VOLUME NINETY ONE

THE ALKALOIDS

VOLUME NINETY ONE

THE ALKALOIDS

Edited by

HANS-JOACHIM KNÖLKER
*Fakultät Chemie
Technische Universität Dresden
Dresden
Germany*

ACADEMIC PRESS
An imprint of Elsevier

Academic Press is an imprint of Elsevier
50 Hampshire Street, 5th Floor, Cambridge, MA 02139, United States
525 B Street, Suite 1650, San Diego, CA 92101, United States
125 London Wall, London, EC2Y 5AS, United Kingdom

First edition 2024

Copyright © 2024 Elsevier Inc. All rights are reserved, including those for text and data mining, AI training, and similar technologies.

Publisher's note: Elsevier takes a neutral position with respect to territorial disputes or jurisdictional claims in its published content, including in maps and institutional affiliations.

No part of this publication may be reproduced or transmitted in any form or by any means, electronic or mechanical, including photocopying, recording, or any information storage and retrieval system, without permission in writing from the publisher. Details on how to seek permission, further information about the Publisher's permissions policies and our arrangements with organizations such as the Copyright Clearance Center and the Copyright Licensing Agency, can be found at our website: www.elsevier.com/permissions.

This book and the individual contributions contained in it are protected under copyright by the Publisher (other than as may be noted herein).

Notices
Knowledge and best practice in this field are constantly changing. As new research and experience broaden our understanding, changes in research methods, professional practices, or medical treatment may become necessary.

Practitioners and researchers must always rely on their own experience and knowledge in evaluating and using any information, methods, compounds, or experiments described herein. In using such information or methods they should be mindful of their own safety and the safety of others, including parties for whom they have a professional responsibility.

To the fullest extent of the law, neither the Publisher nor the authors, contributors, or editors, assume any liability for any injury and/or damage to persons or property as a matter of products liability, negligence or otherwise, or from any use or operation of any methods, products, instructions, or ideas contained in the material herein.

ISBN: 978-0-443-29554-6
ISSN: 1099-4831

For information on all Academic Press publications
visit our website at https://www.elsevier.com/books-and-journals

Publisher: Zoe Kruze
Acquisitions Editor: Jason Mitchell
Editorial Project Manager: Palash Sharma
Production Project Manager: James Selvam
Cover Designer: Greg Harris

Typeset by STRAIVE, India

Contents

Contributors   vii
Preface   ix

1. **Structural variety and pharmacological potential of naphthylisoquinoline alkaloids**   1
   Doris Feineis and Gerhard Bringmann

   1. Introduction   4
   2. Dioncophyllaceae and Ancistrocladaceae: Woody lianas from tropical Africa and Asia   9
   3. Structural diversity and classification of naphthylisoquinoline alkaloids   44
   4. Isolation of naphthylisoquinoline alkaloids and assignment of their full absolute stereostructures   64
   5. *Ancistrocladus abbreviatus* from Coastal West Africa, a rich source of structurally diverse naphthylisoquinoline alkaloids   101
   6. *Ancistrocladus abbreviatus*, a creative producer of naphthylisoquinoline alkaloids with unusual molecular scaffolds   165
   7. Unprecedented biosynthesis of naphthylisoquinoline alkaloids—A novel acetate-polymalonate pathway to isoquinoline alkaloids in plants   231
   8. Chemo-ecological interactions and localization of naphthylisoquinoline alkaloids in intact plants   249
   9. Summary and outlook   274
   10. Tables of the naphthylisoquinoline alkaloids and related compounds isolated from the West African liana *Ancistrocladus abbreviatus*   278

   Acknowledgments   369
   References   370

Cumulative Index of Titles   411
Index   423

Contributors

Gerhard Bringmann
Institute of Organic Chemistry, University of Würzburg, Würzburg, Germany

Doris Feineis
Institute of Organic Chemistry, University of Würzburg, Würzburg, Germany

Preface

In Volume 91 of *The Alkaloids*, Doris Feineis and Gerhard Bringmann give an exhaustive overview of the recent developments in the area of naphthylisoquinoline alkaloids. Bringmann's group also contributed the two preceding articles on this class of natural products for *The Alkaloids*: Chapter 3 in Volume 29 (published in 1986) and Chapter 4 in Volume 46 (1995). Thus as editor, I am very pleased that the protagonist, who has been working in this field for more than four decades, provides his expertise and experience in this review article. The authors comprehensively describe the botanical details of the Dioncophyllaceae and Ancistrocladaceae plant families, the classification of naphthylisoquinoline alkaloids, their occurrence, isolation, structural elucidation, biogenesis, chemical synthesis, and biological activities.

Hans-Joachim Knölker
Technische Universität Dresden, Dresden, Germany

CHAPTER ONE

Structural variety and pharmacological potential of naphthylisoquinoline alkaloids

Doris Feineis and Gerhard Bringmann*

Institute of Organic Chemistry, University of Würzburg, Würzburg, Germany
*Corresponding author: e-mail address: g.bringmann@uni-wuerzburg.de

Contents

1.	Introduction	4
2.	Dioncophyllaceae and Ancistrocladaceae: Woody lianas from tropical Africa and Asia	9
	2.1 Dioncophyllaceae	12
	2.2 Ancistrocladaceae	18
	2.3 Cultivation of Ancistrocladaceae plants in the greenhouse, clonal propagation, and production of cell cultures	35
	2.4 Cultivation of further African and Asian *Ancistrocladus* species	42
3.	Structural diversity and classification of naphthylisoquinoline alkaloids	44
	3.1 *C,C*-Coupled monomeric naphthylisoquinoline alkaloids and their structural landmarks	44
	3.2 *N,C*-Coupled naphthylisoquinolines, a small subclass of unique alkaloids equipped with a stereogenic *N*-iminium-*C*-aryl axis	53
	3.3 Naphthylisoquinoline dimers and their different subclasses	56
	3.4 Free, non-coupled isoquinoline and naphthalene metabolites related to naphthylisoquinoline alkaloids	61
	3.5 Species-specific production of naphthylisoquinoline alkaloids in Ancistrocladaceae and Dioncophyllaceae	63
4.	Isolation of naphthylisoquinoline alkaloids and assignment of their full absolute stereostructures	64
	4.1 Isolation of naphthylisoquinoline alkaloids and assignment of their constitutions	65
	4.2 Assignment of the relative configuration at the stereogenic centers, C-1 versus C-3, by 1D and 2D NMR spectroscopy	74
	4.3 Assignment of the relative configuration at centers versus axis by NOESY measurements	75
	4.4 Assignment of the absolute configuration at the stereogenic centers by oxidative degradation	79

The Alkaloids, Volume 91
ISSN 1099-4831
https://doi.org/10.1016/bs.alkal.2024.03.001

Copyright © 2024 Elsevier Inc.
All rights are reserved, including those for text and data mining, AI training, and similar technologies.

4.5	Assignment of the absolute axial configuration by experimental ECD spectroscopy in combination with quantum-chemical ECD calculations	80
4.6	Assignment of absolute axial configurations online, by LC-ECD coupling	85
4.7	Exemplified for dioncophylline A: An overview on the repertoire of methods for the assignment of the full absolute stereostructure of naphthylisoquinoline alkaloids	88
4.8	Further phytochemical studies on Dioncophyllaceae and Ancistrocladaceae plants	100

5. *Ancistrocladus abbreviatus* from Coastal West Africa, a rich source of structurally diverse naphthylisoquinoline alkaloids   101

5.1	Naphthylisoquinoline alkaloids with a 5,1'-coupling site and their structural diversity	104
5.2	Ancistrobrevine B and related 5,8'-coupled naphthylisoquinoline alkaloids	119
5.3	Dioncophylline A and related 7,1'-coupled naphthylisoquinoline alkaloids from four different subclasses	122
5.4	Dioncophyllidine E, only the second example of a 7,3'-coupled dioncophyllaceous naphthylisoquinoline alkaloid discovered in nature	145
5.5	Ancistrobrevines A, H, I, and J, and related 7,8'-coupled naphthyltetra- and -dihydroisoquinoline alkaloids	150
5.6	A unique series of fully dehydrogenated naphthylisoquinoline alkaloids from the roots of *Ancistrocladus abbreviatus*	153
5.7	*Ancistrocladus abbreviatus* as a rich source of naphthylisoquinoline alkaloids—Summary and outlook	161

6. *Ancistrocladus abbreviatus*, a creative producer of naphthylisoquinoline alkaloids with unusual molecular scaffolds   165

6.1	A near-complete series of atropo-diastereomeric naphthylisoquinoline dimers, discovered in the roots of *Ancistrocladus abbreviatus*	167
6.2	Ancistrobreviquinones A and B, novel-type naphthylisoquinoline alkaloids with an *ortho*-quinoid unit in the naphthalene part	185
6.3	Ancistrobrevinium A, the first—and so far only—Cationic C,C-coupled naphthylisoquinoline alkaloid known from nature	189
6.4	1-*nor*-8-*O*-Demethylancistrobrevine H, the only known example of a naphthylisoquinoline alkaloid lacking the otherwise generally present methyl function at C-1	192
6.5	Ancistrosecolines A–F, the first ring-opened naphthylisoquinoline alkaloids, discovered in the roots of *Ancistrocladus abbreviatus*	195
6.6	The naphthylisoindolinones ancistrobrevolines A–D, unprecedented ring-contracted naphthylisoquinoline alkaloids	205
6.7	Highly oxygenated naphthoquinones from solid callus cultures of *Ancistrocladus abbreviatus*	215

7. Unprecedented biosynthesis of naphthylisoquinoline alkaloids—A novel acetate-polymalonate pathway to isoquinoline alkaloids in plants   231

7.1	Biosynthesis of isoshinanolone, plumbagin, and droserone in Ancistrocladaceae and Dioncophyllaceae	236

7.2 Biosynthesis of dioncophylline A following an unprecedented
acetate-polymalonate pathway to isoquinoline alkaloids   238
7.3 Biosynthesis of dioncophylline A from late-stage dihydro- and
tetrahydroisoquinoline precursors   240
8. Chemo-ecological interactions and localization of naphthylisoquinoline
alkaloids in intact plants   249
  8.1 Chemo-ecological interactions   250
  8.2 NMR imaging of organs and seeds of *Ancistrocladus* species and
localization of naphthylisoquinoline alkaloids in intact plants   260
  8.3 In vivo localization of naphthylisoquinoline alkaloids by non-invasive
FT-Raman microspectroscopy   267
  8.4 Chemodiversity, storage, and ecological functions of
naphthylisoquinoline alkaloids—Summary
and outlook   273
9. Summary and outlook   274
10. Tables of the naphthylisoquinoline alkaloids and related compounds isolated
from the West African liana *Ancistrocladus abbreviatus*   278
Acknowledgments   369
References   370

Abstract

Naphthylisoquinoline alkaloids are a fascinating class of natural biaryl compounds. They show characteristic mono- and dimeric scaffolds, with chiral axes and stereogenic centers. Since the appearance of the last comprehensive overview on these secondary plant metabolites in this series in 1995, the number of discovered representatives has tremendously increased to more than 280 examples known today. Many novel-type compounds have meanwhile been discovered, among them naphthylisoquinoline-related follow-up products like e.g., the first *seco*-type (i.e., ring-opened) and ring-contracted analogues. As highlighted in this review, the knowledge on the broad structural chemodiversity of naphthylisoquinoline alkaloids has been decisively driven forward by extensive phytochemical studies on the metabolite pattern of *Ancistrocladus abbreviatus* from Coastal West Africa, which is a particularly "creative" plant. These investigations furnished a considerable number of more than 80—mostly new—natural products from this single species, with promising antiplasmodial activities and with pronounced cytotoxic effects against human leukemia, pancreatic, cervical, and breast cancer cells. Another unique feature of naphthylisoquinoline alkaloids is their unprecedented biosynthetic origin from polyketidic precursors and not, as usual for isoquinoline alkaloids, from aromatic amino acids—a striking example of biosynthetic convergence in nature. Furthermore, remarkable botanical results are presented on the natural producers of naphthylisoquinoline alkaloids, the paleotropical Dioncophyllaceae and Ancistrocladaceae lianas, including first investigations on the chemoecological role of these plant metabolites and their storage and accumulation in particular plant organs.

1. Introduction

During the past decades, naphthylisoquinoline alkaloids[1–7] like dioncophylline A (**1a**)[a] and its atropo-diastereomer 7-*epi*-dioncophylline A (**1b**),[5,6,8,9] ancistrolikokine E$_3$ (**2**),[10] and ealamine A (**3**)[11] (Fig. 1) have been in the focus of intense research efforts, since they constitute structurally (in particular stereochemically) challenging, biosynthetically unprecedented, and pharmacologically promising natural compounds. As indicated by the name, they all consist of a naphthalene and an isoquinoline moiety linked by a *C,C*-biaryl axis. The structural variety of naphthylisoquinoline alkaloids is further enlarged by the occurrence of dimeric representatives[3] displaying unique molecular architectures like jozimine A$_2$ (**4a**)[3,12] and

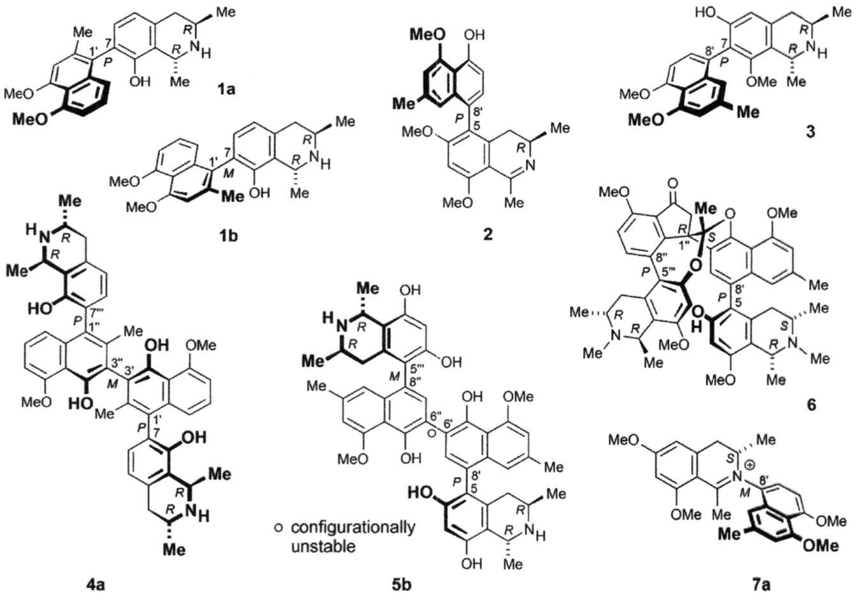

Fig. 1 Representatives of the three main structural subtypes of naphthylisoquinoline alkaloids: *C,C*-coupled monomeric compounds such as dioncophylline A (**1a**) and its atropo-diastereomer 7-*epi*-dioncophylline A (**1b**), ancistrolikokine E$_3$ (**2**), and ealamine A (**3**), dimeric representatives such as jozimine A$_2$ (**4a**) and michellamine B (**5b**), and fused dimers with a complex molecular architecture like spirombandakamine A$_1$ (**6**), as well as *N,C*-coupled alkaloids like ancistrocladinium A (**7a**).

---

[a] For naphthylisoquinoline alkaloids appearing as spectroscopically distinguishable rotational isomers, a numbering system is applied in which the two atropo-diastereomers are denoted as **a** and **b**, like e.g., for dioncophylline A (**1a**) and 7-*epi*-dioncophylline A (**1b**, see Fig. 1). The same applies to atropo-enantiomers (i.e., for naphthylisoquinoline alkaloids that have no stereogenic centers).

michellamine B (**5b**),[3,5,13,14] and the further cyclized dimer spirombandakamine A$_1$ (**6**)[3,15] (Fig. 1). A third subfamily of naphthylisoquinolines, also showing a high degree of structural originality, comprises compounds with an unprecedented stereogenic *N*-iminium-*C*-aryl axis between the isoquinoline part and the naphthalene moiety, like e.g., ancistrocladinium A (**7a**).[2,16] Besides different coupling positions of their two molecular halves, the alkaloids vary by their stereochemical features in the isoquinoline subunit, by the degree of hydrogenation in the heterocyclic isoquinoline ring system, and by their *O*- and *N*-methylation patterns.[1–7] Most of the naphthylisoquinoline alkaloids show the fascinating phenomenon of atropisomerism due to bulky substituents close to the biaryl axis.[1–6,17]

Over 280 naphthylisoquinoline alkaloids have so far been isolated,[1–7] all of them exclusively from plant species belonging to the Ancistrocladaceae[18–25] and Dioncophyllaceae[26] families, which are indigenous to the evergreen forests of West, Central, and East Africa, India, Sri Lanka, and Southeast Asia. For a long time, these two paleotropical plant families received only limited attention, because most of the lianas are fairly rare, and their natural habitats are not easily accessible, and nowadays often severely endangered. For this reason, most of the species are botanically only incompletely known, so that their taxonomic treatment is still problematic.[18–25] Recent field work in Southeast Asia[24] and Central Africa[25] even hinted at the existence of further new, as yet undescribed species.

With the first isolation of naphthylisoquinoline alkaloids in the early 1970s by the pioneer in this field of research, T. Govindachari,[27,28] and with the discovery of promising bioactivities, in particular the protective effects of michellamine B (**5b**) against the HI virus, as found by M. Boyd et al.,[13,14] in the 1990s, Ancistrocladaceae and Dioncophyllaceae lianas have become quite popular. Since then, about 20 naphthylisoquinoline-producing plant species have been investigated more thoroughly for their secondary metabolites.[1–7] Among them were long-known species such as *Triphyophyllum peltatum*[5–7] (Dioncophyllaceae) from Ivory Coast and Ancistrocladaceae lianas such as *Ancistrocladus tectorius*[1] from the Chinese Hainan Island, *Ancistrocladus ealaensis*[3,4,11] and *Ancistrocladus likoko*[3,4,10] from Central Africa, but also newly discovered taxa like the Cameroonian species *Ancistrocladus korupensis*[3,5,7,13,14,21] and *Ancistrocladus benomensis*,[1,22] which is endemic to the Southwest-Malaysian highland. Botanically yet undescribed *Ancistrocladus* lianas from the Congo Basin have been phytochemically studied in more detail, too.[2–4,12,15,16,29–35] Some of the naphthyliso-quinoline alkaloids isolated from these plants have attracted attention as potential chemotherapeutic agents for the treatment of severe and widespread tropical diseases such as malaria, Chagas' disease, African sleeping sickness, and leishmaniasis.[1–4,7,36–50] Other representatives of the alkaloids strongly inhibit multiple myeloma,[4,51] leukemia,[1–4,51–54] cervical,[3,4,31,55,56] and pancreatic cancer cells[3,4,10,11,30,31,56,57] and may thus be considered as new lead structures for drug development.

More recently, through extensive phytochemical studies on *Ancistrocladus abbreviatus* from Coastal West Africa, further new fascinating aspects regarding the structural variability of naphthylisoquinoline alkaloids have emerged.[54–61] The synthetic repertoire of this liana, which is in the focus of the present review, was found to be excessively diverse, giving rise to complex mixtures of secondary metabolites, among them numerous new representatives. Without a particular regioselectivity or stereocontrol, the plant produces a large number of monomeric alkaloids equipped with a tetrahydro- or a dihydroisoquinoline subunit,[4,5,56–58,61–65] or possessing a non-hydrogenated isoquinoline portion[54,60] such as ancistrobrevine H (**8**),[57] 5-*epi*-dioncophyllidine C$_2$ (**9**),[56] and ancistrobreveine D (**10a**)[54] (Fig. 2).

The discovery of the first 1-unsubstituted naphthylisoquinoline, 1-*nor*-8-*O*-demethylancistrobrevine H (**12**),[55] lacking the otherwise generally present methyl group at C-1, along with the formation of unique

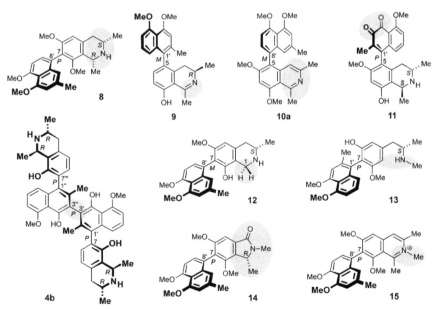

Fig. 2 The structural diversity of secondary metabolites from the West African liana *Ancistrocladus abbreviatus*. The degrees of hydrogenation in the isoquinoline portions in **8–10a** and unprecedented structural modifications in the alkaloids **11–15** and in the dimer **4b** are marked in gray.

follow-up metabolites such as the ring-opened (i.e., *seco*-type) alkaloid ancistrosecoline A (**13**)[55] (Fig. 2), highlights the outstanding synthetic "creativity" of *A. abbreviatus*. Likewise intriguing is the occurrence of unprecedented ring-contracted naphthylisoindolinones such as ancistrobrevoline A (**14**)[59] (Fig. 2) in this plant species. The unusual molecular scaffolds of the metabolites of *A. abbreviatus* are further exemplified by the first quinoid naphthylisoquinolines consisting of a 3,4-naphthoquinone portion coupled to a tetrahydroisoquinoline subunit like e.g. ancistrobreviquinone A (**11**),[56] and by unprecedented alkaloids like ancistrobrevinium A (**15**),[60] the only example of a cationic *C,C*-coupled naphthylisoquinoline alkaloid known from nature. Furthermore, *A. abbreviatus* is the so far only species that was found to produce a near-complete series of atropisomeric dimers (four out of six possible stereoisomers) possessing three consecutive chiral biaryl axes, among them the new C_2-symmetric quateraryl jozibrevine A (**4b**)[58] (Fig. 2).

For the isolation and structural elucidation of naphthylisoquinoline alkaloids, in particular for the unambiguous attribution of the absolute configuration at the stereocenters and stereogenic axes, a wide repertoire of chromatographic, spectroscopic, degradative, synthetic, and computational methods have been used, some of which were specifically adapted or even newly developed for this class of compounds.[1–3,5,66–79] Besides conventional methods (like, *i.a.*, MS, IR, and UV), the most important techniques for a reliable assignment of complete stereostructures are one- and two-dimensional nuclear magnetic resonance spectroscopy (NMR),[1,3,66–68] electronic circular dichroism (ECD) spectroscopy in combination with quantum-chemical ECD calculations,[71–75] and a ruthenium-mediated oxidative degradation procedure[1,3,69,70] to give known, easy-to-analyze amino acids.

This review also deals with the unprecedented biosynthetic origin of naphthylisoquinoline alkaloids following a novel pathway to isoquinolines.[80–83] Within the large family of more than 2500 known plant-derived di- and tetrahydroisoquinolines,[4,84–88] these are the first—and as yet only—isoquinoline alkaloids that are not formed from aromatic amino acids,[84,85,87,89–94] but from acetate/malonate units.[80–83] The molecular halves of naphthylisoquinoline alkaloids—the isoquinoline part and the naphthalene portion—are built up separately, yet via joint polyketide intermediates, and are then coupled to each other at a late stage of the biosynthetic route, by phenol-oxidative cross-coupling.[95,96]

An exciting, albeit much less explored aspect concerns the interactions of the Dioncophyllaceae and Ancistrocladaceae with their natural environment and, in particular, the chemo-ecological functions of naphthylisoquinoline alkaloids and related compounds to cope with abiotic and biotic stress and to mediate interactions with herbivores, pathogens, or neighboring plants. The diversity of bioactivities of different naphthylisoquinoline alkaloids, among them fungicidal effects against plant-pathogenic fungi,[5,97] antifeedant and growth-retarding activities against herbivorous insects,[98,99] but also pronounced larvicidal[100,101] and molluscicidal[102,103] properties, suggests that these secondary metabolites play an important role in defense-mediating processes. First information has been acquired regarding the transport and storage of naphthylisoquinoline alkaloids in distinct plant organs and tissues. Non-invasive investigations by NMR micro-imaging permitted to visualize the water distribution in plant organs, thus providing fascinating insight into the anatomy of roots and leaves,[104–107] and to track, non-invasively, early stages of germination and seedling growth.[107] By means of NMR chemical-shift imaging (CSI) measurements[104–107] and FT-Raman microspectroscopy,[108–110] an accumulation and storage of naphthylisoquinoline alkaloids and related compounds in special compartments was observed.

Since the appearance of the last major review[5] on naphthylisoquinoline alkaloids in this series in 1995, with about 60 representatives recognized at that time, this field of research has developed tremendously in many respects. Meanwhile a much higher number of new alkaloids have been isolated, among them compounds displaying unprecedented molecular architectures. Moreover, the above-mentioned ultimate proof of a fundamentally novel biosynthetic origin of isoquinoline alkaloids in plants from polyketide precursors, and the discovered significant biological activities have given this topic an additional future-oriented impact. In 2015, an overview[7] on mono- and dimeric naphthylisoquinoline alkaloids was published, with a main focus on the bioactivities of these unique natural products. A comprehensive article on dimeric naphthylisoquinoline alkaloids,[3] presented by the authors' group in 2019, summarizes the most important results regarding the isolation, structural elucidation, total synthesis, and bioactivities of this fascinating, rapidly growing class of plant metabolites. In 2021, an overview appeared on N,C-coupled naphthylisoquinoline alkaloids,[2] with special emphasis on the assignment of their full stereostructures, the development of synthetic approaches toward these compounds and related synthetic analogues, and on their bioactivities. A further review,[79] published in 2022, reports on the conceptual approaches successfully developed for the regio- and atropo-selective synthesis of naphthylisoquinoline alkaloids.

Most recently, a major article[1] has been presented by the authors' group dealing with phytochemical, phylogenetic, and taxonomic investigations on six *Ancistrocladus* lianas from Asia with special emphasis on their botanical peculiarities, the diversity of their metabolite profiles, and their difficult and (in part) still uncertain taxonomic treatment.

This review will complement the above-referenced articles by providing a detailed overview on the botanical peculiarities of the Dioncophyllaceae and Ancistrocladaceae plant families (Section 2), followed by a summary regarding the classification and structural variability of mono- and dimeric naphthylisoquinoline alkaloids and their occurrence in different plant species (Section 3). The repertoire of efficient and reliable methods for the isolation and structural elucidation of these bi- and quateraryl natural products will be described thoroughly in Section 4. In Sections 5 and 6, attention will be drawn to the outstanding position of the West African liana *Ancistrocladus abbreviatus* as a particularly rich source of bioactive and structurally most diverse naphthylisoquinoline alkaloids, many of them displaying unprecedented structural features that have been detected in no other naphthylisoquinoline-producing plant. These two sections are complemented by Section 10 showing a complete list of all the naphthylisoquinoline alkaloids discovered in *A. abbreviatus*, categorized according to their coupling sites and their affiliation to the various subgroups of these natural products. Furthermore, the unusual molecular framework of naphthylisoquinoline alkaloids triggered investigations on their unprecedented biosynthesis via the acetate-polymalonate pathway as described in Section 7. Finally, Section 8 deals with the chemo-ecological functions of naphthylisoquinoline alkaloids and their accumulation in particular plant organs, and Section 9 gives a short summary and outlook on the key results presented in this review and on some challenging tasks for the future.

2. Dioncophyllaceae and Ancistrocladaceae: Woody lianas from tropical Africa and Asia

The Dioncophyllaceae and Ancistrocladaceae are tropical scandent shrubs or woody lianas with tendril-like modified shoots provided with typical circinate woody hooks as climbing devices. The names of the two plant families reflect their botanical peculiarity: "Dioncophyllaceae" literally means "with double-hooked leaves,"[26,111,112] while the term "Ancistrocladaceae" comes from the Greek words "ankistron" (= hook) and "klados" (= shoot).[18–20,24,25,113] The Dioncophyllaceae consist of only three monotypic genera: *Triphyophyllum*,[26,111,112,114–121] *Dioncophyllum*,[26,111,112,114,115,122–125] and *Habropetalum*.[26,111,112,114,115] Their occurrence is restricted to the coastal evergreen rain forests of tropical West Africa.[26,111,112,115,116] *Ancistrocladus* Wall. is the only genus of the Ancistrocladaceae plant family. It comprises 18 acknowledged species[19,20,22,23] and is characterized by its disjunct habitats in the paleotropics, with main centers of distribution in West, Central, and East Africa, India, Sri Lanka, and in Southeast Asia (Fig. 3).[18–25,113]

For many decades, the systematic position of the Ancistrocladaceae and Dioncophyllaceae has been discussed controversely, and the presumed interfamiliar relationships and botanical classifications have thus widely varied in the past.[20,22,113] Comparison of DNA sequence data due to the plastid *rbc*L gene and *18S* rDNA, but also comparative sequencing of the chloroplast gene *mat*K and the flanking *trn*K intron region as well as

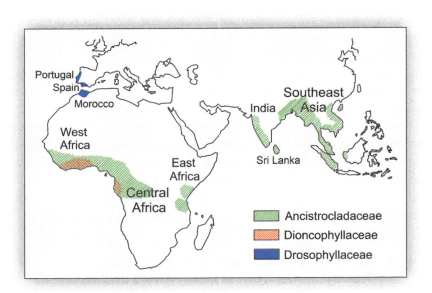

Fig. 3 Geographic distribution of Dioncophyllaceae and Ancistrocladaceae in Africa and Southeast Asia. For comparison, see the area of distribution (Portugal, Spain, and Morocco) of the most closely related carnivorous Drosophyllaceae[126] plant family (see Fig. 4), which consists of a single monotypic genus (*Drosophyllum*). According to a cladistics analysis of *mat*K gene sequences, it occupies a sister position to a clade comprising the Dioncophyllaceae and Ancistrocladaceae[127–129]—but it does not produce naphthylisoquinoline alkaloids, only related, nitrogen-free naphthalene derivatives.[126,130,131]

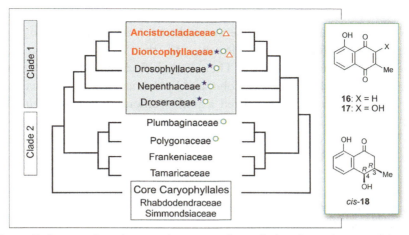

Fig. 4 Cladograms based on single-gene analysis of plastid *matK*, summarizing the phylogenetic relationships of the paleotropical plant families Dioncophyllaceae and Ancistrocladaceae among the taxa of Caryophyllidae s.l. (*sensu lato*). The two, highly congruent, tree topologies[133] presented here are based on different *matK* data sets: studies by Meimberg et al.[127] and Cuénoud et al.[128] gave rise to the strict-consensus tree shown on the left; investigations by Hilu et al.[129] provided the strict-consensus tree illustrated on the right. Plant families containing naphthoquinones such as plumbagin (**16**) or droserone (**17**) and tetralones like *cis*-isoshinanolone (*cis*-**18**) are labeled with a green circle (○), naphthylisoquinoline producing families are highlighted in red and with a red triangle (△), and families possessing carnivorous species are marked with an asterisk (★).

combined analyses based on plastid *mat*K, *atp*B, *rbc*L, and nuclear *18S* rDNA have more recently permitted a reliable phylogenetic placement of the Ancistrocladaceae and Dioncophyllaceae. As shown in Fig. 4, the two plant families occupy a linking position within the Caryophyllidae s.l. (sensu lato), as a sister clade of the carnivorous Drosophyllaceae, in a near phylogenetic vicinity to the likewise carnivorous Nepenthaceae and Droseraceae, along with a second, closely related clade combining the Plumbaginaceae, Polygonaceae, Tamaricaceae, and Frankeniaceae.[127,128,132,133] The strong relationships between the Ancistrocladaceae and the Dioncophyllaceae, as indicated by chloroplast gene *mat*K and *rbc*L phylogenies,[127–129,132–137] are confirmed by several anatomical, morphological,[114,115,138–140] and palynological[141] properties. Furthermore, the Ancistrocladaceae and Dioncophyllaceae share common habitats in West Africa.

A most convincing argument for the close phylogenetic relationships of the Ancistrocladaceae and Dioncophyllaceae and a strong support for their placement within the Caryophyllales is, chemotaxonomically, the presence of naphthylisoquinoline alkaloids found exclusively in these two small plant families.[1–7] These metabolites are, thus, excellent phylogenetic markers for species belonging to the Ancistrocladaceae and Dioncophyllaceae.[1–6,22,24,25,130] Their botanical classification among the taxa of Caryophyllidae s.l. is corroborated by the occurrence of structurally and biosynthetically closely related naphthoquinones like plumbagin (**16**) and droserone (**17**) and tetralones like *cis*-isoshinanolone (*cis*-**18**) (Fig. 4) in Ancistrocladaceae and Dioncophyllaceae—and in most families of the two mentioned clades.[1,5,130,131,135]

2.1 Dioncophyllaceae

Triphyophyllum peltatum (Hutch. & Dalz.) Airy Shaw[26,111,112,114–121] is the most widespread species of the Dioncophyllaceae. The plant predominantly grows in the evergreen rainforests of Sierra Leone, Liberia, and Ivory Coast (Fig. 5a). As a liana, it climbs trees to heights up to 70 m by means of double-hooked leaves (Fig. 5g). This botanical peculiarity was even selected as a motive on a stamp of Ivory Coast in 1985 (Fig. 5b), as part of a series of three stamps dedicated to rare indigenous plant species.[142]

Only two other—very rare—species belong to the Dioncophyllaceae: *Habropetalum dawei* (Hutch. & Dalz.) Airy Shaw,[26,111,112,114,115] mainly found on sandy ground along the coastal border regions of Sierra Leone and Liberia, and *Dioncophyllum tholloni* Baillon,[26,111,112,114,115,122–125] growing in primary and secondary rainforests of the Western Coast of Central Africa in Gabon and in the Republic of the Congo (also known as Congo-Brazzaville).

For a long time, the taxonomic position of the Dioncophyllaceae was doubtful. The history of this notable plant family started in 1890 with the first description of the (at that time) new species, *Dioncophyllum tholloni* (and its even new genus), by Baillon.[122] He regarded the plant as a link between the Bixaceae and Passifloraceae. A short time later, Warburg (1893)[123] and Gilg (1908)[124] classified *Dioncophyllum* as a genus of the plant family Flacourtiaceae, while Sprague (1916)[125] and Hallier (1923)[143] noted significant morphological links to the carnivorous Nepenthaceae. In 1951, Airy Shaw[26] reinvestigated the genus *Dioncophyllum* and was the first to describe each of the three known species, *Triphyophyllum peltatum* (Hutch. & Dalz.) Airy Shaw, *Dioncophyllum tholloni* Baillon, and *Habropetalum dawei* (Hutch. & Dalz.) Airy Shaw, as belonging to three independent monotypic genera (*Triphyophyllum*, *Dioncophyllum*, and

Fig. 5 (a) Geographic distribution of *Triphyophyllum peltatum*, and (b) the typical double-hooked leaves of this peculiar liana pictured on a stamp issued from Ivory Coast in 1985. (c,d) *T. peltatum* flowering in the greenhouse of the Botanical Garden of the University of Würzburg: The plant produces branchlets with hooked leaves bearing small white to pale-beige buds and flowers, which appear directly at the branchlet apex. (e) Opening young fruits exposing the developing seeds to air long before maturity and showing a rapid formation of large papery disks (fully 5–12 cm in diameter), which let them disperse. (f) Young plant in its natural habitat at the Taï National Park in the South West of the Ivory Coast before starting to climb into the canopy of high trees by producing new leaves with much larger internode lengths, right after the carnivorous phase. (g) Mature plants forming short leaves with a pair of recurved hooks helping the climbing lianas to attach to branches and twigs in the rain forest. (h,i) *T. peltatum* showing "part-time carnivory," based on glandular carnivorous leaves solely in the juvenile phase, here digesting small invertebrates such as mosquitoes or spiders (see white arrows). (j) *T. peltatum* during the transition from the rosetted growth of a juvenile plant to elongated growth of a young liana. *Photographs: c,d,g*—J. Schlauer; *f*—H. Bringmann; *h,i*—A. Irmer; *e,j*—H. Rischer.

Habropetalum) combining them into the newly generated family Dioncophyllaceae. Today, the family is part of the order Caryophyllales, which comprises the Nepenthaceae, Drosophyllaceae, and Droseraceae, and the closely related family Ancistrocladaceae (Fig. 4).[127–129,132–137] It is noteworthy that phylogenetic relationships according to *mat*K sequences within the Dioncophyllaceae revealed *Dioncophyllum* and *Triphyophyllum* to form a strongly supported subclade (showing a high bootstrap topology of 92%), with *Habropetalum* placed in a position as a sister taxon at the base.[127] The extremely restricted occurrence of *H. dawei* and some distinct morphological characteristics of this species (e.g., thin petals) gave rise to the assumption that the plant may be a relict of an earlier flora, supporting the base position of this genus within the clade. The two remaining genera *Dioncophyllum* and *Triphyophyllum* were found to be anatomically more similar to each other than to *Habropetalum*.[127]

Airy Shaw[26] considered the Dioncophyllaceae to be one of the "strangest groups of plants to be found in the vegetable kingdom" due to some unusual, even bizarre, botanical features. Dioncophyllaceae lianas form different kinds of leaves during the different phases of their lives. *Dioncophyllum thollonii* and *Habropetalum dawei* possess two types of leaves: simple laminate ones and leaves that are hooked at the apex.[26,111,112,115] *Triphyophyllum peltatum* even has trimorphic leaves (*Triphyophyllum* means "three kinds of leaves," derived from the Greek words "triphyes" = "of threefold form," and "phyllon" = "leaf"): simple recurved leaves on short shoots (Fig. 5f and j), filiform glandular leaves (Fig. 5h and i) that develop terminally on short juvenile shoots, and hooked leaves (Fig. 5c and g) on mature climbing shoots.[26,111,112,114–121] *Triphyophyllum peltatum* has attracted special attention regarding its complex life cycle showing the phenomenon of "part-time carnivory."[26,112,116–121,144–146] At the end of the juvenile growth period, just before it starts climbing, the liana forms leaves with stalked glands and reduced lamina that capture and digest invertebrates (Fig. 5h).[116–121] A further botanical peculiarity of all of the Dioncophyllaceae is the formation of seeds with large, paper-thin membraneous wings (Figs. 5e and 6a), which are the only seeds in the plant kingdom that are larger than the fruits from which they stem. Unlike most other angiosperm fruits, they open prior to seed maturity, forming umbrella-shaped winged seeds. The Dioncophyllaceae plants in general have therefore been called "secondary gymnosperms,"[26,146] although they are far from being related e.g., to the conifers. After germination, the cotyledon tips remain in the seed and the growing hypocotyl lifts the seed above the ground (Fig. 6b and c).

Triphyophyllum peltatum seems to be carnivorous solely in its juvenile phase,[116–121,144–146] while carnivory has so far never been observed in the two other Dioncophyllaceae genera (*Dioncophyllum, Habropetalum*),[26,111,127,132,133] nor in any of the Ancistrocladaceae[20,127,132,133] taxa. Their closest phylogenetic neighbors outside the Dioncophyllaceae family, however, are carnivorous, too, like the Droseraceae (with sticky leaves or spring traps),[147] the Drosophyllaceae (with sticky leaves),[126] or the Nepenthaceae (with pitchers).[148] In its juvenile phase, *T. peltatum* forms rosettes of undivided lanceolate leaves with obtuse to acute leaf tips (Figs. 5j and 6d and h). Just before the young plants attain a certain size (ca. 25–40 cm), at the onset

Fig. 6 Cultivation of *Triphyophyllum peltatum* in the greenhouse (Botanical Garden of the University of Würzburg) (top) and production of cell cultures and ex vitro propagation of *T. peltatum* (bottom). (a) Mature seeds, the only seeds in the plant kingdom that are larger than the fruits from which they stem ("secondary gymnosperms"). (b) Germination of seeds sown on a hydroculture substrate (Seramis®). (c) Seedling of *T. peltatum* just after germination, with the growing hypocotyl lifting the seed up (of which, for practical reasons, the outer part was cut off beforehand) and the cotyledons bent sideward at their base; the primary leaves sprouting from a narrow slit between the cotyledons, where they are attached to the hypocotyl. (d) A juvenile plant, 5 months after germination. (e) Induction of callus formation starting from sterile shoots (see f) as the parent material[150]; the axenic cell cultures thus obtained were used for phytochemical studies[151,152] and biosynthetic investigations.[80–83,153] (f) In vitro shoot cultures established on Anderson medium[154] starting from sterile seedlings obtained from surface-sterilized seeds, sown under aseptic conditions on an artificial medium. (g) Use of these sterile shoots to generate aseptic newly rooted plants; for further cultivation, the young plants were grown on sterile Seramis® substrate. (h) For acclimatization to insterile greenhouse conditions, the plants were transferred to small pots filled with Seramis® and grown at 95% humidity and 25 °C with constant light. *Photographs: a,c–h*—A. Irmer; *b*—K. Wolf.

of the rainy season, carnivorous leaves are formed, which have a more or less reduced lamina and a prolonged midrib equipped with glandular emergences (Fig. 5i). These glands secrete a sticky mucilage and capture and digest small invertebrates such as mosquitoes or spiders (Fig. 5h).[116,118,120,121] The close homology of the stalked and sessile glands on the trapping organ of *T. peltatum* to those of *Drosophyllum lusitanicum* (Drosophyllaceae),[115] and the presence of endogenous proteolytic enzymes[149] in the secretions of these glands led to the hypothesis of carnivory in *T. peltatum* very early already.[116,120,121,144] Although nutrient deficiency has been associated with carnivory in *T. peltatum* already by Marburger[120] (1979) and Green et al.[121] (1979), phosphorus starvation has been identified only recently to be essential for the induction of the carnivorus syndrome.[119] Winkelmann et al.[119] (2023) showed that shoots of *T. peltatum* cultivated in vitro on medium suffering from depletion of major nutrients such as nitrogen, potassium, or phosphorus immediately developed pronounced visual symptoms such as chlorotic small leaves, and finally browning and dying of older leaves. Remarkably, only the shoots growing on phosphorus deficiency media started to produce glandular leaves perfectly resembling those described for plants in their natural habitats. Liquid medium allowing to better define the phosphorus content of the medium revealed that nutrient solutions with 0.2 mM phosphorus or lower resulted in the formation of glandular leaves,[119] thus suggesting that in the part-time carnivorous plant *T. peltatum* intracellular phosphorus concentrations seem to serve as a signal when dropping below a critical threshold. Apparently, episodes of soil phosphorus depletion trigger the formation of carnivorous leaves with secreting glands permitting *T. peltatum* to replenish its P stocks by digesting captured invertebrates.[116,118–121] Right after the carnivorous stage, the new elongated shoots, now equipped with considerably grown internodes (Fig. 5f), start climbing into the high canopy of the rain forest, attached to the supporting trees by leaves that have two hooks curling toward the leaf base at either side of the leaf tip (Fig. 5g).[116,117]

In 1997, the authors' group received viable seeds of *T. peltatum* from Ivory Coast, collected by a close scientific partner and highly renowned plant expert, late Prof. L. Aké Assi.[146] To obtain these extremely rare and difficult-to-collect seeds was a most lucky circumstance. It took great efforts to identify fruiting plants just in time during the rainy season, yet facilitated by experience from our earlier collections in 1989. Traveling to the localities of *T. peltatum* already is most troublesome, and it is a challenging task to collect the seeds in the dangerous heights of the forest canopy before they are detached by the wind and rapidly disappear into the dense ground vegetation. In the greenhouse of the Botanical Garden of the University of Würzburg, these seeds were sown on sandy potting soil as the substrate (Fig. 6b). Germination (ca. 50%) was observed after a few weeks to several months, yielding plantlets[146] (Fig. 6c and d), which grew well for years allowing the authors' group to study more closely the life cycle of *T. peltatum*[116,117] under defined greenhouse conditions. The observations included the unique carnivorous stage[116–118] and the ensuing transition from the rosetted growth of a juvenile plant to elongated-shoot formation, along with the production of new leaves, which were separated by internodes of much larger length. Finally, the plants even started flowering and fruiting.[116,117] Some fruits, produced from manually pollinated flowers, developed seeds (Fig. 5e). While the beige-red glabrous flowers (Fig. 5c and d) do not look too spectacular,[26,116,117] the seeds of *T. peltatum* are unique within the plant kingdom. The large, disk-shaped seeds (5–12 cm in diameter) of *T. peltatum* are borne on prolonged, thickened and even woody funiculi (seed stalks) attached to the center of the disks (Fig. 5e).[26,116,117,146] The embryo is located close to the point of attachment in a thickened ring of white endosperm (approximately 1 cm in diameter), the rest of the seed consists of a thin, papery wing, which is slightly striated radially (Fig. 6a).[26,116,117,146] In the rainforest, the detached seeds can fly over large distances until they reach the floor and start germination.

Owing to the limited availability of viable seeds or living plant material, a clonal-propagation protocol,[150] including rooting and weaning of *T. peltatum* was developed, aiming at a sustainable cultivation of this rare plant species in the greenhouse. Reasons for the necessity of a rapid micropropagation in vitro were the slow growth and propagation of *T. peltatum* in the greenhouse,[116–118,146] and, in particular, the dramatically shrinking natural habitat of the liana due to deforestation and plantation expansions.[116] Consequently, it was of special importance to cultivate plants in the greenhouse as an ecologically agreeable alternative to take care of this unique, fascinating plant, which is severely threatened in its natural environment.

Viable seeds of *T. peltatum* (Fig. 6a) were surface-sterilized and sown under aseptic conditions on an artificial medium giving rise to sterile seedlings,[118,146] from which in vitro organ cultures (e.g., shoot cultures) were successfully established on Anderson medium.[118,146,150] These sterile shoots (Fig. 6f) were used as the parent material for inducing callus cultures[151–153] (Fig. 6e) and for clonal propagation.[118,119,146] The production of alkaloid-producing callus cultures was a real milestone, permitting for the first time to investigate in depth the exceptional biosynthesis of naphthylisoquinoline alkaloids from acetate-malonate units[80–83] (see Section 7).

For the generation of rooted plants, suitable for cultivation outside the flasks under greenhouse conditions,[118,119,146,150] phytohormone combinations such as 1-naphthalene acetic acid (NAA) and thidiazuron were added to the medium, but only about 25% of the shoots showed good root growth. Better results were obtained by supporting the phenomenon of spontaneous rooting by keeping the shoots on depleted medium for an extended time exceeding the usual 4-week subculture interval. Aseptic rooted plants were transferred to sterile Seramis® substrate (Fig. 6g), while the rootless shoots were again transferred to fresh Anderson medium to continue the sterile organ culture.[118,150] Rooting of *T. peltatum* is generally a very slow process, and after 3 months the roots were still rarely longer than 1 cm.

The small (and still sterile) plants (Fig. 6g) were cultivated for several weeks under the same conditions as previously applied to the shoots, and were then transferred into small pots filled with Seramis® as the substrate (Fig. 6h). This permitted acclimatization to greenhouse conditions, which was successfully accomplished 3–4 months later.[118,150] The plants did not require any anti-transpirant coating if the weaning process was done slowly, keeping the plants under conditions of high relative humidity (ca. 95%).[150] Initially, the young plants were grown without any fertilizer (to avoid salt stress), and, after 2–3 months, nutrient supply was added resulting in the formation of slightly enlarged leaves (ca. 5 cm long).[150] During summer time, the growth proceeded very slowly, whereas from the beginning of fall, the growth rate and the leaf size increased drastically, and the plants started to produce leaves with a length of up to 10 cm.[118] Although in vitro propagation[118,119,150] according to the protocol described above furnished living plants[117–119,146,150] and alkaloid-producing cell cultures[80–83,151–153] continuously over years, the cultivation and propagation of this extremely delicate tropical liana remains very difficult up to now, being far away from a routine process.

2.2 Ancistrocladaceae

The monogeneric family Ancistrocladaceae comprises only one single genus, *Ancistrocladus* (former synonyms: *Bembix*, *Wormia*, or *Bigamea*), with presently 18 accepted species[20,22,23] out of ca. 30 binominals listed in different plant surveys from Asia and Africa.[18–23] The main centers of distribution are West and Central Africa, as well as Southeast Asia, with outliers in East Africa, India, and Sri Lanka.[18–25,113] A reliable taxonomic treatment of the genus *Ancistrocladus* is still lacking due to incomplete herbarium collections, and, in particular, due to the fact that for some of the species (even for accepted ones!) flowers or fruits are still unknown.[18–25] Moreover, some of the taxa are quite rare, several of them are not easily accessible, also due to the difficult political situation in the respective countries.

Ancistrocladus plants, in particular those from Southeast Asia, have been known for more than 200 years. In 1829 Wallich[155] was the first to introduce the name *Ancistrocladus* into the botanical literature describing four species (*A. heyneanus*, *A. vahlii*, *A. extensus*, and *A. wallichii*) from India, Sri Lanka, Burma (today Myanmar), Malaysia, and the Andaman Islands.[20] Twenty years later Planchon[156] (1849) recognized the Ancistrocladaceae as a separate plant family consisting of one genus with—at that time—seven species (namely *A. vahlii*, *A. extensus*, *A. pinangianus*, *A. griffithii*, *A. sagittatus*, *A. wallichii*, and *A. heyneanus*) from all over Southeast Asia and India.[20,156] In the following decades botanists classified the genus *Ancistrocladus* into various other families, mostly into the Dipterocarpaceae, because of the similar fruits.[20] In 1868 Oliver[157] published a first report on *Ancistrocladus* plants from Africa, namely on *A. guineënis* from Nigeria, as a member of the Dipterocarpaceae. It took nearly 80 years until further *Ancistrocladus* species from West and Central Africa were described,[19–21,23,25,158–165] and the only two ones known so far from East Africa, *A. robertsoniorum*[18–20,166–168] and *A. tanzaniensis*,[19,20,169] have even been documented quite recently, at the end of the past century.

A first synopsis on the Ancistrocladaceae was presented by Gilg[170] in 1925, followed by a detailed account on the typification of 25 binominals based on a survey of local floras, provided by Gereau[18] in 1997, with 11 valid species recognized for Africa, and 12 for Southeast Asia. Nearly 50 years earlier, van Steenis (1948)[171] summarized all Asian species under *A. tectorius* (Lour.) Merr., because of difficulties distinguishing vegetative material, thus giving rise to a taxonomic classification of most of the *Ancistrocladus* collections in Southeast Asia as *A. tectorius*. Recent comparative phylogenetic studies based on the nuclear internal transcribed spacers (ITS) of molecular rDNA, on the *trn*K intron of cpDNA, and on inter-simple sequence repeat (ISSR) fingerprint analyses strongly indicate a species differentiation within this taxon.[24] The isolation of structurally most different naphthylisoquinoline alkaloids from various plant collections,[1,22,24] even though all assigned as "*A. tectorius*," strongly support the presence of further, yet unrecognized *Ancistrocladus* species throughout Southeast Asia.

In 2000, a first synoptic revision of the African taxa was presented by Cheek,[19] with 13 species recognized, followed by a comprehensive revision of the whole genus *Ancistrocladus* by Taylor et al.[20] in 2005 with 16 accepted species, 11 from Africa (Table 1) and 5 from Asia (Table 2). Based on an extensive examination of herbarium collections, reporting in detail on each of the taxa with their type specimens and on their distribution, phenology, habitat, and conservation status, this updated account of the genus *Ancistrocladus* constitutes a comprehensive compendium summarizing the most important results from botanical research. Recently, two new species, *Ancistrocladus benomensis* Rischer & G. Bringmann[22] from Malaysia and *Ancistrocladus ileboensis* Heubl, Mudogo & G. Bringmann[23] from the Congo Basin, have been recognized, with phytochemical contributions from the authors' group.

Table 1 African species of the genus Ancistrocladus.

No.	Species	Author(s)	Typus/herbarium	Locus classicus [type]	Habitat and distribution	Reports in the literature
1a	*Ancistrocladus abbreviatus* Airy Shaw subsp. *abbreviatus*	Airy Shaw[159]	F. C. Deighton 2589/K, designated in 1950 by H. K. Airy Shaw[160]	Sierra Leone, Southern Province: Njala (January 5, 1933)	Wet evergreen forests on river banks and river islands, at 130–160 m alt., Guinea, Sierra Leone, Liberia, Ivory Coast, Ghana, Nigeria, Cameroon[19,20]	Refs. 18–20, 138, 159, 160, 162, 186, 187
1b	*Ancistrocladus abbreviatus* Airy Shaw subsp. *lateralis* Gereau	Gereau[20]	J. A. Emwiogbon 344/FHI-72100	Nigeria, Anambra State: East Central State, Awka District, Mamu Forest Reserve (November 6, 1973)	Wet or swamp forests, seasonal swamp, at 5–200 m alt., Nigeria[20]	Ref. 20
2	*Ancistrocladus barteri**	Scott Elliot[163]	G. F. Scott Elliot 4860/K, designated in 1950 by H. K. Airy Shaw[160]	Sierra Leone, not located as to region: Mount Gonkwi, Duunia Talla Hills (February 16, 1892)	Lowland wet forests, coastal bluffs, along rivers, from 0–95 m alt., Senegal, Guinea-Bissau, Sierra Leone, Liberia to Ghana[19,20]	Refs. 18–20, 138, 159–163
3	*Ancistrocladus congolensis**	J. Léonard[164,165]	J. Louis 11171/BR	Democratic Republic of the Congo (DR Congo), Province Orientale: Forestier Central, Environs de Yangambi, Yafunga sur Zaïre, près d'Isangi (September 8, 1938)	Swamp and riverine forests, with an area ranging from the DR Congo and the Republic of the Congo to Gabon[19,20]	Refs. 18–20, 138, 141, 164, 165, 188

4	*Ancistrocladus ealaensis** J. Léonard[164,165]	J. Léonard 1057/BR	Democratic Republic of the Congo (DR Congo), region of Equateur, Forestier Central, Environs de Eala, Rivière Yuli (November 23, 1946)	Freshwater swamp forests and periodically inundated riverine forests, at 20–400 m alt., Central African Republic, DR Congo, Republic of the Congo, Gabon[19,20]	Refs. 18–20, 141, 164, 165, 188
5	*Ancistrocladus ileboensis* ● Heubl, Mudogo & G. Bringmann[23]	V. Mudogo & G. Bringmann No. 57	Democratic Republic of the Congo (DR Congo), Bambange, north of the town Ilebo (March 16, 2000)	Dry land and secondary forests, at 420–465 m alt., Province Kasaï-Occidental (DR Congo)[23]	Ref. 23
6	*Ancistrocladus grandiflorus** Cheek[19]	M. Cheek & P. Ndumbe 3915/K	Cameroon, South-West Province: near Limbe, Mabeta-Moliwe Forest (November 11, 1992)	Wet primary and old secondary forests, at 60–940 m alt., Cameroon[19,20]	Refs. 19, 20
7	*Ancistrocladus guineënsis** Oliver[157]	G. Mann 2234/K	Nigeria, Cross River State: Old Calabar (1859–1863)	Evergreen vegetation, on open roadsides and along rivers, at 80–600 m alt., Nigeria, Cameroon to Gabon[19,20]	Refs. 18–20, 157, 159–163, 172
8	*Ancistrocladus korupensis** D. W. Thomas & Gereau[21]	R. E. Gereau, D. W. Thomas, F. Namata & E. Jato 5180/MO4568407	Cameroon, SW Province: Ndian Division, 0.6 km East of the confluence of Mededibe (Moliba) & Ndian (Mana) Rivers, 5°02′N, 8°53′E, 90 m (March 4, 1993)	Lowland evergreen rainforests, at 50–360 m alt., Southeast Nigeria to Southwest Cameroon[19–21]	Refs. 18–21, 189, 190

Continued

Table 1 African species of the genus *Ancistrocladus*.—cont'd

No.	Species	Author(s)	Typus/herbarium	Locus classicus [type]	Habitat and distribution	Reports in the literature
9	*Ancistrocladus letestui**	Pellegrin[174]	G. M. P. C. Le Testu 8627/P	Gabon, Estuaire Province, Région de Lastoursville, Région de Monts Iboundji (December 27, 1930)	Lowland or submontane evergreen rainforests, at 125–950 m alt., Nigeria to Gabon and Republic of the Congo[19,20]	Refs. 18–20, 138, 174, 188
10	*Ancistrocladus likoko**	J. Léonard[164,165]	J. Léonard 1850/BR	Democratic Republic of the Congo (DR Congo), Orientale Province: Forestier Central, Rivière Boamba, Environs de Yangambi (July 1948)	Along rivers and in swamp forests, at 430–470 m alt., DR Congo[19,20]	Refs. 18–20, 141, 164, 165
11	*Ancistrocladus pachyrrhachis* reduced to taxonomic synonymy under *A. barteri*.[20]	Airy Shaw[160]	J. T. Baldwin Jr. 5839/K	Liberia, Montserrado County: Monrovia, sandy area (May 30, 1947)	Lowland evergreen forests, Liberia—the two collections known were made from the same single plant, no further plants have ever been found again, despite an intense search[19,20]	Refs. 18–20, 160, 162
12	*Ancistrocladus robertsoniorum**	J. Léonard[166,167]	S. A. Robertson 3687/BR	Kenya, Coast Province: Kwale District, Buda Forest, 60 m alt. (September 7, 1983)	Moist evergreen coastal forests, at 25–300 m alt., Southeastern Kenya[19,20]	Refs. 18–20, 113, 166–168

13	*Ancistrocladus tanzaniensis**	Cheek & Frimodt-Moller[169]	C. Frimodt-Moller, H. Ndangalasi & Joker *s.n.* in TZ-537/K	Tanzania, Iringa Region: Iringa Rural District, Uzungwa Scarp Forest Reserve, 8°21′S, 35°58′E, 1100 m alt. (December 15, 1997)	Wet evergreen forests of Uzungwa Mountains, at 1100–1310 m alt., South-Central Tanzania[19,20]	Refs. 19, 20, 169
14	*Ancistrocladus uncinattus* reduced to taxonomic synonymy under *A. guineënsis*[20]	Hutchinson & Dalziel[173]	P. A. Talbot & Mrs. P. A. Talbot 3108/K	Nigeria, Akwa-Ibom State, Eket District (1912–1913)	Lowland evergreen rainforests, Nigeria, Southeast Nigeria—the plant has not been found again since the type collection, despite botanical survey work in the 1990s,[19,20] but was, in 2013, reported to have been re-discovered in the Cross River National Park in Nigeria[185]	Refs. 18–20, 160–162, 173, 185

In 2000, a synoptic revision of the African taxa of the genus *Ancistrocladus* was presented by Cheek,[19] with an identification key and detailed information on taxonomy, distribution, and ecology of 13 species recognized at that time. In 2005, Taylor et al.[20] published a new revision of the genus *Ancistrocladus* Wall., accepting 11 species of lianas (marked with an asterisk, *) from tropical West, Central, and East Africa. Two African *Ancistrocladus* lianas have been reduced to taxonomic synonymy: *Ancistrocladus pachyrrhachis* Airy Shaw (entry no. 11) under *A. barteri* Scott-Elliot, and *A. uncinatus* Hutch. & Dalz. (entry no. 14) under *A. guineënsis* Oliv. (see also column "Remarks"). In 2010, a new species from the Ilebo region in the South of the Central Congo Basin (DR Congo), *A. ileboensis* Heubl, Mudogo & G. Bringmann (entry no. 5, marked with a circle, ●), was described and illustrated.[23]

Table 2 Southeast Asian species of the genus *Ancistrocladus*.

No.	Species	Author(s)	Typus/herbarium	Locus classicus [type]	Habitat and distribution	Reports in the literature
1	*Ancistrocladus attenuatus**	Dyer[191]	J. W. Helfer s.n. 724/K, designated in 1997 by Gereau[18]	India and Burma (today Myanmar), Tenasserim and Andaman Islands (date not given)	In mixed evergreen scrub and bamboo forests, at 50 m up to 150 m alt., India, Bangladesh, Myanmar, Andaman Islands[20,191,204]; the hard stems are used by indigenous people of the Andaman Islands for arrow production[204,205]	Refs. 18, 20, 170, 191, 196, 204–211
2	*Ancistrocladus benomensis* ●	Rischer & G. Bringmann[22]	H. Rischer & K. Wolf s.n./M, No. KL4932	Malaysia, Pahang, Gunung Benom, Kerau Game Reserve (April 12, 2000)	So far, only known from the type location Gunung Benom, Malaysia[22]	Ref. 22
3	*Ancistrocladus carallioides* reduced to taxonomic synonymy under *A. tectorius*[20]	Craib[192]	A. F. G. Kerr 5053/K	Thailand, Northern Region, Nan, Doi Tiu (March 9, 1921)	Thailand: Doi Tiu Nan, ca. 900 m alt. (type collection), in secondary forests[20,192]	Refs. 18, 20, 192, 212
4	*Ancistrocladus cochinchinensis* reduced to taxonomic synonymy under *A. tectorius*[20]	Gagnepain[193]	L. Pierre 16/P, designated in 1997 by R. E. Gereau[18]	Vietnam, Bin Dinh, in montibus, 300 m alt. (March, 1867)	Vietnam,[20,193,213,214] Province Bien-hoa; Vietnamese name: Trung quân nam bộ and Trường quân nam bộ[214]	Refs. 18, 20, 213–215

5	*Ancistrocladus extensus* taxonomic synonym of *A. tectorius*[18,20]	Wallich ex Planchon[156]	Wallich Numer. List 1052.1.1829/K	Burma (today Myanmar), probably Mon State: Amherst, Province Martaban, today District Thaton	Wallich Numer. List 1052.1 from Amherst, a British station in Burma; the plant material was presumably collected by Wallich himself in evergreen lowland forests, 30–60 m alt.[20,205] Lower Burma, Andaman Islands[206,209,212,216]: the stems are used by indigenous people of the Andaman Islands for arrow production[206]	Refs. 18, 20, 156, 170, 191, 196, 201, 202, 205, 206, 209, 211, 212, 216
6	*Ancistrocladus griffithii**	Planchon[156]	W. Griffith *s.n.*/K	Burma (today Myanmar), Tenasserim, Mergui	Swamp forests and muddy river banks,[209] to 350 m alt, India, Andaman Islands, Lower Myanmar, Thailand, Cambodia, Vietnam[20,205,209]	Refs. 18, 20, 156, 170, 191, 196, 202, 205, 209, 211, 212
7	*Ancistrocladus hainanensis* reduced to taxonomic synonymy under *A. tectorius*.[18,20]	Hayata[194]	Z. Katsumada *s.n.*/K designated in 2005 by Taylor et al.[20]	China, South Central, Hainan Island, Ngai District, Fung Leng (August 29, 1932)	Hainan Island, PR China[18,20,201,217]	Refs. 18, 20, 194, 201, 217
8	*Ancistrocladus hamatus**	(Vahl) Gilg[195–197]	J. Koenig *s.n.*/C	Ceylon (today Sri Lanka), près de Bigamen	Moist evergreen forests, at 100–150 m, Sri Lanka[20,170,191,211,218,219]	Refs. 18, 20, 138, 156, 170, 191, 195–197, 202, 205, 211, 218–224
9	*Ancistrocladus harmandii* reduced to taxonomic synonymy under *A. tectorius*[20]	Gagnepain[193]	B. F. J. Harmand 989/P	Laos, Koukou: Bassin du Sé-moun (Mun River) (January 1877)	Type collections in evergreen forests of Laos[20]	Refs. 18, 20, 193

Continued

Table 2 Southeast Asian species of the genus *Ancistrocladus*.—cont'd

No.	Species	Author(s)	Typus/herbarium	Locus classicus [type]	Habitat and distribution	Reports in the literature
10	*Ancistrocladus heyneanus**	Wallich ex J. Graham[198]	Wallich Numer. List 7262.1832/K	Locality unknown, but presumably India	Seasonal to evergreen forests, low elevations in the Western Ghats mountains of Southwest India,[20,205,209,210] occurrence also noted for the Indian states Karnataka, Kerala, Maharashtra, and Tamil Nadu[210]	Refs. 18, 20, 138, 141, 155, 156, 170, 191, 196, 198, 202, 205, 209–211, 223–232
11	*Ancistrocladus pentagynus* = *Durandea* Planchon (Linaceae)[18,20,143,171]	Warburg[203]	—	Not designated	Excluded species[18,20,143,171]	Refs. 18, 20, 143, 171, 203
12	*Ancistrocladus pinangianus* reduced to taxonomic synonymy under *A. tectorius*[18,20]	Wallich ex Planchon[156]	G. Porter *s.n.*/K in Wallich Numer. List 1054	Malaysia, Pilau Pinang; Penang	Evergreen forests of Malaysia (type location): very common on sandy hills near the sea, more rarely inland; occurrence also noted for Singapore, Sumatra[199,233]	Refs. 18, 20, 156, 170, 191, 196, 199, 202, 211, 212, 233
13	*Ancistrocladus sagittatus* = *Tetramerista glabra* Miquel (Tetrameristaceae)[18,20,171,191,202]	Wallich ex Planchon[156]	Wallich Numer. List 1055.1829	Singapore (locality unknown)	Excluded species[18,20,171,191,202]	Refs. 18, 20, 138, 171, 191, 202, 205
14	*Ancistrocladus stelligerus* name of dubious identity[20]	Wallich ex A. DC.[202]	Wallich Numer. List 1053/K	Burma (today Myanmar), Province Martaban	The description of *A. stelligerus* was possibly based on a mixed collection since parts of the circumscription are consistent with the ones of *A. griffithii* (fruit, leaves), *A. attenuatus* (leaves), and *A. tectorius*; and most of the specimens are doubtful[20]	Refs. 20, 196, 202

15	*Anistrocladus tectorius** syn. *Bembix tectoria* Loureiro[200]	(Loureiro) Merrill[200,201]	J. de Loureiro s.n./BM	Vietnam (locality unknown)	Lowland to premontane evergreen but rather dry forests, at 0–1525 m alt., growing on red basaltic, granitic, or sandy to clayey soils; natural habitats located in Myanmar, Thailand, Cambodia, Laos, Hainan Island (PR China), Vietnam, Malaysia, Singapore, Andaman and Nicobar Islands, Sumatra and Borneo of Indonesia[20,171,204,205,210]; stems used by indigenous people of the Andaman Islands for arrow production[210]	Refs. 18, 20, 138, 171, 200, 201, 204, 205, 208, 210, 213, 230, 231, 233–237
16	*Ancistrocladus thwaitesii* taxonomic synonym of *A. hamatus*[18,20]	Tieghem[196]	G. H. K. Thwaites 1600/P	Ceylon (today Sri Lanka), locality unknown	Van Tieghem[196] (1903) published the names *Ancistrocladus thwaitesii* and *Bigamea thwaitesii*, both names referring to the *Ancistrocladus* species endemic to Sri Lanka[18,20]	Refs. 18, 20, 196, 205, 220
17	*Ancistrocladus vahlii* taxonomic synonym of *A. hamatus*[18,20]	Arnott[195]	J. Koenig s.n./C	Ceylon (today Sri Lanka)	Vahl[197] (1810) published the name *Wormia* Vahl, based on *Wormia hamata* Vahl, referring to the Sri Lankan *Ancistrocladus* species; reported to occur in the central and southern regions of Sri Lanka[18,20,218,219]	Refs. 18, 20, 156, 170, 191, 195–197, 202, 205, 211, 218–221

Continued

Table 2 Southeast Asian species of the genus Ancistrocladus.—cont'd

No.	Species	Author(s)	Typus/herbarium	Locus classicus [type]	Habitat and distribution	Reports in the literature
18	*Ancistrocladus wallichii* name of dubious identity[20]: not clear due to various incorrect applications, in particular for *A. griffithii* and *A. attenuatus*	Planchon[156]	F. de Sylva *s.n.*, in Wallich Numer. List 1052.2.1829/K	India/Bangladesh (border region), Pundua—the locality has disappeared from modern maps, but was a town located in the Ganges Valley near the border of Bengal and Assam (today known as Bangladesh and Assam State, India)[20,205]	Wallich Numer. List 1052.2 from Pundua, a British station in Northeastern India, collected by Wallich's collaborator Francis da Sylva, described by Planchon as *A. wallichii* Planch. The type specimens, however, lack flowers and fruits, thus they cannot unambiguously be identified as any of the recognized species.[20] Furthermore, the plant was reported to occur in mixed evergreen forests up to 150 m on the Andaman and Nicobar Islands (very rare),[204,210] in Bangladesh, and in Myanmar[205,210,212]	Refs. 18, 20, 138, 156, 170, 191, 196, 204, 205, 208–212, 231

In 2005, Taylor et al.[20] published a comprehensive revision of the genus *Ancistrocladus* Wall., accepting only five species of lianas (marked with an asterisk, *) from Southeast Asia, neglecting *A. stelligerus* Wall. ex A.DC. (entry no. 14) and *A. wallichii* Planch. (entry no. 18), and considering *A. thwaitesii* Tiegh. (entry no. 16) and *A. vahlii* Arn. (entry no. 17) as synonymous to *A. hamatus* (Vahl) Gilg (entry no. 8). Six Asian *Ancistrocladus* lianas were reduced to taxonomic synonymy under *A. tectorius* (Lour.) Merr. (entry no. 15): *A. extensus* Wall. ex Planch. (entry no. 5), *A. pinangianus* Wall. ex Planch. (entry no. 12), *A. cochinchinensis* Gagnep. (entry no. 4), *A. harmandii* Gagnep. (entry no. 9), *A. hainanensis* Hayata (entry no. 7), and *A. carallioides* Craib (entry no. 3). Two species, *A. pentagynus* Warburg (entry no. 11) and *A. sagittatus* Wallich ex Planchon (entry no. 13), were excluded considering the fact that these are species of the genera *Durandea* (Linaceae) and *Tetramerista* (Tetrameristaceae). In 2005, a new species from Malaysia was described and illustrated (see column "Remarks"): *A. benomensis* Rischer & G. Bringmann[22] (entry no. 2, marked with a circle, ●), mentioned in the revision of the genus *Ancistrocladus* Wall. by Taylor et al.[20] as a note added in proof).

Ancistrocladus usually occurs in evergreen rainforests (lowland or submontane), in swamps and seasonally inundated riverside forests, sometimes close to the coast, along roadsides, and also in cultivated areas.[18–25,113] All species are woody lianas, climbing into the forest canopy, with a series of recurved to spiraling hooks serving as climbing devices (Fig. 7a). In their juvenile stage, they start as monopodial, erect, self-supporting saplings (Fig. 8c), then developing to become sympodial lianas in the adult phase, with climbing stems and smaller lateral branches that bear the typical hooks as well as terminal rosettes of leaves and inflorescences (Fig. 8a and b).[18–25,113] The leaves are simple, alternate, and spirally arranged, varying from elliptic or obovate to very

Fig. 7 Typical features that characterize the monotypic genus *Ancistrocladus* Wall. (Ancistrocladaceae). Photos (a–g) were taken from African and Asian individual *Ancistrocladus* plants cultivated in the greenhouse of the Botanical Garden of the University of Würzburg. Photo (h) shows *A. ealaensis* in its natural habitat in Eala Botanical Garden located near the town of Mbandaka (Province Équateur, Democratic Republic of the Congo). (a) Leafy hooked twigs ("*Ancistrocladus*"), here from the West African plant *A. abbreviatus*. (b) Detail of the leaf surface of *A. tectorius* from Laos, with well-developed venation and a large epidermal pit. (c) Inflorescence of the new Congolese liana *A. ileboensis*: cyme with more than 50 buds emerging from a leaf rosette. (d) Red-colored inflorescence of *A. benomensis* from Peninsular Malaysia with flowers in buds. (e) Cyme of the Indian species *A. heyneanus* with a young flower possessing contorted petals at the beginning of anthesis. (f) The West African species *A. abbreviatus* flowering: cincinnate inflorescence with fully developed flowers, showing five brilliant-red petals (but also cream- or pink-colored petals can be observed). (g) Mature fruit of *A. ileboensis*, being a mono-seeded nut with five irregularly formed "wings," derived from enlarged sepals. (h) *Ancistrocladus* species inhabiting swamp forests such as *A. ealaensis* (Congo Basin), producing fruits with small wings, shorter than the fruit body. *Photographs: a—A. Irmer; b–d,f,g—I. Kajahn; e—M. Dreyer; h—B. K. Lombe.*

Fig. 8 *Ancistrocladus cochinchinensis* Gagnep. (Ancistrocladaceae), endemic to the evergreen tropical forests of South Central Vietnam (area Ba Nang, Province Da Nang), in its natural habitat. (a) The large woody liana *A. cochinchinensis*, equipped with apical hooks, which enable the plant to climb through the surrounding vegetation to the level of the tree canopy, attaining lengths of more than 20 m. The main stems are sparingly branched and carry smaller extra-axillary lateral stems (i.e., branchlets) equipped with several woody hooks in a single plane. (b) The leaves on the main stems are widely spaced, whereas the leaves of lateral branches are crowded in rosettes. (c) *Ancistrocladus* plants start growing as erect, self-supporting saplings in their juvenile stage; when they reach maturity, climbing stems are formed from elongated shoots that bear robust leaves as shown in (b). *Photographs*: M. Dreyer.

narrowly oblanceolate (Fig. 8). They often possess epidermal pits (Fig. 7b) containing a single wax-secreting trichome. These pits are located on both surfaces of the leaves and sometimes on the sepals. The leaves and sepals bear mostly tiny, but also large flat glands of various forms, the function of which is still unknown.[20]

Several inflorescences (Fig. 7c, d, and g) are borne on lateral branchlets, from extra-axillary positions below the apex, bearing some hooks, but without bractal leaves.[20] Cymes emerging from a leaf rosette produce about 30–60 flowers (Fig. 7c and d). There is one exception: the inflorescences of *A. abbreviatus* (Fig. 7f) are typically solitary and are located very close to the branchlet apex, thus appearing to be terminal. The flowers are usually subtended by one to several bracts, and are bisexual and actinomorphic, with a partially to fully inferior ovary. They possess five oblong to obovate sepals, nearly equal in size, frequently bearing small pits and glands. There are also five petals (equal in size and shape), varying from pale-green, white, yellow, orange, or pink to red (Fig. 7e and f), with some color variations within the same species (e.g., the petals of *A. abbreviatus* were observed to vary from pale pink to brilliant red). At anthesis the petals seem to remain generally erect while the sepals usually spread widely (Fig. 7e).[19,20,113] Most of the plants possess ten stamens, which are arranged in one or two whorls (Fig. 7f). In some species, the filaments are generally equal in length, but in several species they are dimorphic. In most species, the anthers are similar in size and comprise two thecae, which are usually separated by a well-developed connective. The ovary is mostly semi-inferior, crowned in the flower by a disk and a short columnar structure, and both of them persist in the fruit. The basal stylar column bears three styles or, rarely, only one style as observed in the Congolese species *A. ealaensis*.[20]

Ancistrocladus is also notable for its unusual nut-like, mono-seeded fruits (Fig. 7g and h).[18–25,113] Most of the species have the fruiting sepals extended into wings several times longer than the body of the fruit, well-suited for wind dispersal (Fig. 7g). The elongated papery wings apparently also serve to orient the nut-like fruits in an upright position, upon landing on the forest floor, for germination. Swamp forest species, by contrast, such as e.g., the Central African liana *A. ealaensis*, have fruits with the sepaline fruiting wings shorter than the fruit body (Fig. 7h). They can thus be dispersed by floating in the water.[19,20,164,165] The seeds are solitary, relatively large (5–8 mm of diameter), and distinctive in their deeply ruminate endosperm, which is covered with a thin-textured brown testa. The endosperm is starchy and the embryo is straight.[20]

2.2.1 Ancistrocladaceae from West, Central, and East Africa

The two main centers of distribution of the Ancistrocladaceae family in Africa are the west and the center of the continent, with two outliers in East Africa.[18–21,23,157–174] In total, 15 species (including one subspecies) have so far been described for Africa (see Table 1), among these were nine species from the Gulf of Guinea as one center of diversity, where seven known species occur between the Cross and Sanaga Rivers, *viz.*, *A. guineënsis* Oliv.,[157,161,162,172] *A. uncinatus* Hutch. & Dalz.,[173] *A. abbreviatus* Airy Shaw[159] and its subspecies, *A. abbreviatus* Airy Shaw *subsp. lateralis* Gereau,[20] *A. korupensis* D.W. Thomas & Gereau,[21] *A. grandiflorus* Cheek,[19] and *A. letestui* Pellegr.,[174] the latter also expanding to Gabon and the Republic of the Congo. *Ancistrocladus abbreviatus* Airy Shaw is widely distributed in Upper Guinea, where *A. barteri* Scott-Elliot[160,163] is quite common, and *A. pachyrrhachis* Airy Shaw[160] has been collected there, too. From previous field work in the Democratic Republic of the Congo

(DR Congo) involving scientists from the authors' group, from the University of Munich, and from the University of Kinshasa, the Congo Basin turned out to be another major center of diversity. Besides the four well-known species, *A. letestui* Pellegr.,[174] *A. congolensis* J. Léonard, *A. likoko* J. Léonard, and *A. ealaensis* J. Léonard,[164,165] a new species, *A. ileboensis* Heubl, Mudogo & G. Bringmann, was discovered in the Southern Congo Basin, and was botanically described for the first time in 2010.[23] Moreover, phytochemical studies on *Ancistrocladus* plants collected from different locations in the West, South, and Central Congo Basin led to the discovery of some yet undescribed *Ancistrocladus* species. These *Ancistrocladus* plants produce unique patterns of naphthylisoquinoline alkaloids,[2–4,12,15,16,29–35] which differ significantly from those of the five accepted Central African taxa.[2–4,10,11,51–53,175–184] The findings gave rise to the assumption that through an increased screening of *Ancistrocladus* plants from different localities in Central Africa, further yet unknown taxa may be discovered. All the *Ancistrocladus* lianas known so far were collected no further east than 25°E in the center of the Congo Basin, except for the two additional species from East Africa, *A. robertsoniorum* J. Léonard,[19,20,166–168] endemic to Kenya, and *A. tanzaniensis* Cheek & Frimodt-Møller,[19,20,169] which was found in submontane coastal forests in Tanzania.

Table 1 summarizes all the species names of the African taxa listed previously in the botanical literature. The synoptic revision exclusively on the African taxa presented by Cheek[19] in 2000 describes an identification key and gives detailed information on the taxonomy, distribution, and ecology of 13 species. In the comprehensive revision by Taylor et al.[20] in 2005 covering the entire Ancistrocladaceae family, 11 valid species and a newly described subspecies of *A. abbreviatus* restricted to Southern Nigeria (previously listed under *A. barteri* by Scott-Elliot, then attributed to *A. guineënsis* by Airy Shaw) were recognized from Africa. Two African *Ancistrocladus* lianas were reduced to taxonomic synonymy, viz., *Ancistrocladus pachyrrhachis* Airy Shaw under *A. barteri* Scott-Elliot, and *A. uncinatus* Hutch. & Dalz. under *A. guineënsis* Oliv., since their isotypes do not differ significantly from collections of *A. barteri* and *A. guineënsis*, respectively.[20] From *A. pachyrrhachis* only two collections (from the same plant!) have been reported, while, despite several targeted searching expeditions, no second plant specimen of this species has as yet been found elsewhere.[19,20] *Ancistrocladus uncinatus* is only known from a single type collection performed about 100 years ago; yet, from the general appearance of its leaves and its inflorescence fragment, there were no significant differences compared to *A. guineënsis*.[19,20]

In 2013, a South African group[185] from veterinary science reported on a collection of plant material assigned as *A. uncinatus* (which would be the first discovery after 100 years!)[19,20] from the Cross River National Park in Nigeria, and on the detection of 35 volatile compounds from roots, leaves, and stems using gas chromatography in combination with mass spectrometry. The interest in these extracts resulted from significant activities of a decoction of the respective plant against the African swine fever (ASF) virus, but they were also found to exert a strong cytotoxic effect on primary cells used in the assay.[185] The compounds responsible for the antiviral activities of this Nigerian liana, possibly naphthylisoquinoline alkaloids, are still unknown. Likewise unexplored is the alkaloid pattern of that possibly re-discovered species *A. uncinatus* and, thus, it has as yet not been possible to compare the naphthylisoquinoline profile of this liana with those of the accepted *Ancistrocladus* species *A. guineënsis*, *A. korupensis*, and *A. letestui* also known to occur in Nigeria.[18–20] The identity and species status of *A. uncinatus* thus still remains unclear and should be investigated by tracking down the plant and performing in-depth phytochemical studies.

2.2.2 Ancistrocladaceae from India, Sri Lanka, and Southeast Asia

About 18 *Ancistrocladus* species have been so far described for Asia (Table 2), although they have meanwhile been reduced to a smaller number.[18,20,22,171] Most of them are found in dry or wet, lowland or submontane, evergreen forests in Southeast Asia, ranging from Malaysia, Singapore, Sumatra, and (in part) Borneo, through China (Island of Hainan), Vietnam, Laos, Cambodia, Thailand, Myanmar, and the Indian Andaman and Nicobar Islands, to the Indian Subcontinent and Sri Lanka. The taxonomy of the Asian *Ancistrocladus* plants is still difficult. Most of the taxa are known only from relatively few field observations, some of the species are quite rare, and some type collections lack flowers and mature fruits, so that there is still a great uncertainty in the identification and delimitation of species.

In 1925, Gilg[170] published a first synopsis on the taxonomy of the genus *Ancistrocladus*, comprising eight species (seven from Asia, only one from Africa, *A. guineënsis*). In 1997, based on a survey of local floras from Africa and Asia, Gereau[18] presented an annotated checklist of *Ancistrocladus* species, comprising 17 Asian species names, among them 12 valid species recognized by the author for Southeast Asia: *A. attenuatus* Dyer,[191] *A. carallioides* Craib,[192] *A. cochinchinensis* Gagnep.,[193] *A. extensus* Wall. ex Planch.,[156] *A. griffithii* Planch.,[156] *A. hainanensis* Hayata,[194] *A. hamatus* (Vahl) Gilg,[195–197] *A. harmandii* Gagnep.,[193] *A. heyneanus* Wall. ex J. Graham,[198]

A. pinangianus Wall. ex Planch.,[156,199] *A. tectorius* (Lour.) Merr.,[200,201] *A. wallichii* Planch.,[156] and *A. thwaitesii* Tiegh.[196] (probably a synonym of *A. hamatus*),[18,20] neglecting *A. stelligerus* Wall. ex A. DC.,[202] and considering *A. vahlii* Arn.[195] as a synonym of *A. hamatus*.[18,20] Gereau[18] excluded *A. pentagynus* Warb.[203] and *A. sagittatus* Wall. ex Planch.[156] since these plants are, without any doubt, members of other plant families.[18,20]

Given the great difficulties in distinguishing vegetative material of the various Southeast Asian *Ancistrocladus* plants, van Steenis (1948)[171] had previously suggested a completely different taxonomic concept. He reduced all of the collections of *Ancistrocladus* in Southeast Asia to the taxon *Ancistrocladus tectorius* (Lour.) Merr.,[201] which had originally been described from Vietnam (1790) by De Loureiro as *Bembix tectoria*, being the first known *Ancistrocladus* species from this region.[200] Since then, most of the Southeast Asian *Ancistrocladus* collections have been uniformly classified as *A. tectorius*.

This taxonomic treatment was also accepted by Taylor et al. (2005),[20] yet with two exceptions: *A. griffithii*[156] (distributed from India through Myanmar, Thailand, Cambodia to Vietnam) and *A. attenuatus*[191] (described for India, Myanmar, and the Andaman Islands) were recognized as two distinct species, significantly independent from each other and from *A. tectorius*. Thus, five *Ancistrocladus* species were, at that time, recognized as valid taxa from India, Sri Lanka, and Southeast Asia.[20] Meanwhile the number of accepted species from Asia has increased to a total of six taxa, because in Central Malaysia, in Gunung Benom, Pahang, located in the Kerau Game Reserve, a new taxon, *A. benomensis* Rischer & G. Bringmann,[22] was discovered.

Plants under the joint name *Ancistrocladus tectorius* were found to show a large chemodiversity with respect to the naphthylisoquinoline alkaloids isolated from different collections all over Southeast Asia,[1–3,22,238–255] giving rise to the assumption that this taxon might, in reality, consist of several distinct species, even though morphologically similar. Another hint at the existence of further distinguishable species not yet recognized taxonomically might be deduced from the fact that the huge area of distribution of *A. tectorius* consists of most different climatic regions ranging from humid in the south to semi-humid in the north, even with drought periods for several months.[20,24] Samples collected from different locations all over Southeast Asia were investigated more closely using sequence data from the cpDNA *trn*K intron and the ITS region of molecular rDNA, and by ISSR fingerprints.[24] This comparative study clearly revealed that the large diversity of naphthylisoquinoline alkaloids reported for *A. tectorius*[1–3,238–255] is indeed paralleled by a high genetic variability,[24] thus indicating that the current taxonomic concept for *Ancistrocladus*,[20,22] with only four recognized species in Southeast Asia, *viz.*, *A. attenuatus*, *A. griffithii*, *A. benomensis*, and *A. tectorius*, was insufficient. The molecular approach to taxonomy also strongly supported the status of *A. griffithii* and *A. benomensis* as independent species and indicated that *A. pinangianus*, *A. cochinchinensis*, and some other *Ancistrocladus* collections from Southern Thailand and Laos may possibly deserve the status of being species of their own, too.[24] These findings make a thorough morphological re-examination, accompanied by more-in-depth phytochemical and genetic studies, highly rewarding.

2.3 Cultivation of Ancistrocladaceae plants in the greenhouse, clonal propagation, and production of cell cultures

2.3.1 Cultivation and propagation of the Indian liana Ancistrocladus heyneanus

Among the Asian Ancistrocladaceae plants, *Ancistrocladus heyneanus* Wall. ex J. Graham (Fig. 9) from India was the first intensely investigated species, not only with respect to its naphthylisoquinoline pattern,[1,27,28,256–267] but also concerning the manifold unusual facets of *Ancistrocladus* lianas in general. It is a large climbing shrub with hooked lateral branchlets up to 30 cm long, each bearing several hooks and a cluster of leaves (Fig. 9a).[18,20,191,198,202,205,208–210,223–232] The taxon is restricted to seasonal and evergreen forests in low elevations in the Western Ghats of Peninsular India. Its huge area of distribution comprises the Indian states Maharashtra, Goa, Karnataka, Tamil Nadu, and Kerala.[18,20,205,209,210,223,229–232]

In the Botanical Gardens of the Universities of Münster and Würzburg, the authors' group successfully cultivated *A. heyneanus* from fruits.[1,5,224,225] Some of the plants grew to a height of more than 4 m and developed the characteristic hooked branches, and, at the age of 3 years, even started flowering and fruiting (Fig. 9b–d).[1,5,224,268]

The availability of fully matured seeds, collected right before the monsoon, was recognized to be a fundamental prerequisite for a successful cultivation of *A. heyneanus* in the greenhouse (Fig. 9e–h). Although the seeds rapidly lost their viability, first encouraging results were obtained by growing this sensitive liana from fruits on various soil mixtures (soil/sand/peat). The breakthrough for a more reliable and reproducible cultivation of the plants came from the use of hydroculture substrates.[1,5,224,268] When sowing the seeds in pure river sand (temperature 30–33 °C, relative humidity 90–100%), permanently keeping them moist, germination was observed to take place after 3–4 weeks (Fig. 9f). Under

Fig. 9 *Ancistrocladus heyneanus*, endemic to the Western Ghats of India, cultivated in the greenhouse of the Botanical Garden of the University of Würzburg. (a) A 3-year old plant with leaves arranged in rosettes. (b) Flower at anthesis, fully developed, with the view on five pink petals; ovary fully inferior with three white styles. (c,d) Inflorescence emerging from a leaf rosette and bearing immature fruits with five irregularly formed "wings" (enlarged sepals). (e) Seedling of *A. heyneanus* just after germination from the mono-seeded nut, with a diameter of ca. 1–2 cm (since deprived of its sepals). (f) Well-developed young seedlings grown in pure river sand 4 weeks after germination with a rate of ca. 90%. (g) A plantation of prospering specimens (here only presented as a detail out of ca. 400 plantlets) of *A. heyneanus* at the age of about 6 months. (h) Adaption of single plants at a size of about 10 cm to a hydroculture substrate. *Photographs: a,c*—M. Dreyer; *b*—A. Irmer; *d*—B. Wiesen; *e–h*—K. Wolf.

these conditions, a large number of ca. 400 healthy and well-growing green seedlings were obtained (Fig. 9g).[1,5,224,225,268]

At a size of about 10 cm, the young plants were singularized and adapted to hydroculture substrate (granulated clay, 8–16 mm) (Fig. 9h). From this time on, they were irradiated for 13 h per day using Na lamps (distance approx. 1 m, max. 12.000 lux, controlled by a photocell), and were sprinkled with rainwater twice per hour, maintaining a relative humidity of 70–90% (controlled by a hygrostate). Among the growth parameters, a sufficiently high substrate water temperature (night: approx. 22 °C; day: ca. 27 °C) proved to be of crucial importance. The plants were fertilized via the roots and leaves twice a week using a commercially available fertilizer.[224,225] The availability of living plants permitted fascinating insight into the anatomy of *A. heyneanus*, and provided interesting results on the accumulation and storage of naphthylisoquinoline alkaloids in plant organs (Section 8), by the use of non-invasive analytical techniques such as FT-Raman microspectroscopy[108] or NMR microscopy, the latter also in combination with NMR chemical-shift imaging (CSI) measurements.[104–107]

Moreover, aseptic plants and alkaloid-producing callus cultures used for biosynthetic studies (see Section 7) and for in vitro propagation were established.[269,270] Induction of callus formation (Fig. 10) was best achieved by placing nodal stem segments horizontally on 1/5 Murashige and Skoog (1/5 MS) medium[271] supplemented with 6-benzylaminopurine (BAP), 1-naphthalene acetic acid (NAA), and 2,4-dichlorophenoxyacetic acid (2,4-DPA). This combination of phytohormones was also found to successfully promote a rapid development of undifferentiated callus material.

As presented in Fig. 11, variation of subcultivation conditions resulted in the production of morphologically differentiated calli.[269,270] A significant tendency toward root formation was observed when regenerating calli were grown on media supplemented with $0.3\,\mathrm{mg\,L^{-1}}$ of kinetin and $4.0\,\mathrm{mg\,L^{-1}}$ of NAA (Fig. 11a). Calli cultivated on solid medium containing $2\,\mathrm{mg\,L^{-1}}$ of BAP and $0.01\,\mathrm{mg\,L^{-1}}$ of NAA showed a significant organogenesis of shoots (Fig. 11b). When adding this combination of phytohormones to liquid medium, however, translucent red-colored calli were produced (Fig. 11c). As already mentioned above, the best results for callus

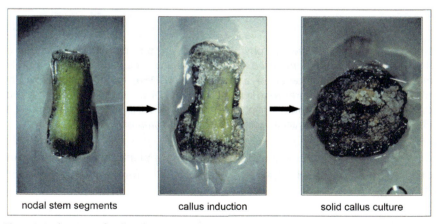

Fig. 10 Induction of callus formation within 6 weeks starting from sterile nodal stem segments of *A. heyneanus*.[270] The calli were cultivated on a modified Murashige and Skoog (MS) medium,[271] with addition of $0.5\,\mathrm{mg\,L^{-1}}$ of 6-benzylaminopurine (BAP), $0.5\,\mathrm{mg\,L^{-1}}$ of 1-naphthalene acetic acid (NAA), $0.5\,\mathrm{mg\,L^{-1}}$ of 2,4-dichlorophenoxyacetic acid (2,4-DPA), $30\,\mathrm{g\,L^{-1}}$ of sucrose, and $2\,\mathrm{g\,L^{-1}}$ of Gelrite for solidification of the medium. The cultures were kept in darkness at $24\pm2\,°\mathrm{C}$ and were repeatedly subcultured for at least 6 months. *Photographs:* H. Rischer.

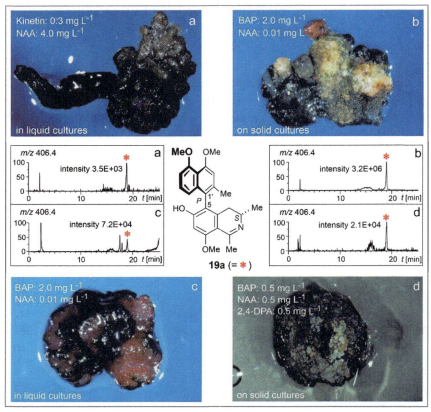

Fig. 11 Influence of different phytohormones on the morphology of calli of *A. heyneanus* cultivated on modified 1/5 Murashige and Skoog (1/5 MS) liquid or solid media,[270,271] and identification of ancistrocladinine (**19a**) in the calli samples by HPLC-ESI-MS/MS. The cultures were kept in the dark at $24 \pm 2\,°C$ and were monthly sub-cultured for at least 6 months; 1-naphthalene acetic acid (NAA), 6-benzylaminopurine (BAP), 2,4-dichlorophenoxyacetic acid (2,4-DPA). *Photographs:* H. Rischer.

development were achieved on solid medium supplemented with a mixture of BAP, NAA, and 2,4-DPA ($0.5\,\text{mg}\,L^{-1}$ for each component) (Fig. 11d).[269,270]

HPLC-ESI-MS/MS experiments on the callus material thus obtained clearly revealed the presence of the naphthyldihydroisoquinoline alkaloid ancistrocladinine (**19a**).[270] The identity of **19a** was proven by its retention time and the characteristic product ion spectrum. Although usually only low quantities of ancistrocladinine (**19a**) were detected in the calli, significant differences concerning the concentrations of **19a** in the samples were monitored depending on the conditions used for callus propagation. As deduced from the intensities of the protonated pseudomolecular ion m/z 406.4 $[M+H]^+$ of **19a** in the mass chromatograms of the calli samples (Fig. 11, center), the regenerating calli grown on solid media supplemented with $2\,\mathrm{mg\,L^{-1}}$ of BAP and $0.01\,\mathrm{mg\,L^{-1}}$ of NAA (Fig. 11b) were found to produce the highest quantities of **19a**. The lowest concentrations of **19a** within this series of samples were detected in calli with a tendency toward root formation (Fig. 11a). The quantities of ancistrocladinine (**19a**) measured in callus samples obtained from the two remaining cultivation approaches (Fig. 11c and d) were nearly identical.[270]

Online investigations of the naphthylisoquinoline alkaloid pattern of suspended callus cultures of *A. heyneanus* using HPLC-ESI-MS/MS led to the identification of further naphthylisoquinoline alkaloids.[270] But, similar to the results described above for ancistrocladinine (**19a**), they only occurred as minor or even trace metabolites.[270] In general, the cultivation of tissue cultures from roots or leaves of *A. heyneanus* was found to impose stress on the cells, which triggered the formation of high quantities of plumbagin (**16**) and *cis*-isoshinanolone (*cis*-**18**),[268,269] (for the structures of **16** and *cis*-**18**, see Fig. 4), instead of the anticipated production of naphthylisoquinoline alkaloids. The occurrence of abundant amounts of these two bicyclic compounds was used to study more closely their biosynthetic origin by feeding differently ^{13}C-labeled acetate to the cell cultures[268,269] (see Section 7).

Callus production on modified solid 1/5 Murashige and Skoog (1/5 MS) media containing $2\,\mathrm{mg\,L^{-1}}$ of BAP and $0.01\,\mathrm{mg\,L^{-1}}$ of NAA led to the formation of red-colored meristems[270,271] (Fig. 12a), which continuously increased in size within a period of 6 months (Fig. 12b), finally developing sprouting shoots[270] (Fig. 12c). During this time, the calli were transferred to new medium every 4 weeks. Single shoots were detached when reaching a length of approximately 1 cm, and were separately cultivated on media supplemented with $0.04\,\mathrm{mg\,L^{-1}}$ of TDZ and $0.01\,\mathrm{mg\,L^{-1}}$ of NAA (Fig. 12d).[270] Unfortunately, the capability of these callus cultures to regenerate decreased over time, and finally got lost after a period of 3 years.

2.3.2 Cultivation and propagation of the West African liana Ancistrocladus abbreviatus

Ancistrocladus abbreviatus Airy Shaw is a relatively widespread, common shrub (Fig. 13b and c) growing in the lowland wet evergreen and swamp forests on river banks and river islands in a quite large area, extending from Guinea to Cameroon.[18–20] Previously treated as *A. barteri* by Scott-Elliot,[163] it was recognized as an own species in 1950 by Airy Shaw.[159,160]

Fig. 12 Regeneration of shoots from callus cultures of *A. heyneanus*.[270] (a) Formation of red-colored meristematic cells in callus tissue 4 weeks after their subcultivation on modified 1/5 Murashige and Skoog (1/5 MS) solid medium[270,271] containing 2 mg L^{-1} of 6-benzylaminopurine (BAP) and 0.01 mg L^{-1} of 1-naphthalene acetic acid (NAA). (b) Proliferation of meristem tissue in monthly subcultured callus cultures containing BAP and NAA. (c) Development of sprouting shoots after 6 months of subcultivation. (d) Detached axillary shoots grown on media supplemented with 0.04 mg L^{-1} of thidiazuron (TDZ) and 0.01 mg L^{-1} of NAA. *Photographs:* H. Rischer.

Following the protocol established for the cultivation of *A. heyneanus*,[224,225] the authors' group managed to raise young seedlings of *A. abbreviatus* obtained from the Parc de Taï (Ivory Coast) to maturity in the greenhouse of the Botanical Garden of the University of Würzburg (Fig. 13c). In single cases, the plants even climbed up to a height of more than 6 m, supported by means of recurved to spiraling woody hooks as typical of the Ancistrocladaceae (Fig. 13e). At the age of ca. 3 years, the plants even started to form blossoms (Fig. 13d and f). As mentioned earlier in this section, the occurrence of inflorescences directly located at the branchlet apex, thus appearing to be terminal (Fig. 13d), is quite unusual within the genus since all other known *Ancistrocladus* species produce several buds and flowers on lateral branchlets.[20]

The authors' group likewise managed to cultivate *A. abbreviatus* starting from viable fruits from Ivory Coast provided by late Prof. L. Aké Assi in 1997.[107,187] Investigations by ^{1}H NMR microscopy and NMR chemical shift-imaging (CSI) experiments permitted a non-destructive monitoring and visualization of the germination process of seeds of *A. abbreviatus*.[107]

Structural variety and pharmacological potential of naphthylisoquinoline alkaloids 41

Fig. 13 (a) Geographic distribution of *A. abbreviatus* in West Africa and photographs of individual plants in their natural habitat (b) and of those cultivated (c–f) in the greenhouse of the Botanical Garden of the University of Würzburg. (b) A young erect plant with leaves arranged in terminal rosettes observed in wet evergreen forests of the West Ivory Coast. (c) Adult flourishing plant in the greenhouse with large epidermal pits lying flat on the surface of the leaves as typical of the genus *Ancistrocladus*.[20] (d) Inflorescence with flower at anthesis, located very close to the branchlet apex, thus appearing to be terminal. (e) A characteristic hooked branch. (f) Flower at anthesis, fully developed, here with 5 brilliant-pink petals (but also cream- or red-colored petals can be observed), equal in size, and with 10 white stamina in 1 whorl, filaments shortly fused at their rather thickened bases, dimorphic in length, ovary fully inferior with 3 white styles. *Photographs: b*—L. Aké Assi; *c,d,f*—M. Dreyer; *e*—A. Irmer.

This non-invasive NMR method provided images of the structures of the living embryos in the seeds and made it possible to track the formation of roots and shoots on the newly formed seedlings over time[107] (Section 8).

Although a high number of young plants of *A. abbreviatus* developed to give prospering adult specimens (Fig. 13c) in the greenhouse, a reliable and reproducible availability of living plants, as required for biosynthetic or comparative morphological studies, still remained difficult, due to the fact that propagation by cuttings was limited by a low rooting percentage.

Thus, it was a real breakthrough that a rapid clonal propagation of *A. abbreviatus*[187] was achieved by growing nodal stem segments on appropriate media under aseptical conditions, as presented in Fig. 14. These segments had initially been obtained from three seedlings successfully grown from sterile seeds in vitro on autoclaved Seramis® clay pearls (germination rate: three out of seven). From these 3 original plants, 81 shoot cultures were produced within 1 year by repeated use of the lower parts of the stems as explants for clonal propagation.[187,270] Within 1 month the explants showed

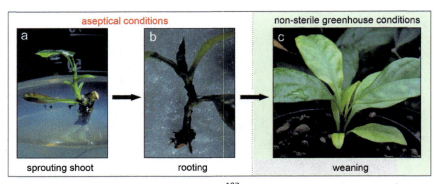

Fig. 14 Clonal propagation of A. abbreviatus.[187] (a) Sprouting shoot from a single nodal stem segment. (b) Rooted plant before weaning. (c) Rooted young plant potted in Seramis® successfully introduced to the greenhouse. *Photographs: a–c*—H. Rischer.

swollen nodes, which developed into shoots when transformed to medium without phytohormones. Healthy plants with many short hairy roots suited for weaning were finally obtained by putting the shoots to sterile Seramis® soaked with sterile water. Rooted plants (about 5 cm of size) became acclimatized to non-sterile greenhouse conditions and developed well in a propagation chamber over 3 months. They were then transferred to the greenhouse to reach appropriate final growing conditions (relative humidity of 80%, 30 °C).[107,187,225,270]

From sterile leaf material of *A. abbreviatus*, alkaloid-producing callus cultures[151,272] (Fig. 15) were established using Murashige and Skoog (MS)[271] medium. The content of naphthylisoquinoline alkaloids in the calli, however, was quite low.[151] Solidification of the medium by stepwise increase of the concentration of calcium and magnesium salts (necessary for gelling), led to more aerobic conditions for the calli. This favored the formation of naphthoquinones as the major metabolites, among them a series of new highly oxygenated compounds,[151] which had not been detected in the intact plants of *A. abbreviatus* (Section 6.7).

2.4 Cultivation of further African and Asian *Ancistrocladus* species

Not described here in detail, although worth mentioning, is the successful cultivation of further African and Asian *Ancistrocladus* plants. Among them were *A. griffithii*[156] from Thailand, *A. tectorius*[200,201] from Laos, *A. barteri*[160,163] from Ivory Coast, *A. congolensis*[164,165] from the Democratic Republic of the Congo, *A. robertsoniorum*[166–168] from Kenya (Fig. 16a), and the two newly described

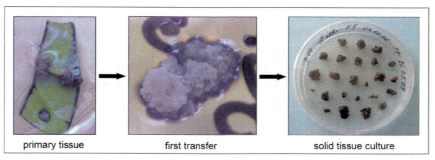

Fig. 15 Solid callus cultures obtained from fully developed lanceolate leaves of A. abbreviatus.[151,272] The calli were cultivated on a modified Murashige and Skoog (MS) medium,[271] with full strength of microelements and organics, but 1/5 of macroelements, except for Ca^{2+} and Mg^{2+} ions (using 1/4). The medium was supplemented with $0.5\,mg\,L^{-1}$ of 2,4-DPA, $0.5\,mg\,L^{-1}$ of BAP, $0.5\,mg\,L^{-1}$ of NAA, 3% sucrose, 0.25% Gelrite, $100\,mg\,L^{-1}$ of glutathione, and $4.5\,mg\,L^{-1}$ of polyvinylpolypyrrolidone (PVPP). The cultures were kept in the dark at $24 \pm 2\,°C$ and were monthly subcultured for at least 6 months. *Photographs:* T. Noll *and* A. Irmer.

Fig. 16 *Ancistrocladus* species cultivated in the greenhouse of the Botanical Garden of the University of Würzburg. (a) A 3-year old specimen of *A. robertsoniorum* (from Kenya), with a main stem bearing large, robust leaves. (b) Lateral hooked branch of *A. ileboensis* (Southern Congo Basin) with crowded leaves and an inflorescence bearing buds, flowers at anthesis, and a young fruit. (c) Lateral branch of a 3-year old plant of *A. benomensis* (from Malaysia) with inflorescences. *Photographs: a,c*—M. Dreyer; *b*—A. Irmer.

species, *A. ileboensis*[23] (Fig. 16b) from the Southern Congo Basin and *A. benomensis*[22] (Fig. 16c) from Malaysia.

Ancistrocladus ileboensis Heubl, Mudogo & G. Bringmann (Fig. 16b) was discovered by the authors' group in 2000 in the Democratic Republic of the

Congo, near the village Bambange north of the town Ilebo along the Kasaï River, close to its junction with the Sankuru River.[23] It was the first *Ancistrocladus* species found in the South-Central Congo Basin, because all the other Central African species described so far are distributed in the northwestern part of the Democratic Republic of the Congo, thus the type locality of *A. ileboensis* was clearly separated from that of *A. likoko*, *A. congolensis*, and *A. ealaensis*.[20,164,165] There was also a delineation to *A. letestui*,[20,174] which occurs in Gabon and in the Republic of the Congo, but has to this day never been found in the Democratic Republic of the Congo.[20,23] The new species occurs mainly in dry land and secondary forests at ca. 420–465 m above sea level. In the greenhouse of the Botanical Garden of the University of Würzburg, the authors' group managed to cultivate some specimens of this newly discovered Congolese plant, which developed flowers and even fruits (Fig. 7c and g).

In Central Malaysia, in Gunung Benom, Pahang, located in the Kerau Game Reserve, the new taxon *A. benomensis* Rischer & G. Bringmann (Fig. 16c) was discovered. This submontane liana, which was found to occur at an altitude of 500–1180 m above sea level on the slopes of Gunung Benom ("benom" is an old Mon-Khmer word for "mountain"), was described for the first time in 2005.[22] The authors' group successfully cultivated some specimens from young seedlings to adult plants (including flower development, see Fig. 7d) in the greenhouse of the Botanical Garden of the University of Würzburg.

3. Structural diversity and classification of naphthylisoquinoline alkaloids

3.1 C,C-Coupled monomeric naphthylisoquinoline alkaloids and their structural landmarks

In the early 1970s, Govindachari's pioneering work on the chemical constituents of Indian plants led to the discovery of a novel, unprecedented secondary metabolite, the naphthylisoquinoline alkaloid ancistrocladine (**20a**).[27,28,256,257,259] It was the first member of this unique class of naturally occurring biaryl compounds, isolated from the tropical liana *Ancistrocladus heyneanus*. It is still, even today, one of their most prominent representatives[1,5,79] (Fig. 17). Ancistrocladine (**20a**) is characterized by the presence of two stereogenic centers in the tetrahydroisoquinoline moiety and a rotationally hindered biaryl axis, which connects the naphthalene portion

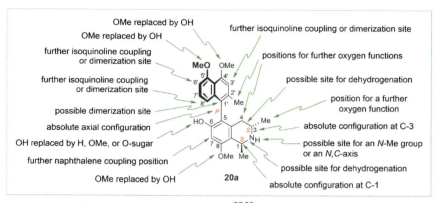

Fig. 17 Exemplified for ancistrocladine (**20a**)[27,28]: structural features of naphthylisoquinoline alkaloids responsible for the broad variety of compounds produced by plants belonging to the Ancistrocladaceae and Dioncophyllaceae families. Further variations arise from the—less frequent—occurrence of quinone entities in the naphthalene part, and from the presence of 1-*nor*- or *seco*-analogues, or from isoindolinone units in the heterocyclic moiety (see Section 6).

to the isoquinoline part. Virtually all of the naphthylisoquinoline alkaloids that have so far been isolated consist of a 4,5-dioxy-2-methylnaphthalene[b]<foot_source> building block (**A** in Fig. 20) and a 1,3-dimethyl-8-oxy- or 1,3-dimethyl-6,8-dioxyisoquinoline moiety (**B** in Fig. 20), the latter usually occurring as a di- or a tetrahydroisoquinoline. They show a unique, polyketide-derived substitution pattern, not only in the naphthalene portion, but also in the isoquinoline part, with a methyl group at C-3 and a *meta*-oxygenation pattern at C-6 and C-8. This heterocyclic portion may additionally be equipped with an *N*-methyl group[1–7] or/and with oxygen functions at the C-3 methyl group,[1,252,273] sometimes also at C-4,[5,6,274,275] and can then even have three stereogenic centers. The oxygen substituents are either free hydroxy functions or methoxy groups (or, in rare cases, they bear sugar entities[276]). In most of these compounds, a total of three stereogenic elements are present, two stereocenters and the biaryl axis, which can be a chiral entity, too, if it is rotationally hindered by the presence of four substituents in the *ortho*-positions. It can be configurationally stable even in the presence of only three substituents—if they are

---

[b] For an easier comparison of naphthylisoquinoline alkaloids with different coupling positions, a 2-methyl-4,5-dioxy substitution pattern is throughout applied in the numbering of the naphthalene moiety, regardless of the coupling site.

sufficiently bulky.[1–3,5,6,17] The molecular diversity of the about 280 naphthylisoquinoline alkaloids[1–7] that have so far been isolated from Dioncophyllaceae and Ancistrocladaceae is attained by a broad variation of most of these structural features (see Fig. 17).

Naphthylisoquinolines show an impressive chemodiversity[1–7]—and yet, the plants that produce the alkaloids seem to follow strict synthetic principles, which are of significant chemo- and geotaxonomic relevance for a reliable botanical and phylogenetic classification of Ancistrocladaceae and Dioncophyllaceae species.[1,5,6] As an example, the West African Dioncophyllaceae solely produce naphthylisoquinolines with an R-configuration at C-3, and always lacking an oxygen function at C-6 ("Dioncophyllaceae type") like e.g., dioncophylline A (**1a**)[5,6,8,9] (Fig. 18). The alkaloid patterns of the East African and Southeast Asian *Ancistrocladus* species, by contrast, are characterized by a (nearly) exclusive production of 3S-configured and 6-oxygenated representatives ("Ancistrocladaceae type") like e.g., ancistrocladine (**20a**)[1,27,28] (Fig. 17) or ancistrobertsonine D (**21**)[277] (Fig. 18). This finding clearly reveals distinct chemo- and geotaxonomic relationships of the only two known taxa from East Africa, *A. tanzaniensis* (from Tanzania)[278,279] and *A. robertsoniorum* (from Kenya),[277,280] with the Asian species such as *A. heyneanus* (from India),[1,27,28,256–267] *A. hamatus* (from Sri Lanka),[1,281,282] *A. griffithii* (from Thailand),[1,283,284] or *A. tectorius* (distributed all over Southeast Asia[1–3,238–255]).

The West African lianas *A. abbreviatus*[5,54–65] and *A. barteri*[5,285,286] (from Ivory Coast) are, with respect to their constituents, chemotaxonomically closely related to the Dioncophyllaceae, since they produce, besides the (expected) Ancistrocladaceae-type alkaloids, also a considerable number of Dioncophyllaceae-type representatives. From these two plant species, Ancistrocladaceae-Dioncophyllaceae mixed-type naphthylisoquinolines, like e.g., ancistrobrevine C (**22**)[64] (Fig. 18), with an oxygen function at C-6 and the R-configuration at C-3, have furthermore been isolated, hence classified as hybrid-type alkaloids. The West African liana *A. abbreviatus*, which is in the focus of this review, is even the only species that does not merely produce all these three subtypes, but even a fourth imaginable one. Besides Ancistrocladaceae-, Dioncophyllaceae-, and hybrid-type compounds, the plant contains the first and only ever found inverse hybrid-type alkaloid dioncoline A (**23**)[58] (Fig. 18), which has been isolated from the roots, thus demonstrating that *A. abbreviatus* occupies a particular taxonomic

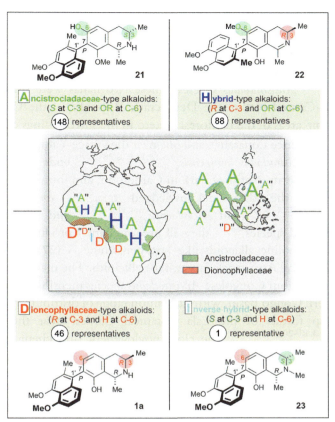

Fig. 18 Ancistrocladaceae-, Dioncophyllaceae-, and hybrid-type compounds, and one single inverse hybrid-type alkaloid, all serving as chemo- and geotaxonomic markers in Ancistrocladaceae and Dioncophyllaceae plants, exemplified by selected representatives such as ancistrobertsonine D (**21**),[277] ancistrobrevine C (**22**),[64] dioncophylline A (**1a**),[5,6,8,9] and dioncoline A (**23**).[58] The map illustrates the occurrence of the four principal subclasses of naphthylisoquinoline alkaloids in Asian and African Ancistrocladaceae plants and in Dioncophyllaceae lianas from Coastal West Africa. For the definition of "A" and "D," see Fig. 19.

position within the genus *Ancistrocladus*. Similar to the Dioncophyllaceae liana *Triphyophyllum peltatum*, also *A. abbreviatus* and *A. barteri* are endemic to West Africa and share a common range of distribution,[5,6,20,26] which corroborates their role as chemo- and geotaxonomical links between the Ancistrocladaceae and the Dioncophyllaceae.

A geographical dependence of the occurrence of alkaloids with distinct structural features also applies to the *Ancistrocladus* lianas indigenous to

Central Africa. Two botanically so far undescribed species[16,29,30] were found to produce solely Ancistrocladaceae-type naphthylisoquinoline alkaloids such as ancistrocladine (**20a**) (Fig. 17) or ancistrobertsonine D (**21**) (Fig. 18), reminiscent of the alkaloid patterns of the Asian[1] and East African[277–280] *Ancistrocladus* species. These two lianas were discovered in the North-Central part of the Democratic Republic of the Congo, one in the swamp areas of the rainforest Yeteto in the surroundings of the town Ikela[16,29] (Province Équateur), and one near the village Yafunga[30] (Province Orientale). *Ancistrocladus korupensis*,[3,5,13,14,276,287–290] endemic to the rain forests of the Korup National Park in Cameroon, and four Congolese species, namely the three acknowledged taxa *A. congolensis*,[3,177,178] *A. likoko*,[3,4,10,53,175,176] and *A. ealaensis*,[3,4,11,179–182] and a botanically as yet unidentified species,[3,4,31] they all predominantly produce hybrid-type naphthylisoquinoline alkaloids like korupensamine A (**27**)[5,287] (Fig. 20), along with typical Ancistrocladaceae-type alkaloids. The main center of distribution of the four Congolese *Ancistrocladus* species mentioned above is the northwestern Congo Basin, with a common range of distribution in the rain and swamp forests of the Eala Botanical Garden near the town of Mbandaka.[3,18–20,31,164,165] The close phylogenetic and chemotaxonomic relationships between these four Congolese *Ancistrocladus* species to the Cameroonian liana *A. korupensis*[18–21] are also evident from the fact that they all contain michellamine-type dimers[3,5,13,14,31,53,176,178,180,181,288,289] (see Section 3.3). Two of the Central African species, *A. ileboensis*[3,23,51,52,183] from the South-Central Congo Basin and a botanically unexplored taxon[3,12] from the North-Central Congo Basin, collected in the Tshuapa District near the town Ikela (Province Équateur, Democratic Republic of the Congo), can be regarded as geo- and chemotaxonomical links to the Ancistrocladaceae and Dioncophyllaceae from West Africa. They combine, to a large degree, the "synthetic repertoire" of the two tropical plant families, and produce naphthylisoquinoline alkaloids of the Ancistrocladaceae, Dioncophyllaceae, and hybrid types.[3,12,51,52,183]

Within the genus *Ancistrocladus*, the Malaysian species *A. benomensis*[22] occupies a special chemotaxonomic position. The plant shows an alkaloid profile[1,273,291] substantially different from those of all the other known[1,20,22] Asian species. From the leaves of this plant, four non-hydrogenated naphthylisoquinoline alkaloids like e.g., *ent*-dioncophylleine A (**24b**), were isolated, all of them unexpectedly lacking an oxygen function at C-6.[291] Within the classification as Dioncophyllaceae- or Ancistrocladaceae-type alkaloids, the presence or absence of the 6-oxygen function is given

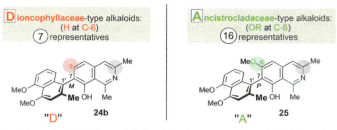

Fig. 19 Tentative classification of fully dehydrogenated naphthylisoquinoline alkaloids as Dioncophyllaceae- ("D," red) or Ancistrocladaceae-type ("A," green) compounds, here exemplified for *ent*-dioncophylline A (**24b**)[291] and ancistrobreveine A (**25**).[54] The presence or absence of a 6-oxygen function as a criterion is given priority over that of the absolute configuration at C-3 (here not applicable, therefore underlaid in gray).

priority over the—here not applicable—absolute configuration at C-3 (i.e., constitutional criteria over stereochemical ones). Thus, these four compounds were tentatively categorized as Dioncophyllaceae-type naphthylisoquinolines[1,5] (see the denotation "D" in Figs. 18 and 19), although, due to their lack of a stereocenter at C-3, it can only be assumed that they are really derived from precursors with an *R*-configuration at C-3 or, possibly, from 3*S* (which would make them hybrid-type compounds, see below). Apart from these four metabolites, all other naphthylisoquinoline alkaloids isolated from *A. benomensis* belong to the Ancistrocladaceae type,[1,273] in complete agreement with the chemotaxonomic rule mentioned above. Vice versa, fully dehydrogenated naphthylisoquinoline alkaloids possessing an oxygen function at C-6 such as ancistrobreveine A (**25**),[54] isolated from the roots of *A. abbreviatus*, were classified as Ancistrocladaceae-type compounds[1,5] (see the "A" in Figs. 18 and 19).

Summarizing, the "Ancistrocladaceae/Dioncophyllaceae" classification[1,5] clearly marks a chemotaxonomic borderline between these two families, and it strongly implies relationships within the genus *Ancistrocladus* in Southeast Asia and East Africa on the one hand, and in Central and West Africa on the other.

Besides the differences in their stereochemical features—regarding the hydrogenation degrees in the heterocyclic isoquinoline ring and the oxygenation and *O*- or *N*-methylation patterns in the two molecular halves—it is mainly the different positions of the biaryl linkage that contribute to the broad structural variability of naphthylisoquinoline alkaloids, by significantly determining their molecular shapes. According to their biosynthetic formation through regio- and stereoselective phenol-oxidative coupling,[95,96] the two molecular portions, which are separately formed beforehand,[80-83] can be linked either in *ortho-* or in *para-*positions relative to the phenolic oxygen functions, i.e., via C-1', C-3', C-6', or C-8' in the naphthalene moiety **A** and via C-5 or C-7 in the isoquinoline part **B** (see Fig. 20). Of the eight resulting possible *C,C*-coupling types, seven have already been identified in nature, typical examples being ancistrobenomine A (**26**)[1,273] (5,1'-coupling), korupensamine A (**27**)[5,287] (5,8'-coupling), ancistrotanzanine A (**28**)[1,249,278] (5,3'-coupling), dioncophyllinol B (**29**)[275] (7,6'-coupling), dioncophylline E (**30a/b**)[292] (7,3'-coupling), ancistrogriffine A (**31**)[1,284] (7,8'-coupling), and dioncopeltine A (**32**)[5,6,293] (7,1'-coupling).

3.1.1 5,8'-Coupling
Among the secondary metabolites of Central African *Ancistrocladus* species, 5,8'-linked alkaloids like korupensamine A (**27**)[5,287] are by far largest group, comprising many monomeric[5,10,29-31,35,51,53,175-177,179,181,287,288] compounds and most of the dimers[5,13-15,31-35,53,176,178,181,182,288] found so far (see Section 3.3). In Asian species, on the contrary, this 5,8'-coupling mode is quite rare, and only five such compounds[1,245,249,251] have as yet been detected. Interestingly, the metabolite profiles of *A. likoko*[3,4,10,53,175,176] and one single further Congolese *Ancistrocladus* taxon,[3,4,31] which is botanically yet undescribed, comprise naphthylisoquinoline alkaloids whose biaryl axis is located between C-5 and C-8', exclusively. The alkaloid pattern of that unidentified liana, however, is characterized by the production of michellamine-type dimers[3,31] as the main constituents (see also Section 3.3), whereas *A. likoko* is a rich source of monomeric alkaloids.[10,53,175,176]

3.1.2 5,1'-Coupling
The by far highest number of compounds isolated from Asian species are 5,1'-coupled naphthyltetra- and dihydroisoquinolines structurally related to the widely occurring alkaloid ancistrocladine (**20a**).[1,27,28] But also non-hydrogenated representatives of this coupling type have been identified, such as ancistrobenomine A (**26**)[1,273] (Fig. 20), which has an otherwise rarely found hydroxymethylene group at C-3.

3.1.3 7,3'- and 7,8'-Coupling
Representatives of the less frequently occurring 7,3'- and 7,8'-coupling types such as dioncophylline E (**30a/b**)[292] and ancistrogriffine A (**31**)[1,284] (Fig. 20), were discovered in *Ancistrocladus* species from Asia and

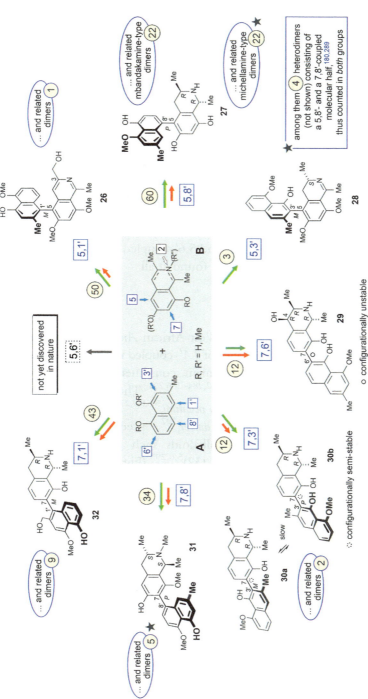

Fig. 20 A selection of naphthylisoquinoline alkaloids representing the C,C-coupling types detected in plants of the families Dioncophyllaceae (red arrows) and Ancistrocladaceae (green arrows). In most cases, the biaryl axis is configurationally stable and constitutes an additional element of chirality, exceptions being, e.g., dioncophyllinol B (**29**),[275] whose axis can freely rotate, and dioncophylline E (**30a/b**),[292] which occurs as a pair of configurationally semi-stable, slowly interconverting atropo-diastereomers.

Africa,[1,3–5,11,28,51,54,55,57,59–61,180,239,240,253,254,264,265,276,279,283,284,291] but also in Dioncophyllaceae[82,292] lianas. The isolation of a series of 7,8′-linked naphthylisoquinoline alkaloids (13 representatives, among them 3 dimers[3,180]) in the Congolese species *A. ealaensis*[3,11,180] was unprecedented, because in most of the other taxa that produce 7,8′-coupled alkaloids, at best four such compounds had been detected.[1,51,54,55,57,59,60,250,264,276,283,284,289]

3.1.4 5,3′- and 5,6′-Coupling

Alkaloids of the 5,3′-coupling type are extremely rare in nature. Prior to the isolation of two further representatives from the Chinese liana *A. tectorius*,[1,249] ancistrotanzanine A (**28**) (Fig. 20) from the East African species *A. tanzaniensis*[278] had been the only naphthylisoquinoline alkaloid showing this unusual biaryl linkage; and 5,6′-coupled naphthylisoquinoline alkaloids have as yet even never been found at all.

3.1.5 7,1′-Coupling

The alkaloid patterns of Asian and West African *Ancistrocladus* species are characterized by a significant number of 7,1′-coupled naphthylisoquinolines, in some cases they are even the main constituents of the respective plants.[1,4,5,54,55,58,59,62,64,65,251,254,260,262,284–286,291] The three West African Dioncophyllaceae lianas, *Triphyophyllum peltatum*,[5,6,8,9,274,293–296] *Dioncophyllum thollonii*,[68,292] and *Habropetalum dawei*,[297] predominantly produce 7,1′-coupled naphthylisoquinoline alkaloids such as dioncophylline A (**1a**)[5,6,8,9] (Fig. 18) or dioncopeltine A (**32**)[5,6,293] (Fig. 20) (ca. 15 representatives), all of them—as mentioned above—belonging to the subclass of Dioncophyllaceae-type compounds. From the West African *Ancistrocladus* taxa *A. abbreviatus*[5,54,58,59,62,63] and *A. barteri*[5,285] as well as from two further *Ancistrocladus* species[12,51,52,183] endemic to the Central Congo Basin mainly Dioncophyllaceae-type alkaloids of this coupling type have been isolated, and even naphthylisoquinoline dimers[12,52,58,183] consisting of dioncophylline-A-related monomers were discovered in some of the plants (see Section 3.3). In a sharp contrast to the Dioncophyllaceae plants, however, they also contain 7,1′-coupled compounds that belong to the Ancistrocladaceae and hybrid types.[5,12,51,54,55,57–59,64,65,286] In East African[277–280] and in Central African[3,4,30,177] *Ancistrocladus* lianas (except for *A. ileboensis*[3,51,52,183]), however, 7,1′-linked naphthylisoquinoline alkaloids have only rarely been found.

3.1.6 7,6'-Coupling

Six Dioncophyllaceae-type naphthylisoquinoline alkaloids of the less frequently occurring 7,6'-coupling type,[5,6,275,298] as occurring e.g., in dioncophyllinol B (**29**)[275] (Fig. 20), have so far been identified in *T. peltatum*, predominantly in leaf material. Because of the presence of only two *ortho*-substituents next to the biaryl axis, they possess a configurationally unstable axis that can freely rotate at room temperature. From the Chinese liana *A. tectorius*, three further 7,6'-linked, yet Ancistrocladaceae-type, naphthylisoquinoline alkaloids have been isolated.[1,248,249] In comparison to the Dioncophyllaceae-type representatives of this coupling type such as **29**, the presence of an additional methoxy group at C-6 hinders the free rotation about the axis to some extent, so that the 7,6'-coupled Ancistrocladaceae-type alkaloids of *A. tectorius* are configurationally semi-stable at the central *C,C*-linkage.[1,248,249]

Depending on the respective *C,C*-coupling type, and on the presence or absence of an oxygen function at C-6, the steric hindrance next to the axis may vary drastically, resulting in a broad energetic range of atropisomerization barriers. Thus, naphthylisoquinoline alkaloids with most different configurational stabilities were isolated from Ancistrocladaceae and Dioncophyllaceae plants. Among them were biaryls with rapidly rotating biaryl axes as e.g., in dioncophyllinol B (**29**)[275] (Fig. 20), or compounds with configurationally semi-stable axes, giving rise to pairs of more or less slowly interconverting atropo-diastereomers like e.g., for dioncophylline E (**30a**) and its rotational isomer, 7-*epi*-dioncophylline E (**30b**)[292] (Fig. 20). The vast majority of the naphthylisoquinoline alkaloids, however, show complete configurational stability at the axis (even up to the decomposition temperature of the molecule), like e.g., in the case of ancistrocladine (**20a**).[1,5,17,27,28,259]

3.2 *N,C*-Coupled naphthylisoquinolines, a small subclass of unique alkaloids equipped with a stereogenic *N*-iminium-*C*-aryl axis

The spectrum of possible coupling types was further enlarged by the discovery of a novel (albeit still small) subclass of naphthylisoquinoline alkaloids with a high degree of structural originality, the first *N,C*-linked naphthylisoquinolinium salts (natural counter anion yet unknown). Recent accounts[1,2] have provided an overview on the key results regarding these *N,C*-coupled alkaloids, on their occurrence in Asian and African *Ancistrocladus* species, on the elucidation of their stereostructures, on their bioactivities, and on their total synthesis.

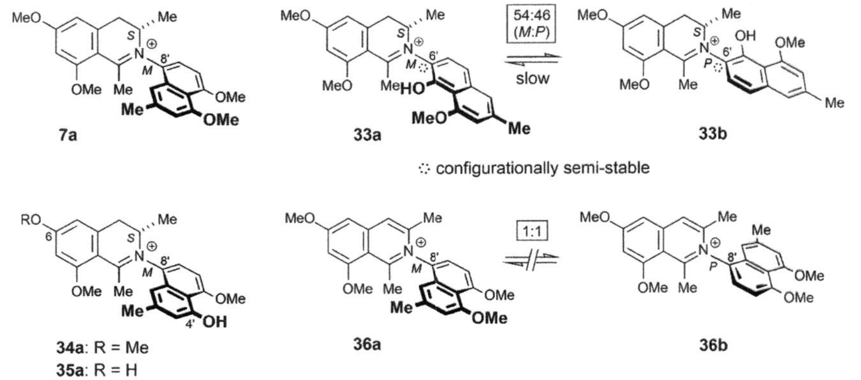

Fig. 21 N,C-Coupled naphthylisoquinoline alkaloids isolated from a broad variety of Asian and African *Ancistrocladus* species: ancistrocladinium A (**7a**),[16] 4′-O-demethylancistrocladinium A (**34a**),[299] 4′,6-O,O-didemethylancistrocladinium A (**35a**),[299] and ancistrocladinium B (**33a/b**).[16] Structure of ancisheynine (**36a/b**),[266,267] a fully racemic trace metabolite found so far only in *A. heyneanus*.

Among these structurally unusual natural products are ancistrocladinium A (**7a**)[1,2,16] and its mono- and bisphenolic analogues **34a** and **35a**[1,2,299] (Fig. 21), all of them displaying an N,8′-linked axis. Ancistrocladinium B (**33a**) and its rotational isomer **33b** (Fig. 21) are based on an N,6′-coupling type, showing a slow rotation about the—now sterically less hindered—central N,C-axis, so that they occur as a mixture of two configurationally semi-stable atropo-diastereomers.[1,2,16] Only four such naphthyldihydroisoquinoline alkaloids (plus their atropo-diastereomers) have so far been identified in nature, all of them belonging to the Ancistrocladaceae type.[1,2] N,C-linked naphthylisoquinoline alkaloids, although structurally not so diverse, are quite widespread, and they have meanwhile been found to be part of the alkaloid patterns of most of the Asian and some of the Central African taxa.[1,2,16,30,181,299] They have, however, not yet been discovered in any Dioncophyllaceae plant.

From the aerial parts of *A. heyneanus*, the likewise N,C-coupled isoquinolinium alkaloid ancisheynine (**36**)[1,2,74,266,267] (Fig. 21) was isolated. This compound has so far been detected exclusively in this Indian *Ancistrocladus* species, and only as a trace constituent, which raised the question if it is a genuine natural product. Online HPLC-ECD analysis of fresh plant parts of *A. heyneanus*, using a chiral Chiralcel® OD-H column, assisted by reference material of **36** obtained by total synthesis,[1,2,74,267] revealed that **36** does occur in the plant, and indeed as a racemate, thus—in view of its configurational stability, evidencing that the formation of ancisheynine (**36a/b**) may proceed in a non-atroposelective—and thus possibly non-enzymatic—way.[1,2,74] Owing to the promising inhibitory activities of ancisheynine (**36a/b**), ancistrocladinium A (**7a**), ancistrocladinium B (**33a/b**), and related compounds[1,2,42–44] against *Leishmania donovani*, the pathogen of visceral leishmaniasis, intense research efforts were initiated, focussing on the total synthesis[1,2,267,300] and on structural modifications[1,2,43,301–303] of these unusual natural products for the development of new drug candidates.

In the twigs and stems of *A. tectorius* collected in the Ledong region of the Chinese island Hainan, two additionally cyclized—and, thus, even pentacyclic—*N,C*-coupled naphthylisoquinoline alkaloids, named ancistrocyclinones A (**37**) and B (**38**)[1,2,255] (Fig. 22), were discovered. They were the first (and so far only!) representatives of a new subtype of naphthylisoquinoline alkaloids exhibiting a molecular framework reminiscent of that of berberine alkaloids, like e.g., berberine (**39**)[304–307] itself (Fig. 22). The two ancistrocyclinones and the berberines are nice examples documenting the "chemical creativity" of nature, here for achieving a biosynthetic convergence[83] of pathways leading to polycyclic isoquinoline alkaloids. While berberine alkaloids (like all the other plant-derived isoquinoline alkaloids!) are formed from aromatic amino acids,[89–94] naphthylisoquinoline alkaloids are built up from acetate/malonate units exclusively, via joint polyketide

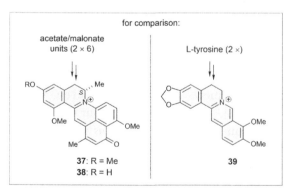

Fig. 22 Ancistrocyclinones A (**37**) and B (**38**),[255] unprecedented pentacyclic *N,C*-coupled naphthylisoquinoline alkaloids from *A. tectorius*, with a molecular framework similar to that of the tetracyclic isoquinoline alkaloid berberine (**39**)[304]—yet arising from totally different biosynthetic pathways. Identical structural parts are underlaid in pale blue-gray.

precursors,[80–83] and, in the case of **37** and **38**, apparently by *N,C*-analogous phenol-oxidative coupling and further oxidative condensation.[1,2,255] Thus, berberine (**39**) and the ancistrocyclinones A (**37**) and B (**38**), although structurally so similar, originate from completely different biosynthetic routes.

3.3 Naphthylisoquinoline dimers and their different subclasses

The third subfamily of naphthylisoquinoline alkaloids are the respective dimers, with meanwhile more than 50 known representatives. This group comprises an entirely novel type of quateraryl alkaloids with four aromatic systems, up to seven stereocenters and up to three consecutive chiral biaryl axes or *C,C*-bonds derived thereof. A major review[3] dealing, in more detail, with the unprecedented molecular architectures of such dimeric naphthylisoquinoline alkaloids and their occurrence in Asian and African *Ancistrocladus* species, appeared in 2019. It focusses on their isolation, structural elucidation, and total synthesis, including their pharmacological properties.

In many of the naphthylisoquinoline dimers (23 representatives),[1,3,5] the monomeric portions are coupled to each other via C-6′ of both naphthalene portions, i.e., via the least-hindered positions, so that the central biaryl axis can freely rotate and is, hence, not an additional element of chirality (see Fig. 23, top). Among these 6′,6″-coupled dimers, there are constitutionally symmetric compounds, consisting of two 5,8′- or two 7,8′-coupled naphthylisoquinoline portions like e.g., michellamine B (**5b**)[3,5,13,14] or ancistrogriffithine A (**41**),[1,3,283,284] but also "mixed," constitutionally unsymmetric representatives are known, like ealapasamine C (**40**),[3,180] a product of a 5,8′-coupled monomer cross-linked to a 7,8′-coupled one. All these 6′,6″-coupled dimers have exclusively been isolated from the Cameroonian liana *A. korupensis* (seven dimers)[3,5,13,14,288,289] and from four Congolese species, namely *A. congolensis* (seven representatives),[3,178] *A. likoko* (six examples),[3,53,176] *A. ealaensis* (four dimers),[3,180,181] and from a botanically as yet undescribed *Ancistrocladus* plant (six dimers[3,31]). Only one single Asian species, *A. griffithii* from Thailand, has been found to produce a 6′,6″-coupled dimer, ancistrogriffithine A (**41**).[1,3,283,284] In the 1990s, some of the compounds, in particular michellamine B (**5b**), were in the focus of intense research efforts (even including extended preclinical studies in the US), because **5b** was found to exhibit strong anticytopathic activities against different HIV strains, acting against both, early and late stages of the viral life cycle.[3,5,13,14,308,309]

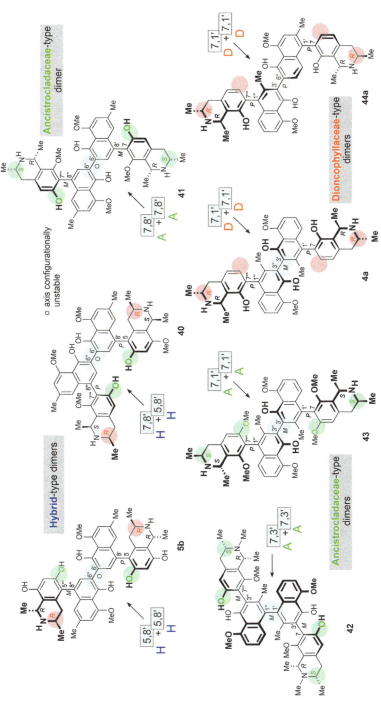

Fig. 23 Constitutionally and stereochemically unprecedented naphthylisoquinoline dimers with structural features of the Ancistrocladaceae-, Dioncophyllaceae-, and hybrid-type naphthylisoquinoline subclasses. Prominent examples are michellamine B (5b),[13,14] ealapasamine C (40),[180] and ancistrogriffithine A (41),[283,284] with two stereogenic outer biaryl axes and a central binaphthalene axis that is not rotationally hindered, while the shuangancistrotectorines D (42) and A (43),[12] jozimine A$_2$ (4a),[254] and jozilebomine A (44a)[183] possess three consecutive chiral biaryl axes.

From the Chinese liana *A. tectorius* five constitutionally symmetric naphthylisoquinoline dimers have been isolated.[1,3,254] Their monomeric halves are linked via sterically more constrained positions in the naphthalene moieties, so that, different from the 6′,6″-dimerized michellamine-type compounds, their central binaphthalene axis constitutes an additional stable element of chirality. Consequently, these quateraryls, like, e.g., the 3′,3″-coupled shuangancistrotectorine A (**43**)[254] or the 1′,1″-dimerized shuangancistrotectorine D (**42**)[254] (Fig. 23, bottom) were the first naphthylisoquinoline alkaloids possessing three consecutive chiral axes and, hence, exhibiting a total of up to seven stereogenic units. They consist of two fully identical, i.e., homomorphous naphthylisoquinoline halves (both belonging to the Ancistrocladaceae type), and are thus C_2-symmetric.[1,3,254]

From *A. ileboensis*,[1,3,52,183] *A. abbreviatus*,[1,58] and a botanically yet undescribed Congolese *Ancistrocladus* species,[1,3,12] a series of seven further dimeric naphthylisoquinoline alkaloids were isolated, likewise characterized by the presence of three consecutive chiral biaryl axes. In contrast to the shuangancistrotectorines[1,254] (Fig. 23), however, these dimers are R-configured at C-3 and lack an oxygen function at C-6 in both of their two isoquinoline subunits. They are, thus, the first naturally occurring Dioncophyllaceae-type naphthylisoquinoline dimers and consist of two 7,1′-coupled naphthylisoquinoline monomers related to dioncophylline A (**1a**).[1,3,12,52,58,183] The compounds are coupled via the C-3 positions in the two naphthalene units, as in jozimine A$_2$ (**4a**),[1,3,12] or their monomeric halves are connected through an unprecedented 3′,6″-coupling in the central binaphthalene core as in jozilebomine A (**44a**)[1,3,183] (Fig. 23). Thus, different from the C_2-symmetric jozimine A$_2$ (**4a**)[3,12] and from the structurally closely related shuangancistrotectorine A (**43**),[1,254] jozilebomine A (**44a**)[1,3,183] is constitutionally highly unsymmetric. Jozimine A$_2$ (**4a**) displays excellent in vitro antiplasmodial activities in the nanomolar range against the NF54 strain of the malaria parasite *Plasmodium falciparum* (sensitive to all known drugs), but exerts an extremely weak cytotoxicity, leading to an outstanding selectivity index (ratio of cytotoxicity of **4a** against rat skeletal myoblast L6 cells versus its antiparasitic effect) of ca. 11,400.[1,3,12] According to the TDR/WHO guidelines,[310] this dimer can thus be considered as a promising lead compound.

Further structurally unique dimers such as mbandakamine A (**45**) (Fig. 24), featuring a highly unsymmetric 6′,1″-coupling at the central binaphthalene axis, were discovered in a botanically as yet unidentified *Ancistrocladus* species endemic to the swamps and rain forests near the town Mbandaka in the Northwestern part of the Democratic Republic of the

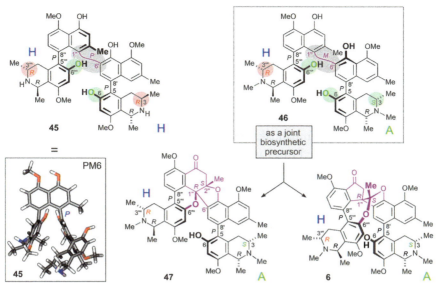

Fig. 24 Mbandakamine A (**45**)[32] and mbandakamine B₂ (**46**),[15] two largely unsymmetrical dimeric naphthylisoquinoline alkaloids with a rotationally highly hindered 6′,1″-coupled central biaryl axis, and two structurally likewise unprecedented dimers, cyclombandakamine A₂ (**47**)[33] and spirombandakamine A₁ (**6**),[15] displaying complex—and rigid—architectures, with a series of consecutive condensed rings, presumably arising from **46** by oxidation-induced cyclization reactions in the plants. The PM6-optimized structure of **45** visualizes the u-turn shape structures of the mbandakamines. Compounds **46**, **47**, and **6** belong to the rare subgroup of such mixed dimers, consisting of an Ancistrocladaceae- (A) and a hybrid-type (H) naphthylisoquinoline portion.

Congo.[3,32] Mbandakamine A (**45**) consists of two highly similar, yet differently O-methylated 5,8′-coupled monomeric halves. They are linked via the sterically strongly constrained 1″-position of one of their naphthalene units. The central axis is, thus, located in the *peri*-position of one of the naphthalene moieties neighboring one of the outer axes, so that, although the 6′-position is not highly shielded, this coupling type gives rise to a rotationally hindered central binaphthalene axis. The 6′,1″-coupling of the central biaryl linkage leads to an extremely high steric load, with one of the as yet highest degrees of steric hindrance at the central axis ever found for naphthylisoquinoline dimers isolated from nature.[3,32] In contrast to all other dimers,[3] which have a more or less linear molecular basic structure, mbandakamines possess a more u-turn like shape (see Fig. 24), entailing a close proximity of the outer ring systems to each other, which explains their high tendency to undergo further follow-up intramolecular linkage reactions (see below).[3,15,33,34] Besides mbandakamine A (**45**) itself, a whole series of further such mbandakamine-type dimers have been identified in the Central African liana *A. ealaensis*[181] and in another botanically as yet unknown *Ancistrocladus* species[35] from the Central Congo Basin, collected near the town Ikela, so that in total eight such mbandakamine-type naphthylisoquinoline alkaloids have so far been discovered.[3,32,35,181]

From the leaves of the as yet undescribed *Ancistrocladus* species from the Mbandaka region mentioned above, four structurally even more fascinating dimers related to the mbandakamines, but with a rigid architecture of eight consecutive condensed rings and a total of eight stereogenic elements—six centers and two axes—have been isolated.[3,33,34] These unusual plant metabolites, like cyclombandakamine A_2 (**47**)[33] (Fig. 24), are the first examples of twofold oxygen-bridged dimeric naphthylisoquinoline alkaloids with an unprecedented pyran-cyclohexenone-dihydrofuran sequence. A whole series of even 7 such natural products have been discovered in the Congolese liana *A. ealaensis*,[182] so that a total of 11 naturally occurring cyclombandakamines are known to date.[3,33,34,182]

Further in-depth isolation work on the leaves of that unknown *Ancistrocladus* species from the Mbandaka region furnished three additional, likewise structurally unique naphthylisoquinoline dimers, with an unprecedented spiro-fused-cage-like molecular framework, such as spirombandakamine A_1 (**6**)[3,15,34] (Fig. 24). These alkaloids possess a five-membered carbon ring and five- and seven-membered oxygen heterocycles, and—similar to the cyclombandakamines[3,33,34,182]—they have six stereogenic centers and two chiral axes.[3,15,34] Another remarkable structural feature of two of the spirombandakamines[15] refers to the fact that they (like their presumed precursor, mbandakamine B_2, **46**)[15] belong to a small subgroup of ten heterodimers[3,15,31,33,34,178,288] that are built up from one Ancistrocladaceae-type molecular half and one hybrid-type naphthylisoquinoline (Fig. 24). A vast majority of michellamine-, jozimine-A_2-, and mbandakamine-type dimers (no less than 47 compounds), by contrast, consist of 2 monomeric halves exclusively belonging to the same subclass of naphthylisoquinoline alkaloids, i.e., both portions are Ancistrocladaceae-compounds (6 compounds),[1,3,254,283] both Dioncophyllaceae-representatives (7 compounds),[1,3,12,52,58,183] or both hybrid-type compounds (34 examples).[1,3,5,13,14,31–35,53,176,178,180–182,288,289] So far, no dimers have been found that are formed from a Dioncophyllaceae-type monomer in combination with an Ancistrocladaceae- or a hybrid-type naphthylisoquinoline. Thus, the molecular halves of all dimers known from nature are always both 6-oxygenated, or—as in the case of the Dioncophyllaceae-type dimers—both lack an oxygen function at C-6 in both of their two isoquinoline subunits.

The cyclo- and spirombandakamines apparently originate as follow-up products from a cascade of oxidative cyclization reactions of respective "open-chain" naphthylisoquinoline dimers.[3,15,33,34] The identification of mbandakamine B$_2$ (**46**)[15] (Fig. 24) in the plants substantiates the assumption that dimers such as cyclombandakamine A$_2$ (**47**)[33] and spirombandakamine A$_1$ (**6**)[15] (Fig. 24) should arise from **46** as a joint precursor, by epoxidation in the highly strained C-1″–C-2″ region and subsequent ring opening, as a result of the unprecedented u-turn shaped structures of the mbandakamines. These follow-up cyclization reactions lead to a significant decrease of steric hindrance—and, simultaneously, to a substantial increase of complexity and rigidity.[3,15,33,34,182]

3.4 Free, non-coupled isoquinoline and naphthalene metabolites related to naphthylisoquinoline alkaloids

All of the mono- and dimeric naphthylisoquinoline alkaloids isolated so far from Ancistrocladaceae and Dioncophyllaceae plants have in common that always a naphthalene portion ("**N**") and an isoquinoline building block ("**IQ**") are cross-linked together ("**N–IQ**").[1–7] Isoquinolines are never homo-coupled to give dimeric isoquinolines ("**IQ–IQ**"), and—with the exception of some few binaphthalenes like triphyoquinol A$_2$ (**54**)[311] (Fig. 25), which were identified in the root cultures of *Triphyophyllum peltatum*—nearly never naphthalenes to naphthalenes ("**N–N**"). Likewise never detected were combinations of two isoquinolines linked to one naphthalene ("**IQ–N–IQ**" or "**IQ–IQ–N**"), or two naphthalenes coupled to one isoquinoline ("**N–IQ–N**" or "**N–N–IQ**"). Thus, the biosynthetic formation of naphthylisoquinoline alkaloids is highly specific. The two aromatic portions, in turn, arise from identical open-chain polyketide precursors,[80–83] followed by an apparently immediate, mostly atropo-selective phenol-oxidative cross-coupling[95,96] with subsequent partial or total O-methylation. This straightforward reaction sequence[80–83] is in agreement with the fact that free (i.e., not naphthalene-linked) tetrahydroisoquinolines ("**IQ**") (Fig. 25, top) like phylline (**48**)[73,297] are formed far less frequently in Dioncophyllaceae and Ancistrocladaceae plants than the complete naphthylisoquinoline alkaloids. *Ancistrocladus korupensis* produces the quaternary tetrahydroisoquinolinium salt gentrymine B (**50**),[290,312] which, despite its *trans*-configuration at C-1 versus C-3, is assumed to originate from the likewise naphthalene-devoid co-occurring, yet *cis*-configured *N*-monomethylated gentrymine A (**49**).[287] This was convincingly plausibilized by the biomimetic semisynthesis of **50** from **49**, which occurred with the predicted

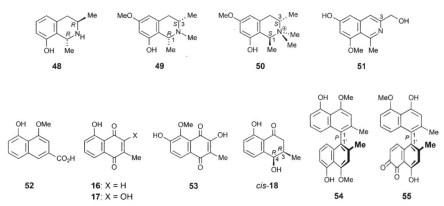

Fig. 25 Secondary metabolites related to the hetero- or isocyclic molecular portions of naphthylisoquinoline alkaloids isolated from Ancistrocladaceae and Dioncophyllaceae plants. The upper row shows free, non-coupled isoquinoline derivatives such as **48–51**, while the lower row displays naphthalene congeners like the naphthoic acid **52**, naphthoquinones like plumbagin (**16**), droserone (**17**), and ancistroquinone A (**53**), the tetralone cis-**18** (as a reduced analogue of the naphthoquinone **16**), the binaphthalene **54**, and the naphthylnaphthoquinone **55**, as an example of an oxidized follow-up product of a binaphthalene.

stereoinversion at C-1. Indeed, N-methylation of authentic (yet synthetically prepared[312]) gentrymine A (**49**), with its 1R,3S-configuration, using methyl iodide resulted in a spontaneous epimerization at C-1 in situ to give the more stable natural trans-isomer **50** in high yields.[312] This also explains the unexpected 1S,3S-array, which is so far unprecedented in *A. korupensis*.[5,287,290] All other trans-tetrahydroisoquinolines in that plant are 1R,3R-configured.[3,5,13,14,287–289] Furthermore, fully dehydrogenated analogues like anciscochine (**51**),[313] with its methyl group at C-3 being hydroxylated (Fig. 25, top), have been identified in Asian *Ancistrocladus* species.

Also free (i.e., not isoquinoline-linked) naphthalenes ("**N**") and follow-up products (Fig. 25, bottom) like e.g., ancistronaphthoic acid B (**52**),[179] have been isolated from Dioncophyllaceae and Ancistrocladaceae plants, mainly originating from oxidation reactions as a consequence of chemical, physical, or biotic stress. They are apparently formed from the free naphthalene part of naphthylisoquinolines, to give the naphthoquinone plumbagin (**16**), which may then be further oxygenated, leading to droserone (**17**),[1,5,81–83,151,153,269] or to compounds with an even higher oxygenation degree like e.g., ancistroquinone A (**53**).[314] In the plants, they occur along with related tetralones, like e.g., cis-isoshinanolone (cis-**18**), which results from the reduction of plumbagin (**16**).[1,5,83,151,269]

Less frequently, isoquinoline-devoid naphthalenes form dimers like triphyoquinol A$_2$ (**54**),[311] or they are further oxygenated to give naphthyl-naphthoquinones like triphyoquinone A (**55**).[311]

3.5 Species-specific production of naphthylisoquinoline alkaloids in Ancistrocladaceae and Dioncophyllaceae

Intense phytochemical studies on more than 20 *Ancistrocladus* species[1-7] (among them 6 botanically yet unidentified plants[12,15,16,29-35]) clearly revealed a species-specific production of mono- and dimeric naphthylisoquinoline alkaloids, thus demonstrating the important role of these secondary metabolites as useful phytochemical markers for an unequivocal species differentiation and for a clear attribution of taxonomic relationships within the genus *Ancistrocladus*. Already the classification of naphthylisoquinoline alkaloids of a plant under investigation as belonging to the Ancistrocladaceae type (3*S*, 6-O*R*), the Dioncophyllaceae type (3*R*, 6-H), or the hybrid type (3*R*, 6-O*R*) significantly indicates chemo- and geotaxonomic relationships between different plant species from Asia and Africa (see Section 3.1). The occurrence of distinct biaryl coupling types in a single plant taxon in combination with the presence of mono- and dimeric naphthylisoquinoline alkaloids likewise supports botanical and phytochemical relationships between certain plant species (see Section 3.1). The detection of non-hydrogenated naphthylisoquinoline alkaloids has significantly contributed to the taxonomic classification of *Ancistrocladus* species in Asia and Africa, since only few such compounds occur in the plants.[1,5,53,54,176,252,273,274,291,299] A quite high number of such metabolites (seven alkaloids, each), however, were identified in the new Malaysian taxon *A. benomensis*[1,273,291] and in *A. abbreviatus*[54] from Ivory Coast.

The alkaloid profiles of many of the Ancistrocladaceae and Dioncophyllaceae plants are dominated by the presence of naphthyl*tetrahydro*isoquinolines that possess a thermodynamically more stable *trans*-configuration at C-1 versus C-3, while *cis*-configured diastereomers (unless *N*-methylated) have so far been detected far less frequently.[1-7] The lower numbers of *cis*-isomers isolated from the plants does not necessarily reflect their rather rare occurrence in nature, because they might simply have gone lost during isolation, possibly due to their pronounced reactivity toward oxidants, leading to a quite rapid conversion to the corresponding dihydroisoquinolines already by air oxygen.[1,5,315] Furthermore, *cis*-configured naphthyltetrahydroisoquinolines are sensitive toward an acid- or base-catalyzed epimerization at C-1, thus leading to the more stable *trans*-compounds.[1,5,315] The instability of the *cis*-diastereomers is in a sharp contrast to the higher stability of the respective *N*-methyl analogues and of the *trans*-isomers. *N*-Methylated naphthylisoquinoline alkaloids[1–7] have been isolated from nearly all of the investigated Ancistrocladaceae and Dioncophyllaceae species (except for *Triphyophyllum peltatum*[5,6]). In some of the *Ancistrocladus* species they even appear in large numbers and quantities like, e.g., in *A. abbreviatus*,[5,55,57,59,60,62,65] *A. likoko*,[53,175,176] or *A. tectorius*.[1,239,244,249–251,253]

Regarding the occurrence of naphthyl*dihydro*isoquinolines, a pronounced number of representatives belonging to the Ancistrocladaceae-type or the hybrid-type subclass have been found in Asian *Ancistrocladus* taxa,[1] in the East African species *A. tanzaniensis*,[278,279] in *A. abbreviatus*[56,57,61,64] from Ivory Coast, and in some of the Central African taxa, among them *A. ealaensis*,[179,181,182] *A. likoko*,[10,53,176] and one botanically as yet undescribed Congolese *Ancistrocladus* species.[31] Dioncophyllaceae-type naphthyldihydroisoquinolines, by contrast, have very rarely been identified in nature. So far, only three such compounds are known, two of them discovered in the leaves of *A. abbreviatus* (see Sections 5.1 and 5.4).[5,6,56,61]

In conclusion, research on mono- and dimeric naphthylisoquinoline alkaloids has tremendously developed in many respects during the past decades (for the latest comprehensive reviews, see Refs. 1–3, 79), now permitting to attribute phylogenetic and chemotaxonomic relationships within the Ancistrocladaceae species as well as between Ancistrocladaceae and Dioncophyllaceae lianas, and to delineate Ancistrocladaceae and Dioncophyllaceae from Drosophyllaceae (see Fig. 4 in Section 2). A more in-depth insight into the high chemodiversity of naphthylisoquinoline alkaloids has become possible by the fruitful interplay of a wide repertoire of chromatographic,[1,3] spectroscopic,[1,3,66–68] degradative,[69,70] synthetic,[76–79] and computational[71–75] methods. Some of the analytical tools used for the isolation and structural elucidation of these plant metabolites have even been designed specifically for this purpose.[1,3,5,67,69,70,73–75] The repertoire of efficient and reliable methods for the isolation and structural elucidation of naphthylisoquinoline alkaloids will be presented in Section 4.

4. Isolation of naphthylisoquinoline alkaloids and assignment of their full absolute stereostructures

The broad range of structural features of the naphthylisoquinoline alkaloids, as presented in the previous section, poses a thrilling analytical challenge. The great structural diversity of the compounds,[1–7] originating from the various substitution patterns and the different coupling positions of the two molecular halves, the—usually stereogenic—biaryl axes, and the stereocenters in the isoquinoline moiety, all this necessitates the application of a broad spectrum of efficient methods for their structural elucidation. This section gives a short introduction into basic principles for a rapid phytochemical screening of plant extracts[1–3,31,67,68,73,283,297] and describes efficient chromatographic procedures for the isolation of the alkaloids. It also presents the spectroscopic (NMR, MS, ECD),[1,3,5,66–68,71–75] chemical (e.g., oxidative degradation[1,3,69,70] and semisynthesis[1,5,76–79]), and computational tools (quantum-chemical ECD calculations[71–75]) that have been applied (and, where required, even newly designed[1,3,5,66–79]) for the reliable assignment of the constitutions and of the relative and absolute configurations, i.e., of the complete stereostructures of naphthylisoquinoline alkaloids and related compounds.

4.1 Isolation of naphthylisoquinoline alkaloids and assignment of their constitutions

Naphthylisoquinoline alkaloids have been isolated from roots, stem bark, twigs, and leaves of Dioncophyllaceae and Ancistrocladaceae lianas[1–7] and from the respective cell cultures.[80–83,151,311] Typical isolation procedures[1–3] involve the exhaustive extraction of the air-dried powdered plant material with methanol (neutral or slightly acidic), followed by maceration of the crude extracts with water (to permit precipitation of chlorophylls) and further processing by liquid–liquid partition using *n*-hexane and dichloromethane to remove undesired non-polar compounds. Consecutive fractionation and purification of the dichloromethane layers by liquid–liquid and/or solid–liquid extraction, applying high-speed countercurrent chromatography (HSCCC) or fast centrifugal partition chromatography (FCPC), column chromatography (CC) on normal-phase silica gel, solid-phase extraction (SPE) on C_{18} reversed-phase silica gel, or ionic or gel permeation chromatography provides a series of alkaloid-containing subfractions. Metabolite-enriched fractions are resolved by semipreparative reversed-phase HPLC, assisted by LC–MS to search for masses hinting at the presence of naphthylisoquinolines and related compounds, finally affording pure naphthylisoquinoline alkaloids for structural and stereochemical investigations and for biological testing.

Combination of HPLC with mass spectrometry (MS) by the use of soft-ionization techniques, such as electrospray ionization (ESI) in positive mode, is a powerful analytical device to screen for the occurrence of mono- and dimeric naphthylisoquinoline alkaloids in crude plant extracts.[1–3,12,15,16,30–35,51–61,67,68,73,180–183,254,283,284,297,299,316] Monomeric compounds typically give monoprotonated molecules ($[M+H]^+$, m/z: 360–440) as the most abundant ions. Naphthylisoquinoline dimers usually display twofold protonated molecular ions ($[M+2H]^{2+}$, m/z 360–440), along with monoprotonated species ($[M+H]^+$, $m/z > 724$), thus the dimeric naphthylisoquinoline alkaloids can easily be distinguished from the monomers.[1,3,31,35,73,254,283,316]

Besides MS, IR, and UV analysis, the manifold 1D and 2D NMR methods are the by far most important tools for the structural elucidation of naphthylisoquinoline alkaloids.[1–3,5,6,66] The exact positions of ^{1}H and ^{13}C NMR signals give valuable insight regarding the molecular framework of naphthylisoquinoline alkaloids and their special structural features. Typical chemical shifts in the ^{1}H NMR spectrum indicative of the constitution of a naphthyl-1,3-dimethyltetrahydroisoquinoline are summarized in Fig. 26 and are as follows:

➢ two doublets resonating around δ_H 0.9–1.7 with a coupling constant of about 6 Hz, each corresponding to three protons, evidencing the presence of methyl groups at C-1 and C-3;
➢ a three-proton singlet at δ_H 2.0–2.5, revealing the presence of a methyl group at C-2′ in the naphthalene part;

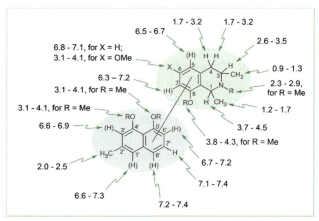

Fig. 26 Characteristic ^{1}H NMR data (δ in ppm) of monomeric C,C-coupled naphthyltetrahydroisoquinoline alkaloids.

> the signals of two aliphatic methin units, H-1 (q, δ_H 3.7–4.5) and H-3 (m, δ_H 2.6–3.5);
> the signals of two diastereotopic protons at C-4 (δ_H 1.7–3.2): Upfield-shifted signals (δ_H 1.7–2.2) hint at a close proximity of the protons at C-4 to the biaryl axis and, thus, give evidence that the naphthalene portion is not linked to C-7, but to C-5 in the isoquinoline moiety. Vice versa, "normal" chemical shifts for the two protons at C-4 (δ_H 2.80–3.20), i.e., with no shielding effect to be seen, indicate that there is no naphthalene part nearby; such values thus indicate that the biaryl axis is not linked to C-5, but to C-7 in the isoquinoline subunit;
> a downfield-shifted three-proton singlet (δ_H 2.3–2.9), as typical of an N-methyl function (not to be confused with the three-proton singlet of the methyl substituent at C-2' in the naphthalene part): The presence of a methyl substituent on the nitrogen atom can easily be assigned by its NOESY correlations to all neighboring protons in the isoquinoline subunit, here to H-1, Me-1, H-3, and Me-3;
> up to four singlets around δ_H 3.1–4.1, each corresponding to three protons, evidencing the presence of up to four methoxy groups, which are located in the naphthalene unit at C-4' and/or C-5', and in the isoquinoline portion at C-6 and/or C-8;
> a sharp signal at around δ_H 9.0–9.5 (not always detectable), hinting at a hydrogen-bridged hydroxy-methoxy array at C-4'–C-5' or C-5'–C-4' of the naphthalene part;
> five or six aromatic protons resonating at δ_H 6.3–7.5: Five protons are observed in the case of Ancistrocladaceae-type naphthylisoquinoline alkaloids, whereas the presence of six aromatic protons is indicative of a Dioncophyllaceae-type representative. The spin pattern of the aromatic protons can give a first hint at the coupling type of a respective naphthylisoquinoline (as an example, in the case of 5,1'-coupled Ancistrocladaceae-type alkaloids, a spin pattern of three doublets of doublets and two singlets is monitored). An unambiguous assignment of the coupling positions in the naphthalene and isoquinoline subunits can, however, only be achieved by specific Nuclear Overhauser and Exchange Spectroscopy (NOESY) and by Heteronuclear Multiple Bond Correlation (HMBC) interactions.

In the naphthalene part, four coupling positions are known, viz. C-1', C-3', C-6', and C-8' (see Fig. 27, left). Assuming a biosynthetic origin of the biaryl axes by a phenol-oxidative coupling of the two molecular halves, only these four positions are possible, all of them located either *ortho* or *para* to an

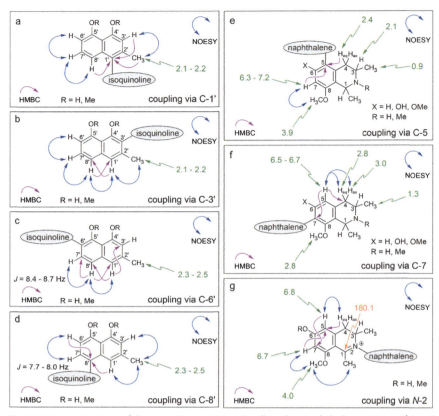

Fig. 27 Determination of the coupling position (a–d) in the naphthalene part and (e–g) in the isoquinoline subunit, by typical ^{1}H NMR chemical shifts (δ in ppm), coupling constants, and by diagnostically valuable HMBC (single purple arrows) and NOESY (double blue arrows) interactions.

oxygen substituent. The question whether the coupling position of the biaryl axis is located in the methyl-substituted ring (i.e., at C-1′ or C-3′) (Fig. 27a and b) or in the other one (at C-6′ or C-8′) (Fig. 27c and d), can easily be answered by the pattern of the aromatic protons and by analysis of the chemical shift of the 2′-Me signal. The anisotropic effect caused by an aromatic substituent (here the isoquinoline residue) results in highfield-shifted signals for the protons near the axis. Hence, the 2′-Me signal, appearing at δ_H 2.3–2.5 in 6′- and 8′-coupled naphthalenes (for some examples, see Refs. 11, 13, 14, 29–35, 51, 53, 175–182, 249, 264, 275, 287–289), is shifted to δ_H 2.1–2.2 in 1′- and 3′-coupled ones (for selected examples, see Refs. 8, 30, 56, 57, 61, 250, 252, 254, 265, 278, 284, 292) and is further shifted to δ_H 1.7–1.9 (i.e., the lowest δ_H values ever reported in naphthylisoquinolines) if the methyl group at C-2′ is flanked by even two axes.[1,3,12,52,58,254] This was the case for 3′,3″-coupled dimers like e.g., shuangancistrotectorines A (**43**)[1,254] and D (**42**)[1,254] or jozimine A$_2$ (**4a**)[3,12] (see Fig. 23 in Section 3) and their atropisomers.[52,58] The differentiation whether the coupling site is C-1′ (Fig. 27a) or C-3′ (Fig. 27b) can sometimes be achieved only by HMBC and NOESY measurements (for some examples, see Refs. 30, 51, 57, 61, 250, 251, 254, 278, 279, 292). Coupling positions at C-6′ (Fig. 27c)[249,275,298] and C-8′ (Fig. 27d) (for some illustrative examples, see Refs. 11, 13, 14, 29–35, 51, 175–182), on the other hand, may be distinguished from each other by the J value of 7′-H and by analyzing the neighboring proton in the other *peri*-position (viz., at C-1′).[275] The higher double-bond character of the C-7′–C-8′ linkage leads to slightly increased coupling constants (J = 8.5 Hz) between H-7′ and H-8′ (in 6′-coupled naphthalenes) compared to those between H-7′ and H-6′ (J = 8.0 Hz) in 8′-coupled ones. Thus, a value of J = 8.5 Hz indicates the axis to be located at C-6′.[275]

A coupling of the naphthalene subunit at C-5 (Fig. 27e) of the isoquinoline part (compared to a linkage via C-7) is easily recognized by the high-field shift of the H-4 signals from ca. δ_H 2.6 to ca. δ_H 2.1 and of the Me-3 peak from δ_H 1.3 to δ_H 0.9 (for some examples, see Refs. 29–35, 51, 53, 57, 66, 176, 250, 251, 278). If the naphthalene substituent is located at C-7 (Fig. 27f), by contrast, it is the signal of a methoxy substituent at C-8 that is high-field shifted (here from δ_H 3.9 to δ_H 3.4) (for illustrative examples, see Refs. 11, 54, 55, 59, 60, 66, 239, 240, 251, 264, 265, 275–277, 279, 295). The positions of the methoxy groups, in turn, can be deduced from their chemical shifts if the coupling type is known, otherwise HMBC and NOE measurements have to be performed.

The detection of one additional aromatic proton, as established by NOE correlations in the series of protons and substituents in the isoquinoline subunit and confirmed by specific HMBC interactions of H-5 to C-7, and by HMBC effects of both H$_{eq}$-4 and H-7 to C-5, supports the presence of an *N*-iminium–*C*-aryl axis (Fig. 27g) in the naphthylisoquinoline alkaloid.[1,2,16,299] The screening of extracts by LC-UV will already give a first hint that the compound is an *N,C*-coupled naphthyldihydroisoquinolinium alkaloid, because in the UV spectrum of such compounds the second maximum is bathochromically shifted by 30–40 nm, in comparison to the UV curves monitored for *C,C*-linked naphthylisoquinolines (typically around 305 nm).[1,2,16,299]

During the past years, intense phytochemical studies on Dioncophyllaceae and Ancistrocladaceae plants have furnished a plethora of naphthyltetrahydroisoquinoline alkaloids like e.g., ancistrobrevine H (**8**)[57] (Fig. 28a),

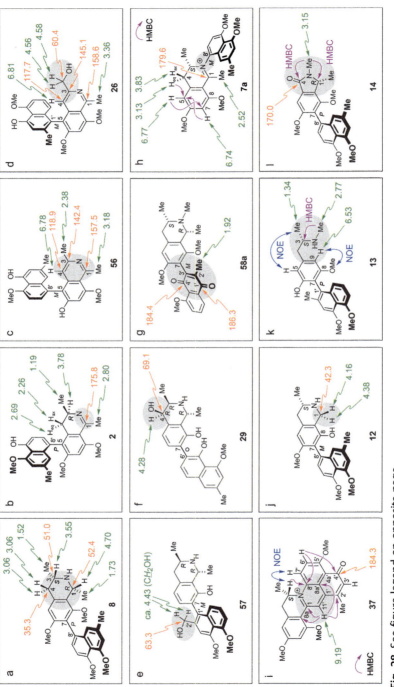

Fig. 28 See figure legend on opposite page.

showing a high structural diversity due to different coupling sites and manifold variations in their O- and N-methylation patterns.[1–7] As presented in Fig. 28, further possible constitutional features of naphthylisoquinoline alkaloids have to be taken into account, as obvious from the chemical shifts of the respective compounds in ^{1}H and ^{13}C NMR:

> the lack of the H-1 quartet (which normally appears around δ_H 4.0 in naphthyltetrahydroisoquinolines), the downfield shift of the protons of the methyl group at C-1, resonating at δ_H 2.2–2.7, along with the changed multiplicity (a singlet instead of the doublet around δ_H 1.4 as typical of a naphthyl*tetrahydro*isoquinoline), and the downfield chemical shift of the ^{13}C NMR resonance of C-1 at around δ_C 170–180, evidencing the presence of a naphthyl-1,3-dimethyl*dihydro*isoquinoline such as ancistrolikokine E$_3$ (**2**)[10] (Fig. 28b);

> an additional singlet (δ_H 6.8–7.6), together with the loss of the diastereotopic geminal protons at C-4 and the simultaneous absence of two doublets of the aliphatic methyl groups at C-1 and C-3 around δ_H 1.1 and δ_H 1.9, the missing quartet of H-1 around δ_H 4.4, and the lacking multiplet of H-3 in the region between δ_H 3.2 and 4.0, are indicative of a naphthylisoquinoline alkaloid like e.g., ancistrolikokine G (**56**)[176] (Fig. 28c) possessing a non-hydrogenated isoquinoline moiety, which is confirmed by significant downfield shifts of the ^{13}C NMR resonances of C-1, C-3, and C-4;

Fig. 28 ^{1}H (in dark green) and ^{13}C (in orange) NMR chemical shifts (δ in ppm), typical of naphthylisoquinoline alkaloids displaying special structural features. NMR chemical shifts (δ in ppm), typical of (a) naphthyl*tetrahydro*isoquinolines like ancistrobrevine H (**8**),[57] (b) naphthyl*dihydro*isoquinolines like ancistrolikokine E$_3$ (**2**),[10] (c) non-hydrogenated alkaloids like ancistrolikokine G (**56**),[176] alkaloids with (d) a hydroxymethylene function at C-3, as in ancistrobenomine A (**26**),[273] and of compounds with special features such as (e) a hydroxymethylene function at C-2', as in habropetaline A (**57**),[296] (f) a secondary hydroxy group at C-4, as in dioncophyllinol B (**29**),[275] or (g) a naphthoquinone portion, as in ancistrotectoquinone A (**58a**)$^\nabla$.[253] Further representatives with unusual molecular scaffolds are N,C-coupled naphthylisoquinolines like (h) ancistrocladinium A (**7a**)[16] and (i) the further cyclized—and, thus, even pentacyclic—ancistrocyclinone A (**37**)[255], the (j) 1-*nor*-alkaloid 1-*nor*-8-O-demethylancistrobrevine H (**12**),[55] and the two representatives ancistrosecoline A (**13**)[55] and ancistrobrevoline A (**14**)[59] of the novel, recently discovered subgroups of (k) *seco*-type and (l) ring-contracted naphthylisoquinoline alkaloids. $^\nabla$Ancistrotectoquinone A (**58**) occurs as a pair of configurationally semi-stable and, thus, slowly interconverting atropo-diastereomers, **58a** and **58b**. Here, arbitrarily, only the *M*-atropo-diastereomer **58a** is shown.

- some of the fully dehydrogenated naphthylisoquinoline alkaloids such as ancistrobenomine A (**26**)[1,252,273] (Fig. 28d) possess a hydroxymethyl group at C-3, as obvious from the lack of resonances for the methyl group at C-3, and from the downfield shifted ^{13}C NMR signal at δ_C 60.4–61.4 typical of the hydroxymethyl function and, due to the nearby axial chirality of **26**; the multiplicity of the two—hence diastereotopic—geminal methylene protons can result in their appearance as a pair of doublets with a large coupling constant, and give rise to a negative signal in the DEPT-135 spectrum;
- like the methyl group at C-3, also the likewise benzylic Me-2′ substituent in the naphthalene subunit can be oxygenated to give a hydroxymethyl group as in the case of habropetaline A (**57**)[296,297] (Fig. 28e); the remaining—now diastereotopic—two protons of that previous methyl group give signals at around δ_H 4.5, which may appear as a pair of doublets, but can also be fused to give a (then usually broad) singlet;
- some few naphthylisoquinoline alkaloids like e.g., dioncophyllinol B (**29**)[275] (Fig. 28f) exhibit a secondary hydroxy function at C-4; the signal of the remaining proton at C-4 will then appear as a (down-field shifted) doublet at δ_H 4.3;
- in a few cases, alkaloids like ancistrotectoquinone A (**58**)[1,253] (Fig. 28g) have been isolated which, besides the "usual" isoquinoline subunit, possess a rare naphthoquinone portion (instead of a naphthalene unit). This can easily be recognized from the appearance of two carbonyl signals at around δ_C 180–186, the extremely upfield-shifted signal of Me-2′ resonating at δ_H 1.7–1.9, and the lack of one aromatic proton compared to the spectra of all other monomeric alkaloids;
- the detection of one additional aromatic proton, as established by NOE correlations in the series of protons and substituents in the isoquinoline subunit and confirmed by specific HMBC interactions of H-5 to C-7, and by HMBC effects of both H$_{eq}$-4 and H-7 to C-5, supports the presence of an N-iminium-C-aryl axis in the respective naphthylisoquinoline alkaloid, shown exemplarily for ancistrocladinium A (**7a**)[1,2,16,299] (Fig. 28h); furthermore, in comparison to the UV spectra monitored for C,C-coupled naphthylisoquinolines (typically around 305 nm), the second maximum of N,C-linked naphthylisoquinolines is bathochromically shifted by 30–40 nm, to appear around 335 nm;[2,16,299]
- some of the N,C-coupled alkaloids such as ancistrocyclinone A (**37**)[1,2,255] (Fig. 28i) show a pentacyclic molecular framework, exhibiting

six protons in the aromatic region, of which one was significantly downfield-shifted, resonating at δ_H 9.2. Further NMR data distinguishing **37** from usual naphthylisoquinoline alkaloids were the lack of the signal for the methyl group at C-1, and the appearance of a carbonyl resonance at δ_C 184.3. HMBC couplings of H-6' to C-8' and C-4' established the carbonyl carbon to be C-4', while HMBC cross-peaks from the downfield-shifted proton at δ_H 9.2 to C-1 and C-8a in the isoquinoline unit, and to C-8a' and C-2' in the naphthalene moiety, assigned that proton to be H-11', thus revealing the dihydroisoquinolinium unit to be linked to the naphthalene moiety by a C—C bond from Me-1 to C-11';

➤ a lack of the—otherwise generally present—methyl group at C-1 (whose ^{1}H NMR signal normally is at δ_H 1.6–1.8), the appearance of a signal at ca. δ_C 42 in the DEPT-135 spectrum, and the presence of two diastereotopic protons (along with the ones at C-4), evidence that the usual methyl group at C-1 is now replaced by a second hydrogen and, thus, indicate a 1-*nor*-alkaloid such as 1-*nor*-8-*O*-demethylancistrobrevine H (**12**)[55] (Fig. 28j);

➤ the occurrence of six instead of five aromatic protons, an unusual NOESY interaction between the methyl group at C-3 and the aromatic proton at C-5, a *meta*-coupled proton at C-9, and—in the case of the presence of an *N*-methyl function—a likewise unprecedented HMBC interaction of that aliphatic *N*-methyl group to C-3, hint at the presence of a *seco*-type naphthylisoquinoline alkaloid such as ancistrosecoline A (**13**)[55] (Fig. 28k);

➤ A downfield-shifted signal in the ^{13}C NMR spectrum at around δ_C 170, typical of a carbonyl group, a downfield-shifted three-proton singlet characteristic of an *N*-methyl function, HMBC interactions from that *N*-methyl group to the aliphatic carbon atom C-1 and to the carbonyl C-atom hint at a ring contraction of the usual six-membered isoquinoline entity due to the loss of C-3 of a normal naphthylisoquinoline alkaloid and, thus, at the presence of an alkaloid such as ancistrobrevoline A (**14**)[59] (Fig. 28l) consisting of a naphthalene portion and an isoindolinone subunit, i.e., with the six-membered heterocycle contracted to become a five-membered ring;

➤ finally, homo- and heterodimeric naphthylisoquinoline alkaloids can be identified by their (nearly) double masses in MS when using mild ionization techniques (mono-ionization), and by ^{1}H NMR, due to the lack of one aromatic proton (lost during the coupling step). A detailed report on the repertoire of methods applied for the assignment of the complete stereostructures of naphthylisoquinoline dimers is presented in a more recent major review,[3] which appeared in 2019.

4.2 Assignment of the relative configuration at the stereogenic centers, C-1 versus C-3, by 1D and 2D NMR spectroscopy

With the constitution of a new alkaloid established, the assignment of its relative configuration at centers and axis is the next issue to be addressed. The conformation of the nitrogen-containing ring of the tetrahydroisoquinoline subunit is usually determined by the Me-3 group, which preferentially adopts an equatorial position (see Fig. 29). For 1,3-dimethyl-1,2,3,4-tetrahydroisoquinolines, the relative array of the methyl groups may be deduced from the chemical shift of the H-3 signal. This peak, which is sometimes difficult to detect due to its low intensity and its extended multiplicity, typically appears above δ_H 3.7 (usually at δ_H 3.7–4.2) for *trans*-configured naphthylisoquinolines[5,8,63,245,248,275,285,294,295] (Fig. 29a and b), while it is below δ_H 3.5 (usually at δ_H 3.0–3.5) for the respective *cis*-diastereomers[5,11,51,53,57,181,275] (Fig. 29c and d). Traces of acid in the sample (as sometimes resulting from the trifluoroacetic acid used in the preceding HPLC run) may lead to a (partial) *N*-protonation and, thus, to a shift to lower field, which may pretend a *trans*-configuration. Thus, in each case, for a reliable assignment of the relative configuration of the two methyl groups at C-1 and C-3, NOESY measurements are inevitable to clarify the situation.

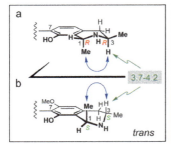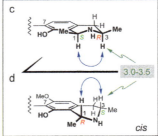

Fig. 29 Relative configurations of the two stereogenic centers at C-1 and C-3 in the tetrahydroisoquinoline portions of naphthylisoquinoline alkaloids, as deduced from NOESY interactions (double blue arrows), and further supported by the chemical shifts (δ in ppm) monitored for H-3. NOESY cross-peaks between Me-1 and H-3 are indicative of a *trans*-configuration like 1*R*,3*R*, as in (a), or 1*S*,3*S*, as in (b), and those between H-1 and H-3 evidence the presence of *cis*-configured compounds like 1*S*,3*R*, as in (c), or 1*R*,3*S*, as in (d).

4.3 Assignment of the relative configuration at centers versus axis by NOESY measurements

For naphthylisoquinoline alkaloids in which the naphthalene moiety is not too far away from the chiral part of the tetrahydroisoquinoline portion, the elucidation of the axial configuration relative to the stereogenic centers is usually achieved by specific NOE interactions across the biaryl axis.[1,3,5,6,17,57,63,66] NMR analysis of the stereochemical relationships between the spin systems in the naphthalene subunit and protons with a defined stereochemical arrangement in the isoquinoline moiety (viz., the protons at C-1 and/or C-4 and/or the three hydrogen atoms of the C-methyl group at C-1) provides valuable information about the configuration at the biaryl axis relative to the centers. Correlation of that relative axial stereoarray to the well-defined absolute configurations in the isoquinoline portion (established by ruthenium-mediated oxidative degradation,[1,3,69,70] see Section 4.4) permits assignment of the absolute configuration at the biaryl axis, and finally gives rise to the full absolute stereostructure of the isolated compound.

NOESY or ROESY spectroscopy in combination with the oxidative degradation procedure has become a reliable and widely used standard method for the stereochemical attribution of the absolute axial configuration of naphthylisoquinoline alkaloids, in particular in the case of 5,1′- and 5,8′-linked representatives.[1,3,14,15,30–35,53,56,57,63,175–182,250,252,259,287,288] This is of special importance because these two coupling types comprise the by far largest group of secondary metabolites (more than 110 compounds) isolated from Dioncophyllaceae and Ancistrocladaceae lianas,[1–7] among them a quite remarkable number of representatives displaying promising activities against pathogens causing tropical diseases[15,34–37,180,181,250,251,287] and/or against malignant cells.[10,30,31,53,56,57,176,181]

In many cases, strong NOE effects between the isoquinoline and naphthalene subunits of 5,1′- and 5,8′-coupled naphthylisoquinoline alkaloids can be observed, because of the spatial proximity (≤ 3.5 Å) of the relevant protons on both sides.[3,5,53,56,57,63,66,176,250,287] As an example, the axis-relevant NOE cross-peaks of H-8′ with H-4$_{eq}$ and of Me-2′ with H-4$_{ax}$ in the 5,1′-coupled alkaloid ancistrobrevine F (**59a**)[57] (Fig. 30a), isolated from the roots of the West African species *A. abbreviatus*, were complementary to those of its co-occurring atropo-diastereomer, 5-*epi*-ancistrobrevine F (**59b**),[57] which displayed specific NOE interactions between H-8′ and H-4$_{ax}$ and between Me-2′ and H-4$_{eq}$ (Fig. 30b). The correlation between H-8′ and the equatorial proton at C-4 in ancistrobrevine F

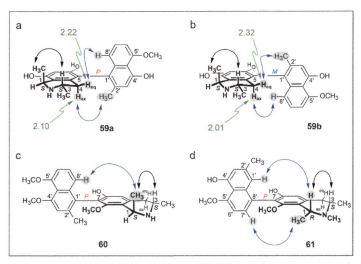

Fig. 30 Determination of the relative configuration of naphthylisoquinoline alkaloids at centers versus axis using NOESY experiments. Illustrated are specific NOESY interactions defining the relative configuration at C-3 versus C-1 within the tetrahydroisoquinoline portions (double black arrows) and the—likewise relative—configurations at the axes (double blue arrows) of the 5,1′-coupled (a) ancistrobrevine F (**59a**) and (b) its *M*-atropo-diastereomer, 5-*epi*-ancistrobrevine F (**59b**),[57] (c) of the 7,1′-coupled ancistectorine B₁ (**60**),[250] and (d) of the 7,8′-coupled ealamine H (**61**).[11]

(**59a**) established the two spin systems to be on the same side of the isoquinoline plane, which, given the known 3*S*-configuration (as assigned by oxidative degradation), evidenced H-4$_{eq}$ to be on the upper side of the isoquinoline and, thus, revealed the biaryl axis to be *P*-configured.[57] This stereochemical assignment of **59a** was further corroborated by the complementary NOE interaction between the axial proton at C-4 and Me-2′, being both underneath, i.e., on the bottom side of the isoquinoline plane (Fig. 30a). In the case of 5-*epi*-ancistrobrevine F (**59b**), where the NOE correlations across the axis were opposite, an *M*-configuration was analogously deduced (Fig. 30b).[57]

In agreement with the hence established 1*S*,3*S*,*P*-array for **59a** and the 1*S*,3*S*,*M*-configuration for **59b**, the equatorial proton at C-4 was upfield shifted in the case of **59a** (δ_H 2.22) as compared to the analogous signal in **59b** (δ_H 2.32). Vice versa, the axial proton at C-4 was upfield shifted for **59b** (δ_H 2.01) as compared to that of **59a** (δ_H 2.10).[57] In general, the protons that are closer to the larger naphthalene portion expectedly are more strongly influenced by its ring-current effect and thus show higher-shifted resonances (Fig. 30a and 30b).

For naphthylisoquinoline alkaloids with a 7,1'- or a 7,8'-coupling site, in which the relevant protons of the two ring systems are quite distant from each other (in some cases up to 5 Å), optimized long-range NOE measurements have to be applied for the elucidation of the axial configuration relative to the stereogenic centers.[11,12,51,52,54,58–60,66,180,183,250,254,264,285] As exemplarily presented for the 7,1'-coupled alkaloid ancistectorine B$_1$ (**60**),[250] which is a minor metabolite from *A. tectorius*, a NOESY correlation between Me-1 and H-8' established that these two spin systems were on the same side of the isoquinoline plane (Fig. 30c). This was, taking into account the 1*S*-configuration assigned by oxidative degradation, in agreement with the *trans*-configuration of the tetrahydroisoquinoline portion as deduced from NMR, the upper side of the isoquinoline, thus giving rise to the attribution of the stereoarray at the axis as *P*-configured.[250] The combination of the results from NMR with those from the oxidative degradation likewise permitted an elucidation of the absolute axial configuration of some representatives of 7,8'-linked naphthylisoquinoline alkaloids, among them ealamine H (**61**),[11] a minor constituent of the Congolese liana *A. ealaensis*. Based on the absolute 1*R*,3*S*-stereoarray in the tetrahydroisoquinoline portion of **61**, long-range NOESY cross-peaks between Me-1 and H-7' and between H-1 and H-1' (Fig. 30d) revealed the small part of the axis-bearing naphthalene-ring (C-6' and C-7') to be below the tetrahydroisoquinoline plane. The larger part of the naphthalene ring system (here the one with the methyl-substituted ring unit) had to be located above the plane, which evidenced the biaryl axis of ealamine H (**61**) to be *P*-configured.[11]

A pivotal prerequisite for the assignment of the relative configuration of naphthylisoquinoline alkaloids at centers versus axes by NOESY or ROESY spectroscopy is the presence of a spin system in the heterocyclic part that has a stereochemically known and well-defined spatial orientation above or below the isoquinoline plane, and that is sufficiently near to protons in the naphthalene part. The method is thus difficult to apply to naphthyldihydroisoquinoline alkaloids, because here C-1 is "flat" (since sp^2-configured). It is particularly tricky for 7,1'- or 7,8'-coupled ones, where the stereocenter at C-3 is too far away for permitting significant NOE long-range interactions. As exemplarily demonstrated for some 7,8'-linked alkaloids isolated from the West African liana *A. abbreviatus*,[57] this problem was solved by diastereoselective conversion of the dihydroisoquinoline subunit into the corresponding *cis*-configured tetrahydroisoquinoline. The newly generated "auxiliary" stereocenter at C-1—much closer to the naphthalene part than C-3—now provided decisive NOE interactions between the isoquinoline and the naphthalene half, and thus a reliable attribution of the axial configuration (for a detailed report on this stereoanalytical device, see Section 5.5).[57] In other cases, specific NOE interactions from a spin system in the naphthalene part with one of the two diastereotopic protons at C-4 can be analyzed (for illustrative examples, see Refs. 29, 30, 35, 53, 56, 175–183, 251, 252), but the method is sometimes hampered by an insufficient resolution of the signals of these two protons, H-4$_{ax}$ and H-4$_{eq}$.[11,31,51,57,82] The method may fail even completely for other representatives if the diagnostically relevant NOE interactions are too weak.[82,251,276,278,289]

A complementary approach is the analysis of the chiroptical properties of the alkaloids. For this purpose, ECD investigations (for compounds with a new molecular framework in combination with quantum-chemical ECD calculations[15,16,55,59,71–75,182,183,253–255,267,278,292,300]) are the method of choice, since the ECD spectra of two atropo-diastereomeric naphthylisoquinoline alkaloids like e.g., ancistrobrevine F (**59a**) and 5-*epi*-ancistrobrevine F (**59b**),[57] are usually nearly opposite to each other (Fig. 31). Their ECD behavior is mainly influenced by the contributions of the main chromophores, here the chiral biaryl system of naphthylisoquinolines, over those of the stereogenic centers in the tetrahydroisoquinoline portion. A detailed introduction to the application of ECD investigations for the elucidation of the axial configuration of naphthylisoquinoline alkaloids will be given in Section 4.5.

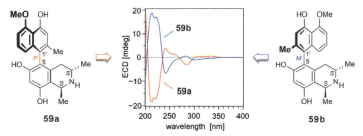

Fig. 31 Nearly opposite ECD spectra of ancistrobrevine F (**59a**) and 5-*epi*-ancistrobrevine F (**59b**), although the two compounds have the same configuration at the stereogenic centers, C-1 and C-3, in the tetrahydroisoquinoline subunit, and are, thus, diastereomers.

4.4 Assignment of the absolute configuration at the stereogenic centers by oxidative degradation

For the elucidation of the absolute configuration at the stereogenic centers C-1 and C-3 of naphthylisoquinoline alkaloids possessing a tetrahydro- or a dihydroisoquinoline subunit, a highly efficient chemical procedure has been developed in the authors' group, based on a smooth ruthenium-mediated oxidative degradation reaction.[1,3,69,70] This reliable and sensitive method (Fig. 32) works routinely down to submilligram quantities (ca. 0.1–0.5 mg) of the analyte. The periodate oxidation transforms the tetra- or dihydropyridine heterocycle into the simple and easy-to-analyze chiral amino acids alanine ("Ala," mainly formed from C-1) and 3-aminobutyric acid ("3-ABA," exclusively derived from C-3), which thus carry the direct information about the absolute configuration at the two—now separately analyzable—stereocenters of the investigated alkaloids.[1,3,69,70] Oxidative degradation of N-methylated naphthylisoquinoline alkaloids analogously provides N-methylalanine and N-methyl-3-aminobutyric acid, usually together

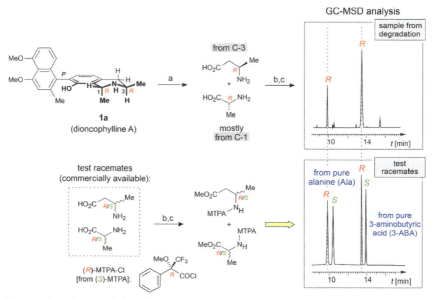

Fig. 32 Elucidation of the absolute configuration at the stereogenic centers in the di- and tetrahydroisoquinoline portions of naphthylisoquinoline alkaloids by ruthenium-mediated oxidative degradation and analysis of the resulting chiral amino acids by gas chromatography with mass-selective detection (GC-MSD) of the Mosher derivatives, here exemplified for dioncophylline A (**1a**). Reaction Conditions: (a) RuCl$_3$, NaIO$_4$; (b) SOCl$_2$, MeOH; (c) (R)-MTPA-Cl ("Mosher's chloride").

with their oxidatively N-demethylated products alanine and 3-aminobutyric acid.[1,3,70] After conversion of the obtained amino acids with SOCl$_2$/MeOH to give the respective methyl esters and Mosher-type derivatization with (R)-α-methoxy-α-trifluoromethylphenylacetyl chloride [(R)-MTPA-Cl, prepared from (S)-MTPA], the absolute configurations of the respective naphthylisoquinolines can be assigned by gas chromatography on an achiral dimethylpolysiloxane-coated silica capillary column coupled to a mass-selective detector (GC-MSD). Chromatographic comparison of the Mosher derivatives of the alkaloid sample with the respective pure derivatives of the authentic amino acids of known configuration (from a test racemate) finally provides direct information about the absolute configuration at the chiral centers.[1,3,69,70]

It must, however, be pointed out that in contrast to the reliable stereo-chemical attribution at C-3, as deduced from the formation of 3-aminobutyric acid, the one at C-1 may be less unambiguous, in particular for *cis*-configured tetrahydroisoquinolines, which are rapidly oxidized to give the respective 3,4-dihydroisoquinolines,[1,3,5,315] thus losing the genuine stereo-information at C-1. An additional problem may arise from the fact that the "northern part," covering C-3, is, to a small degree, also degraded further down to alanine, so that this amino acid does not only result from C-1, but also from C-3 and, thus does not give fully reliable information about the configuration at C-1.[1,3,70] This is the reason why stereochemical investigations at C-1 should always be accompanied by the NOE-based assignment of the relative configuration at C-1 versus C-3, whose absolute configuration is reliably known from the degradation.

The particular value of the oxidative degradation procedure lies in the fact that due to the chiroptical predominance of the biaryl chromophore, it is the only method that gives immediate access to the absolute configuration at the stereogenic center(s), and that it can be performed at a sub-mg level. Otherwise, a reliable stereochemical assignment might only be achieved by a—quite demanding—stereoselective total synthesis[1–3,5,76–79] of the investigated molecules, or by an X-ray structure analysis of a heavy-atom derivative, which is, however, possible only in rare cases.[287,317]

4.5 Assignment of the absolute axial configuration by experimental ECD spectroscopy in combination with quantum-chemical ECD calculations

ECD spectroscopy distinguishes between stereoisomers that show a sufficient chiroptical differentiation, i.e., usually between enantiomers, because their ECD curves are fully opposite.[71–75,318–323] Since the chiroptical behavior of a compound with a chiral axis and stereogenic centers is usually dominated by the biaryl chromophore (in the presence of which the ECD contributions of the centers are hardly to be seen), the ECD spectra of atropo-diastereomeric naphthylisoquinoline alkaloids will usually give near-opposite ECD curves (see Fig. 31).[1–3,5,16,31,32,51,52,57,58,61,62,71–75,250,287,292] Thus, the axial configuration of a new naphthylisoquinoline alkaloid may be established simply by comparison of its ECD spectrum with that of a structurally related known representative of the same coupling type, regardless of the configurations at the stereocenters. If, however, the compounds to be compared do not resemble each other enough and/or the chiroptical similarity is not convincing, quantum-chemical calculations of ECD spectra are an elegant (sometimes even the only) way to address this problem, also for other natural and synthetic products.[71–75,318–330] In contrast to the interpretation of ECD spectra by the use of ECD rules such as the Octant Rule[319,331] or the Exciton Chirality Method,[318–323,326,329] this theoretical approach is not limited by any structural restrictions.

The use of the Octant Rule is restricted to chiral ketones possessing a rigid carbon skeleton displaying a well-defined conformational behavior and having no further chromophores. The applicability of the Exciton Chirality Method depends on the fulfillment of particular structural preconditions such as the presence of a molecule with two electronically similar, if possible (near-)identical chromophores, which must be in a defined spatial orientation to each other. But even then the Exciton Chirality Method can give wrong results.[332] For this reason, the direct application of the Exciton Chirality Method to genuine naphthyltetrahydroisoquinoline alkaloids, with their most different molecular portions, is not reliable.[17,254,333] Therefore, such naphthyltetrahydroisoquinoline alkaloids like e.g., dioncophylline A (**1a**)[5,6,8,9] (Fig. 33) first have to be modified by dehydrogenation. This has to be done under mild conditions, in order to avoid isomerization at the biaryl axis (and thus racemization), which is a concrete risk in the case of **1a**, due to the presence of only three *ortho* substituents next to the biaryl axis. Indeed, catalytic dehydrogenation with Pd/C in refluxing toluene provided the optically active, enantiomerically nearly pure non-hydrogenated naphthylisoquinoline dioncophylleine A (**24a**), with the axis being the only remaining element of chirality in the molecule, possessing two extended, structurally and electronically more similar, bicyclic chromophores.[5,8,333,334] Application of the Exciton Chirality Method permitted a reliable attribution of the absolute axial configuration of **24a**

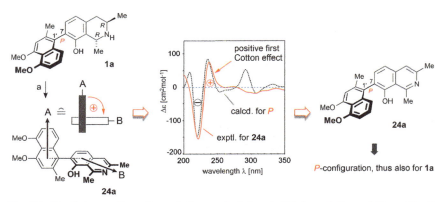

Fig. 33 Absolute configuration of dioncophylleine A (**24a**), and thus of its synthetic precursor, dioncophylline A (**1a**), by applying the Exciton Chirality Method, and by quantum-chemical ECD calculations. Reaction Conditions: (a) Pd/C (5%), toluene, 120 °C, 8%.

(and thus of **1a**), with a positive couplet in the experimental ECD spectrum, i.e., a first positive and a second negative Cotton effect, indicating the biaryl axis to possess a "positive chirality," here meaning that it is *P*-configured. This assignment was further confirmed by quantum-chemical ECD calculations showing a good agreement of the predicted ECD spectrum of **24a** with the experimental ECD curve, in particular for the crucial positive ECD couplet around 240 nm (Fig. 33).[5,6,8,333,334]

Quantum-chemical ECD calculations have become an efficient and reliable tool for the determination of the absolute configuration of naphthylisoquinoline alkaloids, because there is usually no necessity of chemically introducing additional chromophores, and the predictions can be performed for virtually any novel alkaloid,[1–3,11,15,16,33,55,59,71,74,75,183,254,255,267,278,292,300] with axial and/or central chirality, even including naphthylisoquinoline dimers,[1,3,15,33,183,254] with their complex molecular scaffolds.

The chiroptical behavior of a molecule strongly depends on the molecular orientation of its chromophores, which, in turn, is a consequence of the molecular flexibility, more than for any other spectroscopic method. Therefore, detailed investigations of the conformational behavior are indispensable for the computational simulation of an ECD curve. Hence, for the reliable calculation of a high-quality ECD spectrum it will not be sufficient to simply consider the global minimum, but one must take into account *all* possible conformational species that may influence the overall chiroptical behavior of a chiral molecule.[71–75,333] In many cases, the exploration of the conformational space of naphthylisoquinolines or related chiral compound was done by a conformational analysis (CA) with Boltzmann-weighting of the relevant conformers,[71–75,333] or a molecular-dynamics (MD) approach[71,333] was applied, usually for more flexible molecules like e.g., for dioncophylline A (**1a**).[335]

For the conformational analysis of naphthylisoquinolines, a set of reasonable conformers of the molecular framework of a presumed stereoisomer was generated manually by taking into account the most relevant conformational freedom degrees of flexible substituents. These minimum structures were then further refined, in most cases at a molecular-mechanics or at a semiempirical level (AM1, PM3, PM6, PM7),[336–339] aiming at the identification of all minimum geometries that are significantly populated at ambient temperature.[71] The conformers thus found were further optimized by means of the density functional theory (DFT), by using Becke's hybrid exchange correlation B3LYP[340,341] functional, and by application of a split-valence double-zeta basis set such as 6-31G*.[342] For all conformers thus obtained within an energetic cut-off of 3 kcal/mol above the global minimum, the single ECD spectra were then calculated and added up following the Boltzmann statistics, i.e., according to their percentage within the conformational population as a result of their energetic contents.[71,72,74,75,333]

In the case of more flexible molecules, the absolute configuration was assigned by performing molecular-dynamics (MD) simulations instead of Boltzmann-weighted CA calculations,[71,335] using different force fields like TRIPOS[343–345] or MM3.[346–349] Within this approach, the different geometries for the calculation of the single ECD spectra were extracted from the trajectory of the MD simulation. Averaging these single spectra over the time trajectory eventually provided the MD-based theoretical overall ECD spectrum.[71,72,333]

For the generation of reliable wave functions for the ground state and the excited states, together with the corresponding energies, semiempirical approaches like CNDO/S-CI,[350,351] ZINDO/S-CI,[352,353] or OM2[354] Hamiltonians with a configuration interaction (CI) expansion[355] have frequently been used. Higher-level methods have also been applied, like the time-dependent density functional theory (TDDFT)[320,321,324,356,357] or the DFT/MRCI[358] method, which additionally use a multi-reference (MR) wave function in the CI expansion covering single and double excitations. The TDDFT calculations usually involve BLYP and B3LYP/6-31G* functionals[321,329,341,359,360] and a TZVP[361] basis set (i.e., a triple-zeta valence basis set including polarization functions).

The simulated ECD curves are then compared with the experimental ones, aiming at an attribution of the (mostly axial) absolute configuration of the respective naphthylisoquinoline alkaloid. Before this comparison, a correction step for the compensation of systematic computational errors has to be performed. For this purpose, the theoretical overall UV spectrum is calculated in parallel, based on the same conformational analysis, and is then compared with the experimental UV curve, and the shift necessary to match the peaks of the two graphs, is then used as a wavelength calibration by UV comparison. This "UV correction" procedure is easier and more reliable than doing it directly on the ECD spectra because of the usually larger number of peaks of the ECD curve, also with respect to the fact that the latter can have positive *and* negative signs.[71,72,74,333,357]

In some cases, the combination of experimental ECD spectroscopy with quantum-chemical ECD calculations was even the only possible approach to determine the absolute axial configuration of a newly isolated naphthylisoquinoline alkaloid, in particular in the case of novel-type structures like e.g., ancistrotanzanine A (**28**)[1,249,278] with its unprecedented 5,3′-coupling type (Fig. 34). This naphthyldihydroisoquinoline[1,278] was isolated for the first time from the leaves of *Ancistrocladus tanzaniensis*,[20,169] a newly discovered liana from the wet evergreen submontane forests of the southern part of the Uzungwa Mountains National Park in Tanzania. Attempts to assign the configuration at the biaryl axis relative to that of the stereogenic center at C-3, by looking for specific long-range NOESY interactions between H_{eq}-4 and Me-2′, failed because of the overlap of the largely isochronous signals of these two diastereotopic protons at C-4.[278] Therefore, the absolute axial configuration of **28** had to be deduced from a combination of experimental and computational ECD investigations. For this purpose, the molecule was submitted to a conformational analysis applying the Tripos force field,[343,345] followed by a further refinement utilizing the semiempirical AM1[336] method. These calculations were arbitrarily started with the *M,S*-isomer, affording 48 conformers, all appearing within 3 kcal/mol above the global minimum. For all of the conformers, the corresponding ECD spectra were calculated by using the semiempirical CNDO/2S[350,351] method. The single ECD curves thus obtained were then added up following the Boltzmann statistics, i.e., according to their energetic contents, along with the above-described UV correction,[71,72] to give the theoretical overall ECD curve for *M,S*-**28** (Fig. 34, center). In a similar manner, the ECD properties of the *P*-atropo-diastereomer were computed, too (Fig. 34, left). The spectrum calculated for *M,S*-**28** revealed a good

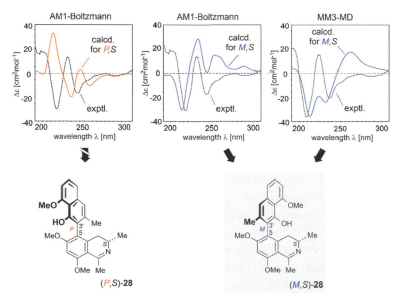

Fig. 34 Assignment of the absolute axial configuration of ancistrotanzanine A (**28**),[278] by comparison of the measured ECD curve with the ECD spectra computed using the semiempirical CNDO/2S Hamiltonian method, based on the AM1-Boltzmann approach (left and center), or applying a MD simulation using the MM3 force field (right).

agreement with the measured one, whereas the theoretical ECD curve for the (*P*,*S*)-atropo-diastereomer was almost opposite, thus permitting attribution of the absolute axial configuration of ancistrotanzanine A (**28**) as *M*.[1,278] This assignment was confirmed by a molecular-dynamics (MD) simulation, which was carried out for *M*,*S*-**28** at a virtual temperature of 500 K using the MM3 force field.[278] From the 500 ps trajectory of motion, 1000 structures were extracted, for which the single ECD spectra were calculated, again by applying the semiempirical CNDO/2S method. The computed 1000 single ECD spectra were added up arithmetically to give the theoretical overall ECD spectrum of (*M*,*S*)-**28** (Fig. 34, right), which revealed a good agreement with the measured one, too.[278]

4.6 Assignment of absolute axial configurations online, by LC-ECD coupling

The combination of HPLC with ECD measurements is the most powerful analytical tool for the elucidation of the absolute axial configuration of naphthylisoquinoline alkaloids.[1,2,16,51,61,73–75,82,253,267,283,284,292,297] From the ECD spectrum easily monitored online, the absolute configuration of the respective stereoisomer can be rapidly assigned if experimental chiroptical data—whether taken online or offline—from structurally related molecules of known absolute configurations are available. This approach has been widely used for the structural assignment of a whole series of naphthylisoquinoline alkaloids.[51,54,61,73–75,82,253,284,291,292,297,299] For an interpretation of the ECD spectra of novel-type structures, i.e., when no comparison with known structures is possible, quantum-chemical calculations of the ECD curves of the relevant atropisomers has led to a reliable stereochemical assignment of structurally unprecedented natural products, without the necessity to isolate the respective compounds beforehand.[1–3,73–75,283,284,297] The LC-ECD option has, in particular, proven to be of high value for establishing the absolute axial configuration of alkaloids possessing a configurationally semi-stable biaryl axis like e.g., in the case of the N,C-coupled naphthyldihydroisoquinolinium alkaloid ancistrocladinium B (**33**),[1,2,16,75,300] which occurs in the plant as a pair of slowly interconverting atropo-diastereomers, **33a** and **33b**. The attribution of their absolute axial configurations is an instructive example illustrating the use of the LC-ECD methodology in combination with quantum-chemical ECD calculations, and is presented here in more detail (see Fig. 35).

For LC-ECD investigations, an experimental setup is used consisting of a standard HPLC system coupled in series to an optical detector and a chiroptical one to simultaneously measure the UV and ECD profiles of the analyte.[74,75,283,297] The application of two independently operating detectors provides the opportunity to use two different wavelengths for a most sensitive detection of the analyte at both its UV and its ECD maxima in the same run (Fig. 35). The monitoring of the actual full ECD spectrum of a chiral analyte is accomplished in the stopped-flow mode.[1,2,16,51,54,61,73–75,82,253,267,291,292,297,362] In many cases, the resolution of mixtures of racemic (or scalemic) naphthylisoquinoline alkaloids is best achieved on an HPLC column with a chiral normal-phase adsorbent like e.g., Chiralcel® OD-H (Daicel)[176,252,267,273,291] or Lux® Cellulose-1 (Phenomenex).[54] This has frequently been the method of choice even for atropo-diastereomeric naphthylisoquinoline alkaloids,[62,362] which sometimes behave like enantiomers and cannot be resolved on achiral phases, see e.g., the—otherwise not successful—resolution of N-methyldioncophylline A and its 7-epimer (Fig. 58 in Section 5).[62]

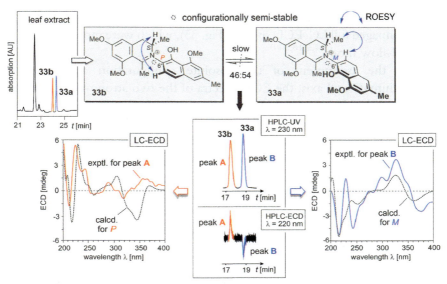

Fig. 35 Stereochemical assignment of the—configurationally semi-stable—two atropo-diastereomers of ancistrocladinium B (**33**),[1,2,16,75] by LC-ECD coupling combined with quantum-chemical ECD calculations using TDDFT (B3LYP/TZVP), and by online-ROESY correlations (double arrows in dark blue) diagnostically indicative of the relative configuration of ancistrocladinium B (**33a**) (peak **B**) and its atropo-diastereomer, *N-epi*-ancistrocladinium B (**33b**) (peak **A**).

The above-mentioned alkaloid ancistrocladinium B (**33**) (Fig. 35) was discovered for the first time in the aerial parts of a botanically yet unknown *Ancistrocladus* species from the Central Congo Basin.[2,16] This alkaloid has meanwhile also been identified in the Indian liana *A. heyneanus*,[1,16] in the Malaysian species *A. benomensis*,[1,16] in *A. tectorius* from the Chinese Hainan Island,[1,2,255] and in yet another, botanically undescribed *Ancistrocladus* plant[30] from the Northern region of the Democratic Republic of the Congo. Despite the configurational semi-stability of **33**, a clear baseline resolution of the two diastereomers **33a** and **33b** was achieved. As outlined in Fig. 35 (center), this succeeded by using a reversed-phase C$_{18}$ HPLC column with an isocratic solvent system consisting of methanol and water (60:40, acidified with 0.05% TFA). A preparative resolution of the two peaks **A** and **B**, however, to give completely pure diastereomers, was not possible due to their gradual interconversion at room temperature.[1,2,16] The ECD analysis of the two configurationally semi-stable atropo–diastereomers of **33** was therefore performed online, expectedly giving rise to two opposite-signed LC-ECD signals and to almost mirror-imaged full LC-ECD spectra (Fig. 35), which were monitored in the stopped-flow mode.[1,2,16,300]

For the assignment of the absolute configuration at the chiral N-iminium-C-aryl axis, the ECD spectra of the two atropo-diastereomers were predicted by quantum-chemical ECD calculations applying TDDFT (B3LYP/TZVP) and compared with the experimental ECD curves.[1,2,16,75] The ECD spectrum of the more rapidly eluting isomer (peak **A**) matched the ECD curve calculated for the *P*-configured atropo-diastereomer **33b**, while the spectrum of the slower isomer (peak **B**) revealed a good agreement with the one computed for the *M*-atropo-diastereomer **33a**. Specific ROESY correlations across the *N,C*-axis recorded by online HPLC-NMR measurements[1,2,16,300] permitted unambiguous assignment of the configuration at the centers relative to the axis. Given the known absolute axial configuration, this clearly indicated the stereogenic center at C-3 to be *S*-configured in both cases. This result was fully corroborated offline, by oxidative degradation of the atropo-diastereomeric mixture **33a/b**,[16] thus confirming the stereochemical assignment of the two atropo-diastereomers as deduced from HPLC-ECD analysis in combination with quantum-chemical ECD calculations.

4.7 Exemplified for dioncophylline A: An overview on the repertoire of methods for the assignment of the full absolute stereostructure of naphthylisoquinoline alkaloids

One of the first naphthylisoquinoline alkaloids investigated more thoroughly by the authors' group was dioncophylline A (**1a**)[5,6,8,9,363,364] (see Fig. 36). Among the meanwhile over 280 naphthylisoquinoline alkaloids discovered in nature, this typical Dioncophyllaceae-type compound, i.e., lacking an oxygen function at C-6 and having the *R*-configuration at C-3, occupies an outstanding role, given its occurrence as a main metabolite in the three Dioncophyllaceae lianas *Triphyophyllum peltatum*,[5,6,8,9] *Dioncophyllum tholloni*,[6,68] and *Habropetalum dawei*,[6,297] but also in various *Ancistrocladus* species from West and Central Africa. Among the Ancistrocladaceae plants that produce dioncophylline A (**1a**) and related alkaloids in substantial quantities are *A. abbreviatus*[5,63] and *A. barteri*[5,285] from Coastal West Africa, and three species from the Congo Basin, *A. letestui*,[184] *A. ileboensis*,[51] and a botanically yet unknown *Ancistrocladus* liana.[12] A broad screening program showed that **1a** displays pronounced

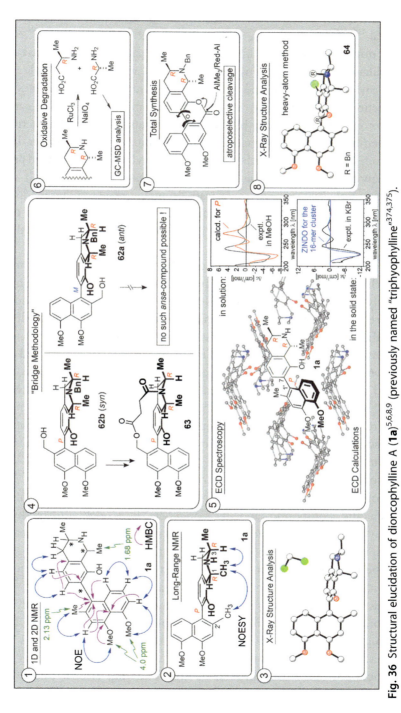

Fig. 36 Structural elucidation of dioncophylline A (**1a**)[5,6,8,9] (previously named "triphyophylline"[374,375]).

bioactivities,[5,7,40,51,97–103,365–373] such as fungicidal,[97] molluscicidal,[102,103] larvicidal,[100,101] insect-feeding deterrency[98] and growth-retarding[98,99] effects (see Section 8.1). Dioncophylline A (**1a**) has also gained interest for its strong antiproliferative activities against breast cancer[365] and toward multiple myeloma[51] and drug-sensitive and multidrug-resistant leukemia cell lines[51,366] (see Section 5.3.2). To gain insight into the mode of action regarding the cytotoxic effects induced by **1a**, studies on its DNA-binding activities were performed providing evidence that **1a** interacts with duplex DNA with moderate affinity.[367] Furthermore, dioncophylline A (**1a**) showed promising activities against *Trypanosoma* and *Plasmodium* parasites, which cause severe tropical diseases such as African sleeping sickness, Chagas' disease, and malaria.[368–372] Moreover, a strong growth-inhibitory potential of **1a** against hematozoan parasites of the genus *Babesia* was demonstrated (see Section 5.3.1).[373]

Dioncophylline A (**1a**) is easily accessible by standard isolation procedures. Besides the two stereocenters at C-1 and C-3, it possesses a configurationally stable biaryl axis.[5,6,8,9] Nonetheless, due to the presence of only three *ortho*-substituents, the axis still has a certain degree of rotational flexibility, thus undergoing a rapid torsional partial rotation. Presumably the same compound, but named "triphyophylline," had previously been isolated from *T. peltatum* by Bruneton and co-workers.[374,375] The structure had been assigned correctly for the constitution, but incorrectly for the stereocenters and for the biaryl axis, which was not recognized as a stereoelement. Its stereostructure was revised by the authors' group according to structural re-investigations on its three stereoelements[5,6,8,66,317,335,363,376,377] (summarized in Fig. 36), and based on its total synthesis,[5,6,9,76,364,378] and "triphyophylline" was renamed into dioncophylline A (**1a**).[5,6,8,9] This renaming was necessary, in particular because of wrong inter-relationships between the compound and its congeners like, e.g., "*O*-methyltriphyophylline,"[379] which actually was not the *O*-methylation product of "triphyophylline."[374,375] Exemplarily for dioncophylline A (**1a**), the repertoire of complementary—in part even newly developed—spectroscopic,[66,67,317,376] chemical,[69,70,363] and computational[71,72,333–335] methods described in this section was applied for the unequivocal confirmation of the assignment of the complete stereostructure of **1a**,[5,6] including the design of a highly efficient methodology[77,79] for its total synthesis.[5,6,9,76,364,378] In detail, the structure of dioncophylline A (**1a**) was elucidated and finally proven by the following arsenal of methods (see Fig. 36), some of which are described in more detail in the sections below:

Box ① constitution[5,6,8,9] by selective HMBC and NOESY interactions;

Box ② relative *trans*-configuration at the stereogenic centers, C-1 versus C-3, by specific NOE interactions, and the relative axial versus central chirality by long-range NOESY correlations[5,6,8,66];

Box ③ constitution and all relative configurations by X-ray crystal structure analysis[5,6,376];

Box ④ relative axial configuration by a remarkably selective, atropisomer-differentiating reaction: specifically only the *syn*-compound **62b** can be bridged by a succinate unit to give **63**, whereas the *anti*-diastereomer **62a** (corresponding to dioncophylline A, **1a**) cannot give a similar *ansa*-compound[5,6,363];

Box ⑤ not shown here, but presented in Fig. 33: Assignment of the absolute axial configuration by experimental ECD spectroscopy, for which dioncophylline A (**1a**) first had to be modified by catalytic dehydrogenation to give dioncophylleine A (**24a**) so that, by enlargement of the isoquinoline chromophore, the Exciton Chirality Method became applicable[5,6,8,333,334];

illustrated here instead: Absolute axial configuration by ECD spectroscopy in combination with quantum-chemical ECD calculations, studied for **1a**, both, in solution[335,377] and in the solid state, including calculations on the ECD spectrum for a 16-molecule cluster taken from the crystal[377];

Box ⑥ absolute configuration at C-1 and C-3 by oxidative degradation[5,6,69,70];

Box ⑦ absolute configuration at C-1 and C-3 by total synthesis starting from D-alanine: for the regio- and stereoselective construction of the biaryl axis, the "lactone method"[76–79] was developed and applied[5,6,9,76,364,378];

Box ⑧ full absolute stereostructure by anomalous X-ray diffraction analysis of the *O,N*-bisbenzylated 5-bromo analogue **64**.[317]

4.7.1 Structural elucidation of dioncophylline A

In the following, 1D and 2D NMR techniques for establishing the absolute stereostructure of dioncophylline A (**1a**)[5,6,8,9,66] are presented in more detail (see Fig. 37). This is, at the same time, an illustrative example regarding the most important aspects of the structural elucidation process, also in view of other naphthylisoquinoline alkaloids. Emphasis is put on the assignment of the constitution, including the coupling position and special structural features, and the relative and absolute configurations at centers and axis.

As deduced from HRESIMS (*m/z* 378.2062, [M + H]$^+$), the molecular formula of dioncophylline A was $C_{24}H_{27}NO_3$. The lack of one oxygen function, as compared to an Ancistrocladaceae-type compound, suggested this alkaloid to be a Dioncophyllaceae-type naphthylisoquinoline.[6,8]

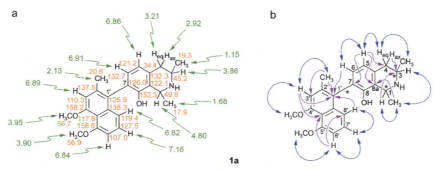

Fig. 37 (a) Selected ¹H (in dark green) and ¹³C (in orange) NMR shifts (in methanol-d_4, δ in ppm) of dioncophylline A (**1a**), and (b) key NOESY (double blue arrows) and HMBC (single purple arrows) interactions indicative of the constitution of **1a**.

Its ¹H NMR spectrum displayed chemical shifts typical of a naphthyltetrahydroisoquinoline, bearing methyl groups at C-1 (δ_H 1.68), C-3 (δ_H 1.15), and C-2' (δ_H 2.13), two methoxy functions (δ_H 3.90 and 3.95), and, as expected for a dioncophyllaceous alkaloid, six aromatic protons (four doublets, one singlet, and one pseudotriplet) (Fig. 37a). The methoxy groups at δ_H 3.90 and 3.95 showed NOESY interactions with H-6' (δ_H 6.84) and H-3' (δ_H 6.89), respectively, evidencing that they were located at C-5' and C-4' in the naphthalene half (Fig. 37b). The third oxygen function thus had to be a free hydroxy group at C-8 in the isoquinoline subunit. With a spin system of three contiguous aromatic protons, giving rise to two doublets (δ_H 6.82 and 6.84) and one pseudotriplet (δ_H 7.16), and from the NOESY correlation sequence {OMe-5' ↔ H-6' ↔ H-7' ↔ H-8'}, which excluded the naphthalene subunit from being connected to the isoquinoline portion via its "methyl-free" ring, the biaryl axis had to be located at C-1'.[6,8] This finding was in agreement with the high-field shift of the signal of Me-2' (δ_H 2.13) and with HMBC long-range couplings from the three aromatic protons H-6 (δ_H 6.91), H-3' (δ_H 6.89), and H-8' (δ_H 6.82) to C-1' (δ_C 125.9), and from Me-2' to C-1' (Fig. 37b). The two diastereotopic protons at C-4 had "normal" chemical shifts (δ_H 2.92 and 3.21), i.e., no shielding effect from a naphthalene part nearby, indicating that the biaryl axis was not linked to C-5, but had to be at C-7 in the isoquinoline moiety. NOESY interactions in the series {H$_{eq}$-4 ↔ H-5 ↔ H-6} established C-7 (δ_C 126.0) as the coupling position, which was further supported by an HMBC interaction from H-5 (δ_H 6.86) to C-7.[6,8]

From a NOESY interaction between Me-1 (δ_H 1.68) and H-3 (δ_H 1.15), the relative configuration of the methyl groups at C-1 and C-3 in dioncophylline A (**1a**) was deduced as *trans* (Fig. 38).[5,6,8] The absolute

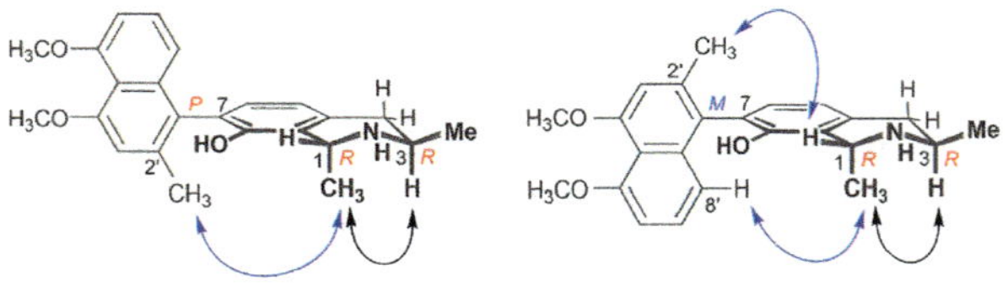

Fig. 38 Assignment of the *trans*-configuration at C-3 versus C-1 (double black arrows) and of the configuration at the biaryl axes of dioncophylline A (**1a**) and its atropo-diastereomer 7-*epi*-dioncophylline A (**1b**) by specific NOE interactions (double blue arrows).

configuration at both of these stereocenters was established as 1R,3R by GC-MSD analysis of the Mosher derivatives of D-alanine (derived from C-1 and, to a small degree, also from C-3) and (R)-3-aminobutyric acid (derived from C-3, exclusively) obtained by ruthenium-mediated oxidative degradation[5,6,69,70] of the alkaloid as described in Section 4.4 (see also Fig. 36, Box ⑥).

For the elucidation of the axial configuration relative to the stereogenic centers in compounds like **1a**, in which the relevant protons of the two ring systems are quite distant from each other (≥ 4 Å), long-range NOE or ROE measurements have to be applied.[5,6,17,66] Atropisomer-specific NOE interactions (Fig. 38) clearly permitted to distinguish between **1a**, in which the methyl groups at C-1 and C-2′ are on the same side of the molecule, and its likewise naturally occurring atropo-diastereomer, 7-*epi*-dioncophylline A (**1b**), in which Me-1 is now close (and thus *syn*) to H-8′, while Me-2′ is on the same side as H-1.[17,66] Thus, based on the NOESY cross-peak found between Me-2′ in the naphthalene part and Me-1 in the isoquinoline portion, and in conjunction with the 1R,3S-stereoarray in the tetra-hydroisoquinoline part as deduced from the result of the oxidative degradation, the absolute axial configuration of dioncophylline A (**1a**) was determined to be P. For 7-*epi*-dioncophylline A (**1b**), by contrast, long-range NOESY correlations between Me-1 and H-8′ and between H-1 and Me-2′ attributed this alkaloid to be M-configured.[17,66]

4.7.2 Absolute axial configuration of dioncophylline A by ECD spectroscopy in combination with quantum-chemical ECD calculations

The assignment of the absolute configuration of dioncophylline A (**1a**) by electronic circular dichroism (ECD) studies was a particular challenge.

It was independently established to be *P* by experimental ECD spectroscopy (after planarization of the isoquinoline part and extension of the chromophore, both achieved by dehydrogenation) with application of the Exciton Chirality Method[5,8,333,334] (see Fig. 33), and by semiempirical ECD calculations of the authentic alkaloid based on the Molecular-Dynamics (MD) approach.[335] Although an unambiguously determined full absolute stereostructure of **1a** with 1*R*,3*R*,7*P*-configuration had already been obtained by NMR investigations[5,6,8,66] in combination with its stereoselective total synthesis from D-alanine,[5,6,9,76,364,378] this alkaloid, with its central importance, still seemed to be a rewarding model for more-in-depth stereochemical studies, also due to the dynamic conformational orientations of the chromophores to each other at the rotationally hindered—yet conformationally still flexible—axis.

Despite its stable configuration (*P*), the conformational dynamics of dioncophylline A (**1a**) has a drastic influence on its ECD spectrum.[377] Partial rotation around the biaryl axis by computationally varying the dihedral angle ABCD at intervals of 5° (30° < θ_{ABCD} < 135°) (Fig. 39a)—yet without changing the *P*-configuration of **1a**—followed by CNDO/S-based ECD calculations on the single conformations thus obtained revealed far-reaching changes of the ECD behavior of dioncophylline A (**1a**) over time, i.e., depending on the current dihedral angle. For some of the torsional geometries, although all possessing the same absolute configuration at the axis and the centers, even fully mirror-like, opposite ECD spectra were predicted (Fig. 39b). These calculations were highly relevant, because

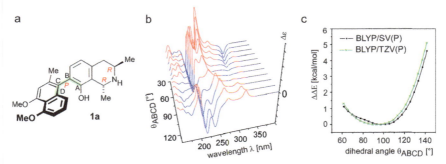

Fig. 39 (a) Dioncophylline A (**1a**) with its absolute stereostructure (ABCD defining the dihedral angle at the biaryl axis). (b) A series of CNDO/S-calculated ECD spectra of the torsional geometries at the axis (red and blue areas correspond to positive and negative $\Delta\varepsilon$ values, respectively). (c) Energy profiles for the axial rotation calculated at the DFT level.

the energy profiles for the axial rotation predicted by calculations at the DFT level (BLYP/6-31G**) indicated a substantial torsional flexibility at the axis of **1a**. This resulted in very similar energies within a quite broad range of dihedral angles, leading to a population of all these conformeric species—but with totally different chiroptical contributions to the macroscopically observed experimental ECD spectrum (Fig. 39c).[377]

Therefore, solid-state ECD spectroscopy was applied to study this flexible compound and its chiroptical behavior in a conformationally "frozen" form, as fixed in the crystal. The transmittance ECD spectrum of the microcrystalline material of dioncophylline A (**1a**) obtained by the KBr-matrix method provided two peaks with positive Cotton effects at 280 and 245 nm, separated by a low-intensity signal at 227 nm.[377] Similar transmittance ECD and UV absorption spectra were obtained, both in the solid state (KBr disk) and in solution (EtOH) (Fig. 40), indicating related conformational patterns.[377]

Based on the coordinates obtained by X-ray diffraction analysis of the single conformer of dioncophylline A (**1a**) found in the crystalline state (for the X-ray structure of **1a**,[376] see Fig. 36, Box ③), ECD calculations were performed by three different approaches, viz., by CNDO/S-CI[350,351] (Fig. 41a), TDDFT[320,321,324,356,357] (Fig. 41b), and DFT/MRCI[358] (Fig. 41c), each by itself permitting unambiguous attribution of the absolute axial configuration of dioncophylline A (**1a**) as P, which again, independently, confirmed the previous assignments.[5,6,8,9,66,317,335,376] The best overall match between theory and experiment was achieved when applying the DFT/MRCI methodology.[377]

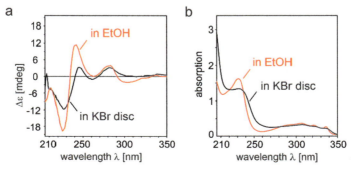

Fig. 40 Experimental (a) ECD and (b) UV spectra of dioncophylline A (**1a**) as measured in solution (EtOH), and in the solid state, prepared as a KBr disc.

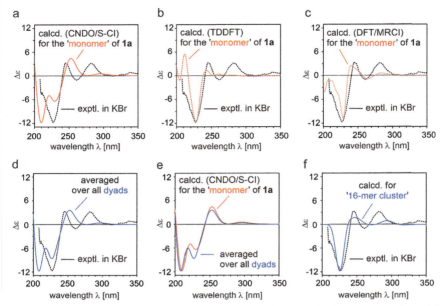

Fig. 41 Comparison of the experimental ECD curve of dioncophylline A (**1a**) in the solid state with the spectra calculated for a single molecule of **1a** by using (a) CNDO-S-CIS, (b) TDB3LYP/TZVP, or (c) DFT/MRCI/SVP. (d) Comparison of the experimental solid-state (KBr matrix) ECD spectrum of **1a** with the curve predicted (CNDO/S-CI) for one single molecule, and (e) with the calculated spectrum obtained as an average over all treated dyads. (f) Comparison of the curve calculated (ZINDO/S-CI) for the cluster of 16 molecules with the experimentally monitored curve of **1a** in the solid state.

Furthermore, ECD computations were performed on a set of molecular dyads (two-molecule arrays) to study more closely neighboring effects that may affect the ECD behavior of **1a** in the crystalline state because of different directions, distances, and angles of the chromophoric units of two individual molecules to each other (isoquinoline to isoquinoline, isoquinoline to naphthalene, and, in particular, naphthalene to naphthalene).[377] These dyads always consisted of a given "central" molecule and a second one from the nearest neighborhood. For investigations on the neighboring effects, one molecule with all its neighbors, one by one, was chosen randomly, since each of the dioncophylline A molecules in the crystal had the same molecular environment.[377] Semiempirical calculations using the CNDO/S-Cl method,[350,351] which had shown a good agreement concerning the predicted wavelengths and the relative intensities for the "monomer" of **1a**, evidenced that the surrounding effects indeed exert a certain influence on the ECD behavior of **1a** in the solid state. A noticeable interdependence between the overlaying naphthalene and isoquinoline moieties was observed, along with remarkably strong intermolecular naphthalene-isoquinoline interactions, giving rise to substantially differing ECD curves for some of the dyads, including curves with opposite signs.[377] Moreover, there were ECD spectra distinctly different from that of the dioncophylline A monomer (data not presented), but the overall effect over all dyads matched the experimental overall spectrum quite well (see Fig. 41d).

The spectrum obtained by arithmetic averaging over the ECD curves of all dyads did not show any substantial change in comparison to the calculated ECD curve of "monomeric" **1a** (Fig. 41e).[377] Similar results were obtained from ECD calculations performed for a significantly larger 16-molecule fragment of the crystal structure of **1a** using ZINDO/S-Cl[352,353] (implemented in ORCA[380]). These calculations provided only minor improvements such as the higher intensity of the peak at about 280 nm (Fig. 41f). From these comprehensive studies it became obvious that although there is indeed a certain effect of the crystalline state on the ECD spectrum of dioncophylline A (**1a**), the consideration of only one single molecule already delivered reliable results for an assignment of the absolute axial configuration of **1a** by means of solid-state ECD spectroscopy.[377]

Still, this work revealed that solid-state ECD measurements might be helpful for an attribution of absolute stereostructures of large chiral molecules, in particular for highly flexible ones, in combination with higher-level quantum-chemical ECD calculations.

4.7.3 Stereoselective total synthesis of dioncophylline A by application of the "lactone method"

For an ultimate proof of the complete stereostructure of dioncophylline A (**1a**), its atropo-selective total synthesis[5,6,9,76,364,378] was performed by applying the "lactone methodology." This concept,[76–79] as developed in the authors' laboratory, is a highly efficient and flexible approach for the atropo-divergent preparation of any of the two rotational isomers of naphthylisoquinoline alkaloids, as desired (here **1a**, or, optionally, its 7-epimer **1b**, see Fig. 42), from late joint precursors. The two formal goals of an atropisomer-selective synthesis—the biaryl bond formation and the asymmetric induction—were achieved consecutively, by ester-type prefixation of the two molecular building blocks, followed by intramolecular coupling, and subsequent atropo-enantio- or -diastereoselective

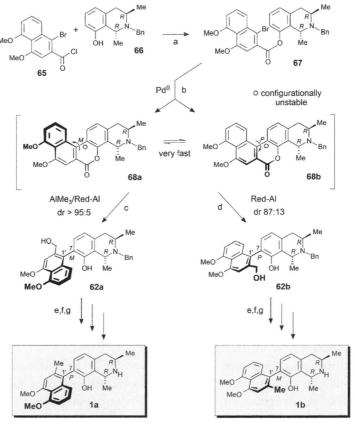

Fig. 42 Atropo-divergent total synthesis of dioncophylline A (**1a**) and its likewise naturally occurring atropo-diastereomer, 7-*epi*-dioncophylline (**1b**), by highly atropisomer-selective ring cleavage of **68** to give **62a** or, if desired, **62b**. Reaction Conditions: (a) NEt$_3$, DMAP, CH$_2$Cl$_2$, 90%; (b) Pd(PPh$_3$)$_2$Cl$_2$, NaOAc, DMA, 130 °C, 75%; (c) AlMe$_3$, THF, r.t., 2 min → Red-Al (= sodium bis(2-methoxyethoxy)-aluminum hydride), −60 to −40 °C, 2 min, then 25 °C, 2N HCl, 0 °C, 90% (for **62a**)[364]; (d) Red-Al, diethyl ether, THF, −60 to −40 °C, 30 min, then 2N HCl, 0 °C, 78% (for **62b**)[364]; (e) (CBrCl$_2$)$_2$, PPh$_3$, CH$_2$Cl$_2$[397]; (f) LAH, THF; (g) H$_2$, Pd/C (10%), MeOH; 83% (for **1a**)[9] or 82% (for **1b**).[9]

cleavage of the lactone auxiliary bridge.[76–79] The method proved to be applicable to the synthesis of a broad spectrum of structurally diverse axially chiral molecules,[76–79,381–396] focussing on a specific substitution pattern that is present in many natural biaryl systems, viz., with a C$_1$ unit and an oxygen function in opposite *ortho*-positions next to the axis.

In view of the occurrence of both, dioncophylline A (**1a**)[5,6,12,51,63,68,73,184,285,297] and its atropo-diastereomer, 7-*epi*-dioncophylline A (**1b**),[5,51,285] as genuine natural products, the authors' group aimed at a directed, atropo-divergent total synthesis of **1a** or, optionally, of **1b** (Fig. 42).[5,6,9,76,364,378] Prefixation of the two aromatic halves of the target molecules, the 1-bromonaphthoic acid chloride **65**[383] and the isoquinoline building block **66**, gave the bromoester **67**, which was subjected to a Pd-catalyzed intramolecular biaryl coupling, affording the lactone-bridged biaryl **68** as an ideal common intermediate for the synthesis of any of the two atropisomeric target molecules. Although already containing the biaryl linkage, and being helically distorted, and thus axially chiral, this intermediate **68** does not occur as stable atropisomers, because its rotational barrier is drastically lowered by the lactone bridge.[9,364] Out of this mixture of the two rapidly interconverting atropo-diastereomers, ring-opening of the lactone bridge in **68** succeeded in an atroposelective manner, even when using achiral hydrogen nucleophiles, so that the configurationally stable atropo-diastereomeric alcohols **62a** (*anti*) and **62b** (*syn*) (see also Fig. 36) were obtained in unequal amounts. As an example, simple hydride transfer reagents, such as e.g., sodium bis(2-methoxyethoxy)-aluminum hydride (Red-Al) in THF gave good atropisomeric ratios in favor of the *P*-configured alcohol **62b** (dr 87:13) (Fig. 42, right). The use of two separate reagents for the complexation and the subsequent reduction step— pretreatment of **68** with trimethylaluminum (which as such does not open the lactone) and addition of this mixture to an excess of the achiral reductant Red-Al—almost exclusively provided the *M*-configured intermediate **62a** (dr 95:5) (Fig. 42, left). Side-chain deoxygenation of **62a** and subsequent *N*-deprotection by standard procedures yielded dioncophylline A (**1a**).[9,364] In a similar manner, **62b** was efficiently converted to the other desired target molecule, 7-*epi*-dioncophylline A (**1b**).[5,6,9,76,364,378]

4.7.4 Constitution and stereostructure of dioncophylline A by X-ray diffraction

For an additional proof of the constitution and for the determination of the conformation and relative configuration of naphthylisoquinoline alkaloids, crystal structure analysis[398,399] may be the method of choice—provided that crystals of appropriate quality are available. Yet, naphthylisoquinoline alkaloids are usually difficult to crystallize without prior chemical derivatization,[5,6,11,12,239,259,274,293,376,400] and only very few X-ray structure analyses have so far succeeded.[5,6,11,12,239,259,274,287,293,317,376,400] In the case of dioncophylline A (**1a**), luckily crystals of suited quality were obtained by slow diffusion of dichloromethane into a solution of **1a** in ethanol at room temperature, giving colorless plates. The crystal structure of dioncophylline A (**1a**), together with one defined molecule of dichloromethane in the crystal, as illustrated in Fig. 36 (Box ③), fully confirmed the anticipated constitution of **1a** and its relative configuration at the three stereoelements (two centers and the axis),[5,6,376] in agreement with the full absolute stereostructure as previously established by NMR[5,6,8,9,66] and ECD[5,6,8,333–335,377] investigations, chemical methods,[5,6,69,70,363] and by stereoselective total synthesis.[5,6,9,364,378] The X-ray structure analysis revealed the heterocyclic tetrahydropyrido ring of **1a** to adopt the anticipated half-chair conformation, and it corroborated the *trans*-orientation of the two methyl groups at C-1 and at C-3 and the relative axial configuration with the methyl group on the naphthalene portion being located on the same side of the molecule as the methyl group at C-1.[376] Thus, the X-ray diffraction analysis firmly corroborated the relative and, in view of the results of the oxidative degradation and the ECD investigations, the full absolute stereostructure of **1a** to be 1R,3R,7P-configured.[5,6]

For the assignment of the complete stereostructure, including the absolute configuration, by crystal structure analysis, anomalous X-ray dispersion (also referred to as "Bijvoet analysis")[398,401] can be applied, provided that the respective molecule contains heavy elements like e.g., bromine or sulfur, therefore also named "heavy-atom method." Although dioncophylline A (**1a**) itself does not possess a heavy atom substituent, an appropriate derivative was easily available by regioselective bromination of **1a** and subsequent *O*- and *N*-benzylation, affording the *O,N*-bisbenzylated 5-bromo derivative **64**.[317] Crystallization of **64** from methanol gave colorless plates suitable for an anomalous X-ray dispersion analysis[317] (Fig. 36, Box ⑧), which here unambiguously confirmed both, the relative and, thus, also the absolute configuration of **1a**.

4.8 Further phytochemical studies on Dioncophyllaceae and Ancistrocladaceae plants

By the fruitful interplay of the arsenal of methods presented in Section 4, phytochemical studies on various Dioncophyllaceae and Ancistrocladaceae lianas furnished a large series of structurally most diverse naphthylisoquinoline alkaloids, thus providing far-reaching insight into the metabolic pattern of these plants, also in view of the formation of minor metabolites and even trace constituents.[1–7] Among the phytochemically most intensely investigated plant species are *Triphyophyllum peltatum* (Dioncophyllaceae)[5,6,8,9,274,275,293–296,298,402] and *Ancistrocladus korupensis*[3,5,13,14,276,287–290] from Cameroon, the two Congolese lianas *A. likoko*[10,53,175,176] and *A. ealaensis*,[11,179–182] the Indian plant *A. heyneanus*,[1,27,28,256–267] *A. tectorius* from the Chinese Hainan Island,[1,243–255] and several botanically yet unidentified *Ancistrocladus* lianas from the Congo Basin[12,15,16,29–35] (all belonging to the Ancistrocladaceae family).

Ancistrocladus abbreviatus from Coastal West Africa has proven to be a creative producer of a broad plethora of naphthylisoquinolines and related alkaloids (so far ca. 80 isolated compounds from this one species alone!),[5,54–65] pursuing a nearly combinatorial—and yet selective!—approach, and is thus an excellent example to illustrate the unique, and still manifold, molecular architectures and fascinating stereochemical features of this class of bi- and quateraryl natural products (see Section 5). Particularly intriguing was the detection of novel-type alkaloids with special, unprecedented structural features, among them root bark and leaf constituents with an *ortho*-quinone functionality in the naphthalene ring[56] and naphthylisoquinolines lacking the otherwise generally present methyl group at C-1.[55] But also *seco*-type alkaloids[55] and even ring-contracted naphthylisoindolinones[59] have been discovered (see Section 6). Some of the compounds displayed promising antiplasmodial activities,[57,58] while others showed pronounced cytotoxic effects against human leukemic,[54,366] pancreatic,[56,57,61] colon,[58] breast,[59,365] lung,[59,60] and HeLa[55,56] cancer cells. This will be the topic of Sections 5 and 6.

5. *Ancistrocladus abbreviatus* from Coastal West Africa, a rich source of structurally diverse naphthylisoquinoline alkaloids

The woody liana *Ancistrocladus abbreviatus* Airy Shaw (Fig. 13 in Section 2) is widely distributed along rivers in lowland moist evergreen forests of Coastal West Africa. It can easily be recognized by its typical hooked branches and its rosettes of leaves appearing on adult climbing stems.[18–20,159,160] Early isolation work on this *Ancistrocladus* species in the 1990s by the authors' group led to the identification of 16 main constituents.[5,62–65] At that time already, the plant attracted attention since the synthetic potential of *A. abbreviatus* appeared as excessively diverse. According to the chemotaxonomical rule[1,5] that had previously been established to distinguish the alkaloids of the Dioncophyllaceae plants from those of the Ancistrocladaceae lianas (see Section 3), *A. abbreviatus* as a representative of the genus *Ancistrocladus* had initially been expected to solely produce typical

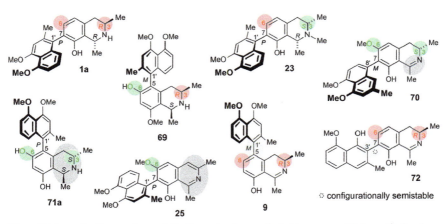

Fig. 43 A selection of naphthylisoquinoline alkaloids exemplifying their four subclasses—Dioncophyllaceae- (6-H, 3R—"red/red"), Ancistrocladaceae- (6-OR, 3S—"green/green"), hybrid- (6-OR, 3R—"green/red"), and inverse hybrid-type (6-H, 3S—"red/green") compounds, and representing the structural variety of alkaloids with a tetrahydro-, dihydro-, or non-hydrogenated isoquinoline portion isolated from *Ancistrocladus abbreviatus*. The structural modifications of interest in **9**, **72**, and in **71a**, **70**, and **25** are labeled in light blue (coupling type) and gray (degree of hydrogenation of the isoquinoline portion).

Ancistrocladaceae-type alkaloids as this had previously been the case, without any exception, for all *Ancistrocladus* species investigated at that time, yet from Asia.[1,5,27,28,238–244,256–262,281,282] The identification of typical Dioncophyllaceae alkaloids, such as dioncophylline A (**1a**)[63] (Fig. 43), was, thus, totally unexpected and necessitated to reconsider the previously strict phytochemical differentiation between the Dioncophyllaceae and Ancistrocladaceae plant families.[5,6] The profile of naphthylisoquinoline alkaloids of *A. abbreviatus* in those early phytochemical studies,[5,62–65] showing an apparently low degree of regio- and stereoselectivity, has, more recently, prompted the authors' group to analyze the metabolite pattern of *A. abbreviatus* in more depths, now additionally focussing on the isolation and structural elucidation of minor and trace compounds.[54–61] For this purpose, an advanced tool of modern chromatographic and spectroscopic methods[1,3,54–61,66,70–75] was compiled and applied. It was based on a meanwhile substantially improved analytical tool box regarding sensitivity and effectiveness since the first phytochemical investigations on *A. abbreviatus* had been performed more than two decades earlier.[5,62–65]

These more recent phytochemical studies[54–61] clearly revealed that *Ancistrocladus abbreviatus* is a creative producer of structurally unique naphthylisoquinoline alkaloids, the majority of which display a 5,1'- or a 7,1'-coupling pattern.[54–59,62–65] Most of the representatives of the 5,1'-linked compounds, like e.g., ancistrobrevine E (**71a**)[57] (Fig. 43), belong to the subclass of Ancistrocladaceae-type compounds,[1,5] i.e., with S-configuration at C-3 and an oxygen function at C-6. The predominant number of the 7,1'-coupled naphthylisoquinolines like dioncophylline A (**1a**),[5,63] by contrast, have an R-configuration at C-3 and no oxygen function at C-6, and are, thus, to be categorized as Dioncophyllaceae-type compounds,[1,5] because the Dioncophyllaceae lianas like *Triphyophyllum peltatum* contain alkaloids with these structural features, exclusively.[5,6,274,275,293–296,402] For this reason and with respect to the fact that *A. abbreviatus*[18–20] shares a common habitat with *T. peltatum*,[26,111,112] this Ancistrocladaceae species can be considered as a chemo- and geotaxonomical link over to the Dioncophyllaceae plant family, thus occupying an exceptional phylogenetic position within the genus *Ancistrocladus*.[5,58]

Moreover, *A. abbreviatus* is one of only four *Ancistrocladus* species (besides *A. abbreviatus*,[5,54–65] it is about *A. barteri*,[5,285,286] *A. ileboensis*,[51,52,183] and an as yet undescribed Congolese *Ancistrocladus* species[12]) that produce both, Ancistrocladaceae- and Dioncophyllaceae-type alkaloids and, additionally, mixed-Ancistrocladaceae/Dioncophyllaceae-type naphthylisoquinolines, i.e., hybrid-type compounds such as ancistrobrevine M (**69**),[58] with an oxygen function at C-6 and an R-configuration at C-3. Even a first inverse hybrid-type alkaloid, dioncoline A (**23**),[58] the as yet only known 3S-configured naphthylisoquinoline that lacks an O-functionality at C-6, was discovered in the plant (Fig. 43). This finding further shows that *A. abbreviatus* obviously does not form its metabolites in a highly regio- and stereoselective manner, but produces a broad plethora of naphthylisoquinoline alkaloids, thus pursuing a nearly combinatorial approach.[5,54–65] It is the only *Ancistrocladus* species known to contain naphthylisoquinoline alkaloids of all four possible subclasses. Remarkably, although the naphthyldi- and tetrahydroisoquinoline alkaloids isolated from *A. abbreviatus* occur in complex regio- and stereoisomeric mixtures, they all have as yet been found to be enantiomerically pure.[5,56–58,61–65]

Ancistrocladus abbreviatus is a rich source of so far nearly 70 discovered naphthylisoquinoline alkaloids, displaying 5 (out of the 7 known)[1–3] different coupling types (5,1', 7,1', 7,3', 5,8', and 7,8').[5,54–65] The by far largest group detected in *A. abbreviatus*, comprising naphthyl*tetrahydro*isoquinolines such as ancistrobrevine E (**71a**),[57] will be presented in this section. But also alkaloids occurring in smaller numbers possessing a *dihydroiso*quinoline subunit, like e.g., ancistrobrevine J (**70**),[57] or a non-hydrogenated isoquinoline portion such as ancistrobreveine A (**25**)[54] will be described in more detail (Fig. 43). The discovery of the latter in *A. abbreviatus* was highly remarkable, since only few naphthylisoquinoline alkaloids like **25** with a rotationally hindered biaryl axis as the only element of chirality had previously been identified in nature.[1,5,53,176,252,273–275,291,299,403] Likewise rare is the occurrence of 5,1'- and 7,3'-coupled dioncophyllaceous naphthylisoquinoline alkaloids in Dioncophyllaceae and Ancistrocladaceae lianas.[5,6,51,292,402] This section reports on the identification of two of these compounds, 5-*epi*-dioncophyllidine C$_2$ (**9**)[56] and dioncophyllidine E (**72**),[61] in the leaves of *A. abbreviatus* (Fig. 43).

Further unusual naphthylisoquinoline alkaloids that have so far been isolated from *A. abbreviatus* exclusively, like e.g., ring-opened (i.e., *seco*-type) naphthylisoquinoline alkaloids[55] or ring-contracted alkaloids possessing a five-membered ring as part of an isoindolinone subunit,[59] will be described in Section 6.

5.1 Naphthylisoquinoline alkaloids with a 5,1'-coupling site and their structural diversity

The by far highest number of naphthylisoquinoline alkaloids isolated from twigs, stems, and roots of *A. abbreviatus* are 5,1'-linked naphthyltetrahydro- and -dihydroisoquinolines (22 compounds),[5,56–58,63] showing a great structural variety of oxygenation and *O*-methylation patterns compared to their parent compounds ancistrocladine (**20a**)[1,27,28,256,257,259] and hamatine (**20b**)[1,28,281,282] (Fig. 44), which were likewise identified in all parts of the plant.[5,63] The majority of these 5,1'-linked compounds (18 representatives)[5,56–58,63] belong to the subclass of Ancistrocladaceae-type compounds, i.e., with *S*-configuration at C-3 and an oxygen function at C-6, but also 3 representatives[56,58] of hybrid-type compounds (3*R*, 6-O*R*) and 1 example[56] of a Dioncophyllaceae-type alkaloid (3*R*, 6-H) were discovered in *A. abbreviatus*.

Naphthylisoquinoline alkaloids structurally closely related to **20a** and **20b** are widely present in Asian and African *Ancistrocladus* species,[1,5,28,242–248,250,252,258,263,273,279–282,284,299,313,403] they even dominate the metabolite pattern of some of these plants significantly like e.g., that of *A. tectorius* from the Chinese Island Hainan (ca. 20 compounds[1,242–248,250,252,254]). In Dioncophyllaceae lianas, by contrast, only one single example of a 5,1'-coupled naphthylisoquinoline alkaloid has so

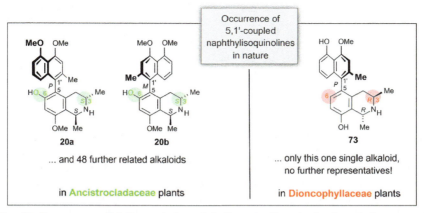

Fig. 44 Occurrence of 5,1'-coupled naphthylisoquinolines in Ancistrocladaceae and Dioncophyllaceae lianas: *Ancistrocladus* species produce a large plethora of alkaloids of this coupling type, whereas dioncophylline C (**73**)[402] is the only 5,1'-linked compound known from Dioncophyllaceae plants, so far found in *Triphyophyllum peltatum*, exclusively.

far been discovered, namely dioncophylline C (**73**) (Fig. 44) from *Triphyophyllum peltatum*.[5,6,402]

Also from a chemotaxonomic point of view, this is an interesting finding, underlining the position of *A. abbreviatus* as a chemo- and geotaxonomic link between the Ancistrocladaceae and the Dioncophyllaceae.[5,58] On the one hand, there is a strong overlap (10 out of 22) of 5,1'-coupled naphthylisoquinolines detected in the roots and stems of *A. abbreviatus* with those found in other *Ancistrocladus* species from Africa and Asia.[5,54,56,57,63] On the other hand, all of the 7,1'-linked Dioncophyllaceae-type alkaloids isolated as main metabolites of *A. abbreviatus*[5,58,62,63] likewise represent major constituents of *Triphyophyllum peltatum*,[5,6,8,9,73] and of the two other Dioncophyllaceae species, *Dioncophyllum thollonii*[68,292] and *Habropetalum dawei*[73,297] (see Section 5.3.1).

5.1.1 5,1'-Linked naphthyltetrahydroisoquinoline alkaloids from the roots of *Ancistrocladus abbreviatus*

Phytochemical studies on air-dried roots, twigs, and stem bark material of *A. abbreviatus* furnished a broad series of 5,1'-coupled naphthyl-*tetrahydro*isoquinolines.[5,57,58,63] Interestingly, a substantial number of them were isolated as atropo-diastereomeric mixtures, while all of the dihydroisoquinolines[56,57] identified so far seem to occur in an atropisomerically pure form

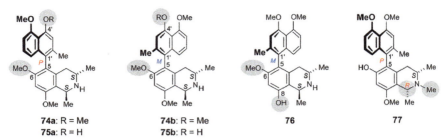

Fig. 45 Six 5,1′-coupled naphthyltetrahydroisoquinoline alkaloids previously identified in related African and Asian *Ancistrocladus* species, now isolated from *A. abbreviatus*: 6-*O*-methylancistrocladine (**74a**) and 6-*O*-methylhamatine (**74b**) and their 4′-*O*-demethyl analogues **75a** and **75b**, and the highly antiplasmodial 5-*epi*-ancistectorine A$_2$ (**76**) (see Section 5.1.2), as well as the *cis*-configured and *N*-methylated ancistrocline (**77**). Structural modifications of these alkaloids compared to their parent compounds ancistrocladine (**20a**) and hamatine (**20b**) (for their structures, see Fig. 44) are underlaid in pale gray.

in the plant (see Section 5.1.4). A total of 16 naphthyl*tetrahydro*isoquinolines of the 5,1′-coupling type have as yet been isolated and structurally elucidated, among them 8 known compounds[5,57,63] (Fig. 45) that had earlier been discovered in related *Ancistrocladus* species[1,5,27–30,51,178,244,245,250,252,281,403] and were now found to be produced by this West African liana, too. While 6-*O*-methylancistrocladine (**74a**)[1,5,28,51,252] and its 5-epimer 6-*O*-methylhamatine (**74b**)[1,5,28–30,51,178,252] are widely distributed in African and Asian *Ancistrocladus* lianas, 6-*O*-methyl-4′-*O*-demethylancistrocladine (**75a**)[1,245] and its atropo-diastereomer, 6-*O*-methyl-4′-*O*-demethylhamatine (**75b**),[1,245] as well as 5-*epi*-ancistectorine A$_2$ (**76**)[1,250] and ancistrocline (**77**)[1,244] had previously been known to occur only in the Southeast Asian liana *A. tectorius*. Being 6-oxygenated and 3*S*-configured, these compounds are typical Ancistrocladaceae-type naphthylisoquinoline alkaloids.

Remarkably, five of the eight new 5,1′-coupled naphthyltetrahydroisoquinoline alkaloids, ancistrobrevines E (**71a**)[57] and F (**59a**),[57] their atropo-diastereomers, 5-*epi*-ancistrobrevines E (**71b**)[57] and F (**59b**),[57] and ancistrobrevine G (**78**)[57] are characterized by a common structural feature, all of them being equipped with a constitutionally and configurationally identical 6,8-dihydroxytetrahydroisoquinoline portion (Fig. 46). They were isolated as minor metabolites from the roots of *A. abbreviatus*, along with two further new compounds, ancistrobrevines K (**80**)[58] and L (**79**).[58] Compound **79** likewise showed a close structural resemblance to representatives of this

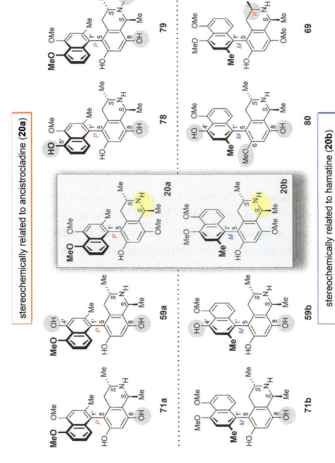

Fig. 46 Eight new 5,1′-coupled naphthyltetrahydroisoquinoline alkaloids from the roots of *A. abbreviatus*, among them seven Ancistrocladaceae-type compounds, all possessing the same tetrahydroisoquinoline subunit, constitutionally and configurationally: ancistrobrevines E (**71a**) and F (**59a**),[57] their atropo-diastereomers 5-*epi*-ancistrobrevines E (**71b**) and F (**59b**),[57] and the ancistrobrevines G (**78**),[57] K (**80**)[58] and L (**79**),[58] discovered along with the only hybrid-type representative (3*R*, 6-OH) within this series of compounds, ancistrobrevine M (**69**).[58]

series possessing a 6-OH/8-OH-substituted isoquinoline subunit, since ancistrobrevine L (**79**)[58] is the *N*-methyl derivative of the *P*-configured ancistrobrevine E (**71a**).[57] Ancistrobrevine K (**80**),[58] by contrast, is the new 8,4′-*O*,*O*-didemethyl-6-*O*-methyl analogue of the *M*-configured parent compound hamatine (**20b**)[1,5,28,281,282]; it might also be addressed as the atropo-diastereomer of 4′-*O*-demethylancistectorine A$_2$, previously isolated as a minor constituent from the twigs of the Chinese liana *A. tectorius*.[252]

Among this unusually large series of 5,1′-coupled naphthyltetrahydroisoquinoline alkaloids discovered in *A. abbreviatus*, ancistrobrevine M (**69**)[58] was the first mixed Ancistrocladaceae/Dioncophyllaceae-type compound of this coupling type, and it was only the second 5,1′-coupled alkaloid at all possessing a *cis*-configured tetrahydroisoquinoline subunit. The only other compound showing this structural feature was the likewise isolated known[1,244] alkaloid ancistrocline (**77**)[404] (Fig. 45), which has an *N*-methylated *cis*-configured tetrahydroisoquinoline subunit. Even more remarkably, ancistrobrevine M (**69**) is the so far only example of a hybrid-type 5,1′-linked naphthyltetrahydroisoquinoline alkaloid ever found in nature.[58]

Oxidative degradation (see Section 4.4) of ancistrobrevine M (**69**)[58] established the absolute configuration at C-3 to be *R*, which, in conjunction with the relative *cis*-configuration at C-1 versus C-3, as determined by NMR, permitted attribution of the stereogenic center at C-1 to be *S*-configured. The absolute axial configuration was assigned to be *M* due to long-range NOESY correlations between H-4$_{eq}$ and H-8′ and between H-4$_{ax}$ and Me-2′, which was confirmed by the fact that the ECD spectrum of **69** was nearly identical to that of the structurally closely related—and likewise *M*-configured—5-*epi*-ancistrobrevine E (**71b**) (Fig. 47a).[58]

In the case of ancistrobrevine K (**80**)[58] (Fig. 47b), however, NOESY cross-peaks of the protons of Me-1 to H-3 and of H-1 to Me-3 proved the methyl groups at C-1 and C-3 to be *trans* to each other. In this case, oxidative degradation evidenced the stereogenic center at C-3 to be *S*-configured, and thus, given the relative *trans*-configuration, to be *S*-configured at C-1, too.[58] On the basis of the absolute 1*S*,3*S*-stereoarray in the tetrahydroisoquinoline subunit thus assigned, NOESY correlations between H-4$_{ax}$ and H-8′ and between H-4$_{eq}$ and Me-2′ (i.e., complementary to those monitored for **69** proved the biaryl axis of ancistrobrevine K (**80**)) to be *M*-configured. This was in agreement with the nearly perfectly matching ECD spectrum of **80** compared to that of the likewise *M*-configured alkaloid **71b**.[58]

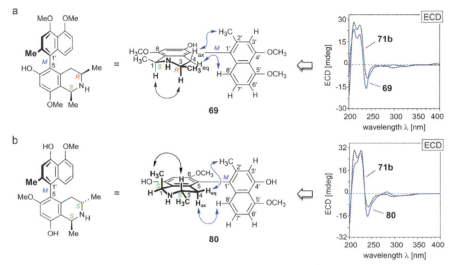

Fig. 47 Assignment of the absolute stereostructures of (a) ancistrobrevine M (**69**) and (b) ancistrobrevine K (**80**) by specific NOESY correlations establishing the relative configurations of **69** and **80** at the stereogenic centers C-1 and C-3 (double black arrow) and at the biaryl axis (double blue arrows). Comparison of the ECD spectra of **69** and **80** with that of the known—and likewise *M*-configured—alkaloid 5-*epi*-ancistrobrevine E (**71b**) (for the structure, see Fig. 46) permitted attribution of the absolute axial configurations of **69** and **80** as *M*.

5.1.2 Antiplasmodial activities of 5,1′-linked naphthyltetrahydroisoquinoline alkaloids

Tropical infectious diseases like malaria, leishmaniasis, Chagas' disease, African sleeping sickness, schistosomiasis, soil-transmitted nematode infections, tuberculosis, and various other parasitic, viral, and bacterial disorders are widely occurring in sub-Saharan Africa and in the MENA region (Middle East and North Africa).[405–409] One of the biggest health problems in many African and Asian tropical and subtropical countries is malaria, because of its high morbidity and mortality.[410–412] According to the WHO statistics, an estimated 1.5 billion malaria incidences and 7.6 million malaria death cases occurred worldwide in the period 2000–2019.[413] Although numerous promising lead compounds to combat malaria have been discovered during the past decade, current treatments suffer from the widespread development of resistance to well-established medical drugs.[414–417] Thus, for a long-term elimination and even eradication of malaria, the discovery and development of novel drugs is urgently required.

Some of the 5,1′-coupled naphthylisoquinoline alkaloids isolated from *A. abbreviatus* exert pronounced activities against the malaria parasite *Plasmodium falciparum*.[57] The most potent compounds were ancistrobrevines E (**71a**) and F (**59a**) and their respective atropo-diastereomers **71b** and **59b**, and ancistrobrevine G (**78**), displaying half-maximum inhibitory concentrations (IC$_{50}$ values) of 0.213 µM (for **71a**), 0.663 µM (for **71b**), 0.928 µM (for **59a**), 0.846 µM (for **59b**), and 0.134 µM (for **78**) against the NF54 strain of *P. falciparum* (sensitive to all known drugs).[57] The by far most potent antiplasmodial agent, however, was 5-*epi*-ancistectorine A$_2$ (**76**),[57] which had initially been discovered in the Chinese liana *A. tectorius*.[250] It revealed a high inhibitory activity against the NF54 strain (IC$_{50}$, 0.0074 µM), and was even found to be about 8.5 times more active than the standard chloroquine (IC$_{50}$, 0.065 µM).[57] Moreover, 5-*epi*-ancistectorine A$_2$ (**76**) showed an excellent selectivity index (SI) of ca. 12,700, exerting only extremely weak cytotoxicity against rat skeletal myoblast (L6) cells.[57] In addition, **76** displayed strong inhibitory effects against the chloroquine-resistant K1 strain of *P. falciparum* (IC$_{50}$, 0.03 µM), with a high SI of >3000.[250] Hence, according to the TDR/WHO guidelines,[310] 5-*epi*-ancistectorine A$_2$ (**76**) can be considered as a promising lead candidate.

The test results indicated that a free hydroxy group at C-8 may favor the antiplasmodial potential of 5,1′-linked naphthylisoquinolines.[57] The two parent compounds of this series of alkaloids, ancistrocladine (**20a**) and hamatine (**20b**), which both have a methoxy function at C-8, showed distinctly lower inhibitory activities, with IC$_{50}$ values of 45 µM (for **20a**) and 8.19 µM (for **20b**), respectively.[57,418]

The potent antiplasmodial activities of the five ancistrobrevines, **71a**, **71b**, **59a**, **59b**, and **78**, and of 5-*epi*-ancistectorine A$_2$ (**76**) proved to be highly pathogen-specific.[57] Against protozoan parasites causing visceral leishmaniasis (*Leishmania donovani*), Chagas' disease (*Trypanosoma cruzi*), or African sleeping sickness (*T. brucei rhodesiense*), very weak or virtually no activities were determined for any of the evaluated 5,1′-coupled naphthylisoquinoline alkaloids.[57]

5.1.3 Preferential cytotoxicities of 5,1′-linked naphthyltetrahydroisoquinoline alkaloids against pancreatic cancer cells

Pancreatic cancer[419–422] is a fatal malignancy, with an extremely poor prognosis and a very low 5-year survival rate of less than 5%. The incidence rate of this deadly disease is dramatically increasing all over the world. Among the solid tumors, pancreatic adenocarcinoma is one of the most aggressive and treatment-resistant human cancers due to impaired drug delivery pathways causing a rapidly progressing ineffectiveness of conventional chemotherapeutic agents.[421–426] In view of the fact that it does not reveal early specific symptoms, the disease often remains unrecognized for a long time. Therefore, the majority of the patients are diagnosed at a timely and spatially advanced metastatic—and thus unresectable—stage.[419–426]

The antiausterity approach[427–431] focusses on the discovery of new anticancer agents that target the capacity of pancreatic cancer cells of tolerating the lack of oxygen and nutrients in the tumor microenvironment.[432–435] Compounds are regarded as antiausterity agents if they show a strong preferential cytotoxicity in nutrient-deprived medium (NDM), by selectively inhibiting the tolerance of pancreatic cancer cells to nutrition starvation, thus leading to cell death.[10,11,30,31,56,57,61,176,183,428–431,436–443] The data obtained by testing the compounds for their preferential cytotoxicity against the PANC-1 cell line in both NDM and standard nutrient-rich medium (Dulbecco's modified Eagle's medium, DMEM) are referred to as preferential cytotoxicity (PC_{50}) values, which is the concentration of 50% cancer cell death in NDM without any toxicity in DMEM.

All of the investigated 5,1′-linked naphthylisoquinoline alkaloids isolated from *A. abbreviatus* displayed moderate to strong preferential cytotoxicities against PANC-1 human pancreatic cancer cells in a concentration-dependent manner, with micromolar PC_{50} values ranging from 45.0 µM down to 7.50 µM. 6-*O*-Methyl-4′-*O*-demethylhamatine (**75b**) and 5-*epi*-ancistectorine A$_2$ (**76**) exerted the most potent effects in NDM, with PC_{50} values of 7.50 µM for **75b** and 11.5 µM for **76** (Fig. 48).[57] The test results suggested that the oxygenation pattern of the isoquinoline subunit plays an important role for the antiproliferative effects of the alkaloids. This became evident from the pronounced cytotoxic activity of 5-*epi*-ancistectorine A$_2$ (**76**), possessing a 6-OMe/8-OH substituted tetrahydroisoquinoline portion, as compared to that of 5-*epi*-ancistrobrevine E (**71b**) (PC_{50}, 45.0 µM), with its two free OH groups at C-8 and C-6.[57] The position of the hydroxy function at C-8 in 5-*epi*-ancistectorine A$_2$ (**76**) obviously had a distinct impact on the antiausterity activity of 5,1′-coupled alkaloids, since hamatine (**20b**), with its opposite arrangement of the substituents at C-6 and C-8 (i.e., 6-OH/8-OMe), exerted only moderate inhibitory effects (PC_{50}, 35.6 µM).[57] The degree of *O*-methylation in the naphthalene part was likewise found to be of importance for the preferential cytotoxicity of naphthylisoquinoline alkaloids with a 5,1′-biaryl linkage. As an example, 6-*O*-methylhamatine

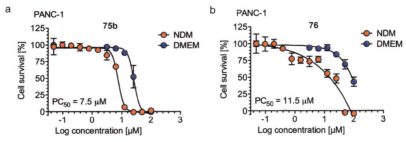

Fig. 48 Preferential cytotoxicities of the two 5,1′-coupled naphthyltetrahydroisoquinolines (a) 6-O-methyl-4′-O-demethylhamatine (**75b**) and (b) 5-*epi*-ancistectorine A$_2$ (**76**) (for their structures, see Fig. 45) against PANC-1 human pancreatic cancer cells in nutrient-deprived medium (NDM) and in nutrient-rich Dulbecco's modified Eagle's medium (DMEM).[57]

(**74b**) (PC$_{50}$, 42.0 μM), with methoxy functions both at C-4′ and C-5′, was about six times less active than its 4′-O-demethyl analogue **75b** (PC$_{50}$, 7.50 μM).[57]

5.1.4 Ancistrobrevidines A–C and related 5,1′-linked naphthyldihydroisoquinoline alkaloids from the leaves of Ancistrocladus abbreviatus

Recent isolation work on the leaves of *A. abbreviatus* has led to the identification of five 5,1′-coupled naphthyl*dihydro*isoquinoline alkaloids,[56] all of them produced in an atropisomerically pure form. This series of minor metabolites comprised three new alkaloids, ancistrobrevidines A–C (**81**, **82**, and **83**),[56] and two known[29,57,252,403] ones. One of the naphthyldihydroisoquinolines, identified earlier in other *Ancistrocladus* species, was the O-permethylated alkaloid 6-O-methylhamatinine (**84**)[29,57,403] (Fig. 49). It was discovered as a main constituent in the Vietnamese liana *A. cochinchinensis*[403] and in a botanically so far undescribed Central Congolese *Ancistrocladus* species.[29] 6-O-Methylancistectorine A$_3$ (**85**) (Fig. 49) was another already known alkaloid, differing from **84** only by its *P*-configuration at the biaryl axis and by a free phenolic hydroxy function at C-4′ on the naphthalene portion. It had been isolated for the first time, along with **84**, from the twigs and stems of *A. tectorius* from the Chinese Hainan Island.[252] With their *S*-configuration at C-3 and an oxygen function at C-6, **84** and **85** belong to the subclass of Ancistrocladaceae-type compounds. This also applies to one of the new naphthyldihydroisoquinoline alkaloids identified in the leaves of *A. abbreviatus*, ancistrobrevidine C (**83**)[56] (Fig. 49), which differs from 6-O-methylhamatinine (**84**) by its free phenolic hydroxy function at C-5′ on the naphthalene subunit.

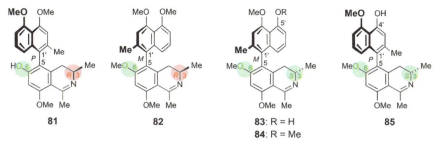

Fig. 49 Naphthyldihydroisoquinoline alkaloids from the leaves of *A. abbreviatus* displaying a 5,1'-biaryl linkage:[56] the hybrid-type compounds ancistrobrevidines A (**81**) and B (**82**), which were isolated along with the Ancistrocladaceae-type alkaloids ancistrobrevidine C (**83**), 6-*O*-methylhamatinine (**84**), and 6-*O*-methylancistectorine A₃ (**85**). The latter two had previously been detected in related Asian *Ancistrocladus* species.[252,403]

The discovery of the ancistrobrevidines A (**81**) and B (**82**) (Fig. 49) in *A. abbreviatus* was unprecedented, since these two compounds are among the very first examples of 5,1'-coupled naphthyldihydroisoquinoline alkaloids found in nature displaying an *R*-configuration at C-3.[56] The only other naturally occurring 5,1'-linked naphthyldihydroisoquinoline possessing such an entity is 5-*epi*-dioncophyllidine C₂ (**9**).[56] This unusual alkaloid has been isolated from the leaves of *A. abbreviatus*, too (see Section 5.1.5). With its lack of an oxygen function at C-6 and its 3*R*-configuration, however, it can clearly be attributed to the subgroup of Dioncophyllaceae-type naphthylisoquinoline alkaloids, whereas the ancistrobrevidines A (**81**) and B (**82**) are typical hybrid-type compounds, being equipped with an oxygen function at C-6 and being 3*R*-configured.[56] The as yet only other known representative of the 5,1'-coupling type belonging to the subclass of hybrid-type alkaloids is the naphthyl*tetrahydro*isoquinoline ancistrobrevine M (**69**)[58] (see Section 5.1.1).

The absolute configurations of the two new ancistrobrevidines A (**81**) and B (**82**) were established by ruthenium-mediated oxidative degradation[1,3,70] (see Section 4.4) in combination with NOESY and ECD measurements.[56] The formation of *R*-3-aminobutyric acid established both alkaloids to be *R*-configured at C-3. The relative and, thus, absolute configuration at the biaryl axis of ancistrobrevidine A (**81**) was attributed to be *P*, by NOESY interactions between H$_{ax}$-4 and H-8' and between H$_{eq}$-4 and Me-2' (Fig. 50a, center). This assignment was confirmed by the fact that the ECD spectrum of **81** was virtually opposite to that of the co-occurring, but *M*-configured 6-*O*-methylhamatinine (**84**) (Fig. 50a, right).[56]

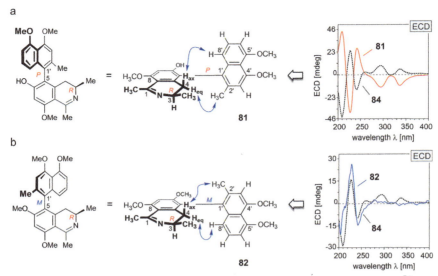

Fig. 50 Assignment of the stereostructures of (a) ancistrobrevidine A (**81**)[56] and (b) ancistrobrevidine B (**82**)[56] by NOESY interactions indicative of their configuration at the biaryl axis relative to the *R*-configured stereogenic center at C-3. Confirmation of the absolute axial configuration of **81** and **82** by comparison of their ECD spectra with those of the known[29,403] related 5,1′-coupled alkaloid 6-*O*-methylhamatinine (**84**) (for its structure, see Fig. 49).

Vice versa, NOESY interactions between H_{ax}-4 and Me-2′ and between H_{eq}-4 and H-8′ revealed the biaryl axis of ancistrobrevidine B (**82**) to be *M*-configured (Fig. 50b, center), which was corroborated by the nearly identical ECD spectra of **82** and **84** (Fig. 50b, right).[56] Hence, the *O*-permethylated 3*R*-configured new alkaloid **82** was the 3-epimer of 6-*O*-methylhamatinine (**84**).

The 5,1′-coupled naphthyldihydroisoquinoline alkaloids presented here showed moderate to strong antiproliferative activities against the HeLa human cervical cancer cell line[56] and against PANC-1 human pancreatic cancer cells in nutrient-deprived medium (NDM)[56] following the antiausterity approach[427–431] (see Section 5.1.3). Within this series of compounds, ancistrobrevidine C (**83**) exerted the strongest cytotoxic effects against the two cancer cell lines in a concentration-dependent manner (Fig. 51). It displayed a half-maximum inhibition concentration (IC_{50}) of 7.2 μM against HeLa cells (Fig. 51a),[56] showing a cytotoxicity comparable to that of 5-fluorouracil (IC_{50}, 13.9 μM), a widely applied therapeutic agent clinically used for cancer treatment.[444–446] Against PANC-1 cancer cells, ancistrobrevidine C (**83**) exhibited potent preferential cell-death inducing

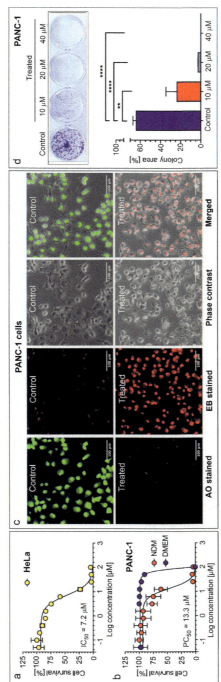

Fig. 51 (a) Impairment of cell viability of human HeLa cervical cancer cells by ancistrobrevidine C (**83**) (for the structure, see Fig. 49). (b) Preferential cytotoxic activity of **83** against PANC-1 human pancreatic cancer cells in nutrient-deprived medium (NDM) and in nutrient-rich Dulbecco's modified Eagle's medium (DMEM). (c) Morphology of PANC-1 cells in the untreated control (top) and in those treated with 20 μM of ancistrobrevidine C (**83**) in nutrient-deprived medium (NDM). Cells were stained with acridine orange (AO) and ethidium bromide (EB) and photographed under the fluorescence (green and red) and phase-contrast modes using an EVOS FL digital microscope (20× objective). (d) Effect of ancistrobrevidine C (**83**) on colony formation by PANC-1 cells. Representative PANC-1 cell colonies (top) treated with different concentrations of **83** and graph (bottom) showing mean values of the area occupied by PANC-1 cell colonies (three replicates). The colony areas were quantified by Image J plugin.[447] ****P < 0.0001, **P < 0.01, when compared with the untreated control group.

cytotoxicity (PC$_{50}$, 13.3 µM) (Fig. 51b) under nutrient-deficient conditions as typical of the hypovascular (austerity) tumor microenvironment, but without causing toxicity in normal, nutrient-rich medium.[56]

Investigations on ancistrobrevidine C (**83**) using the ethidium bromide (EB) and acridine orange (AO) double staining assay showed that the alkaloid significantly affected the morphology of PANC-1 cells under nutrient-deprived conditions (Fig. 51c).[56] AO is a cell-permeable dye, it emits bright green fluorescence in viable cells, whereas EB is permeable only in late apoptotic and necrotic cells, then intercalates into DNA, giving rise to a predominant red fluorescence. The untreated cancer cells, serving as the control (Fig. 51c, top), showed an intact cellular morphology, solely emitting bright green fluorescence, as typical of living cells. Treatment of PANC-1 cells with 20 µM of ancistrobrevidine C (**83**), however, resulted in a red EB fluorescence exclusively, thus proving complete cell death (Fig. 51c, bottom). The EB stained cells displayed dramatic alterations in cellular morphology, as characterized by the rupture of the cell membranes and leakage of cellular contents into the culture medium.[56]

Since pancreatic cancer has a high metastatic potential, leading to a rapid invasion and colonization of distant organs such as stomach, liver, and lungs,[420,424,432,434] the ability of ancistrobrevidine C (**83**) to inhibit PANC-1 colony formation was of great interest, making it worth studying it more closely (Fig. 51d). The cancer cells were exposed to 0 (control), 10, 20, and 40 µM of **83** under nutrient-rich conditions (DMEM) and incubated for 24 h. After replacement of the media with fresh DMEM, the PANC-1 cells were put in a CO$_2$ incubator for further 10 days to allow colony formation. After the end of the experiment, the cells were fixed with 4% aq. formaldehyde, and stained with crystal violet for 15 min. As shown in Fig. 51d, the tumor cells grew aggressively in the control and formed a large number of colonies, which occupied nearly 64% of the total well area.[56] Treatment of PANC-1 cells with ancistrobrevidine C (**83**) for 24 h in DMEM, however, induced a significant inhibition of colony formation.[56] Exposure to 10 µM of **83** already caused a drastic reduction of the occupied colony area by 63% compared to the colony formation in untreated cells (Fig. 51d, bottom), since only 24% of the well area was covered by PANC-1 colonies. Upon treatment with 20 µM of ancistrobrevidine C (**83**), the colony area was reduced by 95%, and cancer cells exposed to 40 µM of **83** even showed complete inhibition of colony formation (Fig. 51d, top).[56] Hence, ancistrobrevidine C (**83**) did not only display preferential cytotoxicity against pancreatic cancer cells in nutrient-deprived medium (NDM) (Fig. 51b), but also significantly suppressed cell migration and colony formation under normal, nutrient-rich conditions in DMEM.[56]

Tackling the migration and invasion of tumor cells and, thus, suppressing the formation of tumor colonies is an important factor to achieve an improved benefit in cancer treatment. Therefore, the inhibitory effect of compound **83** on PANC-1 cell migration was investigated using a quantitative real-time wound-healing assay[56,441] (Fig. 52). For this purpose, the cells were seeded in a two-well culture insert and were allowed to attach overnight to ensure a clean and symmetric wound area. After removal of the insets, live imaging experiments were run at the same time and under identical conditions for the PANC-1 cells incubated either in DMEM alone (control) or in DMEM with 20 or 40 μM of ancistrobrevidine C (**83**). The cells were allowed to migrate for 48 h and a time-lapse video was taken capturing images every 15 min (for the movie, see Ref. 56). While the control (no agent) showed a distinct reduction of the initial wound area by about 50% after 24 h, exposure of the tumor cells to 20 and 40 μM of **83** clearly prevented wound closure (Fig. 52a), resulting in open wound areas of 62 % and 77% compared to the wound area at T_0 (Fig. 52b). The effect was even more pronounced after 48 h. The control group showed complete wound closure, whereas the cells treated with 20 and 40 μM of ancistrobrevidine C (**83**) still had an open wound area of about 25% and 70%, respectively (Fig. 52b).[56]

Hence, ancistrobrevidine C (**83**) did not only display preferential cytotoxicity against pancreatic cancer cells in nutrient-deprived medium (NDM) (Fig. 51b), but also significantly suppressed cell migration and colony formation under normal, nutrient-rich conditions in DMEM and, therefore, has a pronounced antimetastatic and anti-invasion potential.[56] The results suggest that naphthylisoquinoline alkaloids in general, and ancistrobrevidine C (**83**) in particular, can be regarded as promising leads for the search of new anticancer agents based on the antiausterity strategy.

5.1.5 5-epi-*Dioncophyllidine C₂*, only the third example of a 5,1′-linked dioncophyllaceous naphthylisoquinoline alkaloid from nature

Dioncophylline C (**73**)[402] is the only 5,1′-coupled naphthylisoquinoline alkaloid identified in the West African liana *Triphyophyllum peltatum*, which exclusively produces Dioncophyllaceae-type compounds (i.e., with 6-H, 3R). For a long time, it was even the only 5,1′-linked dioncophyllaceous alkaloid at all known from nature—until, some years ago, the closely related naphthylisoquinoline dioncophylline C₂ (**86**) (Fig. 53) was isolated from the recently described[23] Congolese liana *A. ileboensis*.[51] It has been only in 2021

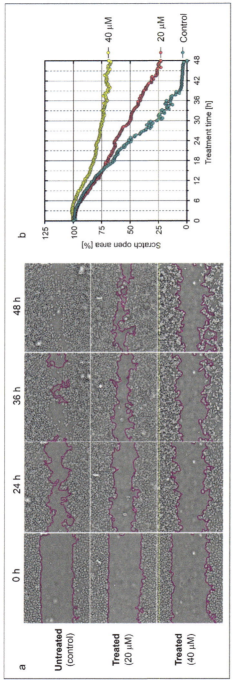

Fig. 52 (a) Suppression of the migration of PANC-1 cells in DMEM by ancistrobrevidine C (**83**) (for the structure, see Fig. 49) in a concentration-dependent manner in a wound-healing assay. Untreated PANC-1 cells (control) and cancer cells treated with compound **83** (20 and 40 μM) were subjected to time-lapse imaging at 15 min intervals for 48 h, for which images taken at 0, 24, 36, and 48 h are shown (for the corresponding real-time movie, see Ref. 56). (b) Quantification of wound healing by measuring the open scratch area for each time point using a Fiji platform.[448]

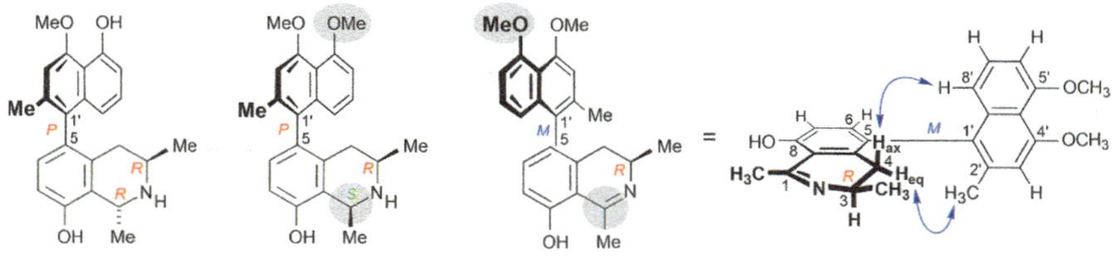

Fig. 53 The only three as yet known naturally occurring Dioncophyllaceae-type naphthylisoquinoline alkaloids exhibiting a 5,1′-coupling type: dioncophylline C (**73**)[402] from *T. peltatum*, dioncophylline C$_2$ (**86**)[51] from *A. ileboensis*, and 5-*epi*-dioncophyllidine C$_2$ (**9**)[56] from the leaves of *A. abbreviatus*; attribution of the relative stereostructure of **9** by specific NOESY interactions. Structural elements in **86** and **9** differing from those of the parent compound **73** are underlaid in gray.

that a third example of a Dioncophyllaceae-type alkaloid displaying the 5,1′-coupling type has been discovered, it was found in the leaves of *A. abbreviatus*, 5-*epi*-dioncophyllidine C$_2$ (**9**).[56] It is the first representative within this small subgroup of alkaloids possessing a dihydroisoquinoline portion and displaying the *M*-configuration at the biaryl axis (Fig. 53). The absolute configuration of **9** at the stereocenter was assigned by the ruthenium-mediated oxidative degradation procedure,[1,3,70] leading to *R*-3-aminobutyric acid, thus establishing the configuration at C-3 to be *R*. The absolute axial configuration was determined to be *M* due to specific NOESY interactions between H$_{eq}$-4 and Me-2′ and between H$_{ax}$-4 and H-8′.[56]

5.2 Ancistrobrevine B and related 5,8′-coupled naphthylisoquinoline alkaloids

Ancistrobrevine B (**87a**) was already known as a main alkaloid of *A. abbreviatus* in early phytochemical studies on this West African liana,[5,63] and has since then been found to likewise occur in the Kenyan species *A. robertsoniorum*[280] and in *A. congolensis*[178] and in a botanically yet undescribed liana[30] from North-Central Africa. Recent investigations on *A. abbreviatus* have now furnished two further 5,8′-coupled naphthyltetrahydroisoquinoline alkaloids as minor compounds, 5′-*O*-demethylancistrobrevine B (**88**)[57] and ancistroealaine D (**89**),[404] along with the likewise 5,8′-coupled fully dehydrogenated naphthylisoquinoline ancistrobreveine D (**10a**)[54] (Fig. 54). 5′-*O*-Demethylancistrobrevine B (**88**) had so far only been identified in *A. abbreviatus*,[57] whereas ancistroealaine D (**89**) had previously been isolated from the Congolese species *A. ealaensis*.[181] Ancistrobreveine D (**10a**) belongs to a unique series of seven alkaloids with a non-hydrogenated isoquinoline

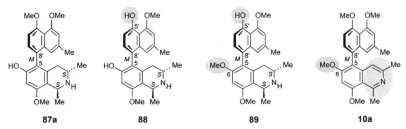

Fig. 54 The only four 5,8′-coupled naphthylisoquinoline alkaloids identified so far in *A. abbreviatus*: ancistrobrevine B (**87a**),[5,63] 5′-*O*-demethylancistrobrevine B (**88**),[57] ancistroealaine D (**89**),[404] and ancistrobreveine D (**10a**).[54] The structural modifications in **88**, **89**, and **10a** compared to their parent compound, ancistrobrevine B (**87a**), are underlaid in light gray.

portion discovered in the roots of *A. abbreviatus*.[54] A more detailed report on these seven naphthylisoquinolines displaying four different coupling types is given separately, in Section 5.6.

The by far largest group of alkaloids isolated from *A. abbreviatus* are 5,1′-linked naphthyltetra- and -dihydroisoquinolines (22 compounds, see Section 5.1), but there are strikingly few 5,8′-coupled compounds. This finding is remarkable with respect to the fact that the alkaloid patterns of most of the Central African *Ancistrocladus* plants are dominated by the presence of a noticeable number of 5,8′-coupled naphthylisoquinolines. These alkaloids are even the main constituents of *Ancistrocladus* species such as *A. korupensis*[5,13,14,287,288] from Cameroon or *A. congolensis*,[3,177,178] *A. likoko*,[3,53,175,176] *A. ealaensis*,[3,179,181,182] and of four botanically undescribed lianas[3,15,29–35] from the Congo Basin. Many of these naphthylisoquinolines are hybrid-type alkaloids, with *R*-configuration at C-3 and an oxygen function at C-6, among them an impressive number of michellamine-type[3,13,14,31,178,181,288] and mbandakamine-type[3,15,32–35,181,182] dimers (see Fig. 24 in Section 3).

Ancistrobrevine B (**87a**)[5,63] (Fig. 55) was the first 5,8′-coupled naphthylisoquinoline alkaloid found in nature. Meanwhile, a broad series of more than 100 such 5,8′-linked biaryl natural products have been discovered in other African and in Asian *Ancistrocladus* species,[1,3–5,13–15,29–35,67,245,251,277,278,280,287,288] so that these compounds currently represent the by far largest subclass of naphthylisoquinoline alkaloids (see Section 3).

NOE interaction between Me-1 and H-3 revealed a relative *trans*-array of the methyl groups at C-1 versus C-3 in **87a** (see Fig. 55a). By oxidative degradation[1,3,70] (see Section 4.4), the absolute configuration at C-3 was established to be *S*, from which, given the relative *trans*-configuration

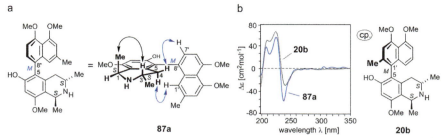

Fig. 55 Assignment of the complete stereostructure of ancistrobrevine B (**87a**)[5,63,66] (a) by specific NOESY interactions defining the relative configuration at C-3 versus C-1 in the tetrahydroisoquinoline portion (double black arrow) and the configuration at the axis relative to the stereogenic centers (double blue arrows) and (b) the absolute axial configuration by comparison of its ECD spectrum with that of the structurally related, but 5,1'-coupled alkaloid hamatine (**20b**).

(as evident from NMR), the absolute configuration at C-1 had to be *S*, too.[63] The absolute axial configuration was established as *M* due the resemblance of the ECD spectrum of ancistrobrevine B (**87a**) to that of its stereochemically closely related, yet 5,1'-coupled isomer hamatine (**20b**) (see Fig. 55b). Although formally possessing a different coupling site as compared to **87a** (which is 5,8'-coupled), the two alkaloids actually differ only by the chiroptically less decisive position of the methyl group on the naphthalene part.[63] This assignment of the absolute axial configuration of **87a** was confirmed by the attribution of the relative configuration by specific NOE interactions across the biaryl axis due to the favorably short distances to be covered in this coupling type, namely between the decisive protons H$_{eq}$-4 and H-7' and between H$_{ax}$-4 and H-1' (see Fig. 55a).[5,63,66] The structural elucidation of ancistrobrevine B (**87a**) was the first example of the attribution of the relative configuration of axes versus centers by 2D NMR spectroscopy in the field of naphthylisoquinoline alkaloids.[5,66] Today, this is a standard method, successfully applied to the stereochemical assignment of the configuration at the biaryl axis of a structurally broad variety of naphthylisoquinoline alkaloids (see Section 4.3).

The total synthesis of ancistrobrevine B (**87a**) (Fig. 56) was achieved by Pd-catalyzed Suzuki-Miyaura cross-coupling of its appropriately modified naphthalene and isoquinoline moieties, with the coupling positions activated by iodine in the tetrahydroisoquinoline part and boronic acid in the naphthalene portion, according to a strategy developed by Hoye et al.[78,449,450] for the synthesis of other 5,8'-coupled naphthylisoquinoline alkaloids. The synthesis started from the 8-*O*-methylated and 6-*O*-benzylated 5-iodotetrahydroisoquinoline **91**,[449,450] which was prepared

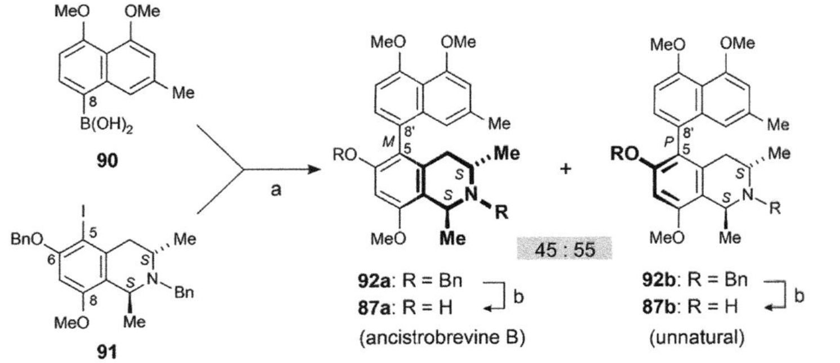

Fig. 56 Total synthesis of ancistrobrevine B (**87a**) by Pd-catalyzed Suzuki-Miyaura cross-coupling starting from the boronic acid **90** and the 5-iodo-activated tetrahydroisoquinoline **91**, giving rise to the natural target molecule **87a** and its atropisomer **87b** in excellent chemical yields, but with a low diastereoselectivity in favor of the non-natural, merely artificial **87b**. Reaction Conditions: (a) Pd(PPh$_3$)$_4$, PhMe, EtOH, satd. aq. NaHCO$_3$, 100–110 °C, 86%, 11% de; (b) H$_2$, Pd/C (10%), MeOH, rt, 98%.

via chiral aziridines (derived from L-alanine) and protected by O,N-dibenzylation, and the naphthalene boronic acid **90**.[449,450] Cross-coupling of **90** and **91** furnished the atropisomeric biaryl products **92a** and **92b** in an excellent chemical yield of 86%.[449,450] The diastereoselectivity, however, was low, with 1:1.25 in favor of the "wrong" atropisomer, **92b** (11% de).[449,450] Apparently, the stereogenic centers in the isoquinoline building block were too far away from the coupling position to provide a satisfactory internal asymmetric induction in the coupling step. Hydrogenolytic O- and N-debenzylation of pure **92a** and **92b** and HPLC resolution on an achiral reversed-phase amino-bonded phase afforded the target molecule ancistrobrevine B (**87a**), along with its atropo-diastereomer **87b** (so far not found in nature), both in a stereochemically pure form.[449,450]

5.3 Dioncophylline A and related 7,1′-coupled naphthylisoquinoline alkaloids from four different subclasses

5.3.1 Dioncophylline A and related Dioncophyllaceae-type naphthylisoquinoline alkaloids

Three of the major metabolites of *A. abbreviatus* were readily identified as the well-known alkaloid dioncophylline A (**1a**)[5,6,63] and its 4′-O-demethyl and N-methyl analogues **93a** and **94a**.[5,58,62] They were isolated together

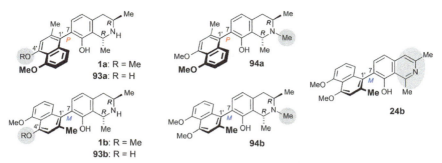

Fig. 57 Dioncophyllaceous 7,1′-coupled naphthylisoquinoline alkaloids from the roots of *A. abbreviatus*[5,54,58,62,404]: dioncophylline A (**1a**), 4′-*O*-demethyldioncophylline A (**93a**), and *N*-methyldionophylline A (**94a**), accompanied by the respective atropo-diastereomers **1b**, **93b**, and **94b**, isolated along with the non-hydrogenated compound *ent*-dioncophylleine A (**24b**), which was atropisomerically pure.

with the respective atropo-diastereomers **1b**, **93b**, and **94b**, respectively (Fig. 57).[5,58,62,404] Dioncophylline A (**1a**) and its congeners **93a** and **94a** (including their atropisomers) have been widely identified as main or minor secondary metabolites in Dioncophyllaceae lianas[5,6,68,292,297] and also in *Ancistrocladus* species[5,12,51,58,62,63,184,285,404] from West and Central Africa. Interestingly, the isolation of all of them from the same plant, as pairs of atropo-diastereomers, has so far been reported only for the title plant, *A. abbreviatus*.[5,58,62,404] A further dioncophyllaceous, yet fully dehydrogenated 7,1′-coupled alkaloid, *ent*-dioncophylleine A (**24b**)[54] (Fig. 57), discovered in the roots of *A. abbreviatus*, will be presented in Section 5.6. This compound had already been known from previous phytochemical investigations on the new Malaysian species *A. benomensis*.[1,291]

Dioncophylline A (**1a**) is a main secondary metabolite of the West African Dioncophyllaceae lianas *Triphyophyllum peltatum*,[5,6,8,9] *Dioncophyllum thollonii*,[68,292] and *Habropetalum dawei*.[297] It likewise dominates the alkaloid pattern of the Central African *Ancistrocladus* taxa *A. letestui*,[184] and *A. ileboensis*,[51] and of a botanically so far undescribed Congolese *Ancistrocladus* species.[12] The three Dioncophyllaceae plants apparently produce **1a** in an atropisomerically pure form,[5,6,8,68,292,297] i.e., without any co-occurring 7-*epi*-dioncophylline A (**1b**), whereas in *A. ileboensis*, dioncophylline A (**1a**) is accompanied by its atropo-diastereomer **1b**,[51] as also here in *A. abbreviatus*.[58,404] Complementary to the situation in Dioncophyllaceae lianas, 7-*epi*-dioncophylline A (**1b**) was found to occur in the West African species *A. barteri* in a stereochemically homogeneous form, i.e., without dioncophylline A (**1a**) being detectable.[5,285]

4′-O-Demethyldioncophylline A (**93a**), differing from **1a** by the presence of a free phenolic hydroxy group on the naphthalene ring, was isolated from the roots and stems of *A. abbreviatus*, along with nearly equal amounts of its atropisomer **93b**.[5,58,404] The latter two alkaloids were also detected in *A. ileboensis* from the Southern Congo Basin, occurring in the plant as major metabolites,[51] whereas in *A. barteri*, 4′-O-demethyl-7-*epi*-dioncophylline A (**93b**) was produced in an atropisomerically pure form, albeit as a minor constituent.[5,285] The two atropo-diastereomers **93a** and **93b** have recently attracted attention, since they are the molecular halves of a unique series of four Dioncophyllaceae-type dimers discovered in the roots of *A. abbreviatus* (see Section 6.1).[58] One of them was jozimine A$_2$ (**4a**) (see Fig. 23 in Section 3), a highly antiplasmodial compound,[12,180] which had previously been isolated from an as yet unknown Congolese *Ancistrocladus* species,[12] and from *A. ileboensis*.[183]

The two main alkaloids of *A. abbreviatus* were *N*-methyldioncophylline A (**94a**) and its atropo-diastereomer, **94b**.[5,62] They were isolated from root and stem bark extracts in nearly equal amounts. These two alkaloids were also identified as major constituents of three other West African plant species, *D. thollonii*,[68,292] *H. dawei*,[297] and *A. barteri*.[285] In the latter, **94a** and **94b** occurred as a 2:1 mixture, whereas *D. thollonii* and *H. dawei* produced the two atropisomers in a ratio of 1:1, similar to the situation in *A. abbreviatus* (see above). The separation of **94a** and **94b**, although being diastereomers, proved to be challenging due to their very similar chromatographic behavior, nearly as if they were enantiomers. It, thus, finally was achieved only by application of racemate-resolution techniques.[5,62] Analytically, HPLC separation succeeded on a chiral phase (Chiralcel OD®) in combination with ECD spectroscopy, giving rise to two baseline-separated peaks, which exhibited two nearly mirror-imaged ECD spectra (Fig. 58).[5,62] Preparatively, **94a** and **94b** were resolved by enhancing the—only weakly developed—diastereomeric character of the atropisomers by derivatization with enantiomerically pure menthoxyacetyl chloride (**95**), as a chiral reagent, followed by resolution of the resulting diastereomeric menthoxyacetic esters by column chromatography (CC) on normal-phase silica gel, and subsequent cleavage of the chiral auxiliary by hydrolysis (Fig. 58).[5,62] Also in some other cases, the resolution of atropo-diastereomeric naphthylisoquinoline alkaloids was possible only by chromatography on a chiral phase, i.e., again by treating these stereoisomers as if they were enantiomers.[1,2,16,299,362]

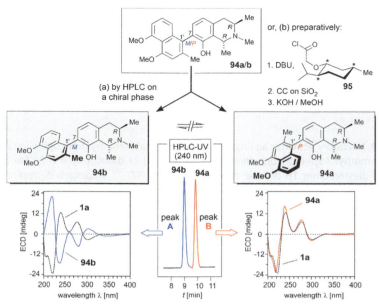

Fig. 58 Resolution of N-methyldioncophylline A (**94a**) and its atropo-diastereomer, N-methyl-7-epi-dioncophylline A (**94b**), by racemate-resolution techniques (mixture of **94a**:**94b**, ca. 1:1). Online chiroptical investigations provided the LC-ECD spectra of the two atropo-diastereomers **94a** and **94b**, and comparison with the ECD curve of dioncophylline A (**1a**) revealed N-methyldioncophylline A (**94a**) to be the more slowly eluting compound, while the epimer **94b** was the faster one. Conditions and Reagents[62]: (a) HPLC on a chiral adsorbent (Chiralcel® OD column, Daicel Chemical Industries Ltd.); (b) enantiopure menthoxyacetyl chloride (**95**, obtained from (−)-menthoxyacetic acid and SOCl$_2$), 1,8-diazobicyclo[5.4.0]undec-7-ene (DBU), rt, 1 h → repeated CC on SiO$_2$ (15–25 mesh), eluent: CH$_2$Cl$_2$–MeOH–NEt$_3$ (1990:10:1, v/v/v) → saponification using 0.05 M methanolic KOH.

Among a series of 7,1′- and 5,1′-coupled genuine and artificial naphthylisoquinolines, N-methyldioncophylline A (**94a**), its atropo-diastereomer **94b**, and dioncophylline A (**1a**) were found to display strong activities against *Babesia canis* in vitro (see Fig. 59a and b), with half-maximum inhibition concentration (IC$_{50}$) values of 1.45 μM for **94a**, 0.14 μM for **94b**, and 0.48 μM for **1a**.[373] Canine babesiosis is a tick–borne disease induced by *Babesia* parasites, which proliferate in erythrocytes by lysing the cells, thus causing a severe, potentially fatal anemia.[451–453] The aromatic diamidine compound imidocarb dipropionate is commonly used for the treatment of dogs suffering from a babesial infection, but its effectiveness is hampered by increasing drug resistance of the protozoa and various adverse side-effects.[452–455] Therefore, alternative chemotherapeutic agents are urgently

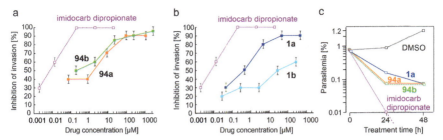

Fig. 59 Inhibitory activities against *Babesia canis* of (a) *N*-methyldioncophylline A (**94a**) and *N*-methyl-7-*epi*-dioncophylline A (**94b**), and (b) of dioncophylline A (**1a**) and its atropo-diastereomer **1b** (for the structures, see Fig. 57).[373] Imidocarb dipropionate[454] was used as the standard. Error bars indicate the range of values derived from three independent experiments. (c) Parasitemia of cells after application of **1a**, **94a**, and **94b**, determined in Giemsa-stained blood smears of the parasites. Control group: 0.05% aq. DMSO. Treatment with imidocarb dipropionate reduced parasitemia after 24 h completely.[373]

needed, in particular with respect to the fact that only a limited number of useful drugs are currently available to combat the disease.[452–454]

The antibabesial effects of *N*-methyldioncophylline A (**94a**), *N*-methyl-7-*epi*-dioncophylline A (**94b**), and dioncophylline A (**1a**) were high, although none of the tested naphthylisoquinolines reached 100% inhibition of parasite infection as achieved by imidocarb dipropionate (IC$_{50}$, 0.07 µM).[373] After application of **94a**, **94b**, and **1a**, the parasitemia of the cells was significantly lower than the one in the control group (treated with 0.5% aq. DMSO, with no compound) at 24 and 48 h after incubation, as determined in Giemsa-stained blood smears of *B. canis* (see Fig. 59c). Furthermore, the strong effects of *N*-methyl-7-*epi*-dioncophylline A (**94b**) and related compounds against *B. canis* correlated with only weak hemolytic effects (<0.7%), whereas compounds with moderate to weak inhibitory activities like e.g., the 5,1′-coupled alkaloid 6-*O*-methyl-4′-*O*-demethylhamatine (**75b**) (Fig. 45), with an IC$_{50}$ value of 830 µM, caused a high degree of hemolysis (11.2%).[373]

Since *Plasmodium* and *Babesia* parasites have similar life cycles and induce comparable typical symptoms of disease such as high fever and anemia due to the lysis of erythrocytes,[456,457] antiplasmodial compounds are regarded as potential candidates for the discovery and development of new potent antibabesial agents.[458–463] However, two of the most active antiplasmodial naphthylisoquinoline alkaloids, jozimine A$_2$ (**4a**)[12,180] (Section 6.1.1) and 5-*epi*-ancistectorine A$_2$ (**76**)[57,250] (Section 5.1.2), with IC$_{50}$ values against *Plasmodium falciparum* in the nanomolar range, and classified as promising antimalarial lead compounds,[12,57,310] exerted only extremely weak antibabesial effects, with high IC$_{50}$ values of 140 µM for **4a** and 615 µM for **76**.[373] Thus, the divergence of the antiplasmodial and antibabesial activities exhibited by **4a** and **76** evidenced a substantial parasite-specificity of the two compounds.

The study also revealed that axial chirality has a strong impact on the inhibition of parasite infection, as seen for the comparison of the activities of dioncophylline A (**1a**) (IC$_{50}$, 0.48 µM) with the atropisomeric 7-*epi*-dioncophylline A (**1b**) (IC$_{50}$, 26.5 µM) (Fig. 59b), or of 6-*O*-methyl-4′-*O*-demethylancistrocladine (**75a**) (IC$_{50}$, 14.0 µM) with its atropodiastereomer 6-*O*-methyl-4′-*O*-demethylhamatine (**75b**) (IC$_{50}$, 830 µM) (for the structures, see Fig. 45).[373]

5.3.2 Dioncophylline A, a promising potential lead for anticancer drug development

Dioncophylline A (**1a**) displayed strong cytotoxic activities at a low micromolar range against various cancer cell lines,[51,365,366,404] indicating that this naphthylisoquinoline alkaloid may have therapeutic potential, and is thus worth more intense investigations as a promising candidate for further drug development as a new anti-cancer agent. It was found to strongly inhibit the viability of multiple myeloma INA-6 cells and of drug-sensitive acute lymphoblastic CCRF-CEM leukemia cells and their multidrug-resistant subline, CEM/ADR5000.[51,366] Dioncophylline A (**1a**) also showed pronounced antiproliferative effects in the hormone-positive and -negative breast cancer models MCF-7 and MDA-MB-231,[365] and displayed significant cytotoxic activities against human cervical HeLa and colon carcinoma (HT-29) cells.[404]

Multiple myeloma (MM)[464–466] is the second-most prevalent hematological malignancy, caused by an excess of abnormal cells forming tumors in multiple locations within the bone marrow. It accounts for about 1% of all cancers and approximately 10% of all hematological cancer disorders. MM primarily affects the elderly and is typically diagnosed between the ages of 65–74 years, with a median overall survival rate of about 6 years.[465] Despite the increased effectiveness of combinatorial first-line chemotherapeutic regimens,[465–468] cure still remains elusive. The majority of patients will eventually relapse and the malignant cells may become drug resistant.[465–468] Therefore, the search for novel therapeutic agents to efficiently target MM still remains an important task.

Dioncophylline A (**1a**), but also 4′-*O*-demethyldioncophylline A (**93a**), isolated from *A. abbreviatus*,[5,58,63] and 5′-*O*-demethyldioncophylline A (**97**),[294] a metabolite of *Triphyophyllum peltatum* (Dioncophyllaceae), exerted pronounced antiproliferative properties against multiple myeloma

INA-6 cells,[51] with effective concentration ranges similar to those of the therapeutically used DNA-alkylating agent melphalan[469] (see Table 3). Normal mononuclear cells from the peripheral blood of healthy donors (PMBCs), by contrast, were not affected by **1a** and its two mono-O-demethylated analogues **93a** and **97** at the tested concentrations.[51] Thus, the excellent—and structure-specific—activity of dioncophylline A (**1a**) (even ten times better than that of melphalan[469]) and of its two mono-O-demethyl analogues demonstrated their promising potential for a further development as potential anti-MM agents.[51]

Table 3 EC$_{50}$ values (µM) of INA-6 multiple myeloma cells and peripheral mononuclear blood cells (PMBCs) treated with dioncophylline A (**1a**) and its 4'- and 5'-O-demethyl derivatives **93a** and **97**, and with further naphthylisoquinoline alkaloids (**96**, **93b**, **98**–**100**, **77**, **20b**, and **20a**), or with melphalan.

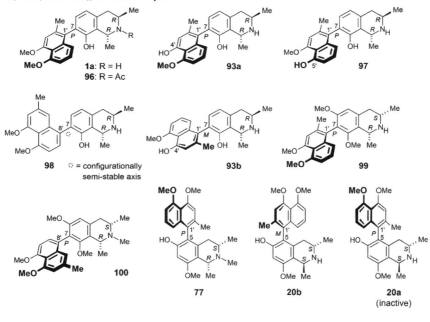

	1a	96	93a	93b	97	98	99	100	77	20b	melphalan[c]
INA-6[a]	0.22	0.8	2.7	16.0	1.5	2.6	4.8	16.0	32.0	32.0	2.0
PBMCs	NR[b]	NR	NR	50	NR	19	NR	35	50	30	3.0

[a]Multiple myeloma cells were treated with different concentrations of compounds **1a**, **96**, **93a**, **93b**, **97**–**100**, **77**, **20a**, and **20b** or with melphalan (positive control).[51,250] Viable cell fractions were determined by annexin V-FITC and propidium iodide (PI) staining using flow cytometry.[51,470] [b]NR: not reached. [c]Positive control. The anti-MM activities of the two most potent compounds, dioncophylline A (**1a**) and its N-acetyl analogue **96**, are underlaid in gray.

Remarkably, the atropo-diastereomer of **93a**, 4′-O-demethyl-7-*epi*-dioncophylline (**93b**), displayed only moderate effects against MM cells,[51] again showing the importance of axial chirality for bioactivities (Table 3). *N*-Acetyldioncophylline A (**96**),[99] an unnatural, merely semisynthetic analogue, was also found to display strong anti-MM activities, without any cytotoxic effects in non-malignant PMBCs.[51] Another 7,1′-coupled, but ancistrocladaceous representative, ancistrocladisine A (**99**)[51] from the Congolese liana *A. ileboensis*, was also found to exhibit high and specific anti-MM activities. Among the other alkaloids with different coupling patterns, the 7,8′-linked naphthylisoquinoline 5′-O-methyldioncophylline D (**98**)[82] from *T. peltatum* showed an anti-MM activity similar to that of 4′-O-demethyldioncophylline A (**93a**), whereas the likewise 7,8′-coupled, yet ancistrocladaceous alkaloid ancistrobrevine A (**100**)[5,57,66] from *A. abbreviatus* exhibited only moderate activities against MM cells, but exerted distinct cytotoxic effects toward normal blood cells.[51] A similar result was obtained for the 5,1′-coupled naphthylisoquinolines hamatine (**20b**) and ancistrocline (**77**), which displayed only weak anti-MM activities, but strong cytotoxicities against PMBCs.[250] The atropo-diastereomer of **20b**, ancistrocladine (**20a**), on the contrary, did not reveal any cytotoxic effect, neither against the cancer cells nor against the normal blood cells,[250] again an impressive example of the impact of axial chirality on bioactivities.

One of the major problems in cancer treatment is multi-drug resistance (MDR), leading to recurrent tumor growth.[471–474] Overexpression of P-glycoprotein (P-gp), an ATP-dependent drug pump, is considered to be the most important factor for the development of MDR. It triggers an increased drug efflux, resulting in a dramatic reduction of the intracellular concentrations of therapeutic agents in tumor cells, thus causing failure of chemotherapy.[475,476] To overcome this phenomenon, there is an urgent need for new effective anticancer drugs for clinical pharmacotherapy.

Dioncophylline A (**1a**), its two likewise naturally occurring regioisomeric analogues, 4′-O-demethyldioncophylline A (**93a**)[5,58] and 5′-O-demethyldioncophylline A (**97**),[294] and an unnatural derivative of **1a**, 8-O-(*p*-nitrobenzyl)dioncophylline A (**101**)[99] significantly inhibited the viability of drug-sensitive human lymphoblastic CCRF-CEM leukemia cells and their multi-drug resistant P-gp-overexpressing subline, CEM/ADR5000 (Table 4).[366] Among the investigated naphthylisoquinolines, the parent compound **1a** was the by far most active agent, exhibiting the lowest IC_{50} values for the two tested leukemia cell lines, 0.46 μM (for CCRF-CEM) and 0.69 μM (for CEM/ADR5000).

Table 4 IC$_{50}$ values (µM) of CCRF-CEM leukemia cells and their multidrug-resistant subline, CEM/ADR5000, treated with dioncophylline A (**1a**) and its 4'- and 5'-O-demethyl congeners, **93a** and **97**, and with the unnatural derivative **101**, or with doxorubicin.

Compound	CCRF-CEM	CEM/ADR5000	Degree of resistance[b]
Doxorubicin	0.14 ± 0.004	81.24 ± 6.32	580
1a	0.46 ± 0.01	0.69 ± 0.04	1.50
93a	5.37 ± 0.59	18.4 ± 0.65	3.43
97	1.75 ± 0.05	5.10 ± 0.14	2.91
101	0.90 ± 0.10	0.94 ± 0.06	1.04

IC$_{50}$ [µM][a]

[a]The lymphoblastic leukemia cells were treated with different concentrations of **1a**, **93a**, **97**, and **101**, or with doxorubicin. Cell viability was assessed by the resazurin assay.[366] Mean values and standard deviation of three independent experiments with each six parallel measurements are shown.
[b]The degrees of resistance were calculated as the ratio of the IC$_{50}$ values of the compounds for CEM/ADR5000 cells and the corresponding IC$_{50}$ values for CCRF/CEM cells.

As outlined in Table 4, compounds **1a**, **93a**, **97**, and **101** affected cell proliferation in CCRF-CEM cells in a concentration range similar to that of the standard antileukemic drug doxorubicin[477,478] (IC$_{50}$, 0.14 µM).[366] In contrast to doxorubicin, however, the four naphthylisoquinolines showed only minimal cross-resistance toward CEM/ADR5000 cells, thus indicating that both drug-sensitive and -resistant cells were inhibited by **1a**, **93a**, **97**, and **101** with similar efficacies.[366] Therefore, these compounds may be considered as promising candidates for the treatment of resistant and refractory tumors.

The atropisomer of **93a**, 4'-O-demethyl-7-*epi*-dioncophylline A (**93b**), and the three 5,1'-coupled alkaloids ancistrocladine (**20a**), hamatine (**20b**), and ancistrocline (**77**), by contrast, showed only moderate to weak cytotoxic effects on the viability of the CCRF-CEM cell line at a concentration of 10 µM (Fig. 60).[51,250] Proliferation of the leukemic cells was inhibited by >30% (for **93b**),[51] and still by ca. 15% in the cases of **20a**, **20b**, and **77**.[250]

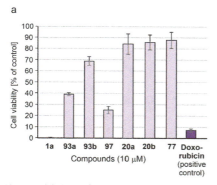 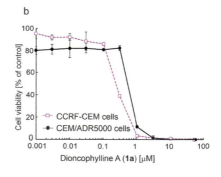

Fig. 60 (a) Growth percentage (%) of drug-sensitive lymphoblastic leukemia CCRF-CEM cells treated with the 5,1'-coupled naphthylisoquinoline alkaloids **20a**, **20b**, or **77**[250] (their structures are shown in Table 3), and with the 7,1'-linked representatives **1a**, **93a**, **93b**, or **97**[51] (for their structures, see Table 3), or with doxorubicin[477,478] as the reference drug at a concentration of 10 μM. (b) Cytotoxic activity of dioncophylline A (**1a**) against CCRF-CEM cells and their multi-drug resistant subline, CEM/ADR5000. The graphic shows mean values and standard deviations of three independent experiments with six parallel measurements each.

The nuclear factor kappa B (NF-κB) signaling pathway is considered to have a profound impact on cancer initiation, promotion, and progression.[479–482] It is also known to play a key role in the development of resistance to current anticancer therapeutics.[481–483] The search for new effective drugs addressing the NF-κB network is thus a promising approach for the treatment of acute myeloid leukemia (AML) and other lethal leukemia types.

Bioinformatic analysis predicted dioncophylline A (**1a**), its two 4'- and 5'-O-demethyl derivatives **93a** and **97**, and the unnatural analogue of **1a**, 8-O-(p-nitrobenzyl)dioncophylline A (**101**), to strongly interact with the NF-κB complex.[366] Molecular docking revealed **1a**, **93a**, and **97** to share the same binding pocket with triptolide, a known[483] NF-κB inhibitor used as a positive control (Fig. 61). Dioncophylline A (**1a**) overlapped with **93a** and **97** in the triptolide binding site (not shown), while 8-O-(p-nitrobenzyl) dioncophylline A (**101**) was located in a different position of the protein (Fig. 61, left).[366] The in silico binding study of the four naphthylisoquinolines to NF-κB was confirmed in vitro, by their inhibitory potential toward the NF-κB network of HEK-Blue™ Null1 cells.[366] Dioncophylline A (**1a**) showed an outstanding NF-κB inhibition, much better than its derivatives, at all concentrations investigated (1, 5, and 10 μM). At a concentration of 1 μM, **1a** lowered the NF-κB activity by more than 50%, and after treatment of the cells with 5 or 10 μM of **1a**, the NF-κB pathway was

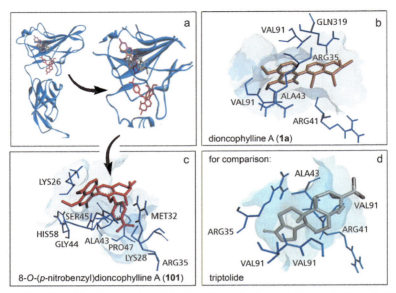

Fig. 61 In silico binding study of (a,c) 8-*O*-(*p*-nitrobenzyl)dioncophylline A (**101**), (b) dioncophylline A (**1a**), and (d) triptolide (positive control), to NF-κB with the docking poses and interacting amino acid residues; each docking was carried out in triplicate.[366]

inhibited nearly completely, comparable to the strong suppression induced by triptolide. 5′-*O*-Demethyldioncophylline A (**97**) exhibited a significant inhibitory effect at a concentration of 10 µM, while its natural regioisomer 4′-*O*-demethyldioncophylline A (**93a**) and the synthetic derivative 8-*O*-(*p*-nitrobenzyl)dioncophylline A (**101**) exerted merely weak activity.[366]

Many anticancer agents derived from natural products, such as paclitaxel, vinblastine, and vincristine, are known to exert growth inhibitory activity on cancer cells by affecting cell cycle regulation during the M phase and by effecting the microtubulin assembly.[484,485] Similar to these known antiproliferative compounds, the cytotoxicity of dioncophylline A (**1a**) toward leukemia CCRF-CEM cells was found to be mediated by cell cycle arrest leading to an increased number of cells in the G2/M phase.[366]

Malignant cells require oxygen and nutrients to survive and proliferate, and therefore need to reside in a close proximity to blood vessels to get access to blood circulation. Suppression of blood vessel formation (angiogenesis) has thus been suggested as a rewarding approach for cancer therapy.[486,487] Dioncophylline A (**1a**) significantly affected angiogenesis in a dose-dependent manner. Treatment of human umbilical vein endothelial cells (HUVECs) with **1a** in three different concentrations (1, 5, and 10 µM) led to a decrease of total tube length and total branching points, accompanied by a reduction of vessel sprouting.[366]

Dioncophylline A (**1a**) was furthermore found to induce autophagy, a process for resolving and recycling proteins and damaged cellular organs,[488,489] while it did not significantly promote apoptosis in CCRF-CEM cells.[366] Thus, autophagy seems to be the more decisive mechanism of cancer cell death caused by dioncophylline A (**1a**) rather than apoptosis or necrosis, similar to the mode of action of the known[490,491] anti-cancer drug rapamycin.

Finally, dioncophylline A (**1a**) displayed strong anti-tumor activity in vivo, in the zebrafish larvae CCRF-CEM cell xenograft model.[366] Studies in zebrafish provide an efficient platform for a rapid testing of safety and efficacy of potential drug candidates in vivo.[492,493] The larvae have a short reproductive cycle and show no immunological rejection to human cells. The transparent fish body easily permits a visual monitoring of tumor development by epifluorescence and confocal microscopy. The zebrafish model has therefore become an established tool for the evaluation of drug toxicity.[493] Microinjection of dioncophylline A (**1a**) up to a concentration of 312 nM did not show any abnormal phenotypes in zebrafish and was therefore considered as NOAEL (No Observed Adverse Effect Level) of **1a**. Treatment of the larvae with 312 nM of **1a** clearly revealed the ability of dioncophylline A to suppress tumor growth in vivo.[366]

Dioncophylline A (**1a**) also showed cytotoxicity against human cervical HeLa and colon carcinoma (HT-29) cancer cells (Fig. 62), displaying IC$_{50}$ values of 2.4 and 5.2 µM, respectively.[404] Its atropo-diastereomer, 7-*epi*-dioncophylline A (**1b**), however, affected the tumor cells only moderately, thus again reflecting the strong impact of axial chirality on the bioactivities of naphthylisoquinoline alkaloids.[404]

Breast cancer[494–496] is currently one of the most frequently diagnosed malignancies and the leading cause of cancer-related deaths among women.

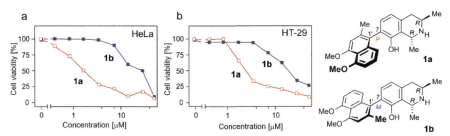

Fig. 62 Cytotoxic activities of dioncophylline A (**1a**) and its atropo-diastereomer **1b** against (a) HeLa cervical cancer cells and (b) HT-29 colon carcinoma cells.

In 2020, more than 2.3 million new cases were diagnosed worldwide, and almost 700,000 patients died from the disease.[496,497] Although mortality from breast cancer has steadily decreased over the past three decades[497–499] as a result of early detection due to mammography screening programs and improved treatment management, breast cancer remains a major threat to women, accounting for almost one-third of new cancer diagnoses in 2022.[498] Since conventional and targeted protocols clinically used for breast cancer therapies still have severe drawbacks such as recurrence of cancer and emergence of drug resistance,[495,496] there is an urgent need for alternative strategies and novel agents to efficiently combat breast cancer.

Three major subtypes are in the focus of breast cancer treatment, with different medicinal implications, based on the presence or absence of molecular markers for estrogen receptors (ER), progesterone receptors (PR), and human epidermal growth factor 2 (HER2). In general, hormone receptor expressing breast cancers have a more favorable prognosis than triple-negative tumors, which do not contain ER nor PR, and do not have HER2. With respect to this fact, triple-negative breast cancer has only limited treatment options, because the cancer cells do not respond to hormonal therapies, nor to HER2 receptor blockers.[494–496,499,500]

Given the pronounced antiproliferative activities of dioncophylline A (1a) against various cancer cell lines,[51,366] as reported above, it seemed rewarding to likewise study the cytotoxic potential of 1a on the cellular proliferation and cell viability of human breast cancer cells (Fig. 63). For this purpose, MDA-MB-231 (hormonal therapy unresponsive)[499] and MCF-7 (hormonal therapy responsive)[500] in vitro breast cancer models were used. MDA-MB-231 is a highly aggressive, invasive, and poorly differentiated triple-negative breast cancer cell line as it lacks ER and PR expression, as well as HER2 amplification.[499] The MCF-7 cell line, by contrast, is ER- and PR-positive, poorly-aggressive, and non-invasive, exerting only low metastatic potential.[500]

Dioncophylline A (1a) showed significant cytotoxic effects against the two tested breast cancer cell lines at a micromolar level in a time- and dose-dependent manner (Fig. 63a–d). Exposure to varying concentrations (1, 3, 5, 7, and 10 µM) of 1a resulted in strong effects on cellular proliferation and cell viability, with IC_{50} values of 1.57 µM in MDA-MB-231 cells and 0.87 µM in MCF-7 breast cancer cells.[365] Thus, in hormone-positive MCF-7 cells the half maximal inhibition concentration of dioncophylline A (1a) was about half as low than the one found in the hormone-negative MDA-MB-231 cancer cells, hinting at a more pronounced sensitivity of MCF-7 cells to the growth inhibition induced by 1a.

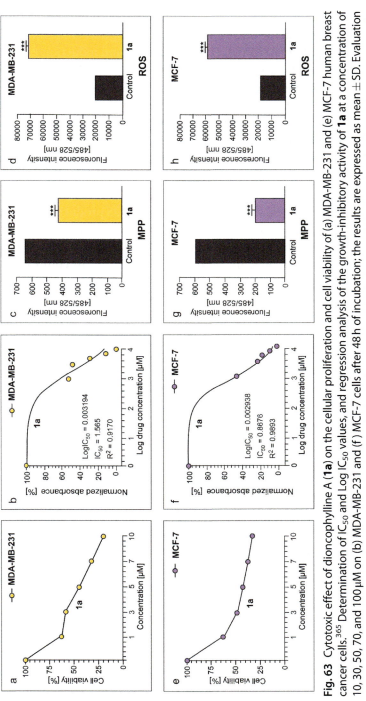

Fig. 63 Cytotoxic effect of dioncophylline A (**1a**) on the cellular proliferation and cell viability of (a) MDA-MB-231 and (e) MCF-7 human breast cancer cells.[365] Determination of IC$_{50}$ and Log IC$_{50}$ values, and regression analysis of the growth-inhibitory activity of **1a** at a concentration of 10, 30, 50, 70, and 100 μM on (b) MDA-MB-231 and (f) MCF-7 cells after 48 h of incubation; the results are expressed as mean ± SD. Evaluation of the potential of **1a** on the disruption of the mitochondrial membrane potential (MMP) of (c) MDA-MB-231 and (g) MCF-7 breast cancer cells, and effect of **1a** on reactive oxygen species (ROS) production in the (d) MDA-MB-231 and (h) MCF-7 breast cancer cell lines; the cells were treated with the respective IC$_{50}$ concentrations of **1a** for 48 h; results are expressed as mean ± SD of the three replicates.

Mechanistic studies revealed that **1a** induced apoptotic cell death, by causing deformations in the nuclear membrane, disruption of the mitochondrial membrane potential (MMP) (Fig. 63e and f), and induction of oxidative stress by elevated reactive oxygen species (ROS) production (Fig. 63g and h) in both ER-positive (MCF-7) and triple-negative (MDA-MB-231) breast cancer cells.[365] Depletion of MPP is one of the early prerequisites of apoptosis.[501] Dioncophylline A (**1a**) induced a significant decrease in MMP at its IC$_{50}$ concentrations of 1.57 µM (in MDA-MB-231 cells, see Fig. 63e) and 0.87 µM (in MCF-7 cells, see Fig. 63f), with a markedly pronounced effect of **1a** toward the mitochondria in MDA-MB-231 cells compared to the one exerted in MCF-7 breast cancer cells.[365]

Many natural products are known to enhance ROS production in cancer cells causing disruption of the redox balance with damage of proteins, lipids, nucleic acids, and cellular organelles, which leads to the activation of the apoptosis pathway and, thus, finally to cell death.[502–505] In comparison to the control group, dioncophylline A (**1a**) significantly increased ROS production in MDA-MB-231 (Fig. 63g) and MCF-7 (Fig. 63h) cancer cells after application of the IC$_{50}$ concentrations of **1a**, with the generation of ROS occurring to a similar degree in both cell lines.[365]

Furthermore, dioncophylline A (**1a**) was studied for its effects on nuclear morphology and apoptosis in MDA-MB-231 and MCF-7 cells using the nucleus-specific Hoechst 33342 staining method. Hoechst 33342 is a cell-membrane-permeable dye that intercalates with DNA and emits bright blue fluorescence in both live and dead cells. As illustrated in Fig. 64a, untreated MDA-MB-231 breast cancer cells (control) displayed a regular cell morphology with intact nuclei, whereas incubation with 1.57 µM of **1a** induced a substantial deformation of the nuclei, with shrinking and membrane blebbing, suggestive of cells undergoing apoptosis. Similar effects were also observed for the MCF-7 cell line (not shown).[365]

The changes in the mitochondrial membrane potential (MMP) was analyzed using the fluorescent potentiometric dye JC-1 (5,5′,6,6′-tetrachloro-1,1′,3,3′-tetraethylbenzimidazolyl-carbocyanine iodide).[365] At a high MMP, the dye accumulates in the organelles and forms aggregates, which, upon excitation, exhibit red fluorescence. Once the MMP is disrupted, JC-1 remains in the cytoplasma in its monomeric form, emitting light of green color. As presented in Fig. 64b, untreated MDA-MB-231 breast cancer cells displayed red fluorescence, typical of intact mitochondria with a high MMP. After treatment with 1.57 µM of dioncophylline A (**1a**), JC-1 showed green fluorescence (along with a reduced red fluorescence)

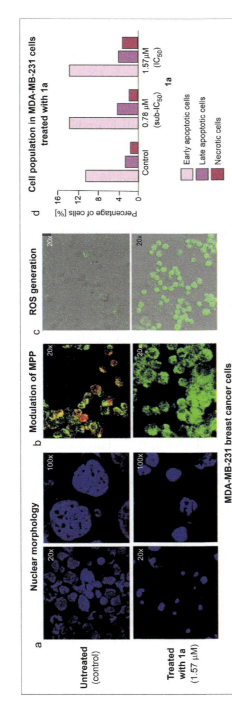

Fig. 64 Cytotoxic effects of dioncophylline A (**1a**) on hormone-unresponsive MDA-MB-231 breast cancer cells at the IC$_{50}$ concentration of 1.57 μM.[365] (a) Changes in nuclear morphology and induction of apoptosis, analyzed by confocal microscopy (resolution: 20× and 100×) after nuclear staining of the cells using the blue-fluorescent Hoechst 33342 dye. (b) Modulation of the mitochondrial membrane potential (MMP), analyzed by confocal microscopy (resolution: 20×); for the detection of changes in MMP, the JC-1 (5,5′,6,6′-tetrachloro-1,1′,3,3′-tetraethylbenzimidazolylcarbocyanine iodide) dye was applied. (c) Production of reactive oxygen species (ROS); modulation of the ROS levels was analyzed by confocal microscopy (resolution: 20×) using the cell-permeable free indicator H$_2$DCFDA (2′,7′-dichlorodihydrofluorescein diacetate) as the staining reagent. (d) Early and late apoptotic, and necrotic cell populations in MDA-MB-231 cells treated with **1a** at the sub-IC$_{50}$ (0.78 μM) and IC$_{50}$ (1.57 μM) concentrations, in comparison to non-treated cancer cells (control).

indicating that the MMP was significantly lowered. The results obtained for MDA-MB-231 and for MCF-7 cells (data not shown) demonstrated that **1a** directly affected the mitochondria of these two breast cancer cell lines and induced apoptosis.[365]

Alterations in the generation of reactive oxygen species (ROS) in MDA-MB-231 and MCF-7 breast cancer cells treated with dioncophylline A (**1a**) were investigated by using H$_2$DCFDA (2′,7′-dichlorodihydrofluorescein diacetate), which is the reduced form of fluorescein.[365] The non-fluorescent H$_2$DCFDA dye passively penetrates the cell membrane, where its acetate groups are cleaved by intracellular esterases and its thiol-reactive chloromethyl group reacts with intracellular glutathione and other thiols. Subsequent oxidation by ROS affords brightly green fluorescent DCF (2′,7′-dichlorofluorescein), which is trapped inside the cells. Dioncophylline A (**1a**) was found to significantly increase the ROS production in MDA-MB-231 cells (Fig. 64c) and in MCF-7 cells (data not shown) after incubation with the IC$_{50}$ concentrations, as compared to non-treated cells, thus corroborating the potential of **1a** to induce ROS-mediated apoptotic cell death in the two breast cancer cell lines.[365]

Another proof of the involvement of apoptosis in MDA-MB-231 and MCF-7 breast cancer cells after application of dioncophylline A (**1a**) was the downregulation of the expression of inactive zymogen pro-caspase-3 (a caspase-3 precursor protein), which was cleaved to produce activated caspase-3. Western blotting (data not shown) revealed that the expression of procaspase protein was lowered in treated cells after 12 h of incubation, in comparison to the non-treated cells (control).[365] This finding is in agreement with previous reports on the potential of plant metabolites to trigger caspase-mediated apoptosis by altering the procaspase levels in breast cancer cells.[503–506]

For a direct evaluation of the apoptotic cell death induced by dioncophylline A (**1a**), double staining of breast cancer cells with annexin V fluorescein isothiocyanate (annexin V FITC) and propidium iodide (PI) was performed.[365] Annexin V binds to phosphatidylserine, which is located on the inner layer of the cell membrane, but switches from the inner surface to the outer one in apoptotic cells. Propidium iodide is only capable of penetrating the disintegrated membrane in apoptotic cells; it enters the cells and intercalates into double-strand DNA. Flow cytometric analysis showed that cancer cell treatment with **1a** at both IC$_{50}$ and sub-IC$_{50}$ (i.e., half of the IC$_{50}$ value) concentrations resulted in an increased number of stained cells, in the early and in the late apoptotic phase.

Moreover, treatment with IC_{50} concentrations of **1a** led to a drastic increase of necrotic cells, whereas no significantly higher number of necrotic cells was observed at sub-IC_{50} concentrations (Fig. 64d).[365]

Summarizing, dioncophylline A (**1a**) is a potent inhibitor of breast cancer cell proliferation at micromolar concentrations, probably by causing apoptosis leading to pronounced alterations in nuclear morphology, along with depletion of the mitochondrial membrane potential (MMP) and an enhanced generation of reactive oxygen species (ROS).[365]

In view of the strong cytotoxicity of dioncophylline A (**1a**), its potential to act as a DNA-binding ligand and, thus, its ability to affect the biological activity of DNA, have been studied to gain further insight into the mode of action of this promising anticancer agent.[367] Photometric and fluorimetric titrations clearly revealed **1a** to intercalate with regular duplex DNA with moderate affinity, exhibiting a binding constant of $K_b = 2.1 \times 10^4 \, M^{-1}$, which was comparable to those reported for well-known DNA-binding alkaloids.[507] Addition of ct (calf thymus) DNA to **1a** in aqueous buffer solution caused significant changes in the UV spectrum of dioncophylline A (**1a**), such as a continuous decrease of the absorption, along with a slight shift of the two UV maxima of **1a** (at 318 and 333 nm) and the formation of a broad red-shifted band.[367] These developments in the spectrum of **1a** on DNA addition—hypochromicity and red shift of the absorption band—are typical features of DNA-binding ligands.[508] Dioncophylline A (**1a**) showed a weak emission with a band maximum at 355 nm. Upon addition of ct DNA, the fluorescence intensity of the emission band decreased steadily, without a significant shift of the emission maximum of **1a**.[367] Fluorescence quenching is a commonly observed effect of DNA-bound ligands resulting from a photo-induced electron transfer from the excited ligand to the DNA base,[508] thus corroborating an association of dioncophylline A (**1a**) with ct DNA.

Linear dichroisms (LD) studies of the DNA-dioncophylline-A complex, performed in a rotating cuvette to align the DNA and its bound ligand along the hydrodynamic field, indicated that, most likely, the naphthalene unit of **1a** is oriented in an intercalative binding mode, coplanar relative to the DNA bases, whereas the isoquinoline part is accommodated in the DNA groove.[367] This proposed intercalation mode of **1a** with DNA was confirmed by titration experiments with synthetic polynucleotides revealing **1a** to bind to DNA sequences rich in adenosine-thymine (AT) and not to regions with a high degree of guanine-cytosine (GC) units, because in the latter case a strong steric repulsion occurs between the groove-bound isoquinoline moiety of **1a** and the amino substituent of guanine.

Moreover, AT-rich sequences possess a slightly wider groove than GC-regions, and thus provide more space for sterically demanding ligands such as dioncophylline A (**1a**).[367]

In view of the DNA-binding properties of **1a**, this effect should be considered as an important feature regarding its cytotoxic activity toward cancer cells.

5.3.3 Ancistrobrevine D and related Ancistrocladaceae-type naphthylisoquinoline alkaloids

From the roots, stems, and leaves of *A. abbreviatus*, three 7,1′-coupled Ancistrocladaceae-type naphthyltetrahydroisoquinoline alkaloids (i.e., with *S*-configuration at C-3 and an oxygen function at C-6) have so far been identified (see Fig. 65).[5,57,65,404] Two of them, ancistrobrevine D (**102a**)[5,57,65] and *N*-methylancistrocladisine A (**104**),[404] are known to occur only in this liana from Coastal West Africa, while ancistrocladisine B (**103**)[57] had already earlier been discovered in the Congolese liana *A. ileboensis*.[51] More recently, two 7,1′-coupled naphthylisoquinoline alkaloids possessing a non-hydrogenated isoquinoline subunit, ancistrobreveines A (**25**) and B (**105**) (Fig. 65), have been isolated from the roots of *A. abbreviatus*, albeit as minor constituents.[54] With an oxygen function at C-6, they were categorized as Ancistrocladaceae-type alkaloids, in contrast to the likewise 7,1′-linked fully dehydrogenated *ent*-dioncophylleine A (**24b**)[54,291] (Fig. 57), which was classified as a Dioncophyllaceae-type compound, since it lacked an oxygen group at C-6. In both cases, the second classification criterion, the configuration at C-3, did not apply due to the sp²-hybridization there (see also Section 5.6).

In *A. abbreviatus*, ancistrobrevine D (**102a**)[5,57,65] was produced in an atropisomerically pure form, i.e., without any co-occurring 7-*epi*-ancistrobrevine D (**102b**), which, in turn, was found to be a major

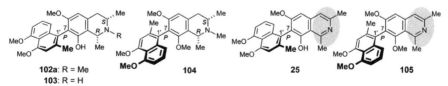

Fig. 65 Ancistrobrevine D (**102a**),[5,57,65] ancistrocladisine B (**103**),[57] and *N*-methylancistrocladisine A (**104**),[404] three 7,1′-coupled ancistrocladaceous naphthylisoquinoline alkaloids occurring in the roots of *A. abbreviatus*, along with the two likewise 7,1′-linked, but fully dehydrogenated alkaloids ancistrobreveines A (**25**) and B (**105**).[54]

constituent of the Vietnamese liana *A. cochinchinensis*. From this plant species, compound **102b** was likewise isolated in an atropo-diastereomerically pure form, together with two related congeners.[403]

The fruitful interplay of NMR spectroscopy (in particular in the analysis of atropisomer-specific long-range NOE interactions) with ECD spectroscopy, oxidative degradation, and the stereoselective total synthesis of **102a** and of all its possible *cis*-configured stereoisomers, permitted reliable assignment of the complete absolute stereostructure of ancistrobrevine D (**102a**).[1,5,57,65] The preparation of atropisomerically pure ancistrobrevine D (**102a**) and 7-*epi*-ancistrobrevine D (**102b**) (Fig. 66) and of their unnatural enantiomers, *ent*-**102a** (1*S*,3*R*,7*M*) and *ent*-**102b** (1*S*,3*R*,7*P*),[1,65] was

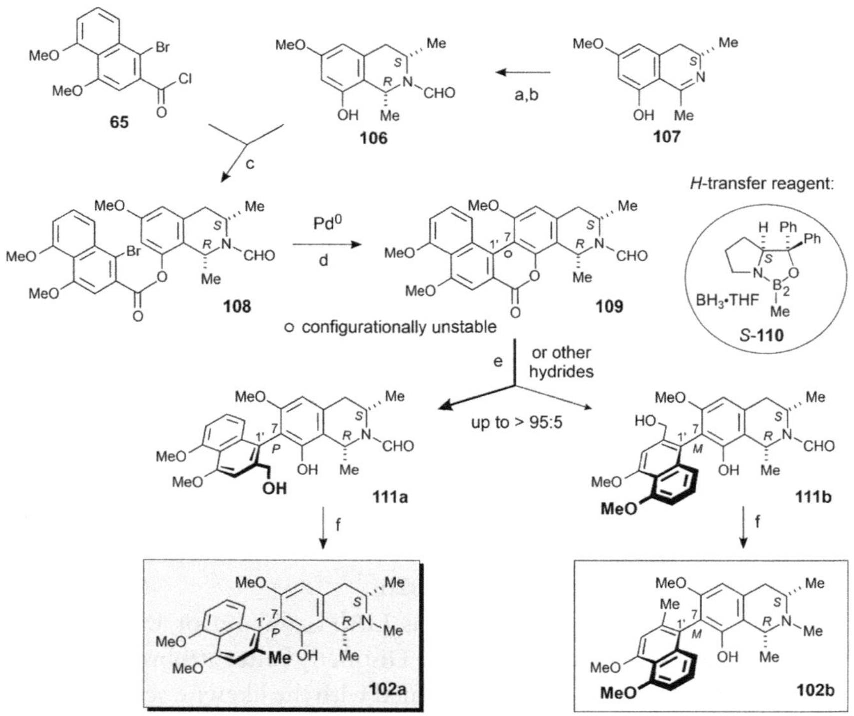

Fig. 66 Atropo-divergent first total synthesis of ancistrobrevine D (**102a**)[1,5,57,65] from *A. abbreviatus* and of its likewise naturally occurring atropo-diastereomer, 7-*epi*-ancistrobrevine D (**102b**),[1,403] as isolated from the Vietnamese liana *A. cochinchinensis*, by following the "lactone concept." Reaction Conditions: (a) NaBH$_4$, MeOH, 0 °C, 80%; (b) *p*-nitrophenyl formate, Hünig's base, DMF, 0–25 °C, 89%; (c) oxalyl chloride, DMF, NEt$_3$, CH$_2$Cl$_2$, 0 °C, 84%; (d) (PPh$_3$)$_2$PdCl$_2$, NaOAc, DMA, 84%; (e) *S*-**110**·BH$_3$·THF, 0 °C, 29%; (f) (CBrCl$_2$)$_2$, PPh$_3$, CH$_2$Cl$_2$,[397] then LiAlH$_4$, THF, reflux, 83% (for **102a**) and 72% (for **102b**).

achieved by applying the "lactone strategy,"[76–79] starting from the known[383] bromonaphthoic acid chloride **65** and the respective enantiopure tetrahydroisoquinoline building blocks **106** and **107**, which were synthesized following well-established protocols.[315] In view of the fact that only ancistrobrevine D (**102a**) and its atropo-diastereomer, 7-*epi*-ancistrobrevine D (**102b**), are genuine natural products, and since all of the decisive steps within these atropo-divergent pathways were basically analogous, only the total synthesis of ancistrobrevine D (**102a**) and 7-*epi*-ancistrobrevine D (**102b**) will be described here, exemplarily, in more detail (see Fig. 66).

Reduction of the dioxygenated monophenolic dihydroisoquinoline **107** with NaBH$_4$ occurred with a good *cis*-diastereoselectivity (dr 93:7).[5,65] Subsequent *N*-formylation provided the suitably *N*-protected 1*R*,3*S*-configured tetrahydroisoquinoline building block **106**. Free, not *N*-methylated 1,3-*cis*-configured tetrahydroisoquinolines are known to be sensitive to oxidation,[1,5,315] being easily transformed to the corresponding dihydroisoquinolines, even just by atmospheric oxygen,[1,5,315] thus requiring cautious handling of the compounds. Reaction of the *N*-formyl-protected tetrahydroisoquinoline **106** with the bromonaphthoyl chloride **65** furnished the bromoester **108**, which was intramolecularly coupled with Pd catalysis to give the biaryl lactone **109** in excellent yield (84%). Atropo-diastereoselective reductive ring-opening using an excess of diborane in the presence of the *S*-2-methyl-oxazaborolidine *S*-**110** delivered the *P*-configured benzylic alcohol **111a** (>95:5).[1,5,65] Alternatively, by the use of the enantiomeric *R*-2-methyl-oxazaborolidine as the catalyst, the configurationally stable *M*-configured alcohol **111b** was obtained almost exclusively (dr = 97:3).[1,5,65] Hydroxy-bromine exchange and reduction by standard procedures eventually yielded the target molecule ancistrobrevine D (**102a**) or, optionally, its atropisomer 7-*epi*-ancistrobrevine D (**102b**).[1,5,65]

Achiral *H*-transfer reagents such as LiAlH$_4$ (60:40) or LiBH$_4$ (62:38) resulted in low asymmetric inductions. Distinctly better atropisomeric ratios in favor of the alcohol **111a** were obtained with the likewise achiral reagents sodium bis(2-methoxyethoxy)aluminum hydride ("Red-Al") (68:32) and lithium tri-*sec*-butylborohydride ("L-Selectride") (73:27).[5,65]

5.3.4 Ancistrobrevine C, one of only four known examples of 7,1′-linked hybrid-type naphthylisoquinoline alkaloids found in nature

Ancistrobrevine C (**22**),[5,64] with its oxygen function at C-6 and the *R*-configuration at C-3, is one of the very rare examples of naphthylisoquinoline alkaloids that belong to the subclass of 7,1′-coupled hybrid-type compounds. At the time when **22** was discovered in the roots and stem bark of *A. abbreviatus* in the early 1990s,[64] it was the first "mixed," Ancistrocladaceae- and Dioncophyllaceae-type alkaloid from nature possessing structural landmarks of both subclasses of compounds. Although such hybrid-type naphthylisoquinoline alkaloids have meanwhile widely been identified in *Ancistrocladus* lianas (no less than 88 compounds!),[1,3,5,10–15,31–35,51,56,58,64,175–178,180–182,286–289] representatives with a 7,1'-biaryl linkage are extremely rare, and only four such compounds have so far been isolated, all from West and Central African *Ancistrocladus* species, exclusively.[5,12,51,64,177,286] Besides the naphthyltetrahydroisoquinoline ancistrocongoline D (**112**)[177] from the Congolese liana *A. congolensis*, three further 7,1'-linked hybrid-type alkaloids were identified. All of them possessed a dihydroisoquinoline subunit, namely 3-*epi*-ancistrocladisine (**114**)[51] from *A. ileboensis* (Southern Congo Basin), ancistrobarterine A (**113**)[5,286] from *A. barteri* (Ivory Coast), and ancistrobrevine C (**22**),[5,64] the title compound of this section (Fig. 67). The latter is not only known to occur in *A. abbreviatus*, but has meanwhile also been found as a minor metabolite in *A. ileboensis*[51] and in a botanically yet unexplored Congolese *Ancistrocladus* species.[12]

The structure of ancistrobrevine C (**22**)[5,64] (Fig. 67) closely resembles that of ancistrobrevine D (**102a**)[5,65] (Fig. 66), differing only by its opposite *R*-configuration at C-3 and the presence of a *di*hydroisoquinoline subunit instead of a *cis*-configured and *N*-methylated *tetra*hydroisoquinoline moiety as occurring in **102a**, but with the same axial configuration. Thus, except for the small variation at C-1, **22** and **102a** belong to the same diastereomeric series with different configurations at the stereogenic center, thus again emphasizing the necessity for a thorough configurational assignment of each of the alkaloids of *A. abbreviatus* individually (instead of relying on mere assumptions, as reported in the literature for similar cases).[238–241,245–249] This example demonstrates that no reliable predictions, just based on analogies, can be made for the stereostructures of naphthylisoquinoline alkaloids from this and some other *Ancistrocladus* plant species.

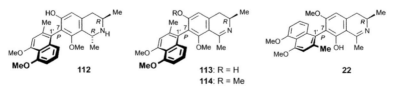

Fig. 67 Ancistrocongoline D (**112**),[117] ancistrobarterine A (**113**),[5,286] 3-*epi*-ancistrocladisine (**114**),[51] and ancistrobrevine C (**22**),[5,12,51,64] the only four representatives of 7,1'-coupled hybrid-type alkaloids (3*R*, 6-O*R*) known from nature.

5.3.5 Dioncoline A, the only known example of an inverse hybrid-type naphthylisoquinoline alkaloid found in nature

Resolution of one of the more polar fractions of the root bark extract of *A. abbreviatus* yielded the 7,1'-coupled naphthylisoquinoline alkaloid dioncoline A (**23**)[58] (Fig. 68). Its absolute stereostructure was established by 1D and 2D NMR methods, ECD spectroscopy, and oxidative degradation. The compound showed NMR signals typical of an *N*-methylated naphthyltetrahydroisoquinoline with six aromatic protons (i.e., no O-functionality at C-6) (Fig. 68a). This finding, together with the presence of only three oxygen atoms, suggested the alkaloid to possess the constitution of a Dioncophyllaceae-type compound.[58] In view of its usual 2'-methyl-4', 5'-dimethoxy-substituted naphthalene portion, the remaining third oxygen function had to be a phenolic hydroxy group at C-8.[58] The constitution of the compound thus resembled that of an alkaloid related to *N*-methyldioncophylline A (**94a**)[5,62] (Fig. 68, bottom, right). The two methyl groups at C-1 and C-3 were, however, *cis* to each other, as established by a NOESY interaction between H-1 and H-3 (Fig. 68a), but no long-range

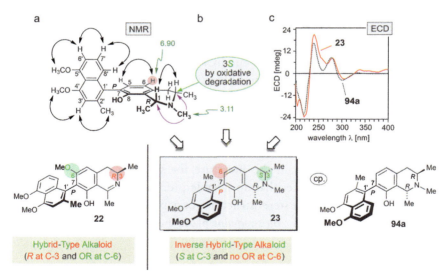

Fig. 68 (a) The first—and as yet only—inverse hybrid-type alkaloid dioncoline A (**23**)[58] with its opposite combination of structural landmarks compared to the hybrid-type compound ancistrobrevine C (**22**).[5,64] Complete stereostructure of **23** as established (a) by specific HMBC (single purple arrows) and NOESY interactions (double black arrows) defining the constitution and the relative configuration in the isoquinoline portion, (b) by oxidative degradation, and (c) by comparison of its ECD spectrum with that of the known[5,62] alkaloid *N*-methyldioncophylline A (**94a**) evidencing the absolute axial configuration as *P*.

interactions indicative of the relative axial configuration were observed between the isoquinoline and the naphthalene subunits. The ECD spectrum matched that of N-methyldioncophylline A (**94a**) nearly perfectly, thus the biaryl axis was attributed to be P-configured (Fig. 68b).[58] Different from **94a**, however, the oxidative degradation procedure[1,3,70] unexpectedly established the absolute configuration at C-3 as S, and thus, in conjunction with the relative *cis*-configuration in the isoquinoline moiety, dioncoline A had to possess the full stereostructure **23**, with a 1R,3S,P-configuration,[58] as shown in Fig. 68. Hence, **23** was, astoundingly, the 3-epimer of N-methyldioncophylline A (**94a**), and not the—rather expected—1-epimer.

As mentioned earlier, the majority of the alkaloids isolated from *A. abbreviatus* were 6-oxygenated and 3S-configured and, thus, typical Ancistrocladaceae-type compounds (49 representatives), or they belonged to the subgroup of Dioncophyllaceae-type naphthylisoquinolines, i.e., devoid of an oxygen function at C-6 and bearing the R-configuration at C-3 (15 representatives). Only four alkaloids identified in this West African liana were categorized as "mixed," Ancistrocladaceae/Dioncophyllaceae hybrid-type compounds (i.e., 6-OR, 3R) like e.g., ancistrobrevine C (**22**) (Fig. 68, bottom, left). Dioncoline A (**23**) was even the very first—and so far the only—example of a novel fourth subclass of alkaloids, showing the opposite characteristics of hybrid-type naphthylisoquinolines: It had no oxygen function at C-6, but was S-configured at C-3, and was thus classified as an "inverse hybrid-type" naphthylisoquinoline alkaloid (Fig. 68). Hence, it occupies a unique position within the ca. 280 known naphthylisoquinoline alkaloids.

5.4 Dioncophyllidine E, only the second example of a 7,3′-coupled dioncophyllaceous naphthylisoquinoline alkaloid discovered in nature

Naphthylisoquinoline alkaloids displaying a 7,3′-coupling site have only rarely been identified in Ancistrocladaceae and Dioncophyllaceae lianas. So far, only 14 such compounds have been detected in nature,[28,61,67,239,240,253,254,261,262,265,279,292] the majority of them (10 alkaloids) belonging to the subclass of Ancistrocladaceae-type naphthylisoquinolines.[67,239,240, 253,254,261,262,265,279] The Chinese liana *A. tectorius* is the by far richest source of 7,3′-linked alkaloids, with no less than eight representatives.[1,3,247,249,253,254] The East African species *A. tanzaniensis* and *A. heyneanus* from India were likewise found to produce some naphthylisoquinolines of this coupling type (three compounds, each).[1,28,261,262,265,279]

Within this small group of 7,3'-coupled naphthylisoquinolines, the structural diversity is remarkable, comprising four monomeric representatives equipped with a tetra- or a dihydroisoquinoline subunit like e.g., ancistrotectorine (**115**)[67,239,247,249,279] and ancistrocladidine (**116**),[28,240, 261,262,265,279] and two dimers, such as shuangancistrotectorine D (**42**)[1,3,254] (Fig. 69), possessing a unique molecular architecture and three consecutive chiral axes. Structurally, the molecular halves of the latter alkaloids are closely related to those of ancistrotectorine (**115**), connected through an unprecedented 1′,1″-coupling in the central binaphthalene core. Likewise unique was the discovery of ancistrotectoquinones A (**58**) and B (**117**)[1,253] (Fig. 69), since these compounds were the first—and so far only—alkaloids possessing a 1,4-naphthoquinone portion coupled to an isoquinoline moiety. In contrast to the monomers **115** and **116** and the dimer **42**, which show complete configurational axial stability,[1,239,240,262,265,279] the ancistrotectoquinones A (**58**) and B (**117**)[1,253] occur as pairs of configurationally semi-stable—and thus slowly interconverting—atropo-diastereomers. This is due to the medium-sized steric hindrance next to the biaryl axis, but also as a consequence of electronic reasons (mesomeric stabilization of the transition state of the axial rotation by the electron-withdrawing effect

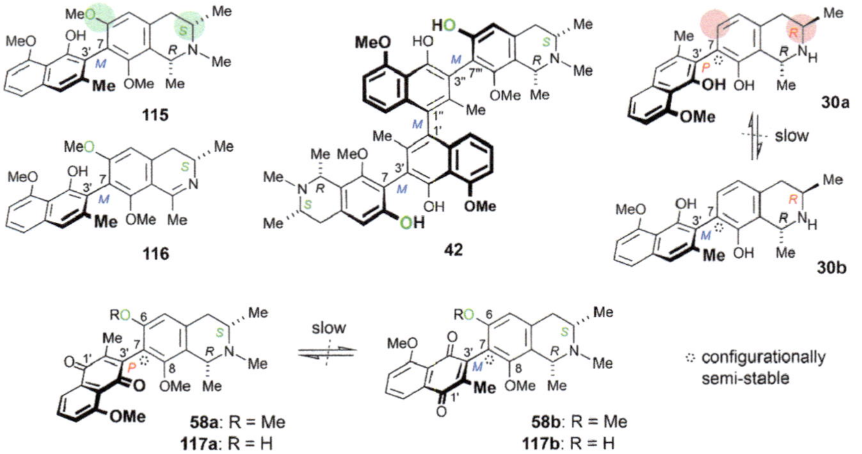

Fig. 69 Structural variability of naturally occurring 7,3′-coupled naphthylisoquinoline alkaloids. Besides the monomeric ancistrocladaceous alkaloids ancistrotectorine (**115**) and ancistrocladidine (**116**), Dioncophyllaceae-type representative such as dioncophylline E (**30a/b**)[292] were discovered, but also dimeric compounds like e.g., shuangancistrotectorine D (**42**)[1,3,254] and quinoid alkaloids like ancistrotectoquinones A (**58**) and B (**117**).[1,253] The characteristic structural landmarks of Ancistrocladaceae- (3S, 6-OR, green/green) and Dioncophyllaceae-type (3R, 6-H, red/red) compounds are labeled accordingly in the structures **115** and **30a**.

of the carbonyl group at C-1'), in combination with the electron-donating oxygen functions at C-6 and C-8.[1,253]

From the Gabonese liana *Dioncophyllum thollonii*, dioncophylline E (**30**)[292] (Fig. 69) was isolated, which was the first—and for a long time the only—example of a 7,3'-coupled dioncophyllaceous naphthylisoquinoline alkaloid, i.e., devoid of an oxygen function at C-6 and possessing the *R*-configuration at C-3. From one of the more polar sub-fractions of a leaf extract of *A. abbreviatus*, a further Dioncophyllaceae-type alkaloid with a 7,3'-biaryl linkage has most recently been isolated, dioncophyllidine E (**72**)[61] (Fig. 70), possessing a dihydroisoquinoline portion instead of a tetrahydroisoquinoline subunit as found in **30**, which is indicative from the

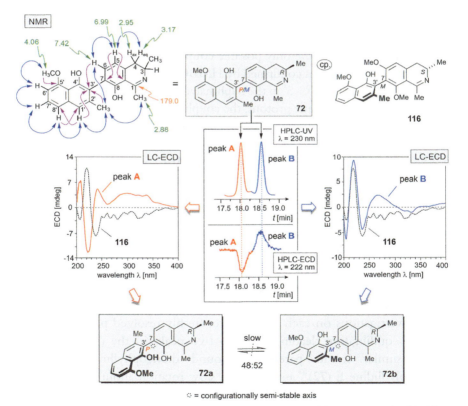

Fig. 70 Selected NMR shifts (in methanol-d_4, δ in ppm) and key NOESY (double blue arrows) and HMBC (single purple arrows) interactions indicative of the constitution of dioncophyllidine E (**72**)[61] and assignment of the absolute axial configuration of its two configurationally semi-stable atropo-diastereomers, **72a** and **72b**, by LC-ECD coupling and subsequent comparison of the online measured ECD spectra of peak **A** (left) and peak **B** (right) with the ECD curve of the *M*-configured and likewise 7,3'-linked known[1,261,262,265,279] alkaloid ancistrocladidine (**116**).

typical downfield shift of the methyl group at C-1 (δ_H 2.88) in the ^{1}H NMR spectrum, the absence of a quartet signal for H-1, which usually resonates at δ_H 4.6,[51,292] and the value of the ^{13}C NMR signal of C-1 (δ_C 179.0) (Fig. 70). The ^{1}H NMR spectrum displayed signals typical of an alkaloid with one methoxy substituent (δ_H 4.06), suggesting the other two oxygen functions to be part of hydroxy groups, as evident from the molecular formula of $C_{23}H_{24}NO_3$. The methoxy group was shown to be located at C-5′, due to NOESY cross-peaks with H-6′ and HMBC interactions of the methyl protons with C-5′. The aromatic region showed the presence of a one-proton system and of groups of two and three adjacent protons, by H,H-COSY experiments.[61] In view of a series of NOESY interactions {H-4$_{eq}$ ↔ H-5 ↔ H-6} and HMBC signals from H-5 (δ_H 6.99) to the quaternary carbon atoms C-7 and C-4, the group of two adjacent protons resonating at δ_H 7.42 (H-6) and 6.99 (H-5) was clearly located at C-5 and C-6 in the isoquinoline moiety, leaving C-7 as the only possible coupling position in this molecular half. This assignment was confirmed by HMBC interactions from H-6 to the quaternary carbon atom C-3′ in the naphthalene part.[61] With a spin system of three neighboring aromatic protons, the coupling position in the naphthalene molecular half had to be in the methyl-bearing ring, either at C-1′ or C-3′. HMBC interactions from H-1′ and Me-2′ to the quaternary carbon atom C-3′ and NOESY cross-peaks in the series {Me-2′ ↔ H-1′ ↔ H-8′ ↔ H-7′ ↔ H-6′ ↔ OMe-5′} (Fig. 70), established C-3′ to be the axis-bearing carbon atom. Oxidative degradation[1,3,70] established the absolute configuration at the stereogenic center at C-3 to be *R*.[61]

The presence of a double set of signals in the NMR spectrum of **72** hinted at its occurrence as a mixture of two configurationally semi-stable atropo-diastereomers, **72a** and **72b**.[61] HPLC analysis, both on an analytical and on a preparative scale, gave two well-separated peaks with distinctly diverging retention times, differing by ca. 1 min (see Fig. 70). Even immediate investigations on each of the pure, freshly resolved peaks showed a gradual equilibration, leading again to two peaks in the chromatogram. Thus, similar to its parent compound dioncophylline E (**30**),[292] dioncophyllidine E (**72**)[61] underwent a slow rotation about the biaryl axis at room temperature due to the presence of only three—and not too bulky—*ortho*-substituents next to the biaryl axis (OH, Me, OH).

Online, however, directly from the respective HPLC peaks, by HPLC-ECD coupling, the ECD spectra of the pure two atropo-diastereomers, **72a** and **72b**, were efficiently monitored (Fig. 70). In accordance with the presence of two alkaloids with opposite configurations at the biaryl axis, the two LC-UV peaks corresponded to the respective atropo-diastereomers **72a** and **72b**, as obvious from their opposite-signed single-wavelength LC-ECD signals at 222 nm and the mirror-imaged full LC-ECD spectra recorded in the stopped-flow mode (Fig. 70).[61] The ECD spectrum of the more rapidly eluting isomer (= peak **A**) was virtually opposite to that of the likewise 7,3′-coupled—yet configurationally stable—naphthyldihydroisoquinoline alkaloid ancistrocladidine (**116**),[1,261,262,265,279] which is known to be *M*-configured. Thus, the absolute axial configuration of the faster eluting atropo-diastereomer was deduced to be *P*, whereas the more slowly eluting peak **B** corresponded to the isomer **72b**, with *M*-configuration at the biaryl axis and an ECD spectrum that was nearly identical to that of ancistrocladidine (**116**).[61]

Dioncophyllidine E (**72**) was found to display strong preferential cytotoxic effects against human PANC-1 pancreatic cancer cells (PC_{50}, 7.4 μM)[61] in a concentration-dependent manner under nutrient-deprived conditions, without exerting toxicity in normal, nutrient-rich Dulbecco's modified Eagle's medium (DMEM), thus following the antiausterity approach[427–431] (see Section 5.1.3). Using the double staining fluorescence assay with ethidium bromide (EB) and acridine orange (AO), the effects of **72** on the cellular morphology of PANC-1 human cancer cells, when exposed to complete nutrition starvation, were investigated. While AO can easily pass the membrane, EB penetrates only when the membrane integrity is lost during the cell death process. Untreated PANC-1 cells (the control) remained intact and emitted bright green fluorescence, exclusively, due to AO. Treatment with dioncophyllidine E (**72**), on the contrary, led to a concentration-dependent increase of EB-stained cells exhibiting a strong red fluorescence, indicating a highly altered morphology of PANC-1 cancer cells.[61]

Since pancreatic cancer cells are highly aggressive and easily migrate to nutrient-rich organs such as liver, stomach, and duodenum to form new cancer colonies,[420,424,432,434] the inhibitory potential of **72** to reduce or even prevent cancer metastasis was tested. PANC-1 cells were exposed to dioncophyllidine E (**72**) at 10, 20, or 40 μM for 24 h in DMEM, then changed to fresh nutrient-rich medium to allow the colonies to grow for another 10 days. While the non-treated control group showed a colony area of 52%, PANC-1 cells incubated with **72** exerted a concentration-dependent inhibition of colony formation, with colony areas of 47% (for 10 μM), 38% (for 20 μM), and 24% (for 40 μM), thus demonstrating the potential of this new dioncophyllaceous naphthyldihydroisoquinoline alkaloid as a possible new agent against pancreatic cancer.[61]

Dioncophyllidine E (**72**) showed virtually no inhibitory effects against the chloroquine-susceptible NF54 strain of the malaria parasite *Plasmodium falciparum* and toward the pathogens causing African sleeping sickness (*Trypanosoma brucei rhodesiense*), Chagas' disease (*T. cruzi*), or leishmaniasis (*Leishmania donovani*). This finding was quite remarkable, because its parent compound, dioncophylline E (**30**), had exhibited strong antiplasmodial activities, with nearly identical IC_{50} values toward the chloroquine-sensitive strain NF54 (60 nM) and the chloroquine-resistant one K1 (58 nM).[292] The inhibitory potential of **30** was very high, within the range of some of the so far most active monomeric naphthylisoquinoline alkaloids like e.g., dioncopeltine A (**32**)[36] or habropetaline A (**57**),[296] and only slightly weaker (by a factor of 5–10) than those of the standard drugs artemisinin and chloroquine.[292]

5.5 Ancistrobrevines A, H, I, and J, and related 7,8′-coupled naphthyltetra- and -dihydroisoquinoline alkaloids

Ancistrobrevine A (**100**)[5,57,66] and its 6-O-demethyl analogue **118**[5,57] (Fig. 71) were the first examples of 7,8′-coupled naphthylisoquinoline alkaloids found in nature. In the course of the early phytochemical studies on this

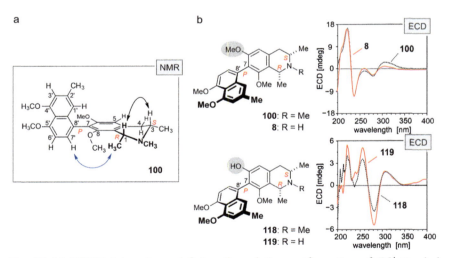

Fig. 71 (a) NOESY interactions defining the relative configuration of 7,8′-coupled naphthyltetrahydroisoquinolines at C-1 and C-3 in the tetrahydroisoquinoline subunit (double black arrow) and at the biaryl axis (double blue arrow), here exemplarily shown for ancistrobrevine A (**100**). (b) Confirmation of the absolute axial configurations of ancistrobrevine H (**8**) and its 6-O-demethyl analogue **119** by comparison of their ECD spectra with those of their corresponding N-methylated analogues **100** and **118**.

West African liana in the 1990s,[5] the two compounds were identified as the main metabolites in the roots and stem bark of *A. abbreviatus*. They possess *cis*-configured tetrahydroisoquinoline portions, here stabilized by an *N*-methyl group. Recent isolation work on the stem bark of *A. abbreviatus* furnished two further 7,8′-coupled alkaloids, ancistrobrevine H (**8**)[57] and its 6-*O*-demethyl derivative **119**,[57] which differed from ancistrobrevine A (**100**) and its analogue **118** only by the missing *N*-methyl group, since **8** and **119** possess a free, *N*-unsubstituted amino function.[57] Oxidative degradation[1,3,70] established the absolute configuration at C-1 and C-3 in the tetrahydroisoquinoline portions of the four compounds to be 1*R*,3*S*.[57,404] Based on the fact that they all possess an oxygen function at C-6, they were thus classified as Ancistrocladaceae-type[1,5] naphthylisoquinolines. The absolute axial configuration of these four 7,8′-linked alkaloids was assigned as *P* by specific long-range NOESY interactions between Me-1 (which is below the tetrahydroisoquinoline plane) and H-7′ (Fig. 71).[57,404] This attribution was further supported by the ECD curves of ancistrobrevine A (**100**) and its 6-*O*-demethyl analogue **118**, which nearly perfectly matched those of ancistrobrevine H (**8**) and 6-*O*-demethylancistrobrevine H (**119**), respectively (Fig. 71).[57]

While the oxidative degradation[1,3,70] provided reliable information about the absolute configuration of the 7,8′-coupled naphthyldihydroisoquinoline alkaloid ancistrobrevine I (**120**)[57] (Fig. 72), NOESY measurements were not applicable for a reliable assignment of the absolute axial configuration of this minor metabolite from the stem bark of *A. abbreviatus*. In view of the lack of a stereogenic center sufficiently near to the biaryl axis of **120** (because of the flat array around C-1, which is sp^2-hybridized), no long-range NOESY or ROESY interactions from spatially differentiated parts of the dihydroisoquinoline moiety over to the naphthalene portion indicative of the relative axial configuration were observed.[57] The problem was solved by installing an additional stereogenic center at C-1, now much closer located to the biaryl axis than C-3.[57] This was achieved by diastereoselective reduction of **120** using NaBH$_4$ to give the tetrahydroisoquinoline **119**, which, as expected, was *cis*-configured, as confirmed by the NOE cross-peak between H-1 and H-3. The hence well-defined newly generated auxiliary center, with Me-1 now located below the plane, permitted unambiguous long-range interactions between Me-1 in the isoquinoline half and H-7′ in the naphthalene part, thus finally leading to a clear assignment of the axial chirality of ancistrobrevine I (**120**) (Fig. 72).[57] Moreover, the semisynthetic product happened to be the—co-occurring—root metabolite 6-*O*-demethylancistrobrevine H (**119**),[57]

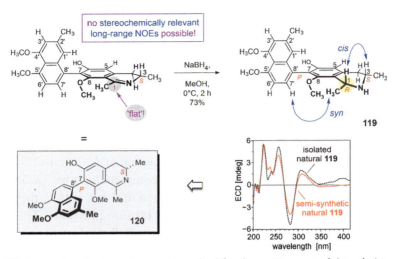

Fig. 72 A new chemical-spectroscopic method for the assignment of the relative axial configuration of 7,8′-coupled naphthyldihydroisoquinoline alkaloids, here exemplified for ancistrobrevine I (**120**),[57]; the new auxiliary stereocenter is underlaid in yellow. The *cis*-configuration at the stereogenic centers C-1 versus C-3 of the semisynthetic product **119** was established by a NOESY interaction between H-1 and H-3. The biaryl axis of **119** was assigned to be *P*-configured due to the decisive NOE cross-peak between Me-1 and H-7′, in conjunction with the 1*R*,3*S*-stereoarray in the tetrahydroisoquinoline subunit, thus evidencing that ancistrobrevine I had the stereostructure **120**. This was corroborated by the ECD spectrum of semisynthetic **119**, which was fully identical with that of the authentic natural product 6-*O*-demethylancistrobrevine H (**119**).[57]

also confirmed by the ECD spectrum of the isolated material, which perfectly matched the ECD curve of the compound obtained semisynthetically (Fig. 72).[57]

A further minor metabolite of *A. abbreviatus*, ancistrobrevine J (**70**),[57] differing from ancistrobrevine I (**120**) only by its OH/OMe substitution pattern in the dihydroisoquinoline portion (6-OMe/8-OH), revealed an ECD spectrum that was nearly identical to that of **120**, thus indicating that these two 7,8′-coupled naphthyldihydroisoquinolines had the same axial orientations (Fig. 73). For formal reasons, however, **70** and **120** had opposite descriptors according to the Cahn-Ingold-Prelog formalism. Again, application of the chemical-spectroscopic method described above permitted an unambiguous attribution of the axial configuration of ancistrobrevine J (**70**) by NMR, thus underlining the general applicability and reliability of this new approach.[57] The introduction of a new stereocenter at C-1 close

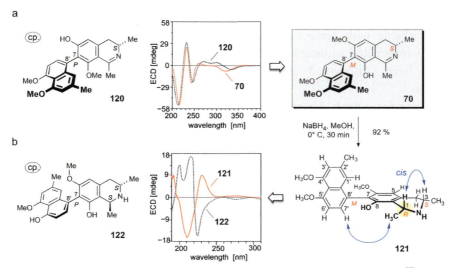

Fig. 73 Assignment of the absolute axial configuration of ancistrobrevine J (**70**)[57] (a) by comparison of its ECD spectrum with that of ancistrobrevine I (**120**),[57] and (b) by generation of an auxiliary stereogenic center at C-1 by stereoselective conversion of **70** to the *cis*-configured tetrahydroisoquinoline **121**. Decisive NOESY interactions between the isoquinoline and the naphthalene subunits of **121** (double blue arrows) and comparison of the ECD spectrum of **121** with that of ancistrogriffine C (**122**)[1,284] evidencing ancistrobrevine J (**70**) to be *M*-configured at the biaryl axis.

enough to the naphthalene ring system, by *cis*-selective conversion of the dihydroisoquinoline **70** into the corresponding tetrahydroisoquinoline **121**, provided a diagnostically significant NOE cross-peak between Me-1 and H-7′, and thus established the axial configuration of **121** as *M*.[57] This assignment was confirmed by the almost mirror-image-like ECD spectrum of **121** compared to that of the structurally closely related alkaloid ancistrogriffine C (**122**),[1,284] which is likewise 7,8′-coupled and—similar to **121**—equipped with a 6-methoxy-8-hydroxy-substituted tetrahydroisoquinoline portion, but *P*-configured at the biaryl axis.[57]

5.6 A unique series of fully dehydrogenated naphthylisoquinoline alkaloids from the roots of *Ancistrocladus abbreviatus*

HPLC-UV guided analysis of an alkaloid-enriched fraction obtained from root bark material of *A. abbreviatus* revealed the presence of a series of minor constituents displaying UV spectra with a third maximum at around 370 nm,[54] characteristic of naphthylisoquinoline alkaloids possessing a non-hydrogenated isoquinoline portion.[53,176,252,273] This finding was quite remarkable with respect to the fact that only a small number of fully dehydrogenated naphthylisoquinoline alkaloids (23 representatives) had so far been discovered in Ancistrocladaceae and Dioncophyllaceae lianas.[1,5,6,53,176,252,273–275,291,403] Naphthylisoquinoline alkaloids with such a molecular entity had been identified as part of the metabolite profiles of some of the Asian *Ancistrocladus* species.[1,252,273,291,403] Besides *A. abbreviatus*, the Congolese liana *A. likoko*[53,176] is the only other African *Ancistrocladus* species that was found to produce non-hydrogenated naphthylisoquinoline alkaloids.

The seven representatives discovered in *A. abbreviatus* displayed four different coupling types (5,1', 5,8', 7,1', and 7,8'),[54] thus significantly differing from the four ones isolated from *A. likoko*, which all belong to the subclass of 5,8'-linked naphthylisoquinolines, exclusively.[53,176] The fully dehydrogenated alkaloids of *A. abbreviatus* comprised five new natural products, ancistrobreveines A–D (**25, 105, 123a/b**, and **10a**)[54] and two previously known compounds, 6-*O*-methylhamateine (**124**)[1,54,403] and *ent*-dioncophylleine A (**24b**)[1,54,291] (Fig. 74). All of these naphthylisoquinolines had in common that they were optically active without the presence of stereogenic centers, but still possessing the rotationally hindered biaryl axis as a stable element of chirality.

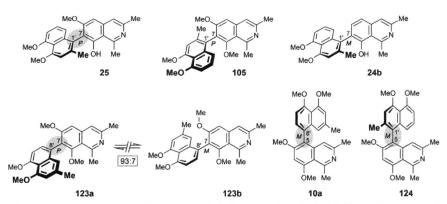

Fig. 74 A unique series of seven naphthylisoquinoline isolated from the roots of *A. abbreviatus* belonging to the small subclass of alkaloids with a non-hydrogenated isoquinoline portion: the new ancistrobreveines A–D (**25, 105, 123a/b**, and **10a**),[54] and the known compounds 6-*O*-methylhamateine (**124**)[54,403] and *ent*-dioncophylleine A (**24b**),[54,291] earlier identified in *Ancistrocladus* species from Southeast Asia. The four different coupling types of these fully dehydrogenated alkaloids—7,1', 7,8', 5,8', and 5,1'—are underlaid in light gray, exemplarily for **25, 123a, 10a**, and **124**.

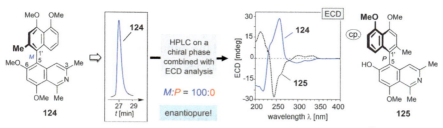

Fig. 75 HPLC analysis of 6-O-methylhamateine (**124**) on a chiral phase (Lux® Cellulose-1), demonstrating that **124** occurs in *A. abbreviatus* in an enantiomerically pure form, and comparison of the ECD spectrum of **124** with that of the known,[259] structurally closely related *P*-configured ancistrocladeine (**125**), semisynthetically prepared by catalytic thermal dehydrogenation of natural ancistrocladine.[259,334]

The 5,1′-linked 6-O-methylhamateine (**124**) (Fig. 75) had previously been discovered as a major metabolite in the leaves of the Vietnamese liana *A. cochinchinensis*.[1,403] The alkaloid exhibited an ECD curve that was virtually opposite to the one measured for the known,[259] likewise fully dehydrogenated and 5,1′-coupled ancistrocladeine (**125**). This metabolite is structurally similar, yet merely synthetic, but *P*-configured (and 6-O-demethylated) naphthylisoquinoline, previously obtained by semisynthesis from ancistrocladine.[259,334] Thus, the axial configuration of **124** was attributed to be *M*. The enantiomeric purity of **124** isolated from *A. cochinchinensis*, however, had never been analyzed.[1,403] The material of **124** isolated from *A. abbreviatus* was found to be enantiomerically pure, as obvious from the HPLC-ECD analysis on a chiral phase (Lux Cellulose-1), giving only one—moreover sharp—peak (Fig. 75).[54] The online ECD spectra from different positions of the peak were all identical, and mirror-imaged as compared to the ECD curve of ancistrocladeine (**125**).[54,259,334] Since in the classification of naphthylisoquinoline alkaloids the presence (or absence) of an oxygen function at C-6 is given priority over the criterion of the absolute configuration at C-3 (which is not applicable for fully dehydrogenated representatives), 6-O-methylhamateine (**124**) was classified as an Ancistrocladaceae-type naphthylisoquinoline alkaloid[1,5] (see also Fig. 19 in Section 3.1).

The second alkaloid with a non-hydrogenated isoquinoline moiety isolated from the roots of *A. abbreviatus* was the 7,1′-coupled Dioncophyllaceae-type naphthylisoquinoline *ent*-dioncophylleine A (**24b**)[54] (Fig. 74). It had earlier been identified in the leaves of the Malaysian species *A. benomensis*, where it occurs as a scalemic mixture, in a ratio of 93:7 of **24b** and its *P*-configured enantiomer dioncophylleine A (**24a**).[1,291] In the case of the material of **24b** obtained from *A. abbreviatus*, on the contrary, HPLC on a chiral phase (Lux® Cellulose-1) in combination with ECD spectroscopy to determine the enantiomeric purity gave only one peak, and the online ECD spectra, measured in intervals, including the left and right slopes of the peak, were all identical, showing that in *A. abbreviatus ent*-dioncophylleine A (**24b**) was thus present in an enantiopure form.[54]

The first new alkaloid discovered in the root bark of *A. abbreviatus*, ancistrobreveine A (**25**)[54] (Fig. 76a), displayed a constitution similar to that of *ent*-dioncophylleine A (**24b**), yet equipped with an additional methoxy group at C-6, thus representing the first fully dehydrogenated natural 7,1′-coupled naphthylisoquinoline alkaloid belonging to the subclass of Ancistrocladaceae-type[1,5] compounds. The ECD spectrum of ancistrobreveine A (**25**) was virtually identical to that of *ent*-dioncophylleine A (**24b**) (Fig. 76a), evidencing that these two alkaloids had the same axial configuration.[54] For formal reasons, however, **25** and **24b** had opposite descriptors, according to the Cahn-Ingold-Prelog notation. HPLC-ECD analysis on a chiral phase (Lux® Cellulose-1) revealed ancistrobreveine A (**25**) to occur in the plant in an enantiomerically pure form.[54] The ECD chromatogram monitored for one single wavelength (here 258 nm, where **25** has a strong negative ECD signal) led to the detection of only one single—and negative—peak (Fig. 76a).

As a second example of a 7,1′-linked Ancistrocladaceae-type alkaloid with a non-hydrogenated isoquinoline part, ancistrobreveine B (**105**)[54]

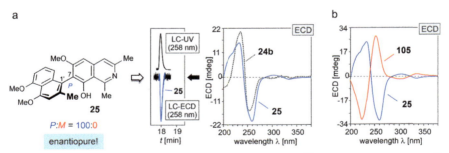

Fig. 76 (a) HPLC-ECD analysis of ancistrobreveine A (**25**)[54] on a chiral phase (Lux® Cellulose-1), revealing that in *A. abbreviatus*, **25** occurs in an enantiomerically pure form, and assignment of the absolute axial configuration by comparison of its ECD spectrum with that of the known[291] *M*-configured *ent*-dioncophylleine A (**24b**; the structure is shown in Fig. 74). (b) Assignment of the absolute axial configuration of the likewise fully dehydrogenated 7,1′-coupled, but *P*-configured ancistrobreveine B (**105**;[54] for the structure, see Fig. 74) by comparison of the ECD curve of **105** with that of **25**.

(Fig. 74) was isolated from the roots of *A. abbreviatus*. This metabolite had nearly the same constitution as **25**, but possessed a methoxy function at C-8 (instead of a hydroxy group as in **25**). The almost mirror–image-like ECD spectra of **105** and ancistrobreveine A (**25**) (Fig. 76b) showed that the two compounds had opposite axial stereoarrays.[54] For formal reasons, however, **105** and **25** had the same *P*-descriptor according to the Cahn–Ingold–Prelog priority rules. Similar to **24b** and **25**, ancistrobreveine B (**105**) was likewise found to be enantiomerically pure by HPLC analysis on a chiral phase (Lux® Cellulose-1) coupled to ECD spectroscopy.[54]

The third new compound within this series of fully dehydrogenated naphthylisoquinoline alkaloids was the 7,8′-coupled and *O*-permethylated ancistrobreveine C (**123**)[54] (Fig. 77), which differed from the 7,1′-linked ancistrobreveine B (**105**) only by the position of the methyl group on the

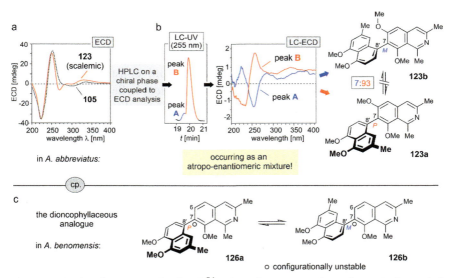

Fig. 77 Ancistrobreveine C (**123**),[54] the first example of a 7,8′-coupled Ancistrocladaceae-type alkaloid found in nature that possesses a non-hydrogenated isoquinoline portion. (a) Assignment of the prevailing absolute axial configuration by comparison of the overall ECD spectrum of the scalemic natural product with that of the structurally related, likewise *P*-configured, but 7,1′-linked ancistrobreveine B (**105**,[54] for its structure, see Fig. 74). (b) HPLC-ECD analysis of **123** on a chiral phase (Lux® Cellulose-1), evidencing that *A. abbreviatus* produces **123** as an atropo-enantiomeric mixture, with the *P*-enantiomer being the by far prevailing isomer in the plant. (c) For comparison: The likewise 7,8′-coupled, but dioncophyllaceous alkaloid dioncophylleine D (**126**)[291] lacking a methyl group *ortho* to its biaryl axis, was found to be present in *A. benomensis* as a racemic mixture of its interconverting enantiomers.

naphthalene portion. The influence of that methyl group on the chiroptical properties of **105** and **123** was assumed as small compared with that of the strong naphthalene chromophore. This permitted to attribute the absolute axial configuration of ancistrobreveine C (**123**) as *P*, since **123** and **105** provided two nearly identical ECD curves (Fig. 77a).[54] Analysis of **123** by HPLC on a chiral phase (Lux® Cellulose-1) revealed a small peak splitting at the rising slope of the UV-detected peak (Fig. 77b). The simultaneously acquired ECD trace monitored at 255 nm clearly gave two peaks (not shown). The detection of a strong positive ECD signal along with a clear negative ECD signal suggested ancistrobreveine C (**123**) to occur as a scalemic mixture in the plant, with the *P*-configured enantiomer, **123a**, being the by far prevalent one.[54] Full LC-ECD spectra directly taken online, in the stopped-flow mode, at the left and the right slopes of the UV peak, indeed furnished nearly mirror-imaged ECD curves (Fig. 77b), thus establishing that here, in *A. abbreviatus*, the two atropo-enantiomers, **123a** and **123b**, occurred in a ratio of 93:7 (*P:M*).[54]

Ancistrobrevine C (**123**) is the first—and so far only—example of a 7,8′-coupled alkaloid with a non-hydrogenated isoquinoline portion that is optically active. The likewise 7,8′-linked and constitutionally nearly identical Dioncophyllaceae-type alkaloid dioncophylleine D (**126**)[1,291] (Fig. 77c), by contrast, differing from **123** only by the lacking methoxy group at C-6, was found to occur as a racemate in the Malaysian liana *A. benomensis*, as a 1:1 mixture of two configuratively unstable, and thus interconverting atropo-enantiomers.

Remarkably, most of the non-hydrogenated alkaloids isolated from the roots of *A. abbreviatus* showed an *O*-permethylated substitution pattern. Another example of such a naphthylisoquinoline was the new 5,8′-coupled ancistrobreveine D (**10a**)[54] (Fig. 78). Its absolute axial configuration was assigned as *M* by comparison of its ECD spectrum with that of the constitutionally identical and likewise 5,8′-linked, but *P*-configured alkaloid ancistrolikokine J₃ (**10b**)[53] from the Congolese species *A. likoko*, i.e., the enantiomer of ancistrobreveine D (**10a**). Again, similar to the case of the 7,1′-coupled ancistrobreveines A (**25**) and B (**105**), and the 5,1′-linked 6-*O*-methylhamateine (**124**), HPLC analysis of **10a** on a chiral phase coupled to ECD spectroscopy resulted in only one single UV signal and only one peak in the ECD trace. The ECD spectra measured in intervals at different positions of the peak were all identical, thus revealing that ancistrobreveine D (**10a**) from *A. abbreviatus* was enantiomerically pure.[54]

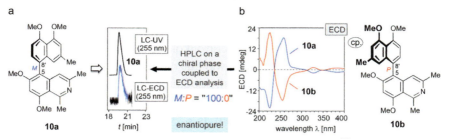

Fig. 78 (a) Occurrence of the 5,8′-coupled ancistrobreveine D (**10a**)[54] in an enantiopure form in the roots of *A. abbreviatus*, as determined by HPLC-ECD analysis of **10a** on a chiral phase (Lux® Cellulose-1). (b) Assignment of the axial configuration of **10a** as *M*, by comparison of its ECD spectrum with that of its enantiomer, ancistrolikokine J$_3$ (**10b**),[53] as previously isolated from *A. likoko*.

Some of the fully dehydrogenated naphthylisoquinoline alkaloids isolated from *A. abbreviatus* like 6-*O*-methylhamateine (**124**) and ancistrobreveine C (**123**),[54] but also related 5,1′- and 5,8′-coupled compounds discovered in the Chinese liana *A. tectorius*[1,252] and in the Congolese species *A. likoko*,[53,176] displayed significant cytotoxic activities against drug-sensitive CCRF-CEM leukemia cells and also against their multidrug-resistant CEM/ADR5000 subline. In view of their strong antiproliferative effects, similar to those exhibited by the well-known anticancer agent doxorubicin,[477,478] some of these non-hydrogenated naphthylisoquinolines may have a substantial therapeutic potential as cytotoxic agents, since they showed a drastically reduced degree of cross-resistance as compared to doxorubicin itself (ca. 580-fold[366]). Most of these alkaloids strongly inhibited the cell viability of CCRF-CEM and CEM/ADR5000 leukemia cells with nearly similar efficacies, revealing only minimal (ca. 1.2- to 6-fold) or even no cross-resistances.[53,54,252] Within the series of the above-presented compounds isolated from *A. abbreviatus*, 6-*O*-methylhamateine (**124**) was the most potent agent, with an excellent half-maximum inhibitory concentration in the low micromolar range against CCRF-CEM cells (IC$_{50}$, 3.95 µM) and against the resistant CEM/ADR5000 subline (IC$_{50}$, 5.52 µM).[54] Considerable antiproliferative activities against these two cell lines were also observed for the 7,8′-coupled ancistrobreveine C (**123**), with nearly identical inhibitory effects determined in CCRF-CEM cells (IC$_{50}$, 12.4 µM) and in CEM/ADR5000 leukemia cells (IC$_{50}$, 12.6 µM).[54] The three 7,1′-coupled alkaloids ancistrobreveines A (**25**) and B (**105**), and *ent*-dioncophylleine A (**24b**) showed only moderate to weak antileukemic activities.[54] The *O*-permethylated 5,8′-coupled ancistrobreveine D (**10a**), differing from 6-O-methylhamateine (**124**) simply by the position of the methyl group in the naphthalene portion, was only moderately active.[54] Its 4′,6-O,O-dimethyl analogue, ancistrolikokine G (**56**)[176] (Fig. 28c in Section 4) from *A. likoko*, by contrast, was one of the most potent compounds among the fully dehydrogenated naphthylisoquinoline alkaloids.[53] It displayed pronounced growth-retarding activities against both, CCRF-CEM and CEM/ADR5000 leukemia cells, with IC_{50} values of 4.73 and 7.73 µM, respectively.[53] The excellent—and structure-specific—antileukemic activities of some of the compounds presented in this paragraph demonstrate their promising potential for a further development as agents to combat cancer cells that are unresponsive toward drugs routinely used for the treatment of malignant disorders.

Most recently, Berthold and van Otterloo[509] have reported a new and concise concept for the asymmetric total synthesis of 5,8′-coupled naphthylisoquinoline alkaloids, based on the construction of the biaryl axis by a Ni-catalyzed Negishi cross-coupling of the naphthalene and isoquinoline building blocks using the bidentate *N,N*-ligand (*S,S*)-**130**. The approach implies an intermolecular coupling of the two molecular precursors, directly providing the final target molecules. Its potential was demonstrated in the atropo-enantioselective first synthesis of the non-hydrogenated alkaloid ancistrolikokine J₃ (**10b**)[53] (Fig. 79). Starting from the appropriately modified precursors **128** and **129**, with the coupling positions activated by a zinc iodide substituent in the naphthalene portion and an iodine atom in the isoquinoline part, Ni-catalyzed cross-coupling

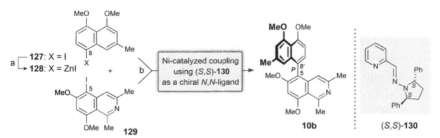

Fig. 79 Enantioselective total synthesis of ancistrolikokine J₃ (**10b**) by Ni-catalyzed Negishi cross-coupling of the zincated naphthalene precursor **128** with the iodinated isoquinoline **129**, using (*S,S*)-**130** as a C₂-symmetric *N,N*-ligand.[509] Reaction Conditions: (a) Zn reagent **128** prepared from the iodinated naphthalene **127**: Zn (2.0 equiv.), DMF (1.0 M), 80 °C, 5 h; (b) coupling on a 0.1 mmol scale: NiCl₂(DME) [dichloro(dimethoxyethane)nickel] (10 mol%), ligand (*S,S*)-**130**[511] (15 mmol%), THF/DMF (1:1, 0.2 M), 60 °C, 18 h, 61%, 96% *ee*.

using the N,N-ligand (S,S)-**130** furnished ancistrolikokine J$_3$ (**10b**) in a good yield of 61% and a high enantiomeric purity of 96% *ee*.[509] Of note, this is the first example successfully addressing the major challenge of a direct coupling of the respective building blocks, the appropriately activated naphthalene and isoquinoline units, to give the targeted alkaloid straightaway.

First attempts by Berthold and van Otterloo[509] making use of the N,N-ligand (S,S)-**130** developed by Lassaletta,[511] but employing the Pd-catalyzed Suzuki-Miyaura cross-coupling approach,[510] had already given the desired alkaloid **10b**, but that reaction had suffered from reproducibility issues, with largely varying chemical yields and stereoselectivities.[509] The Pd/ligand combination was even totally unsuccessful in both Negishi[510] and Kumada[510] cross-coupling reactions. Only when switching the metal catalyst from palladium to nickel under Negishi cross-coupling conditions and employing the N,N-ligand (S,S)-**130**, ancistrolikokine J$_3$ (**10b**) was obtained in consistently good yields and high enantiomeric purities. Remarkably, the approach gave excellent results also when inversely activating the two alkaloid precursors in the coupling positions, i.e., by preparing the zinc reagent from the isoquinoline building block and by introducing a halogen atom (I or Br) into the naphthalene moiety (not shown).[509] The efficiency of the method was also confirmed by the successful atroposelective total synthesis of a series of naphthyl*dihydro*isoquinoline alkaloids,[509] among them the promising agent ancistrolikokine E$_3$ (**2**)[10] (for its structure see Fig. 1), which displays potent cytotoxicity against pancreatic cancer cells.

5.7 *Ancistrocladus abbreviatus* as a rich source of naphthylisoquinoline alkaloids—Summary and outlook

First phytochemical studies on *A. abbreviatus* from Coastal West Africa in the 1990s had led to the isolation and structural elucidation of 16 naphthylisoquinoline alkaloids,[5,62–66] among them prominent Ancistrocladaceae-type compounds such as ancistrocladine (**20a**),[5,63] and the ancistrobrevines A (**100**),[5,66] B (**87a**),[5,63] and D (**102a**)[5,65] (Fig. 80). In addition, well-known Dioncophyllaceae-type alkaloids like dioncophylline A (**1a**)[5,63] and its N-methyl derivative **94a**[5,62] had been isolated, but also the first hybrid-type alkaloid discovered in nature, ancistrobrevine C (**22**),[5,64] had already been identified in those early investigations on *A. abbreviatus* (Fig. 80).

Recent renewed investigations on the metabolite pattern of *A. abbreviatus*, using improved analytical methods, which permitted to focus on minor components and even trace constituents, furnished a plethora of

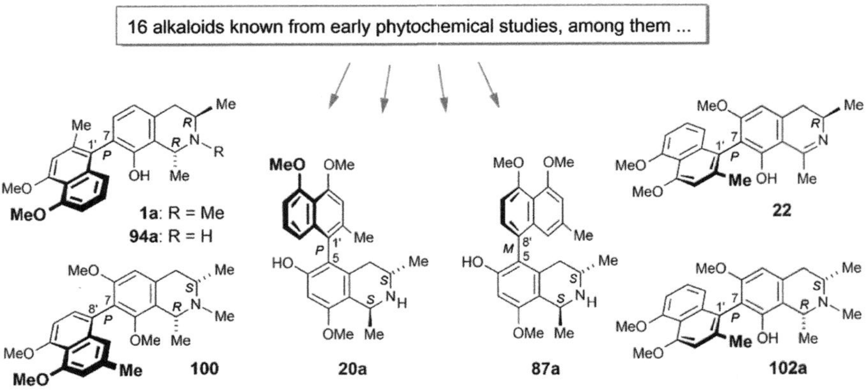

Fig. 80 Seven of the most prominent naphthylisoquinoline alkaloids identified in early phytochemical investigations on *A. abbreviatus*[5,62–66]: dioncophylline A (**1a**), *N*-methyldioncophylline A (**94a**), ancistrobrevine A (**100**), ancistrocladine (**20a**), and the ancistrobrevines B (**87a**), C (**22**), and D (**102a**).

new naphthylisoquinolines with five different coupling types (5,1', 5,8', 7,1', 7,3', and 7,8').[54–61] Subject of the here present Section 5 are the alkaloids equipped with a tetra- or a dihydroisoquinoline subunit, or possessing a non-hydrogenated isoquinoline portion such as ancistrobrevines M (**69**)[58] and J (**70**),[57] dioncophyllidine E (**72**),[61] dioncoline A (**23**),[58] *ent*-dioncophylleine A (**24b**),[54] and ancistrobreveine D (**10a**)[54] (Fig. 81). Today, a total of 51 representatives of these three subgroups of naphthylisoquinoline alkaloids are known to be produced by *A. abbreviatus*, half of these secondary plant metabolites (25 compounds) were new natural products, discovered for the first time in this West African liana.[54,56–58,61–66]

The largest portion of the naphthyl*tetrahydro*isoquinolines discovered in *A. abbreviatus* (33 compounds) possess a 5,1'-biaryl linkage (16 compounds),[5,57,58,63] but also a considerable number of 7,1'-coupled alkaloids have been identified (10 compounds),[5,57,58,62,63,65] whereas only few examples of 5,8'- (3 representatives)[5,57,63] and 7,8'-linked naphthyltetrahydroisoquinolines (4 compounds)[5,57,66] have so far been isolated from the plant. The number of naphthyl*dihydro*isoquinolines found in *A. abbreviatus*, by contrast, was significantly smaller (11 compounds).[56,61,64] Remarkably, this subgroup comprises metabolites with less common structural features, among them the only second and third examples of dioncophyllaceous 5,1'- and 7,3'-coupled alkaloids,[56,61] and the first two 5,1'-linked hybrid-type naphthyldihydro-isoquinolines[56] found in nature. Likewise unusual is the discovery of a series of no less than seven alkaloids with a non-hydrogenated

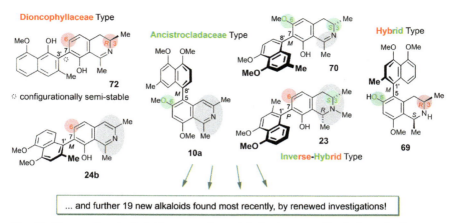

Fig. 81 Six prominent alkaloids possessing a tetrahydro- or a dihydroisoquinoline portion, or a fully non-hydrogenated isoquinoline subunit, most recently discovered in *A. abbreviatus*[54,56,58,61]—a selection (the diverse degree of hydrogenation of the isoquinoline portion is labeled in light gray). The alkaloids were classified as Ancistrocladaceae- (3S, 6-OR—"green/green"), Dioncophyllaceae- (3R, 6-H—"red/red"), hybrid- (3R, 6-OR—"red/green"), and inverse hybrid-type (3S, 6-H—"green/red") compounds.

isoquinoline portion, based on four different coupling types. Prior to their isolation from the roots of *A. abbreviatus*,[54] only three other *Ancistrocladus* species, *A. tectorius* (China),[1,252] *A. benomenis* (Malaysia),[1,273,291] and *A. likoko* (Democratic Republic of the Congo),[53,176] had been known to produce a larger number of fully dehydrogenated naphthylisoquinoline alkaloids, but apart from that their occurrence in nature was rare.

A schematic overview on the presence of naphthylisoquinoline alkaloids in *A. abbreviatus*, based on their differences in the degree of hydrogenation in the isoquinoline moiety, their coupling types, and their classification[1,5] (see Section 3.1) as Ancistrocladaceae-, Dioncophyllaceae-, hybrid-, and inverse-hybrid-type compounds, is given in Fig. 82.

Within the Ancistrocladaceae and Dioncophyllaceae plant families, *Ancistrocladus abbreviatus* occupies an outstanding position. The nearly combinatorial formation of naphthylisoquinoline alkaloids in *A. abbreviatus* appears to be the result of a lack of selectivity never found before in any other plant species belonging to the Ancistrocladaceae and Dioncophyllaceae. As the only known *Ancistrocladus* liana, it contains naphthylisoquinoline alkaloids of all four possible subtypes.[5,54–66] There are even only three other African species from which just three types of compounds have been isolated—Ancistrocladaceae-, Dioncophyllaceae-, and hybrid-type

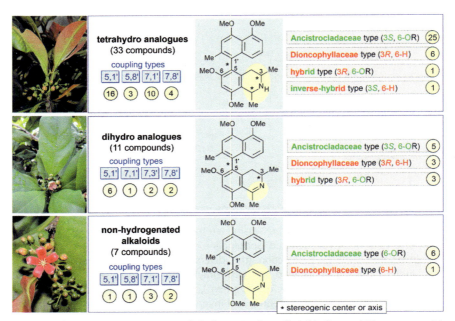

Fig. 82 Schematic overview on naphthylisoquinoline alkaloids (in total 51 compounds) with a tetra- or a dihydroisoquinoline subunit, or possessing a non-hydrogenated isoquinoline subunit, isolated from the roots, twigs, stems, or leaves of *A. abbreviatus*. Photos showing specimens of *A. abbreviatus* with buds and flowers, cultivated in the greenhouse of the Botanical Garden of the University of Würzburg. *Photographs:* A. Irmer *(top and bottom)* and M. Dreyer *(center)*.

alkaloids, but never any inverse-type hybrids. One of these plants is *A. barteri*[5,285,286] from Coastal West Africa, the two others are Central African species, *A. ileboensis*[3,51,52,183] and an as yet undescribed liana[3,12] from the Democratic Republic of the Congo. Thus, *A. abbreviatus* can be regarded as a geo- and chemotaxonomical link,[58] on the one hand over to the mentioned Congolese species and, on the other hand, to the West African *Ancistrocladus* plants such as *A. barteri* and to Dioncophyllaceae lianas, in particular to *Triphyophyllum peltatum*, which exclusively produces Dioncophyllaceae-type compounds.[5,6,58]

The exceptional position of *A. abbreviatus* can also be seen from the even more intriguing discovery of several novel-type alkaloids with special, totally unprecedented structural features.[55,56,59,60] The isolation of these unique natural products from *A. abbreviatus*, their structural elucidation and bioactivities will be the topic of Section 6.

6. Ancistrocladus abbreviatus, a creative producer of naphthylisoquinoline alkaloids with unusual molecular scaffolds

During the past decade, intense isolation work on *Ancistrocladus* plants endemic to the tropical rainforests of tropical Africa and Asia furnished a plethora of mono- and dimeric naphthylisoquinoline alkaloids.[1–4,10–12,15,16,30–35,51–61,176,178,180–183,251–255,299,311] Among them are an impressive number of compounds with unusual structural features,[1–3,12,15,16,32–35,55,56,58–60,180–183,253–255,299,311] enlarging the scope of synthetic originality of the Ancistrocladaceae plant family substantially. In this respect, *Ancistrocladus abbreviatus* from Coastal West Africa is one of the most noticeable species within the *Ancistrocladus* genus, since it produces a broad variety of naphthylisoquinoline alkaloids and related compounds with unprecedented molecular architectures[55,56,58–60] (Fig. 83). Most of

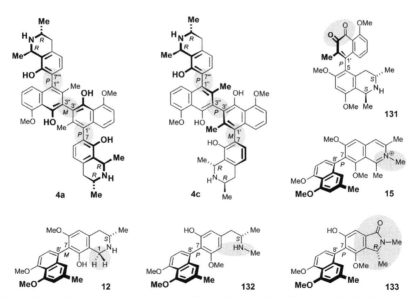

Fig. 83 Secondary metabolites produced by *Ancistrocladus abbreviatus* displaying unusual structural features: the dimers jozimine A$_2$ (**4a**)[12,58] and jozibrevine B (**4c**),[58] the quinoid alkaloid ancistrobreviquinone B (**131**),[56] ancistrobrevinium A (**15**)[60] as the first fully dehydrogenated, *N*-methylated and, thus, cationic C,C-coupled naphthylisoquinoline found in nature, the 1-unsubstituted 1-*nor*-8-O-demethylancistrobrevine H (**12**),[55] the ring-cleaved ancistrosecoline C (**132**),[55] and the naphthylisoindolinone ancistrobrevoline B (**133**).[59]

the structurally thrilling compounds discovered in *A. abbreviatus* have so far not been detected in any other taxon of the Ancistrocladaceae and Dioncophyllaceae families.

In the roots of *A. abbreviatus*, a series of four 3′,3″-coupled Dioncophyllaceae-type naphthylisoquinoline dimers have been discovered, among them the previously already known jozimine A$_2$ (**4a**)[3,12,58,183] and three new alkaloids, such as jozibrevine B (**4c**).[3,58] All of them have the same constitution and identical absolute configurations at the four stereogenic centers, differing from each other only by their axial chirality. This finding was extraordinary with respect to the fact that such a near-complete series of atropo-diastereomers (four out of six possible compounds)[58] had never been found in any *Ancistrocladus* species before. Furthermore, the first quinoid naphthylisoquinoline alkaloids possessing an *ortho*-quinoid entity in the naphthalene part, such as ancistrobreviquinone B (**131**) were detected in the roots of *A. abbreviatus*.[56] Likewise remarkable was the isolation of ancistrobrevinium A (**15**), which is the only example of a cationic *C, C*-coupled *N*-methylnaphthylisoquinoline alkaloid known so far from nature.[60] Completely unexpected and, thus, most spectacular, was the identification of the very first 1-unsubstituted naphthylisoquinoline, 1-*nor*-8-*O*-demethylancistrobrevine H (**12**),[55] lacking the otherwise generally present methyl group at C-1, along with the discovery of structurally intriguing follow-up metabolites. Unprecedented are also the first *seco*-naphthylisoquinolines (six representatives),[55] such as ancistrosecoline C (**132**),[55] in which the tetrahydroisoquinoline ring is cleaved, with loss of C-1. Likewise unique are the new naphthylisoindolinones (four examples),[59] among them ancistrobrevoline B (**133**),[59] which possesses an isoindolinone portion, instead of the usual isoquinoline molecular unit, as a consequence of a ring contraction and loss of the whole C-3/Me-3 entity in the heterocyclic ring system of the parent naphthylisoquinoline alkaloids.

Besides the plant itself, callus cultures of *A. abbreviatus* established from sterile leaf material (see Section 2.3.2) were also found to produce naphthylisoquinoline alkaloids, albeit in small quantities.[151] The cultivation conditions apparently imposed stress to the cells, eventually resulting in a largely increased formation of the naphthoquinones plumbagin (**16**) and droserone (**17**), along with the tetralone *cis*-isoshinanolone (*cis*-**18**). Solidification of the medium, preventing the calli from sinking into the gel, led to more aerobic conditions and favored the additional formation of a series of highly oxygenated naphthoquinones as major metabolites, among them the new compounds ancistroquinone B (**134**) and

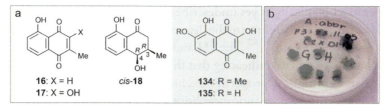

Fig. 84 (a) Oxygenated naphthalene derivatives isolated from (b) solidified callus cultures of *A. abbreviatus*[151]: plumbagin (**16**), droserone (**17**), *cis*-isoshinanolone (*cis*-**18**), and the two newly discovered naphthoquinones ancistroquinone B (**134**) and dioncoquinone B (**135**). *Photograph:* H. Rischer.

dioncoquinone B (**135**) (Fig. 84).[151] This series of highly substituted naphthoquinones was exclusively produced in solidified callus cultures of *A. abbreviatus*, but were not identified in the plants themselves.[151]

The following paragraphs provide a more detailed report on naphthylisoquinoline alkaloids and naphthoquinones with unusual molecular scaffolds discovered in plant organs (see Fig. 83) and cell cultures (see Fig. 84) of *A. abbreviatus*. The focus will in particular be on the elucidation of the complete stereostructures of these new metabolites, on studies regarding their bioactivities, and, for selected compounds, on their partial or total synthesis.

6.1 A near-complete series of atropo-diastereomeric naphthylisoquinoline dimers, discovered in the roots of *Ancistrocladus abbreviatus*

6.1.1 Jozimine A₂ and jozibrevines A–D, Dioncophyllaceae-type naphthylisoquinoline alkaloids, with three chiral axes

In the most polar fraction of the root bark extract of *A. abbreviatus*, LC-UV-MS analysis hinted at the occurrence of four constituents with UV and MS profiles typical of dimeric naphthylisoquinoline alkaloids.[58] All of the compounds showed a mass peak at m/z 725.4 $[M+H]^+$ in combination with a further mass peak at m/z 363.2 $[M+2H]^{2+}$, observed within the very same peak, hinting at the presence of dimers lacking an oxygen function at C-6 in the two molecular halves. The compounds displayed a characteristic UV maximum at ca. 242 nm and a molecular formula of $C_{46}H_{48}N_2O_6$.

One of them, a major constituent within this series of dimeric naphthylisoquinoline alkaloids, was readily identified as the known dimer jozimine A₂ (**4a**),[3,12,58,183] which had earlier been discovered as a trace alkaloid in a botanically as yet undescribed Congolese *Ancistrocladus* liana[3,12] and as a minor metabolite in the leaves of *A. ileboensis* endemic to the Southern Congo Basin.[3,183] The ^{1}H and ^{13}C NMR spectra of **4a** showed only a half set of signals, indicating that this dimer was symmetric. Its optical activity and chiroptical properties excluded the presence of a *meso* compound (i.e., with two enantiomorphous molecular portions), but proved that **4a** was C_2-symmetric consisting of two homomorphous halves.[12,58] The spin pattern of the aromatic protons displayed three contiguous protons in the naphthalene moieties and a two-proton spin system in the tetrahydroisoquinoline portions, thus evidencing the presence of a dimer lacking oxygen functions at C-6 and C-6′′′ in the isoquinoline portion.[12,58] The coupling position in the naphthalene unit, located at C-1′, was deduced from the ROESY cross-peak of H-8′ to the 8-OH group and from HMBC interactions of Me-2′, H-8′, and H-6 with C-1′. The ROESY sequence in the series {H$_{eq}$-4 ↔ H-5 ↔ H-6}, along with the "normal" chemical shifts of the two diastereotopic protons at C-4 (δ_H 2.96 and 3.23), i.e., showing no shielding effect from a naphthalene part nearby, established C-7 to be the axis-bearing carbon atom in the isoquinoline portion. HMBC interactions of H-5 with C-4, C-7, and C-8a confirmed the 7,1′-coupling type (Fig. 85a).[12,58] The two monomeric naphthylisoquinoline halves were connected via C-3′, as deduced from the signal of Me-2′, which was extremely upfield-shifted to δ_H 1.84, revealing that the methyl group was twofold shielded by two naphthalene rings. This assignment was approved by the ROESY signal between Me-2′ and the hydroxy function at C-4′′ monitored in acetone-d_6 (Fig. 85a).[3,12]

From the ROESY interactions between Me-1 and H-3, the relative configuration at the stereocenters at C-1 and C-3 was established as *trans* (Fig. 85b). The absolute configuration at C-3 was assigned as *R* from the ruthenium-mediated oxidative degradation[1,3,70] of the dimer. In view of the relative *trans*-configuration, this finding evidenced C-1 to be *R*-configured, too.[12,58] In combination with the absolute configuration at the two stereogenic centers, long-range ROESY interactions between Me-2′ and Me-1 and between H-1 and H-8′ established the absolute configuration at the two outer axes as *P* (Fig. 85b).[12,58] Hence, jozimine A$_2$ (**4a**) was the homo-coupling product of two identical 7,1′-coupled Dioncophyllaceae-type naphthylisoquinoline monomers, namely 4′-O-demethyldioncophylline A (**93a**) (see Fig. 88).[3,12,58]

The binaphthalene core of jozimine A$_2$ (**4a**) showed a strong structural resemblance to that of shuangancistrotectorine A (**43**) (Fig. 86) from the

Structural variety and pharmacological potential of naphthylisoquinoline alkaloids 169

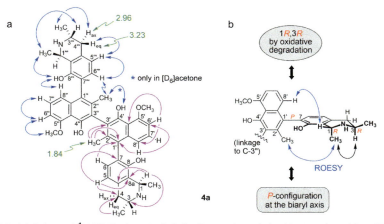

Fig. 85 (a) Selected ^{1}H NMR chemical shifts (in methanol-d_4, δ in ppm) and key ROESY (double blue arrows) and HMBC (single purple arrows) interactions indicative of the constitution of jozimine A$_2$ (**4a**) (including the position of the central axis and the outer ones); (b) absolute configuration of **4a** at the axes and centers in the two (identical) monomeric halves deduced from selective ROESY interactions in combination with the results of the oxidative degradation of **4a**.

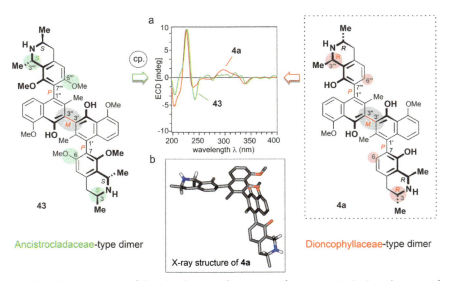

Fig. 86 (a) Assignment of the absolute configuration of jozimine A$_2$ (**4a**) at the central biaryl axis by comparison of its experimental ECD spectrum with that of the related Ancistrocladaceae-type dimer shuangancistrotectorine A (**43**); (b) X-ray structure of **4a** in the solid state (hydrogen atoms, except for those attached to the two nitrogen atoms, have been omitted for reasons of clarity).

Chinese liana *A. tectorius*.[3,12,254] This dimer differed from **4a** by the S-configuration at C-3 and the oxygen function at C-6 in the isoquinoline units. The central axis of **43** and **4a** was flanked by four *ortho*-substituents and was, thus, rotationally hindered. Previous investigations[1,3,254] had shown that the chiroptical properties of such dimeric naphthylisoquinolines with a 3′,3″-linkage are dominated by the orientation of the two naphthalene portions (representing the main chromophores) to each other and, thus, by the configuration at the central biaryl axis, over all other stereogenic elements: the stereogenic centers in the isoquinoline units and even over the two outer axes.[254] Therefore, the absolute configuration at the central biaryl linkage in jozimine A_2 (**4a**) was unambiguously assigned by comparison of its ECD spectrum with that of shuangancistrotectorine A (**43**) (Fig. 86a).[3,12] The ECD spectrum of **4a** was virtually identical to that of **43**, which had been determined to be *M*-configured by spectroscopic and chiroptical methods in combination with quantum-chemical ECD calculations,[1,3,254] thus evidencing that the central axis of **4a** had the same *M*-configuration as in **43**.[3,12]

Jozimine A_2 (**4a**) had been the first Dioncophyllaceae-type dimer discovered in nature.[3,12] It was structurally substantiated by X-ray crystallography (Fig. 86b), confirming the constitution and the relative configuration of **4a**.[12] It was the first—and so far only—naphthylisoquinoline dimer from which crystals suitable for X-ray diffraction analysis have been obtained.

The second naphthylisoquinoline dimer isolated from the roots of *A. abbreviatus* as a minor constituent, exhibited ^{1}H and ^{13}C NMR data that were almost identical to those of jozimine A_2 (**4a**), providing only a half set of signals, thus indicating the presence of yet another spectroscopically symmetric Dioncophyllaceae-type alkaloid.[58] Specific ROESY and HMBC interactions similar to those of **4a** clearly confirmed that this new dimer possessed the same constitution as jozimine A_2 (**4a**). Again, the naphthalene and isoquinoline units were linked to each other by a 7,1′-axis, and the two naphthylisoquinoline monomers were connected via C-3′. The combination of oxidative degradation and long-range ROESY interactions between Me-2′ and Me-1 revealed the new dimer to be *R*-configured at all of the four stereogenic centers, and—similar to jozimine A_2 (**4a**)—to be built up from two 4′-*O*-demethyldioncophylline A (**93a**) moieties.[58] Thus, the new dimer and **4a** only differed by their absolute configurations at the central biaryl axis. The ECD spectrum of the new alkaloid was virtually opposite to that of jozimine A_2 (**4a**, central axis: *M*), evidencing the absolute axial configuration in the binaphthalene core to be *P*.[58] The new naphthylisoquinoline dimer, possessing the full absolute stereostructure **4b** (Fig. 87), was named jozibrevine

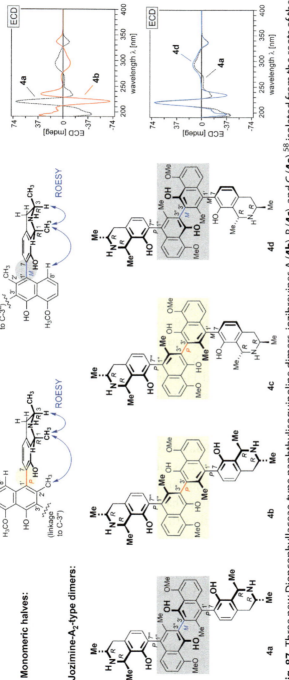

Fig. 87 Three new Dioncophyllaceae-type naphthylisoquinoline dimers, jozibrevines A (4b), B (4c), and C (4e),[58] isolated from the roots of the West African liana A. abbreviatus, along with their parent compound jozimine A$_2$ (4a), known[3,12,183] from previous isolation work on two Congolese Ancistrocladus lianas: attribution of the relative configuration of 4b, 4c, and 4d at the stereogenic centers versus the 7,1'-coupled "outer" biaryl axes in their monomeric halves by ROESY correlations (blue double arrows), and assignment of their absolute configuration at the central 3',3''-linked axis by comparison of their ECD spectra with that of jozimine A$_2$ (4a), here exemplarily presented for jozibrevine A (4b) and jozibrevine C (4d). The dimers possess three consecutive chiral biaryl axes; all of them are coupled via the sterically constrained 3',3''-positions of their two naphthalene units. They all have the same constitution and even the same absolute configurations at the four stereogenic centers, but differ by the chirality of their three biaryl axes.

A, according to its isolation from *A. abbreviatus* and its structural relationship to jozimine A$_2$ (**4a**).[58] Alternatively, it can also be addressed as 3'-*epi*-jozimine A$_2$. Prior to its discovery in *A. abbreviatus*, jozibrevine A (**4b**) had already been known as a semisynthetic compound, obtained along with **4a** by biomimetic phenol-oxidative coupling of 4'-O-demethyldioncophylline A (**93a**) (see Section 6.1.4).[3,12]

The 1D and 2D NMR spectroscopic data of the third minor metabolite discovered in the roots of *A. abbreviatus* again corresponded to a Dioncophyllaceae-type naphthyltetrahydroisoquinoline dimer consisting of two 7,1'-coupled monomers connected by a 3',3''-axis between the two naphthalene units.[58] In contrast to jozimine A$_2$ (**4a**) and jozibrevine A (**4b**), however, the new dimer displayed a full set of ^{1}H and ^{13}C NMR signals, indicating the presence of an unsymmetric compound. The signals in both halves display spectroscopic features similar to those of **4a** and **4b**, suggesting the new metabolite to be composed of two constitutionally identical, yet configurationally different—hence diastereomorphic—monomeric naphthylisoquinolines.[58] ROESY investigations and the chemical degradation established the absolute configurations at C-3, C-3'', C-1, and C-1'' to be all R.[58] Thus, the differences between the two molecular halves had to be due to opposite configurations at the "outer" biaryl axes. This was confirmed by long-range ROESY correlations as outlined in Fig. 87. The interaction between Me-1 and H-8' evidenced an M-configuration at the axis in one half of the new dimer, while a cross-peak between Me-1''' and Me-2'' established the other, second half to be P-configured at the biaryl linkage. Hence, the new dimer was the cross-coupling product of 4'-O-demethyldioncophylline A (**93a**) and its atropo-diastereomer **93b**.[58] Its ECD spectrum was opposite to that of jozimine A$_2$ (**4a**, central axis: M), but virtually identical to that of jozibrevine A (**4b**, central axis: P) (data not shown). Therefore, the new dimer had the full absolute stereostructure **4c**, as presented in Fig. 87. It was thus the 7-*epi*-analogue of jozibrevine A (**4b**) and was henceforth named jozibrevine B.[58]

The fourth naphthylisoquinoline dimer from *A. abbreviatus* was a trace metabolite with a constitution fully identical to those of jozimine A$_2$ (**4a**) and the two jozibrevines A (**4b**) and B (**4c**).[58] The compound revealed a full set of signals in both the ^{1}H and ^{13}C NMR spectra, thus proving it to be unsymmetric. ROESY investigations, the ruthenium-mediated oxidative degradation procedure, and ECD measurements evidenced the new dimer to be most similar to **4a**. The central axis was determined to be M-configured as in **4a** due to the virtually identical ECD spectra of these two dimers,[58] and thus the only difference as compared to **4a** was the configuration at one of the two outer biaryl axes in the isolated alkaloid. Based on the long-range ROESY correlation between H-8′ and the protons of Me-1, it was assigned to be M-configured.[58] Thus, this new Dioncophyllaceae-type naphthylisoquinoline had the full absolute stereostructure **4d** as shown in Fig. 87, and was—similar to jozibrevine B (**4c**)—composed of two diastereomeric alkaloids, **93a** (P-configured) and **93b** (M-configured), but differed from **4c** by its M-configuration at the central axis.[58] The new dimer was, hence, the 3′-epi-analogue of **4c**, but it can likewise be addressed as the 7-epi-analogue of jozimine A$_2$ (**4a**). It was named jozibrevine C.[58]

The discovery of the three new jozibrevines A–C (**4b–d**) along with their parent compound jozimine A$_2$ (**4a**) in the roots of *A. abbreviatus* was particularly thrilling, because such a broad, nearly complete series of atropo-diastereomeric dimers (four out of six possible rotational isomers) had never been found in any *Ancistrocladus* species before.[1,3,58] Joint structural features were their identical constitutions and their absolute R-configurations at the four stereogenic centers in the tetrahydroisoquinoline subunits. The dimers yet differed by their axial chirality (Fig. 87). Two of the dimers, jozimine A$_2$ (**4a**) and jozibrevine A (**4b**), are homo-coupled products of the 7,1′-linked naphthylisoquinoline 4′-O-demethyldioncophylline A (**93a**), diverging from each other only by their stereoarray at the central axis.[58] The jozibrevines B (**4c**) and C (**4d**), by contrast, are formed by cross-coupling of **93a** and its atropisomer 4′-O-demethyl-7-epi-dioncophylline A (**93b**), and are thus unsymmetric dimers, which only differ by their absolute configuration at the central biaryl axis.[58] The two monomeric halves of the four dimers, **93a** and **93b**, have also been found to occur in the roots and stems of *A. abbreviatus*, from which they have been isolated in nearly equal amounts[58] (see Section 5.3.1).

Interestingly, the two dimers jozimine A$_2$ (**4a**) and jozibrevine C (**4d**) had previously also been discovered in stem bark extracts of *A. ileboensis* from the South-Central Congo Basin.[58,183] They were isolated along with yet another such dioncophylleous naphthylisoquinoline dimer, representing the fifth atropo-diastereomer within this series of six possible rotational isomers (Fig. 88). That new alkaloid, named jozibrevine D (**4e**),[52] is C_2-symmetric and is composed of two units of 4′-O-demethyl-7-epi-dioncophylline A (**93b**, biaryl axis: M), which are symmetrically linked via the 3′,3″-positions of their naphthalene portions. The absolute configuration at the central biaryl axis of jozibrevine D (**4e**) was assigned as P, in view of the fact that **4e** and the parent compound jozimine A$_2$ (**4a**, central axis: M) displayed nearly mirror-imaged ECD spectra.[52]

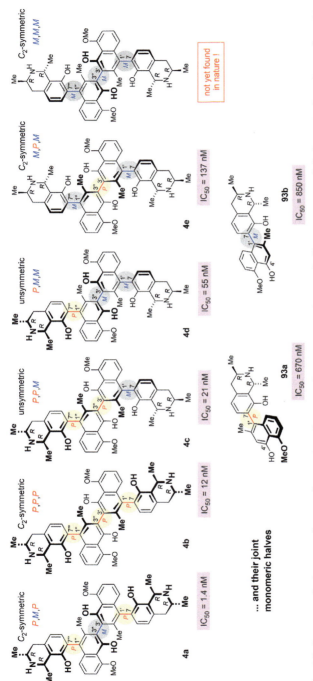

Fig. 88 An unusual, since nearly complete series of atropo-diastereomeric Dioncophyllaceae-type naphthylisoquinoline dimers found in nature, five out of six possible rotational isomers have so far been discovered: jozimine A₂ (**4a**) and the jozibrevines A (**4b**), B (**4c**), C (**4d**), and D (**4e**), all of them exerting pronounced antiplasmodial activities. As shown in Fig. 87, the dimers **4a–d** were identified in the West African liana *A. abbreviatus*,[58] whereas **4a**, **4c**, and **4e** were isolated from the South-Central Congolese species *A. ileboensis*.[52,58,183] A botanically yet undescribed *Ancistrocladus* plant was found to solely produce **4a** as a trace compound.[3,12] Likewise presented are the corresponding monomers, 4′-O-demethyldioncophylline A (**93a**) and its atropo-diastereomer 4′-O-demethyl-7-*epi*-dioncophylline A (**93b**).[58] Half-maximum inhibitory concentration (IC₅₀) values of the compounds against the NF54 strain of *P. falciparum* are underlaid in light pink.

Dioncophyllaceae-type dimers are very rare in nature.[1,3] Besides this series of five atropo-diastereomeric naphthylisoquinoline alkaloids presented in Fig. 88, only two further representatives, jozilebomines A (**44a**) and B (**44b**) (Fig. 89), have as yet been identified, produced as minor metabolites by the Congolese liana *A. ileboensis*.[3,183] They are intrinsically unsymmetric, due to a dissimilar 6′,3″-coupling in their binaphthalene cores. In contrast to **4a–e**, jozilebomines A (**44a**) and B (**44b**) are built up from two constitutionally different, regioisomeric 7,1′-linked naphthylisoquinoline monomers, with 5′-*O*-demethyldioncophylline A (**97**) as one of the molecular units (marked in purple, see Fig. 89) in the two dimers, 4′-*O*-demethyldioncophylline A (**93a**) as the second part in the case of **44a**, and 4′-*O*-demethyl-7-*epi*-dioncophylline A (**93b**) in the case of **44b**.[3,183] The divergence in the OH/OMe patterns dictates the coupling of the two monomeric naphthylisoquinolines via dissimilar 6′- and 3″-positions, leading to the very first natural naphthylisoquinoline dimers with three *ortho*-substituents—two hydroxy groups and one methyl group—next to the central axis, thus making this biaryl linkage configurationally stable and to possess an additional stereogenic element in the dimers **44a** and **44b**.[3,183]

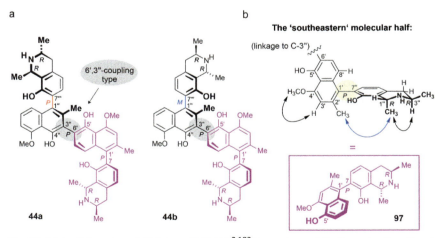

Fig. 89 (a) Jozilebomines A (**44a**) and B (**44b**),[3,183] the only naphthylisoquinoline dimers known so far possessing a 6′,3″-coupled central biaryl axis; and (b) long-range ROESY interaction (double blue arrow) defining the relative configuration at the stereogenic centers versus the axis within the "southeastern" molecular half, which is identical with 5′-*O*-demethyldioncophylline A (**97**, as one molecular unit of **44a** and **44b**, marked in purple).

6.1.2 The antiplasmodial potential of jozimine-A₂-type naphthylisoquinoline dimers

Among all monomeric and dimeric naphthylisoquinoline alkaloids tested so far, jozimine A₂ (**4a**) (Fig. 88) is one of the most potent antiplasmodial compounds, exerting excellent in vitro activities in the low-nanomolar range against chloroquine-sensitive (NF54) and chloroquine- and pyrimethamine-resistant (K1) strains of the malaria parasite *Plasmodium falciparum*, with IC₅₀ values of 1.4 nM (NF54)[3,12] and 16 nM (K1), respectively.[3,180] According to the TDR/WHO guidelines,[310] jozimine A₂ (**4a**) can be considered as a lead compound, since its cytotoxicity (IC₅₀ = 15.9 μM) against rat skeletal myoblast (L6) cells is quite low, giving rise to a very high selectivity index of >11,400. Moreover, the antiplasmodial activities of **4a** seem to be highly pathogen-specific: Against other protozoan parasites, causing visceral leishmaniasis (*L. donovani*), Chagas' disease (*T. cruzi*), and African sleeping sickness (*T. brucei rhodesiense*), **4a** displays only weak inhibitory effects.[3,12] Similar to jozimine A₂ (**4a**), the newly discovered jozibrevines A (**4b**), B (**4c**), C (**4d**), and D (**4e**) (Fig. 88) exhibit strong to good activities in the low nanomolar range against *P. falciparum* and only moderate cytotoxicities against L6 cells.[52,58] The strongest inhibitory effects against the NF54 strain of *P. falciparum* was displayed by jozibrevine A (**4b**).[58] With IC₅₀ values of 12 nM (for **4b**), 21 nM (for **4c**), 55 nM (for **4d**),[58] and 137 nM (for **4e**)[52] (Fig. 88), the antiplasmodial activities of the three jozibrevines, however, are significantly lower (by a factor of ca. 8.6 for **4b**, 15.0 for **4c**, 39.3 for **4d**, and 97.9 for **4e**) compared to that of **4a**. The two jozilebomines A (**44a**) and B (**44b**) (Fig. 89) also exert distinct—and specific—antiplasmodial activities in vitro, with **44a** (IC₅₀ = 43 nM) showing a more potent inhibitory effect against the NF54 strain of *P. falciparum* compared to that of its atropisomer **44b** (IC₅₀ = 102 nM).[3,183] The results clearly suggest a significant structure-dependent specificity and a strong impact of axial chirality on the antiplasmodial activity of jozimine-A₂-type dimers.[1,3,12,52,58,183]

Previous structure-activity relationship studies on a broad series of mono- and dimeric naphthylisoquinolines, including structurally related, merely artificial compounds[76,385,512–520] had revealed the importance of free hydroxy groups for the potency of the alkaloids toward *P. falciparum*. In agreement with this finding, the bisphenolic monomeric halves of the jozimine-A₂-type dimers **4a–e**, **44a**, and **44b**, viz., the alkaloids 4′-O-demethyldioncophylline A (**93a**) (IC₅₀ = 0.67 μM),[58] its 7-epimer **93b** (IC₅₀ = 0.85 μM)[58] (Fig. 88), and the regioisomeric 5′-O-demethyldioncophylline A (**97**) (IC₅₀ = 0.94 μM),[294] each displayed distinctly higher antiplasmodial activities compared to their parent compound dioncophylline A (**1a**) (IC$_{50}$ = 3.81 µM),[369] which possesses only one free hydroxy function. Dimerization of **93a**, **93b**, and **97** (each two free OH groups) to give jozimine-A$_2$-type compounds and thus leading to a doubling of the number of free hydroxy groups even further increased the antiplasmodial activities. The resulting dimers, now equipped with four free hydroxy groups, showed the by far highest effects against *P. falciparum* within this series of compounds.[3,12,52,58,183,520]

Similar to jozimine A$_2$ (**4a**), the jozibrevines A–D (**4b–e**) exerted only moderate to weak activities against *T. brucei rhodesiense* and *T. cruzi*. All of the dimers displayed no inhibitory effects against *L. donovani*. The monomers **93a** and **93b** exhibit virtually no antitrypanosomal or antileishmanial activities.[12,52,58,183]

6.1.3 Jozimine A$_2$, a promising potential lead for anti-cancer drug development

Despite the increased effectiveness of combinatorial first-line chemotherapeutic regimens for the treatment of some common malignancies,[521–523] cervical, pancreatic, and colon cancer as well as leukemia still remain major public health problems.[497,498] Therefore, the search for novel therapeutic agents to efficiently combat the aggressive and metastatic growth of tumors is still an urgent and challenging task. Because of the known pronounced cytotoxic activities of various mono- and dimeric naphthylisoquinoline alkaloids against different types of cancer cells[1–4,10,11,30,31,51–59,61,176,249,252,365,366] (see also Section 5), jozimine A$_2$ (**4a**) and related dimers were investigated for their antiproliferative potential toward human HeLa cervical and PANC-1 pancreatic cancer cells[183] and for their cytotoxic activity against fibrosarcoma (HT-1080) and colon carcinoma (HT-29) cancer cells.[58] Among the evaluated dimers, jozimine A$_2$ (**4a**) exerted the most potent cytotoxic effects against both HeLa and PANC-1 cells (Fig. 90).[183] It also significantly reduced the cell viability of HT1080 fibrosarcoma and HT-29 colon carcinoma cancer cells, whereas jozibrevines A (**4b**) and B (**4c**) did not show any considerable cytotoxic properties (Fig. 91).[58] Jozimine-A$_2$-type dimers were also tested for their growth-inhibitory effects against drug-sensitive acute lymphoblastic CCRF-CEM leukemia cells and their multidrug-resistant CEM/ADR5000 subline.[52] Jozimine A$_2$ (**4a**) displayed strong antiproliferative activities, with half-maximum inhibitory concentration values in the low micromolar range. Pronounced cytotoxicity toward the two leukemia cell lines were also observed for jozibrevines C (**4d**) and D (**4e**), and for jozilebomines A (**44a**) and B (**44b**), but the four dimers did not reach the excellent inhibitory potential of jozimine A$_2$ (**4a**) (Table 5).[52]

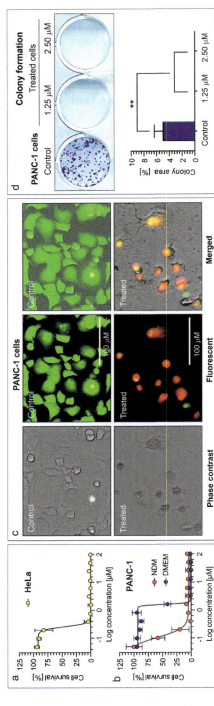

Fig. 90 (a) Impairment of viability of HeLa human cervical cancer cells by jozimine A_2 (**4a**) and (b) preferential cytotoxicity of **4a** against the human PANC-1 pancreatic cancer cell line in nutrient-deprived medium (NDM) and Dulbecco's modified Eagle's medium (DMEM); (c) morphological changes of PANC-1 cells after incubation with 5 μM of jozimine A_2 (**4a**) for 24 h in comparison to non-treated cells (control group). The pancreatic cancer cells were stained with ethidium bromide (EB) and acridine orange (AO) and photographed under fluorescence (red and green) and phase contrast (gray) modes using an EVOS FL digital inverted microscope; (d) effect of jozimine A_2 (**4a**) on colony formation of PANC-1 cells. The graphic shows the mean value of three independent experiments; **$P < 0.01$, when compared with the untreated control group.

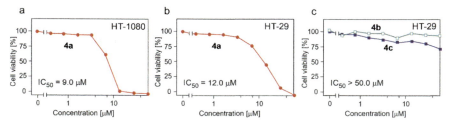

Fig. 91 Impairment of viability of (a) HT-1080 fibrosarcoma and (b) HT-29 colon carcinoma cells by jozimine A$_2$ (**4a**), in a concentration-dependent manner, whereas (c) the related dimers jozibrevines A (**4b**) and B (**4c**) did not show any considerable cytotoxic effects (for the structures of **4a–c**, see Fig. 88).

Table 5 IC$_{50}$ values (μM) of drug-sensitive human lymphoblastic CCRF-CEM and multidrug-resistant CEM/ADR5000 leukemia cells treated with the naphthylisoquinoline dimers **4a**, **4d**, **4e**, **44a**, and **44b**, or with doxorubicin.

Compound	IC$_{50}$ [μM][a] CCRF-CEM	CEM/ADR5000	Degree of resistance[b]
Doxorubicin[366]	0.14 ± 0.004	81.24 ± 6.32	580
Jozimine A$_2$ (**4a**)	0.42 ± 0.02	1.64 ± 0.07	3.92
Jozibrevine C (**4d**)	5.35 ± 0.24	7.59 ± 1.77	1.42
Jozibrevine D (**4e**)	11.5 ± 0.24	16.6 ± 0.75	1.45
Jozilebomine A (**44a**)	4.44 ± 0.21	7.62 ± 1.07	1.71
Jozilebomine A (**44a**)	5.03 ± 0.08	5.16 ± 0.03	1.02

[a]The lymphoblastic leukemia cells were treated with different concentrations of **4a**, **4c**, **4e**, **44a**, and **44b**, or with doxorubicin. Cell viability was assessed by the resazurin assay.[52,366] Mean values and standard deviation of three independent experiments with each six parallel measurements are shown.
[b]The degrees of resistance were calculated by division of the IC$_{50}$ values of the compounds for CEM/ADR5000 cells by the corresponding IC$_{50}$ values for CCRF/CEM cells.

6.1.3.1 Cytotoxicity of jozimine A$_2$ against cervical and pancreatic cancer cells

Jozimine A$_2$ (**4a**) exhibited pronounced cytotoxicity toward HeLa cells in a concentration-dependent manner (Fig. 90a), displaying an IC$_{50}$ value of 0.22 μM[183] and, thus, showing an antiproliferative effect better than 5-fluorouracil (IC$_{50}$ = 13.9 μM), a therapeutic agent widely used for the clinical treatment of cancer.[444–446] Against PANC-1 cancer cells, jozimine A$_2$ (**4a**) exhibited preferential cytotoxicity (PC$_{50}$ = 0.10 μM) under nutrition-deprived conditions (i.e., in nutrient-deficient medium, NDM), but without causing toxicity in normal, nutrient-rich Dulbecco's Eagle's medium (DMEM) (Fig. 90b).[183] As already introduced in Section 5, compounds displaying a selective toxicity to rapidly and aggressively proliferating pancreatic cancer cells at a low state of essential nutrients and oxygen as typical of the hypovascular (austerity) and hypoxic tumor microenvironment,[432–435] but without affecting normal cells, are referred to as antiausterity agents.[427–431] Jozimine A$_2$ (**4a**) significantly inhibited the tolerance of pancreatic cancer cells to nutrition starvation showing a preferential cytotoxicity in a concentration-dependent manner (Fig. 90b), even stronger than that of the positive control arctigenin[428–430] (PC$_{50}$ = 0.83 μM).[183] Thus, this naphthylisoquinoline dimer can be regarded as a promising lead for anticancer drug development based on the antiausterity strategy. Jozimine A$_2$ (**4a**) was further studied for its effects against PANC-1 cell morphology in NDM using the acridine orange (AO)—ethidium bromide (EB) double-staining fluorescence assay.[183] In AO-EB staining, the untreated PANC-1 cells, serving as the control, solely emit the typical bright-green color of the cell-permeable dye AO, as characteristic of living cells with an intact cellular morphology (Fig. 90c, top). The EB reagent, by contrast, is permeable only in dead or dying cells by entering through ruptured cell membranes, giving rise to a predominant red fluorescence. As illustrated by the virtually exclusive red stain (Fig. 90c, bottom), treatment of PANC-1 cells with 5 μM of jozimine A$_2$ (**4a**) caused a complete loss of membrane integrity with membrane blebbing and rounding, followed by disintegration of cell organelles, finally leading to cell death.[183]

Pancreatic cancer cells are highly metastatic, quickly invading distant organs such as liver, stomach, or lungs to form small colonies of cancer cells, which then develop into large tumors.[420–422,424,432] The inhibitory potential of jozimine A$_2$ (**4a**) against colony formation was tested by exposing PANC-1 cells to **4a** at non-cytotoxic concentrations in nutrient-rich DMEM for 24 h.[183] After washing the cells and addition of fresh DMEM (without the test compound), the PANC-1 cells were placed into a CO$_2$ incubator to allow colony formation for 10 days. The area occupied by the colonies was visualized by staining the cells with crystal violet. PANC-1 cells treated with jozimine A$_2$ (**4a**) showed complete inhibition of colony formation even at the lowest tested dose of 1.25 μM, whereas in the absence of **4a** (control), PANC-1 cells grew exponentially to form large numbers of colonies (Fig. 90d).[183]

The jozilebomines A (**44a**) and B (**44b**) likewise exhibited very strong antiproliferative activities against HeLa and PANC-1 cancer cells.[183] The two jozimine-type dimers showed concentration-dependent cytotoxicities against HeLa cervical cells, with IC$_{50}$ values of 1.08 μM for **44a** and 0.68 μM for **44b**, thus displaying slightly lower inhibitory effects compared to jozimine A$_2$ (**4a**) (IC$_{50}$ = 0.22 µM).[183] Following the antiausterity approach, jozilebomine B (**44b**) exerted strong preferential cytotoxicity against PANC-1 cells (PC$_{50}$ = 0.87 µM) under nutrient-deprived conditions, comparable to arctigenin (positive control, PC$_{50}$ = 0.83 µM), while its atropisomer **44a** exhibited slightly lower inhibitory activities (PC$_{50}$ = 2.24 µM). Similar to jozimine A$_2$ (**4a**), **44a** and **44b** severely altered PANC-1 cellular morphology and inhibited colony formation of the cancer cells, when exposed to non-cytotoxic concentrations.[183]

6.1.3.2 Cytotoxicity of jozimine A$_2$ and jozibrevines A and B against fibrosarcoma and colon cancer cells

The P,M,P-configured jozimine A$_2$ (**4a**) exerted pronounced cytotoxicity against HT-29 human colon carcinoma cells in a concentration-dependent manner, exhibiting an IC$_{50}$ value of 12.0 µM after treatment overnight (16 h) (Fig. 91b).[58] The likewise C_2-symmetric P,P,P-configured dimer jozibrevine A (**4b**) and the stereochemically unsymmetric P,P,M-isomer jozibrevine B (**4c**) did not show any considerable cytotoxic properties (IC$_{50}$ > 50 µM) (Fig. 91c).[58] Thus, similar to the aforementioned antiplasmodial activities of the jozimine-A$_2$-type dimers **4a–e** (see Section 6.1.2), their antiproliferative effects toward HT-29 cancer cells again reflected the strong impact of axial chirality on the bioactivities of these quateraryls. Furthermore, jozimine A$_2$ (**4a**) significantly reduced the cell viability of HT-1080 fibrosarcoma cells, displaying an IC$_{50}$ value of 9.0 µM (Fig. 91a).[58]

6.1.3.3 Cytotoxicity of jozimine A$_2$, jozibrevines C and D, and jozilebomines A and B against leukemia cells

Jozimine A$_2$ (**4a**) and the related dimers jozibrevines C (**4d**) and D (**4e**), and jozilebomines A (**44a**) and B (**44b**) were evaluated for their growth-inhibitory activities toward parental drug-sensitive CCRF-CEM leukemia cells and their multidrug-resistant (MDR) P-glycoprotein-overexpressing subline, CEM/ADR5000 (Table 5).[52] The leukemia cells were treated with different concentrations of the respective naphthylisoquinoline dimers in a range of 0.001–100 µM or with the anticancer drug doxorubicin[477,478] as the positive control.

The most potent agent within this series of Dioncophyllaceae-type dimers was jozimine A$_2$ (**4a**). It showed excellent IC$_{50}$ values in the low micromolar range toward CCRF-CEM cells (0.42 µM) and also against the MDR subline CEM/ADR5000 (1.64 µM).[52] CEM/ADR5000 cells, which are highly resistant to the standard drug doxorubicin (ca. 580-fold compared to CCRF-CEM cells), exerted only a low degree of cross-resistance to jozimine A$_2$ (**4a**) (3.9-fold),[52] thus indicating that **4a** might be an efficient inhibitor also against other drug-sensitive and -resistant cancer cells.

Pronounced antiproliferative activities against the drug-sensitive CCFR-CEM and the resistant CEM/ADR5000 leukemia cells were also determined for jozibrevines C (**4d**) and D (**4e**)[52] and for jozilebomines A (**44a**) and B (**44b**) (Table 5). The growth inhibitory potential of **4d**, **4e**, **44a**, and **44b** was distinctly lower than that of jozimine A$_2$ (**4a**). Dose-response curves of the stereochemically unsymmetric dimer jozibrevine C (**4d**) revealed IC$_{50}$ values of 5.35 µM toward CCRF-CEM cells and 7.59 µM against the CEM/ADR5000 subline.

Jozibrevine D (**4e**), built up from two configurationally identical monomeric halves, exerted cytotoxic activities against these two leukemia cell lines two times lower than that of jozibrevine C (**4d**), with IC$_{50}$ values of 11.47 µM against CCRF-CEM cells and 16.61 µM toward the multidrug-resistant CEM/ADR5000 subline.[52]

Jozilebomines A (**44a**) and B (**44b**), consisting of two constitutionally different 7,1'-coupled monomers (Fig. 89), likewise did not reach the excellent growth-inhibitory potential of jozimine A$_2$ (**4a**). In CCRF-CEM cells, an IC$_{50}$ value of 4.45 µM was determined for **44a**, and in CEM/ADR5000 cells the IC$_{50}$ value was found to be 7.62 µM, thus the cytotoxic effects of **44a** were reduced by a factor of over 10 in the drug-sensitive CCRF-CEM cells, and by a factor of nearly 5 in the CEM/ADR5000 subline. Similar results were obtained for jozilebomine B (**44b**), with an IC$_{50}$ value of 5.03 µM against CCRF-CEM cells and 5.16 µM toward the resistant CEM/ADR5000 leukemia cells (Table 5).

The cross-resistance observed for jozibrevines C (**4d**) and D (**4e**)[52] and jozilebomines A (**44a**) and B (**44b**) was low, 1.45-fold (for **4d**), 1.42-fold (for **4e**), 1.71-fold (for **44a**), and 1.02-fold (for **44b**). The IC$_{50}$ values of these four dimers were nearly the same in both cell lines and, thus, in contrast to jozimine A$_2$ (**4a**), they inhibited the cell viability of sensitive and multidrug-resistant leukemia cells with similar efficacies. Again, as already reported for the antiplasmodial activities of jozimine-A$_2$-type dimers (see Section 6.1.2), the absolute configurations at the three stereogenic axes in **4a**, **4d**, **4e**, and in **44a** and **44b** strongly influenced the antileukemic activities of the compounds.[52]

The monomeric halves of jozimine A₂ (**4a**), jozibrevine C (**4d**), and jozibrevine D (**4e**), viz., 4′-O-demethyldioncophylline A (**93a**) and its 7-epimer **93b**, showed only weak (70% cell viability for **93b**) or moderate (40% cell viability for **93a**) effects against the drug-sensitive CCRF-CEM cells at 10 µM.[51] This highlighted the beneficial effect of dimerization and of the resulting overall molecular shape on the cytotoxic potential of these quateraryls.

6.1.4 Semisynthesis of jozimine A₂ and jozibrevine A

The unprecedented molecular scaffolds of 3′,3″-coupled jozimine-A₂-type dimers,[1,3,12,52,58,183,520] with their three consecutive chiral axes and their strong antitumoral[52,58,183] and antiplasmodial[1,3,12,52,58,180,183,520] activities, made these compounds attractive synthetic goals. Previously synthetically prepared 6′,6″-linked dimeric naphthylisoquinoline alkaloids such as the antiviral michellamine A (**5a**),[5,13,14,450,512–514,524–528] widely occurring in *Ancistrocladus* species from Central Africa,[3,5,13,14,176,178] had a significantly less-hindered central biaryl axis, with only two *ortho*-substituents. The central linkage of **5a** had convergently been constructed in a one-step biomimetic oxidative dimerization of its genuine, unprotected monomeric naphthylisoquinoline precursor korupensamine A (**27**), using lead tetraacetate as the oxidizing reagent (Fig. 92).[513,514]

The semisynthesis of jozimine-A₂-type dimers, however, was rewarding with respect to the high steric hindrance to be overcome in the coupling step. Luckily, as described in the following (Fig. 93), phenol-oxidative coupling

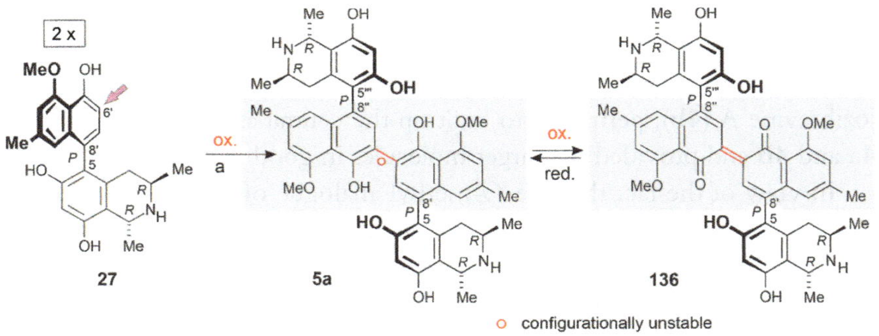

Fig. 92 Smooth, regioselective synthesis of michellamine A (**5a**), by phenol-oxidative coupling of its monomeric precursor, korupensamine A (**27**) using lead tetraacetate as the oxidizing agent, with reversible overoxidation to give the deeply violet-colored diphenoquinone A (**136**). Reaction Conditions: (a) Pb(OAc)₄, BF₃·Et₂O, CH₂Cl₂, 0 °C, 89%.

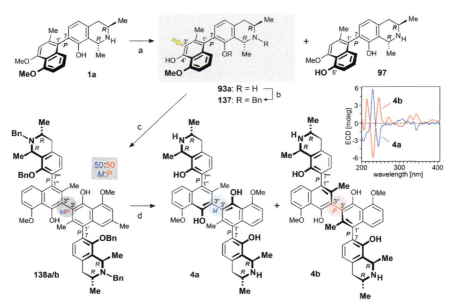

Fig. 93 Semisynthesis of jozimine A₂ (**4a**) and its 3′-atropo-diastereomer, jozibrevine A (**4b**), starting from dioncophylline A (**1a**), by phenol-oxidative coupling of the adequately protected building block N-benzyl-8-O-benzyl-4′-O-demethyldioncophylline A (**137**) using lead tetraacetate as the oxidant. Reaction Conditions: (a) Me₃SiI, CHCl₃, r.t., resolution of **93a** and **97** by preparative HPLC, 49% (for **93a**) and 49% (for **97**); (b) BnBr, Cs₂CO₃, acetone, reflux, 8 h, 92%; (c) Pb(OAc)₄, BF₃·Et₂O, CH₂Cl₂, 0 °C, 10 min, 30%; (d) Pd/C, H₂, EtOH, r.t., 12 h, resolution of **4a** and **4b** by preparative HPLC, 41% (for **4a**) and 44% (for **4b**).

applying the aforementioned lead tetraacetate methodology, yet starting from an adequately protected analogue of 4′-O-demethyldioncophylline A (**93a**), which is the monomeric unit of jozimine A₂ (**4a**) and its 3′-epimer jozibrevine A (**4b**), permitted to built up the central axis of the two dimers **4a** and **4b** and provided the target molecules in good chemical yields.[3,12]

In view of the fact that the O-methyl analogue of **93a**, dioncophylline A (**1a**), is easily available in sufficient amounts by isolation from *Triphyophyllum peltatum* (Dioncophyllaceae)[5,6,8] or from *A. abbreviatus*,[5,63] **1a** was chosen as starting material for a semisynthetic access to **4a** and **4b**. O-Demethylation of isolated natural dioncophylline A (**1a**) using trimethylsilyl iodide in chloroform provided the two regioisomeric alkaloids, 4′- and 5′-O-demethyldioncophylline A (**93a** and **97**) in almost quantitative yield as a 1:1-mixture, which was resolved by preparative HPLC on a C₁₈ reversed-phase column.[12] With respect to the two phenolic oxygen functions at C-8 and C-4′ and the secondary amine group in

4′-O-demethyldioncophylline A (**93a**), undesired by-products had to be expected for the targeted phenol-oxidative coupling step. Selective attachment of O- and N-benzyl protective groups in the isoquinoline portion exclusively, left free only the required hydroxy function at C-4′ next to the desired coupling site. Biomimetic-type phenol-oxidative coupling of the N,O-dibenzyl derivative **137** with lead tetraacetate as the oxidizing reagent finally yielded the two anticipated 3′,3″-linked atropo-diastereomers, **138a** and **138b**, as a 1:1-mixture.[12]

Different from all previous oxidative couplings to sterically less-hindered naphthylisoquinoline dimers such as michellamine A (**5a**), the typical deep-violet color of a transiently over-oxidized diphenoquinone intermediate (see compound **136** in Fig. 92)[5,385,513,514,524,525] was not observed, apparently as a consequence of the high steric demand at the newly generated 3′,3″-linked central axis not permitting a coplanar array of the central binaphthalene unit.[12] The inseparable mixture of **138a** and **138b** was directly submitted to a hydrogenolytic cleavage of the O- and N-benzyl protecting groups, affording the target molecules jozimine A_2 (**4a**) and jozibrevine A (**4b**) (Fig. 93), which were eventually resolved and purified by preparative HPLC. The semisynthesis of **4a** and **4b** firmly corroborated the full absolute stereostructures of the two dimers.[12] Expectedly, the only difference between **4a** and **4b**—their absolute configuration at the central biaryl axes—was clearly evidenced by their ECD spectra, which were virtually opposite to each other (Fig. 93).[12] Given the availability of dioncophylline A (**1a**) by atropo-selective total synthesis following the lactone concept[5,6,9,76,364,378] (see Fig. 42 in Section 4.7.3), the route presented here simultaneously constitutes a formal total synthesis of jozimine A_2 (**4a**) and jozibrevine A (**4b**).

6.2 Ancistrobreviquinones A and B, novel-type naphthylisoquinoline alkaloids with an *ortho*-quinoid unit in the naphthalene part

Metabolic profiling of a less polar fraction of the root bark extract of *A. abbreviatus* led to the discovery of two new compounds possessing an intense orange-yellow color.[56] 1D and 2D NMR measurements suggested the new metabolites to belong to the subclass of 5,1′-coupled naphthylisoquinoline alkaloids, but—in contrast to the series of 5,1′-linked compounds previously identified in the roots, twigs, and leaves[5,54,56,57,63] (see Section 5.1)—the new root bark constituents[56] possessed an unusual naphthalene part (Fig. 94).

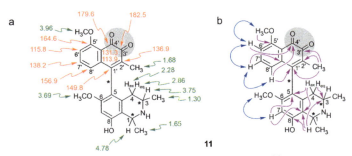

Fig. 94 Structure elucidation of ancistrobreviquinone A (**11**),[56] the first alkaloid with an *ortho*-naphthoquinone coupled to an isoquinoline: (a) selected ^{1}H (in green) and ^{13}C (in orange) NMR data (in methanol-d_4, δ in ppm) and (b) decisive HMBC (purple single arrows) and NOESY (double blue arrows) interactions relevant for the constitution of **11**.

The first of these isolated compounds had a molecular formula of $C_{24}H_{26}NO_5$, corresponding to an m/z value of 408.18034 $[M+H]^+$, as identified by HRESIMS. The IR spectrum displayed a characteristic carbonyl signal at 1678 cm^{-1} and the ^{13}C NMR data showed two downfield-shifted resonances at δ_C 179.6 and 182.5, typical of carbonyl C-atoms, thus indicating the presence of a diketo entity (Fig. 94).[56] The extremely upfield-shifted signal of Me-2' (δ_H 1.68) revealed the methyl group to be in close proximity to the diketo unit and, thus, to be located in the naphthalene portion. HMBC cross-peaks from Me-2' to C-3' (δ_C 182.5) and from H-8' and H-6' to C-4' (δ_C 179.6) evidenced the two keto functions to be *ortho* to each other, suggesting that the new compound possessed a 2'-methyl-5'-methoxy-3',4'-naphthoquinone portion.[56] This was in agreement with the fact that only four protons were monitored in the aromatic part of the ^{1}H NMR spectrum instead of five as observed in the case of all the other monomeric C,C-coupled naphthylisoquinoline alkaloids previously isolated from *A. abbreviatus*.[5,56–58,63–65] The presence of an *ortho*-naphthoquinone subunit was also corroborated by a series of downfield-shifted carbon signals in the naphthalene unit, for example for C-1' (δ_C 149.8 instead of δ_C 120–125), C-5' (δ_C 164.6 instead of δ_C 156–158), C-6' (δ_C 115.8 instead of δ_C 107–110), and C-7' (δ_C 138.2 instead of δ_C 126–128).[56] The new alkaloid possessed two methoxy groups, resonating at δ_H 3.69 and 3.96. They were deduced to be located at C-6 and C-5', due to their characteristic ^{1}H NMR shifts and their NOESY interactions with H-7 and H-6', respectively. The fifth oxygen function thus had to be a free hydroxy group at C-8 (Fig. 94).

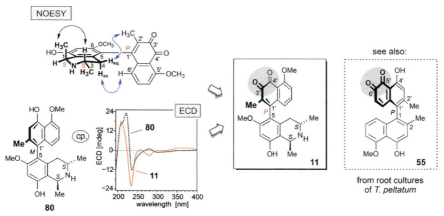

Fig. 95 (a) Attribution of the configuration of ancistrobreviquinone A (**11**)[56] at the biaryl axis relative to the stereogenic centers in the tetrahydroisoquinoline portion, evidenced by NOESY interactions; (b) confirmation of the absolute axial configuration of **11**, by comparison of its ECD spectrum with that of its putative biosynthetic precursor, ancistrobrevine K (**80**),[58] previously likewise isolated from the roots of *A. abbreviatus* and, for comparison, the structure of the known naphthylnaphthoquinone triphyoquinone A (**55**),[311] which likewise possesses an *ortho*-quinone entity (but no isoquinoline portion), isolated from root cultures of the Dioncophyllaceae liana *Triphyophyllum peltatum*.

The relative configuration at C-1 versus C-3 in the tetrahydroisoquinoline unit of the new alkaloid was deduced to be *trans* from specific NOE correlations between Me-1 and H-3. By ruthenium-mediated oxidative degradation, the absolute configuration was assigned to be *S* both at C-1 and C-3 (Fig. 95).[56] Long-range NOESY interactions between H_{eq}-4 (which is above the "plane") and Me-2' and between H_{ax}-4 (which is below) and H-8' suggested that the large naphthoquinone part should be directed down relative to the "plane" of the isoquinoline unit, and the axis should, hence, be *P*-configured. This was further supported by the ECD spectrum of the new alkaloid, which was in accordance with that of the co-occurring, likewise 5,1'-coupled ancistrobrevine K (**80**), despite the different chromophore (naphthoquinone versus naphthalene).[56,58] For formal reasons, however, the two alkaloids had opposite descriptors, according to the Cahn-Ingold-Prelog priority rules. Hence, the new alkaloid possessed the stereostructure **11** as presented in Fig. 95. Referring to its quinoid structure and its discovery in *A. abbreviatus*, it was named ancistrobreviquinone A.[56] Prior to this work, the binaphthoquinone triphyoquinone A (**55**),[311] exclusively produced by root cultures of the

West African liana *Triphyophyllum peltatum* (Dioncophyllaceae), but not found in the plant itself, had been the only other compound also possessing an *ortho*-quinone entity, but—in contrast to **11**—located in the methyl-free ring system (Fig. 95).

Another—likewise dark orange-yellow—new alkaloid, with a molecular formula of $C_{25}H_{27}NO_5$ (*m/z* 422.1956) (as deduced from HRESIMS) was isolated as a minor constituent from the roots of *A. abbreviatus*.[56] The compound showed ^1H and ^{13}C NMR shifts comparable to those of ancistrobreviquinone A (**11**), suggesting the presence of a further 5,1′-coupled quinoid alkaloid, again consisting of a 3′,4′-naphthoquinone coupled to a tetrahydroisoquinoline portion, but—in contrast to **11**—equipped with an additional methoxy function at C-8 in the isoquinoline unit (compared to the free OH group as in **11**). Again, the relative configuration at C-1 versus C-3 was deduced to be *trans* from specific NOE interactions, and the absolute configurations were attributed to be 1*S*,3*S* by ruthenium-mediated oxidative degradation.[56] Like for ancistrobreviquinone A (**11**), long-range NOE interactions across the biaryl linkage established the axis to be *P*-configured. This was in agreement with the nearly identical ECD spectrum of the new alkaloid compared to that of its parent compound **11** (data not shown). The new alkaloid **131** was the 8-*O*-methyl analogue of **11**; it was named ancistrobreviquinone B (Fig. 96).[56]

From the Chinese liana *Ancistrocladus tectorius*, two likewise oxygenated, yet 7,3′-coupled and *para*-quinoid naphthylisoquinoline alkaloids have been isolated, ancistrotectoquinones A (**58**) and B (**117**).[1,253] In contrast to

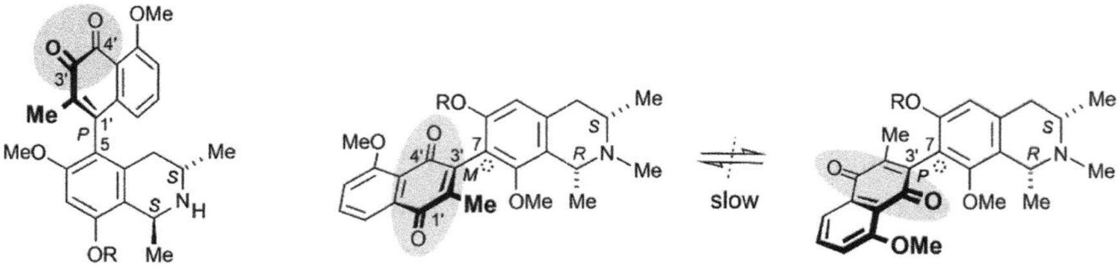

Fig. 96 Quinoid naphthylisoquinoline alkaloids discovered in nature: the new ancistrobreviquinones A (**11**) and B (**131**)[56] from the roots of *A. abbreviatus* and the 7,3′-coupled ancistrotectoquinones A (**58/b**) and B (**117a/b**),[1,253] possessing a 1,4-naphthoquinone portion coupled to an isoquinoline moiety, and occurring as pairs of configurationally semi-stable, slowly interconverting atropo-diastereomers in the Chinese liana *A. tectorius*.

the ancistrobreviquinones A (**11**) and B (**131**),[56] which show complete configurational axial stability, the ancistrotectoquinones A (**58a/b**) and B (**117a/b**) occur as pairs of configurationally semi-stable—and thus slowly interconverting—atropo-diastereomers (Fig. 96), as a consequence of the moderate steric hindrance next to the biaryl axis, but also as a result of mesomeric stabilization of the—nearly coplanar—transition state of axial rotation by the M-effect of the carbonyl group at C-1′.[1,253]

6.3 Ancistrobrevinium A, the first—and so far only—Cationic C,C-coupled naphthylisoquinoline alkaloid known from nature

In the most polar fraction of the root bark extract of *A. abbreviatus*, a fully dehydrogenated alkaloid was discovered.[60] It was found to possess a cationic molecular formula of $C_{27}H_{30}NO_4^+$, as established by HRESIMS (m/z 432.2138 [M]$^+$) and by ^{13}C NMR analysis. The ^1H NMR spectrum exhibited signals for three C-methyl groups (at δ_H 2.27, 2.86, and 3.36), for four O-methyl groups (δ_H 3.33, 3.86, 3.94, and 3.97), and for six aromatic protons (δ_H 6.64, 6.79, 6.97, 7.26, 7.38, and 8.04) (Fig. 97). The spin pattern

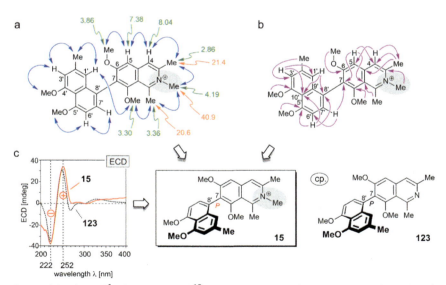

Fig. 97 (a) Selected ^1H (in green) and ^{13}C (in orange) NMR data, and NOESY (blue double arrows) interactions, and (b) HMBC (purple single arrows) of the new cationic C, C-coupled naphthylisoquinoline alkaloid ancistrobrevinium A (**15**); (c) assignment of the absolute axial configuration of **15** by comparison of its ECD spectrum with that of its known[54] *N*-demethylated analogue ancistrobreveine C (**123**).

of the latter, four singlets and two doublets (arising from two protons in *ortho*-position as deduced from the coupling constant, ca. 8.0 Hz), the NOESY interactions between H-4 and H-5, and the HMBC cross-peaks from H-5 and H-7′ to C-7, jointly established a 7,8′-coupling type of the new alkaloid.[60] The absence of resonances for the diastereotopic geminal protons at C-4, and, in turn, the appearance of an additional, downfield-shifted aromatic singlet at δ_H 8.04 suggested the presence of a non-hydrogenated isoquinoline portion. This was in accordance with the lacking quartet of H-1 around δ_H 4.20, and the missing multiplet of H-3 in the region between δ_H 3.20 and 4.00, and with the absence of two doublets for the aliphatic methyl groups at C-1 and C-3 around δ_H 1.10 and 1.90 (see Section 4.1). Such signals would have been expected for a tetrahydroisoquinoline, but were absent for the new alkaloid.[60]

The ^{1}H and ^{13}C NMR spectra were nearly identical to those of the known non-hydrogenated, yet not cationic naphthylisoquinoline ancistrobreveine C (**123**),[54] which had earlier been identified as a trace alkaloid in the roots of *A. abbreviatus* (see Section 5.6). The most significant difference between the NMR data of the new compound, as compared to those of **123**, was the presence of an additional methyl group resonating at δ_H 4.19 and corresponding to a carbon signal at δ_C 40.9, as evidenced by HSQC measurements, suggesting it to be part of an *N*-Me entity.[54,60] This was confirmed by NOESY interactions between the latter and both, Me-1 (δ_H 3.36) and Me-3 (δ_H 2.86), and by HMBC cross-peaks of the protons of this *N*-methyl group to C-1 and C-3 (Fig. 97). Consequently, the new alkaloid was the *N*-methyl analogue of **123**, and, thus, a cationic compound. Its absolute configuration at the biaryl axis was attributed to be *P* based on its ECD spectrum, which showed a clear positive couplet, with a distinct first maximum at 252 nm and a second, negative peak at 222 nm, similar to that of its likewise *P*-configured parent compound ancistrobreveine C (**123**).[54,60] Thus, the new alkaloid had the full stereostructure **15** and was named ancistrobrevinium A.[60] The natural counteranion of **15** is as yet unknown[60]—as for all other cationic naphthylisoquinoline alkaloids, in particular for all *N,C*-coupled representatives.[1,2,16,255,266,267,299] The genuine anion is probably not firmly associated to the alkaloid and may thus be washed out during the isolation process.

In the field of "normal" isoquinolines (which are all derived from aromatic amino acids), some few *N*-methylisoquinolinium alkaloids are known, like the simple isoquinolinium salt **139**[529] and tetradehydroreticuline (**140**)[530,531] (Fig. 98). From such established examples, however, ancistrobrevinium A

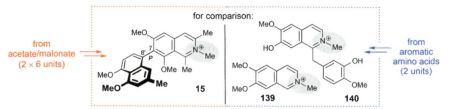

Fig. 98 N-Methylisoquinolinium alkaloids found in nature: ancistrobrevinium A (**15**),[60] built up from acetate-malonate units and the isoquinolinium salt **139**[529] and tetradehydroreticuline (**140**),[530] the latter two originating from aromatic amino acids.

(**15**) differs by its polyketidic origin and its axial chirality. It is the first example of a fully dehydrogenated C,C-coupled N-methylnaphthylisoquinolinium alkaloid found in nature.[60]

The structure of ancistrobrevinium A (**15**) raised the question whether its biosynthetic precursor is the respective N-methyltetrahydroisoquinoline congener, from which **15** would be formed by twofold dehydrogenation, or whether **15** originated from an N-methylation of its already fully dehydrogenated analogue **123**. In the latter case one might wonder if **15** was a genuine natural product or just the spontaneously, non-enzymatically formed N-methylation product of its respective N-unsubstituted analogue, ancistrobreveine C (**123**). Then, however, one would have expected that also other, structurally similar N-methylnaphthylisoquinolinium alkaloids would have been analogously formed in *A. abbreviatus*. But, although a total of six non-hydrogenated—thus non-charged—naphthylisoquinolines had already been isolated from the roots of the liana earlier[54] (see Section 5.6), none of their possible cationic N-methylation products had ever been discovered in *A. abbreviatus*. Also in other, related plant species, fully dehydrogenated naphthylisoquinolines had previously been found[1,5,53,176,252,273–275,291,299,403] (even four examples in the Congolese liana *A. likoko*[53,176]), but never had their cationic N-methyl analogues been detected.

Ancistrobrevinium A (**15**) showed only weak inhibitory activity (IC$_{50}$ = 50.6 µM) against A549 lung cancer cells, compared to the standard drug doxorubicin (IC$_{50}$ = 9.45 µM).[60] The low cytotoxicity of **15** seems to be due, in part, to the positive charge and, thus, polar nature of **15**, which may hinder it from penetration into the cancer cells through the lipophilic phospholipid bilayer. This assumption is in agreement with the ca. two times better IC$_{50}$ value of 24.5 µM for the likewise fully dehydrogenated, but not N-methylated ancistrobreveine C (**123**) as compared to that of **15**.[60]

Summarizing, ancistrobrevinium A (**15**)[60] isolated from the roots of *A. abbreviatus*, is structurally unique within a whole series of more than 280 naphthylisoquinoline alkaloids known so far from nature.[1–7] It combines, for the first time, the structural motifs of a fully dehydrogenated isoquinoline with the presence of an *N*-methyl group. Although several non-saturated naphthylisoquinolines[1,5,53,176,252,273–275,291,299,403] are known, none of them is *N*-methylated. Likewise isolated have been *N*-methylated alkaloids from Ancistrocladaceae and Dioncophyllaceae lianas,[1–7] but they differ from **15** by their heterocyclic portions, which have so far been always fully hydrogenated. This is also true for gentrymine B (**50**)[290,312] (Fig. 99), which is cationic, but it is again a tetrahydroisoquinolinium salt, obtaining its positive charge from a twofold *N*-methylation. This naphthalene-devoid compound is the only *N*-quaternary isoquinolinium alkaloid found in any of the Ancistrocladaceae or related Dioncophyllaceae plants as yet investigated. The only partially or totally unsaturated naphthylisoquinoline alkaloids with a permanent positive charge on the nitrogen are the *N,C*-coupled naphthylisoquinolines (12 as yet known representatives,[1,2,16,255,266,267,299] discovered in various Asian and African *Ancistrocladus* lianas) such as ancistrocladinium A (**7a**)[2,16] (Fig. 99), but they do not have an *N*-methyl group.

6.4 1-*nor*-8-*O*-Demethylancistrobrevine H, the only known example of a naphthylisoquinoline alkaloid lacking the otherwise generally present methyl function at C-1

During the past years, intense phytochemical studies on *Ancistrocladus* species from Asia and Africa led to the discovery of a plethora of naphthylisoquinoline alkaloids.[1–7] Despite their structural diversity, however, all of them have in common that they are equipped with a methyl substituent at C-1. The only exceptions are the *N,C*-coupled ancistrocyclinones A

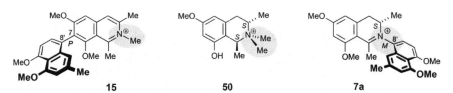

Fig. 99 Cationic metabolites discovered in *Ancistrocladus* lianas from Africa and Asia: the fully dehydrogenated *C,C*-coupled *N*-methylnaphthylisoquinoline alkaloid ancistrobrevinium A (**15**),[60] the *N,N*-dimethylated, i.e., quaternary tetrahydroisoquinoline gentrymine B (**50**),[290,312] and the *N,C*-coupled naphthyldihydroisoquinolinium alkaloid ancistrocladinium A (**7a**).[1,2,16]

missing methyl function at C-1:

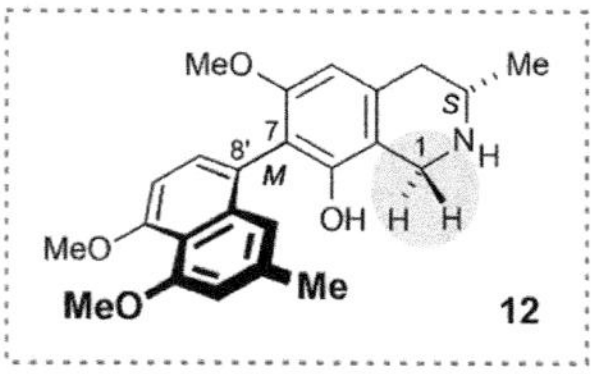

the methyl group at C-1 as part of a polycyclic ring system:

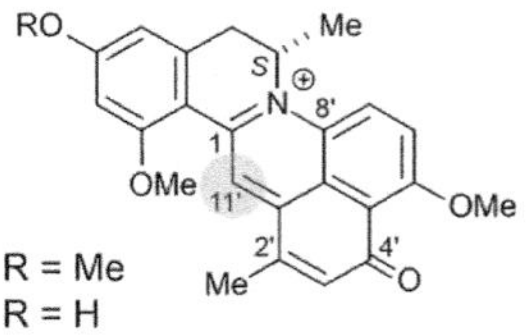

Fig. 100 Naphthylisoquinoline alkaloids differing from all the other over 280 known naphthylisoquinoline alkaloids by lacking a methyl function at C-1: the novel-type 1-*nor*-8-*O*-demethylancistrobrevine H (**12**)[55] from *A. abbreviatus* and the pentacyclic ancistrocyclinones A (**37**) and B (**38**)[255] from *A. tectorius*.

(**37**) and B (**38**)[1,2,255] (Fig. 100) from the Chinese liana *A. tectorius*, where the methyl group at C-1 is now part of a polycyclic ring system.

The discovery of the first naphthylisoquinoline alkaloid in the roots of *A. abbreviatus* lacking the otherwise generally present methyl group at C-1 in its tetrahydroisoquinoline subunit was unprecedented.[55] Being the 1-*nor* derivative of the (yet unknown) 8-*O*-methyl analogue of ancistrobrevine H (**8**),[57] which had likewise been isolated from the roots of *A. abbreviatus*, this new metabolite **12** was named 1-*nor*-8-*O*-demethylancistrobrevine H[55] (Fig. 100).

HRESIMS analysis of the new root metabolite **12** gave a molecular formula of $C_{24}H_{27}NO_4$, corresponding to an m/z value of 394.20144 $[M+H]^+$. 1D and 2D NMR analysis showed signals typical of a 7,8′-coupled naphthylisoquinoline alkaloid equipped with the usual 4′,5′-dimethoxy-2′-methyl substituted naphthalene portion.[55] The 7,8′-coupling site was established by the presence of five aromatic protons, three singlets and an AB spin system of two adjacent protons, and NOESY correlations in the series {OMe-6 ↔ H-5 ↔ H-4_{eq} ↔ H-4_{ax}}, and HMBC cross-peaks from both H-5 and H-7′ to C-7, and from both H-1′ and H-6′ to C-8′ (Fig. 101a). Furthermore, the ^1H NMR data of **12** displayed a three-proton doublet (δ_H 1.51) and a multiplet (δ_H 3.64), characteristic of the CHMe entity at C-3 in the tetrahydroisoquinoline portion.[55] Likewise observed, as usual, were the signals of the diastereotopic geminal protons at C-4 (δ_H 2.96 and 3.18). A most notable difference, compared to other naphthylisoquinoline alkaloids, however, was the appearance of two doublets resonating at δ_H 4.16 and 4.38 (Fig. 101b), indicating the presence of two likewise diastereotopic protons at C-1.[55] This structural peculiarity

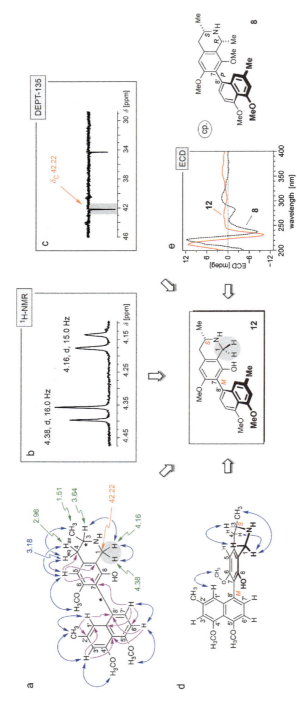

Fig. 101 (a) ^{1}H NMR (in green) and ^{13}C NMR (in orange) data (in methanol-d_4, δ in ppm), key NOESY (double blue arrows) and HMBC (single purple arrows) interactions indicative of the constitution of the 7,8'-coupled 1-*nor*-8-*O*-methylancistrobrevine H (**12**); (b) presence of two diastereotopic protons at C-1 as deduced from their signals at δ_H 4.16 and 4.38 in ^{1}H NMR, (c) confirmed by the signal at δ_C 42.22 in the DEPT-135 spectrum, and (d) by their NOESY interactions with H-3, Me-3, and H-4$_{ax}$; (e) assignment of the absolute axial configuration of **12**, by comparison of its ECD spectrum with that of the structurally closely related, known[57] alkaloid ancistrobrevine H (**8**).

of **12** was highly remarkable, because the isolated metabolite was the first such alkaloid found in nature that lacks the generally present methyl group at C-1 (which is normally monitored around δ_H 1.6–1.8; for comparison, see Refs. 11, 35, 57, 58). In the new compound **12**, the usual CHMe unit at C-1 is replaced by a CH$_2$ unit, resonating in the DEPT-135 spectrum at δ_C 42.22.[55] The position of the two diastereotopic protons at C-1 (Fig. 101c) was further corroborated by their NOESY cross-peaks with H-3, Me-3, and H-4$_{ax}$ (Fig. 101d).[55]

The missing methyl group at C-1 reduced the number of stereogenic elements in **12** to only two compared to conventional naphthyltetrahydroisoquinolines. The absolute configuration of the stereocenter at C-3 was determined to be *S* by ruthenium-mediated oxidative degradation, which afforded *S*-3-aminobutyric acid.[55] The ECD spectrum of **12** was in good agreement with that of the closely related, *P*-configured alkaloid ancistrobrevine H (**8**). Therefore, **12** and **8** had the identical absolute axial configuration, but for formal reasons, according to the Cahn–Ingold–Prelog rules, **12** and **8** had opposite descriptors.[55]

Summarizing, the discovery of the first 1-*nor*-naphthylisoquinoline alkaloid, 1-*nor*-8-*O*-demethylancistrobrevine H (**12**),[55] from *A. abbreviatus* underlines the "synthetic creativity" of this West African *Ancistrocladus* liana. The putative role of **12** as a biosynthetic intermediate in the course of the formation of follow-up metabolites such as *seco*-type naphthylisoquinoline alkaloids from usual representatives is presented in Section 6.5.

6.5 Ancistrosecolines A–F, the first ring-opened naphthylisoquinoline alkaloids, discovered in the roots of *Ancistrocladus abbreviatus*

From the root bark of *A. abbreviatus*, six minor metabolites, exhibiting a novel structural motif, were isolated. All of these compounds have in common that they consist of a naphthalene portion and a *meta*-dioxygenated phenyl ring system, equipped with a chiral, but flexible *N*-methyl-2-aminopropyl side chain. They were the first naphthylphenylisopropylamines discovered in nature. In all these *seco*-type compounds, named ancistrosecolines A–F (**13, 141, 132,** and **142–144**) (Fig. 102), the tetrahydroisoquinoline ring is cleaved, with loss of the entire Me–C-1–H unit.[55] They are typical Ancistrocladaceae-type alkaloids, with an oxygen function at C-6 and *S*-configuration at C-3, belonging to the subclasses of 7,1'- and 7,8'-coupled naphthylisoquinoline alkaloids. Likewise intriguing, from a stereochemical point of view, is the fact that ancistrosecolines E

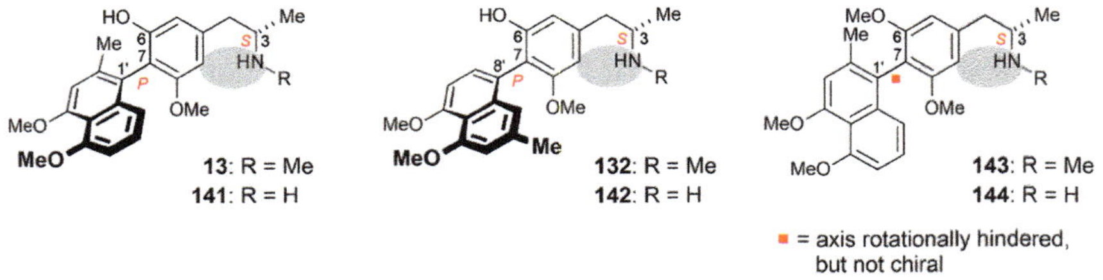

Fig. 102 The first *seco*-type naphthylisoquinoline alkaloids[55] found in nature, named ancistrosecolines A (**13**), B (**141**), C (**132**), D (**142**), E (**143**), and F (**144**), isolated from the roots of *A. abbreviatus* from Coastal West Africa.

(**143**) and F (**144**) are the very first—and so far only—naphthylisoquinoline-derived alkaloids possessing a rotationally hindered, but non-stereogenic biaryl axis. Among the meanwhile over 280 known[1–7] naphthylisoquinoline alkaloids and their natural congeners, **143** and **144** are thus the only ones equipped with oxygen functions at C-6 and C-8 that are not constitutionally heterotopic to each other, but diastereotopic.[55]

HRESIMS analysis of the first new metabolite gave a molecular formula of $C_{24}H_{29}NO_4$, corresponding to a protonated molecular ion peak at m/z 396.21793 $[M+H]^+$. The aliphatic region of the ^1H NMR spectrum showed signals similar to those of a naphthyldihydroisoquinoline (Fig. 103a), with a three-proton doublet (δ_H 1.34) and a downfield-shifted three-proton singlet (δ_H 2.77), as characteristic of the presence of two methyl groups in the (presumed) isoquinoline subunit.[55] Substantially different, however, was the appearance of six aromatic protons (instead of five, as expected in the case of a dihydroisoquinoline) and the observation of an HMBC interaction from the aliphatic methyl group resonating at δ_H 2.77 to C-3 (δ_C 57.7). Such a long-range coupling had never before been detected in any naphthyldihydroisoquinoline alkaloid.[55] The methyl group at C-1 normally displays HMBC cross-peaks to C-1 and C-8a.[10,53,56,57,61,176] Likewise unprecedented was a NOESY correlation between Me-3 (δ_H 1.34) and the aromatic proton (δ_H 6.51) at C-5.[55] As part of a less flexible heterocyclic ring system, the distance between the methyl group at C-3 and H-5 would have been too far for such an interaction to be anticipated. The unexpected HMBC and NOESY cross-peaks, while not fitting with the presence of a rigid dihydroisoquinoline unit (Fig. 103a), gave conclusive evidence of the presence of an open-chain, non-cyclic system in the new alkaloid, as a consequence of the loss of the entire C-1–H–Me entity[55] (Fig. 103b).

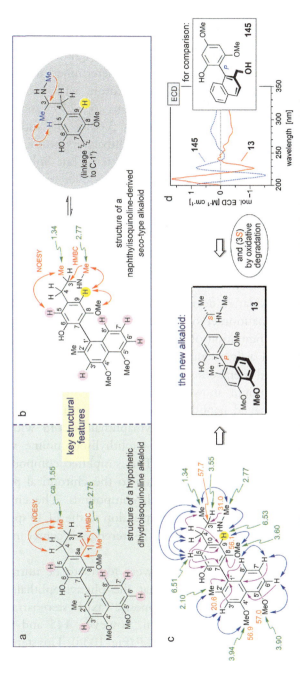

Fig. 103 Decisive 1D and 2D NMR data typical of the structural elements of (a) a naphthyldihydroisoquinoline with five aromatic protons (underlaid in pink) and a rigid heterocyclic dihydropyrido ring system, in comparison to (b) a seco-type alkaloid with six aromatic protons (the "additional" one highlighted in yellow) and with the methyl group at C-3 being part of a flexible *N*-methyl-2-aminopropyl side chain, the isoquinoline ring having been cleaved with concomitant loss of the entire C-1–H–Me entity; (c) ^{1}H NMR (in orange), ^{13}C NMR (in green), HMBC interactions (single arrows in pink), and NOESY cross-peaks (double arrows in blue) establishing the constitution of the naphthalene unit and the *seco*-isoquinoline part of ancistrosecoline A (**13**); (d) assignment of the absolute configuration at the stereogenic center of **13** as *S* by oxidative degradation, and attribution of the absolute axial configuration as *P* by comparison of the ECD spectrum of **13** with that of the related, but stereochemically simplified phenylnaphthalene **145** with known[532,533] absolute configuration.

In the case of such a ring-cleaved *seco*-type structure, the methyl group at C-3 would be part of a flexible side chain equipped with an *N*-methyl group, and the position of the sixth aromatic proton had to be at C-9 (highlighted in yellow in Fig. 103b). This structural assignment was in accordance with the coupling constant of ca. 1.42 determined for H-9 (δ_H 6.53), which was typical of a *meta*-coupled proton, with NOESY interactions of H-9 both with the *N*-methyl protons (δ_H 2.77) and the adjacent methoxy group at C-8 resonating at δ_H 3.60.[55] 1D and 2D NMR analysis revealed the presence of a typical naphthalene portion with the usual 4′,5′-dimethoxy-2′-methyl substitution pattern, but linked to the *seco*-isoquinoline molecular moiety via a biaryl axis over to C-7, as evident from NOESY cross-peaks between H-5 and the two diastereotopic geminal protons at C-4. In the naphthalene half, the coupling position was assigned to be C-1′ due to HMBC interactions from Me-2′, H-8′, and H-3′ to the quaternary carbon atom C-1′ and NOESY correlation sequences in the series {OMe-5′ ↔ H-6′ ↔ H-7′ ↔ H-8′} and {Me-2′ ↔ H-3′ ↔ OMe-4′} (Fig. 103c).[55]

Ruthenium-mediated oxidative degradation[1,3,70] of the new *seco*-type alkaloid, providing the *S*-enantiomer of *N*-methyl-3-aminobutyric acid, established the absolute configuration of the stereogenic center at C-3 to be *S*.[55] A merely empirical assignment of the absolute configuration at the biaryl axis of the isolated metabolite was challenging due to its novel *seco*-type structure, which might show a different chiroptical behavior compared to intact, not cleaved naphthylisoquinolines. Fortunately, the structurally closely related, axially chiral phenylnaphthalene **145** with a chromophore similar to that of the new *seco*-naphthylisoquinoline was available from previous synthetic work.[532,533] That simplified compound **145** was ideally suited for ECD comparison due to the chiroptical predominance of the biaryl chromophore in both compounds. The chiral aromatic system provides a distinctly larger contribution to the ECD spectrum of the new *seco*-type plant metabolite than the stereogenic center in its conformationally flexible side chain.

The ECD curves of the two compounds were found to be mirror-image-like to each other, thus evidencing that the phenylnaphthalene **145** and the ring-opened alkaloid had opposite axial stereoarrays. According to the Cahn-Ingold-Prelog notation, however, **145** and the *seco*-type natural product had the same descriptor, *P* (Fig. 103d). This stereochemical assignment was confirmed by quantum-chemical ECD calculations for the *P*-atropo-diastereomer.[55] The experimental ECD spectrum

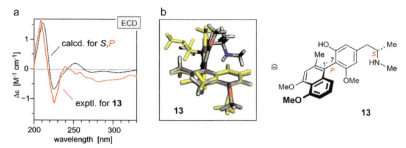

Fig. 104 (a) Confirmation of the absolute axial configuration of ancistrosecoline A (**13**) by the interplay of experimental and computational ECD investigations, by quantum-chemical calculation of the ECD spectrum of the *P*-atropo-diastereomer at the TDωB97XD3/def2-TZVP/B3LYP-D3/def2-TZVP level; (b) matched structures of two DFT-calculated conformers of **13**.

of the new alkaloid nearly perfectly matched the ECD curve computed by TDωB97XD3/def2-TZVP/B3LYP-D3/def2-TZVP calculations (Fig. 104a). The new *seco*-type naphthylisoquinoline alkaloid thus had the full absolute stereostructure **13** as shown in Figs. 103 and 104. It was named ancistrosecoline A.[55]

NMR and MS analysis of a second new metabolite isolated from the roots of *A. abbreviatus* revealed the presence of a likewise 7,1′-coupled *seco*-type naphthylisoquinoline, which, in contrast to ancistrosecoline A (**13**), displayed a free, unsubstituted amino function. Its oxidative degradation afforded *S*-aminobutyric acid and its ECD spectrum was virtually identical to that of the parent compound **13** (for the ECD spectrum of **13**, see Fig. 104a), which established the new metabolite to possess the stereostructure **141**, as shown in Fig. 102. It was henceforth named ancistrosecoline B.[55]

Preparative HPLC of the root bark extract of *A. abbreviatus* furnished two further new trace metabolites, again displaying the typical molecular features of *seco*-type naphthylisoquinoline alkaloids. As already in the case of ancistrosecoline A (**13**), one of the new compounds exhibited a free proton singlet at δ_H 2.76 in the ^{1}H NMR spectrum, thus indicating the presence of an *N*-methyl group, while the other new *seco*-naphthylisoquinoline had a free amino function, as in the case of ancistrosecoline B (**141**).[55] According to NMR, the molecular halves of the new alkaloids were identical to those of **13** and **141**, with a 4′,5′-dimethoxy-2′-methylnaphthalene part and a phenylisopropylamine moiety bearing a free hydroxy group at C-6 and a methoxy function at C-8. 2D NMR measurements revealed that the new *seco*-type compounds differed from **13** and **141** in the position of their biaryl axis. Long-range HMBC interactions from H-1′ and H-6′ to C-8′ and from H-7 and H-5 to C-7 established C-8′ to be the axis-bearing atom in the naphthalene part and C-7 in the *seco*-isoquinoline unit (Fig. 105a).

Oxidative degradation[1,3,70] again determined the configuration at C-3 in the two new alkaloids to be *S*.[55] Their ECD spectra were virtually identical to those of the related 7,1′-coupled ancistrosecolines A (**13**) and B (**141**) (Fig. 105b and c). The two new root metabolites were thus attributed to be *P*-configured at the axis, too, with respect to the fact that, although displaying a different coupling type—7,8′ instead of 7,1′—they actually differed from **13** and **141** only by the chiroptically less decisive position of the methyl group in the naphthalene part. The new *N*-methylated compound was named ancistrosecoline C (**132**) and its *N*-demethylated analogue was given the name ancistrosecoline D (**142**).[55]

Isolation work on the roots of *A. abbreviatus* led to the identification of two further 7,1′-coupled *seco*-type naphthylisoquinoline alkaloids.[55]

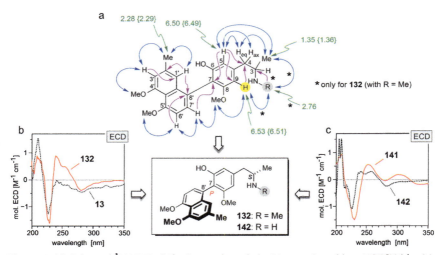

Fig. 105 (a) Selected ¹H NMR shifts (in methanol-d_4, δ in ppm) and key NOESY (double blue arrows) and HMBC (single purple arrows) interactions indicative of the constitution of the 7,8′-linked *seco*-naphthylisoquinolines, ancistrosecolines C (**132**) and D (**142**), with the values of the chemical shifts of **142** that were different from those of **132** given in {}; assignment of the absolute axial configuration of (b) **132** and (c) **142** by comparison of their ECD spectra with those of the related *seco*-compounds, which differ only by their 7,1′-coupling site, namely in the case of **132**, with ancistrosecoline A (**13**) and, in the case of **142**, with the *N*-demethylated ancistrosecoline B (**141**).

The most significant difference of their NMR data compared to those of ancistrosecolines A (**13**) and B (**141**) was the presence of a methoxy group instead of a free hydroxy function at C-6 in the two new compounds. By the oxidative degradation procedure, the absolute configuration at the stereogenic center at C-3 was again established to be *S*. In continuation of this series of *seco*-type naphthylisoquinolines, the new *N*-methylated alkaloid was named ancistrosecoline E (**143**) and its *N*-unsubstituted analogue was given the name ancistrosecoline F (**144**).[55] From a stereochemical point of view, the two compounds occupy an outstanding position. They are the very first naphthylisoquinoline alkaloids whose biaryl axes—although rotationally hindered—are no chiral elements, due to the constitutional symmetry in the phenyl subunit, with the two identical *ortho*-methoxy substituents at C-6 and C-8 next to the biaryl axis. As a consequence of this lack of axial chirality, **143** and **144** only show a weak ECD effect, just arising from the stereocenter at C-3.[55]

All the other representatives of naphthylisoquinolines (ca. 280 known compounds),[1–7] by contrast, possess chiral axes, displaying—without any exception—the phenomenon of atropisomerism. Depending on the absence or presence of bulky substituents next to the biaryl axis, the steric hindrance may vary drastically, resulting in biaryl axes with most different rotational barriers, from rapidly rotating biaryl linkages as in dioncophyllinol B (**29**)[275] (see Fig. 20 in Section 3.1) via configurationally semi-stable ones, as in dioncophyllidine E (**72a/b**)[61] (see Fig. 70 in Section 5.4), to configurationally fully stable axes as in dioncophylline A (**1a**)[5,6,335,363,364,377] (see Fig. 36 in Section 4.7.1) or, even more stable, in ancistrocladine (**20a**)[27,28,257,259] (see Fig. 17 in Section 3.1).

Likewise noteworthy is a second stereochemical peculiarity of ancistrosecolines E (**143**) and F (**144**). The two methoxy groups in **143** and **144** appear at first sight highly similar, if not identical; but being imbedded into a chiral molecule, they are not homo- or enantiotopic, but diastereotopic to each other.[55,534] They should thus give, in principle, two distinct ^{1}H NMR signals. But even at 600 MHz, the two methoxy signals overlapped, giving just one six-proton singlet at δ_H 3.62 (for **143**) and δ_H 3.61 (for **144**). In other, structurally similar, yet synthetic analogues, however, such an—albeit minimal—signal splitting as a consequence of diastereotopicity was indeed observed.[535]

The discovery of *seco*-type follow-up naphthylisoquinoline derivatives such as ancistrosecolines A–F (**13**, **141**, **132**, and **142–144**)[55] (Fig. 102) in the roots of *A. abbreviatus* demonstrates that naphthylisoquinoline alkaloids

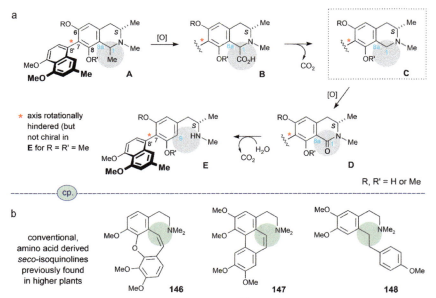

Fig. 106 (a) Proposed biosynthetic pathway[55] to *seco*-type naphthylisoquinoline alkaloids of type **E** by oxidation of the methyl group at C-1 in compound **A** to give the 1-carboxylate **B**, followed by its decarboxylation to the respective 1-*nor*-naphthylisoquinoline (compound **C**) and a further (presumably likewise oxidative) elimination of C-1, possibly via the lactam intermediate **D**; (b) for comparison, see the structures of the "conventional" (i.e., amino acid derived) *seco*-isoquinolines *seco*-sarcocapnine (**146**), *N*-methyl-*seco*-glaucine (**147**), and polysignine (**148**) isolated from higher plants; in **146–148**, in which, however, only the N—C bond was cleaved, but not the C—C bond from C-1 to the phenyl ring (underlaid in light gray-green).

are not just metabolically stable final products, but they can also serve as substrates for further structural modifications. Biosynthetically,[55] *seco*-type naphthylisoquinolines might arise from usual (i.e., still 1-methylated) naphthyltetrahydroisoquinolines like compound **A**. As presented in Fig. 106a, the loss of the methyl group at C-1 in **A** might occur by a stepwise oxidation via the respective 1-carboxylate **B**, followed by decarboxylation to give the 1-*nor*-compound **C**. Subsequent further oxidation at C-1 would result in the formation of the respective lactam (compound **D**), which could then undergo decarboxylation, and would finally provide naphthylisopropylamines of type **E**. The proposed biosynthetic route[55] is in agreement with the isolation of imaginable naphthylisoquinoline precursors related to substrate **A** such as ancistrobrevines A (**100**),[5,57,66] and H (**8**),[57] or their 6-*O*-demethylated analogues **118** and **119**[57] (see Fig. 71 in Section 5.5). In particular, the discovery of the first 1-unsubstituted naphthylisoquinoline, 1-*nor*-8-*O*-demethylancistrobrevine H (**8**)[55] (see previous Section 6.4), convincingly substantiates the postulated biosynthetic pathway (Fig. 106a) to *seco*-type naphthylisoquinoline alkaloids.

This ring opening reaction with concomitant elimination of the Me–C-1–H unit seems to be unique in nature. Some few *seco*-isoquinoline alkaloids[536–541] are already known from higher plants, like *seco*-sarcocapnine (**146**),[539] *N*-methyl-*seco*-glaucine (**147**),[538,539] and polysignine (**148**)[540,541] (Fig. 106b), but they are all amino acid derived, and they all result from a cleavage of the N—C bond, exclusively. Different from the six new *seco*-type ancistrosecolines presented here, the C—C bond from C-1 to the phenyl ring remains intact in **146–148**, leading to the formation of a new conjugated double bond, or—as in the case of polysignine (**148**)—to an alkyl bridge by subsequent hydrogenation.

Ancistrosecoline D (**142**) exhibited pronounced cytotoxicity toward human HeLa cervical cancer cells in a concentration-dependent manner (Fig. 107a), displaying a half-maximum inhibition concentration (IC_{50}) of 11.2 µM[55] and, thus, showing an antiproliferative effect, which was of the same magnitude of activity, compared to the positive control 5-fluorouracil (IC_{50} = 13.9 µM), a therapeutic agent widely used for clinical cancer treatment.[444,542] The related 7,1′-coupled ancistrosecolines A (**13**) and B (**141**), by contrast, exerted only moderate to low cytotoxic properties against HeLa cancer cells, with IC_{50} values of 52.6 µM (for **13**) and 72.4 µM for (**141**).[55]

Cervical cancer[543,544] still is the fourth most common malignancy diagnosed in women worldwide, with about 530,000 new cases and more than 250,000 deaths per year.[497,498] As a consequence of insufficient treatments, most of the global burden of invasive cervical cancer occurs in low- and medium-income countries. In sub-Saharan Africa, where the vast majority of cervical cancer occurs, it is even the main cause of cancer death among women.[544–546] Patients diagnosed with early-stage and only locally advanced tumors can be cured with a combination of surgery, radiation, and chemotherapy, whereas for metastatic or recurrent cervical cancer the prognosis is rather dismal, with a 5-year survival rate of less than 10%.[544,546,547] The discovery and development of new and improved medical drugs is an important task in view of the increasing resistance of cervical cancer cells to currently available agents, resulting in low response rates, which drastically impacts the success of chemotherapy. In particular treatment of the disease in an advanced or recurrent stage is strongly limited and is considered as palliative.[545–547]

Ancistrosecoline D (**142**) can be regarded as a promising candidate for more-in-depth studies on its cytotoxic potential against human HeLa

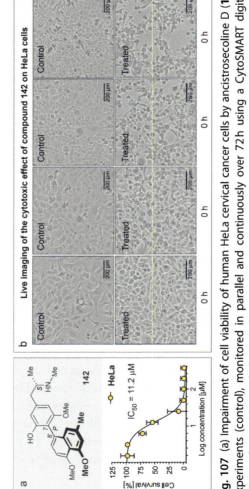

Fig. 107 (a) Impairment of cell viability of human HeLa cervical cancer cells by ancistrosecoline D (**142**); (b) captures of real-time imaging experiments (control), monitored in parallel and continuously over 72 h using a CytoSMART digital imaging system installed within a CO_2 incubator; (c) morphological alterations and induction of apoptosis in HeLa cells treated with 50 μM of **142** and incubated for 72 h. Cells were stained with the Hoechst 33342 dye and photographed under phase-contrast (black/white) and fluorescence (blue) modes using an EVOS FL digital microscope. Nuclei fragmented in the cells are marked with white arrows.

cervical cancer cells. Real-time imaging experiments on HeLa cells treated with 50 µM of **142** revealed this *seco*-type naphthylisoquinoline alkaloid to induce massive nuclei fragmentation, leading to apoptotic cell death.[55] At the end of the experiment, the cancer cells were stained using Hoechst 33342, a DNA-specific and cell-membrane-permeable dye that emits bright-blue fluorescence in both live and dead cells. While untreated HeLa cells (control) displayed a regular morphology with intact nuclei, cancer cells treated with **142** showed massive morphological changes with substantial deformation of the nuclei, leading to apoptotic cell death (Fig. 107c).[55] These promising results warrant further studies aiming at the elucidation of the mode of action of ancistrocecoline D (**142**) and investigations regarding its safety profile.

6.6 The naphthylisoindolinones ancistrobrevolines A–D, unprecedented ring-contracted naphthylisoquinoline alkaloids

Metabolic profiling of an enriched alkaloid fraction of the root bark of *A. abbreviatus* led to the discovery of a new subclass of naphthylisoquinoline-related alkaloids, named ancistrobrevolines A–D (**14**, **133**, **149**, and **150**)[59] (Fig. 108). These four minor metabolites display an unprecedented ring-contracted heterocyclic subunit, in where the heterocyclic part of the isoquinoline portion, as typical of conventional naphthylisoquinolines, is replaced by a five-membered ring, as part of an isoindolinone ring system. This is the consequence of a complete loss of C-3 and the adjacent methyl

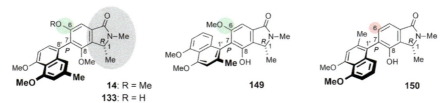

Fig. 108 Ancistrobrevolines A–D (**14**, **133**, **149**, and **150**),[59] the first naphthylisoindolinone alkaloids, which represent a new subtype of naphthylisoquinoline-related natural products, equipped with an unprecedented heterocyclic ring system. Three of these metabolites, discovered in the roots of *A. abbreviatus*, viz., **14**, **133**, and **149**, possess an oxygen function at C-6 (underlaid in green) and are thus to be classified as metabolites derived from Ancistrocladaceae-type naphthylisoquinoline alkaloids, whereas ancistrobrevoline D (**150**) might arise from a dioncophyllaceous metabolite lacking an oxygen substituent at C-6 (underlaid in red). For better comparability, the atom-numbering in the isoindolinone portion follows the one usually applied for naphthylisoquinoline alkaloids.

group.[59] All of these naphthylisoindolinones have in common that they are fully O-methylated in their naphthalene portion, but vary in the coupling position (C-1' or C-8') in the naphthalene unit, and they differ in the stereo-orientation at the biaryl axis. The four ancistrobrevolines are all N-methylated and R-configured at C-1, but they differ by their oxygenation degree at C-6 and their O-methylation at C-8.[59] In terms of the criteria for Ancistrocladaceae- or Dioncophyllaceae-type naphthylisoquinoline alkaloids (see Section 3.1),[1,3,5] the presence or absence of the 6-oxygen function is the only criterion applicable here, not the configuration at C-3, due to the loss of the whole chiral C-3/Me-3 entity in the ring-contracted molecular half. In that respect, three of the new metabolites, namely ancistrobrevolines A (**14**), B (**133**), and C (**149**) might be considered as Ancistrocladaceae-type (rather than hybrid-type) compounds, while ancistrobrevoline D (**150**) might be derived from a Dioncophyllaceae-type (rather than from an inverse hybrid-type) precursor[59] (Fig. 108).

The first isolated compound of this series had a molecular formula of $C_{25}H_{27}NO_5$, as indicated from its sodium adduct in HRESIMS, corresponding to m/z 444.17711 $[M+Na]^+$. 1D and 2D NMR investigations revealed the presence of a naphthalene portion with the usual 4',5'-dimethoxy-2'-methyl substitution pattern, and with C-8' being the axis-bearing carbon atom (Fig. 109a). This became obvious through HMBC interactions from H-1' and H-6' to C-8', from H-7' to C-8a', and from H-3' to C-1' (Fig. 109a), corroborated by NOESY correlation sequences in the series {OMe-5' ↔ H-6' ↔ H-7'} and {OMe-4' ↔ H-3' ↔ Me-2' ↔ H-1'} (Fig. 109b). This assignment was in accordance with the normal, not highfield-shifted signal of Me-2' (δ_H 2.29).[59]

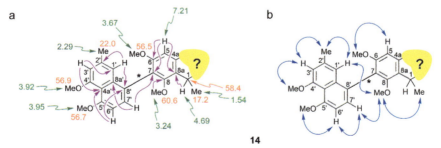

Fig. 109 (a) ^{1}H (in green) and ^{13}C (in orange) NMR shifts (methanol-d_4, δ in ppm), HMBC cross-peaks (single purple arrows), and (b) NOESY interactions (double blue arrows), establishing the constitution of the naphthalene portion and the phenyl part of the naphthalene half of ancistrobrevoline A (**14**), while the novel-type, initially unknown structural motif of **14** is covered by yellow.

The isocyclic benzene ring of the heterocyclic molecular half of the isolated metabolite displayed structural features typical of conventional naphthylisoquinoline alkaloids. Two singlets at δ_H 3.67 and 3.24, each corresponding to three protons, and an aromatic singlet resonating at δ_H 7.21 evidenced the presence of a phenyl unit with methoxy functions at C-6 and C-8 and no substituent at C-5 (Fig. 109a). HMBC interactions from H-5 to C-7 and C-8a, along with a ^4J correlation of C-7 with H-7′ across the biaryl axis, established C-7 to be the axis-bearing carbon atom (see Fig. 109b).[59] Furthermore, the ^{1}H NMR spectrum exhibited chemical shifts characteristic of a CHMe entity at C-1, as obvious from a three-proton doublet (δ_H 1.54) and a quartet signal (δ_H 4.69) in the aliphatic region. The upfield-shifted signal of the methoxy function at C-8 showed NOESY cross-peaks to the protons of that methyl group at C-1 and to two aromatic protons in the naphthalene portion, H-1′ and H-7′, thus confirming the constitution of the phenyl part with the biaryl axis being located at C-7 (Fig. 109).[59]

The structural elements in the molecular part of the new metabolite, yellow overlaid in Fig. 109 and in Structure **I** in Fig. 110, however, substantially differed from those of all other known[1–7] naphthylisoquinoline alkaloids.[59] It was in particular the lack of resonances in the aliphatic region of the ^{1}H NMR spectrum, namely the missing doublet of Me-3 around δ_H 0.9–1.3 and the absent multiplet of H-3 in the region between δ_H 3.2 and 4.0. Likewise not observed were the two geminal protons usually appearing as doublets of doublets in the ^{1}H NMR spectrum of naphthyltetrahydroisoquinoline alkaloids (see Structure **II** in Fig. 110). The detection of a downfield-shifted three-proton singlet at δ_H 3.15 revealed the presence of an N-methyl group, showing a NOESY cross-peak to the methyl group at C-1 and an HMBC interaction to the aliphatic carbon atom C-1 (see Structure **III** in Fig. 110).[59] The appearance of a downfield-shifted signal in the ^{13}C NMR spectrum at δ_C 170.0, typical of a carbonyl function, together with an HMBC interaction from the N-methyl group to the carbonyl C-atom, did not fit with the presence of a six-membered tetrahydropyridinone unit in the new alkaloid (see Structure **IV** in Fig. 110).

The actually observed spectral data, by contrast, suggested that instead of the usual isoquinoline moiety, an isoindolinone part had been formed, with concomitant loss of the whole C-3/Me-3 entity, thus affording a contracted five-membered heterocyclic ring.[59] The eventually proposed Structure **V** (Fig. 110) was in a full agreement with the disappearance of the signal for C-3 in ^{13}C NMR (normally resonating at δ_C 45–53) and with the

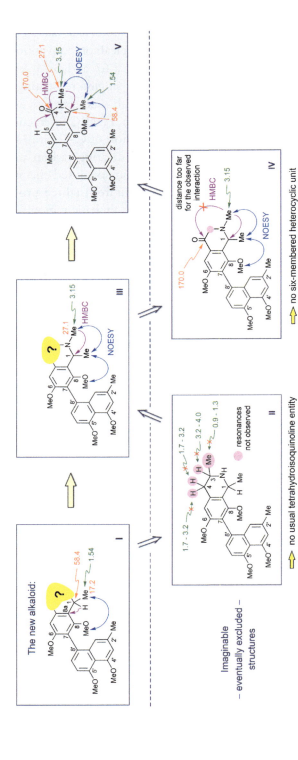

Fig. 110 Decisive ¹H (in green) and ¹³C (in orange) NMR shifts (methanol-d_4, δ in ppm) and key NOESY (double blue arrows) and HMBC (single purple arrows) interactions giving rise, step by step, to the structural assignment of the heterocyclic unit of ancistrobrevoline A (**14**),[59] as presented in Structures **I**, **III**, and **V**. A notable difference compared to usual naphthyltetrahydroisoquinolines was the absence of signals in the ¹H NMR spectrum of **14** expected for the two geminal protons at C-4, for the proton and for the methyl group at C-3 (underlaid in pink in Structure **II**). Likewise not fitting was an imaginable six-membered tetrahydropyridinone ring system as shown in Structure **IV**, since in such a heterocyclic unit the distance between N-Me and C-4 would have been too far for the—actually observed—HMBC cross-peak, thus giving conclusive evidence of a ring contraction with complete loss of C-3 (marked in pink in Structure **IV**), resulting in the formation of the unprecedented isoindolinone subunit of **14** as outlined in Structure **V**.

occurrence of an absorption band at 1673 cm^{-1} in the IR spectrum, as typical of a carbonyl group of an amide function. Hence, the new alkaloid consisted of a conventional naphthalene part and an unprecedented isoindolinone portion, coupled via a 7,8'-linked biaryl axis.[59]

For the assignment of the absolute configuration at C-3 in naphthyltetra- and -dihydroisoquinolines, the ruthenium-mediated oxidative degradation procedure[1,3,70] is routinely used, affording easy-to-analyze chiral amino acids. Luckily, this useful microanalytical device proved to be applicable even in the case of the five-membered ring system of the new naphthylisoindolinone, with cleavage of the strong lactam bond. Here, the degradation procedure resulted in the formation of *N*-methyl-D-alanine, hence establishing the absolute configuration at the stereocenter at C-1 in the isolated alkaloid as *R*.[59]

In view of a possible assignment of the configuration at the axis relative to the stereocenter at C-1, unfortunately, no significant long-range NOESY interactions across the biaryl axis were observed, namely between Me-1 and/or H-1 in the isoindolinone portion and H-1' or H-7' in the naphthalene subunit—apparently due to enlarged distances between the respective molecular entities. Quantum chemical ECD calculations performed for both possible atropo-diastereomers, using the TDωB97XD3/def2-TZVP// B3LYP-D3/def2-TZVP approach, revealed the isolated naphthylisoindolinone to be *P*-configured at the biaryl axis.[59]

As outlined in Fig. 111, the experimental ECD spectrum nearly perfectly matched the curve calculated for the *P,R*-atropisomer, whereas the spectrum computed for *M,R* was virtually opposite to the one monitored for the isolated metabolite. The new naphthylisoindolinone thus had the full absolute stereostructure **14** as presented in Figs. 108 and 111. This first representative of a novel subtype of naphthylisoquinoline-related alkaloids was henceforth named ancistrobrevoline A.[59]

The second new minor metabolite isolated from root bark extracts of *A. abbreviatus* was assigned to be the 6-*O*-demethyl derivative of ancistrobrevoline A (**14**), likewise possessing a 7,8'-linked biaryl axis. Again, the oxidative degradation established the absolute configuration at C-1 as *R*, and the absolute axial configuration was determined to be *P* due to the fact that the ECD spectrum of this second isoindolinone alkaloid matched well that of ancistrobrevoline A (**14**) (not shown). The new alkaloid thus had the full stereostructure **133** (see Fig. 108). It was named ancistrobrevoline B.[59]

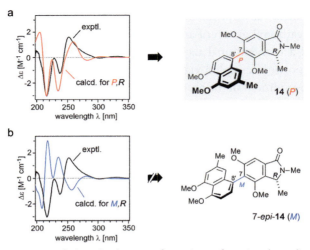

Fig. 111 Assignment of the absolute configuration of ancistrobrevoline A (**14**) by quantum-chemical ECD calculations, using TDωB97XD3/def2-TZVP//B3LYP-D3/def2-TZVP, and comparison of the experimental ECD spectrum of **14** (in black) with the ones calculated (a) for the 1R,7P-isomer (ECD curve in red) and (b) for 1R,7M (ECD spectrum in blue).

The third new compound from the roots of *A. abbreviatus* was a trace metabolite, exhibiting a molecular formula of $C_{24}H_{26}NO_5$ (*m/z* 408.17974), as deduced from HRESIMS and ^{13}C NMR, which was similar to that of ancistrobrevoline B (**133**). 1D and 2D NMR measurements suggested the presence of yet another naphthylisoindolinone alkaloid, but, different from **133**, which is 7,8′-coupled. Its structure was based on a 7,1′-coupling type and was equipped with a methoxy group at C-6 and a free hydroxy function at C-8.[59] HMBC interactions of H-5 to C-4 and C-7 and a NOESY cross-peak between H-5 and MeO-6 in the isoindolinone part and NOESY correlations in the series {H-8′ ↔ H-7′ ↔ H-6′ ↔ MeO-5′} and {MeO-4′ ↔ H-3′ ↔ Me-2} in the naphthalene moiety clearly attributed C-7 and C-1′ to be the axis-bearing carbon atoms (Fig. 112a).[59] The oxidative degradation procedure established the absolute configuration at C-1 as *R*. The ECD spectrum of the new alkaloid was mirror-imaged to the one of ancistrobrevoline A (**14**) (Fig. 112a), thus suggesting that the larger part of the naphthalene portion in the isolated metabolite should be directed down, exhibiting a stereoarray opposite to the ones in **14** and **133**. According to the Cahn-Ingold-Prelog formalism,

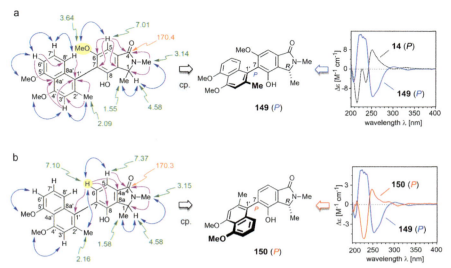

Fig. 112 Elucidation of the full absolute stereostructures of the 7,1′-coupled naphthylisoindolinones ancistrobrevolines C (**149**) and D (**150**): selected ^{1}H (in green) and ^{13}C (in orange) NMR shifts (methanol-d_4, δ in ppm), HMBC cross-peaks (single purple arrows), and NOESY interactions (green blue arrows) of (a) the ancistrocladaceous alkaloid **149** (with the methoxy substituent at C-6 underlaid in yellow) and (b) the dioncophyllaceous metabolite **150** (with H-6 underlaid in yellow), and assignment of their absolute axial configurations, (a) in the case of **149**, by comparison of its ECD spectrum (in blue) with the ECD curve (in black) of ancistrobrevoline A (**14**) and (b) in the case of **150**, by comparison of its ECD spectrum (in red) with that of **149** (in blue).

however, the new naphthylisoindolinone and the ancistrobrevolines A (**14**) and B (**133**) both had the same descriptor, P. Taken together, the new compound had the full absolute stereostructure **149** (see Fig. 112a). It was given the name ancistrobrevoline C.[59]

A further naphthylisoindolinone was detected in *A. abbreviatus*, with a constitution strongly resembling that of ancistrobrevoline C (**149**).[59] Like **149**, it was yet another example of a 7,1′-coupled alkaloid, possessing a 4′,5′-dimethoxy-2′-methylnaphthalene portion and displaying the typical structural features of the five-membered heterocyclic ring in the isoindolinone subunit (Fig. 109b). The ^{1}H NMR spectrum revealed the presence of six aromatic protons, i.e., one more than in the ancistrobrevolines A–C (**14**, **133**, and **149**). The additional aromatic proton was supposed to be located at C-6 as deduced from two adjacent protons resonating at δ_H 7.10 (H-6) and 7.37 (H-5).[59] This assignment was in accordance with the COSY interactions between H-5 and H-6, showing an *ortho*-coupling between each other. This finding indicated that the new metabolite might be derived from a Dioncophyllaceae-type alkaloid,[1,3,5] which is, as described in Section 3.1, characterized by the lack of an oxygen function at C-6 (the second criterion, *R*-configuration at C-3 does not apply here).

Again, the oxidative degradation procedure[1,3,70] delivered the D-enantiomer of *N*-methylalanine, thus assigning C-1 to be *R*-configured. In view of the fact that the ECD spectrum of the new metabolite was virtually mirror-imaged to that of ancistrobrevoline C (**149**), the absolute axial stereoarray had to be opposite, so that, following the Cahn-Ingold-Prelog formalism, the axial configuration had to be *P*. Hence, the new alkaloid, named ancistrobrevoline D,[59] had the full absolute stereostructure **150**, as shown in Fig. 112b.

Some few isoindolinone natural products had already been discovered earlier, derived from usual benzylisoquinoline alkaloids, such as ancistroyagonine (**151**)[539,548,549] and aristolactam I (**152**)[549–551] (Fig. 113a). Biosynthetically, these metabolites are assumed to arise from the corresponding benzyltetrahydroisoquinolines, by 3,4-dioxygenation and subsequent benzilic acid rearrangement, with loss of a carbon atom and ring contraction.[539,548–551]

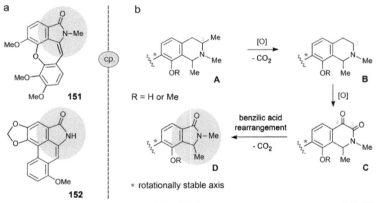

Fig. 113 (a) Aristoyagonine (**151**)[539,548,549] and aristolactam I (**152**),[549–551] two "conventional" benzylisoquinoline-derived alkaloids known from nature, possessing an isoindolinone motif (underlaid in gray) and (b) proposed biosynthetic route[59] from *N*-methylnaphthyltetrahydroisoquinolines (**A**) to ancistrobrevolines (**D**), by oxidative decarboxylation to **B**, and further oxidation to give the diketone **C**, followed by decarboxylation and ring contraction, with complete loss of C-3 and its methyl group.

In an analogous manner,[59] the four new naphthylisoindolinones might be follow-up metabolites formed from the respective *N*-methylated naphthyltetrahydroisoquinoline alkaloids of type **A** (Fig. 113b). Remarkably, the presumable precursors to ancistrobrevolines A (**14**), B (**133**), C (**149**), and D (**150**) are all main metabolites of *A. abbreviatus*, namely ancistrobrevine A (**100**),[5,57,66] its 6-*O*-demethyl analogue **118**,[5,57] ancistrobrevine D (**102a**),[5,65] and *N*-methyldioncophylline A (**94a**)[5,62] (for their structures, see Fig. 80 in Section 5.7).

The first step in the proposed biosynthetic pathway[59] (Fig. 113b) involves removal of the methyl group at C-3 of the naphthylisoquinoline **A** by a stepwise oxidation to the respective carboxylic acid, followed by decarboxylation to give the 3-*nor*-intermediate **B**. Subsequent oxidation should then provide the corresponding 3,4-diketone **C**. As in the case of **151** and **152**,[59] formation of the isoindolinone subunit **D** might finally occur by benzilic acid rearrangement, with loss of CO_2.

Such an elimination of a *C*-methyl group of a naphthylisoquinoline alkaloid had already been observed at C-1, resulting in the formation of 1-*nor*-8-*O*-demethylancistrobrevine H (**12**),[55] likewise occurring in *A. abbreviatus* (see Section 6.4). In a similar way, oxidative follow-up transformations at Me-3[1,252,273] and at C-4[5,274,275] are known to play a role in the biosynthesis of other naphthylisoquinoline alkaloids, examples being ancistrobenomine A (**26**)[273] and dioncophyllinol B (**29**)[275] (see Fig. 20 in Section 3.1).

The excellent growth-inhibitory effects against cancer cells displayed by some of the naphthylisoquinoline alkaloids like dioncophylline A (**1a**)[51,365,366,404] (see Section 5.3.2), but also the remarkable anticancer properties of aristolactam I (**151**) and related non-natural isoindolinone analogues[549–553] stimulated investigations on the cytotoxic potential of the new, polyketide-derived naphthylisoindolinones.

Ancistrobrevolines A (**14**) and B (**133**) exerted only low cytotoxicity against MCF-7 breast adenocarcinoma cells, with antiproliferative activities of 19–35% (for **14**) and 3.7–43% (for **133**) after exposure to varying concentrations (10, 30, 50, 70, and 100 µM) of the agents.[59] Against A549 lung cancer cells, the cytotoxic effects of **14** and **133** were significantly higher, with IC_{50} values of 34.6 µM (for **14**) and 9.05 µM (for **133**)[59] (Fig. 114a). Studies on the immortalized human epithelial cell line MCF-10A showed that ancistrobrevoline B (**133**) exhibited only moderate cytotoxicity on

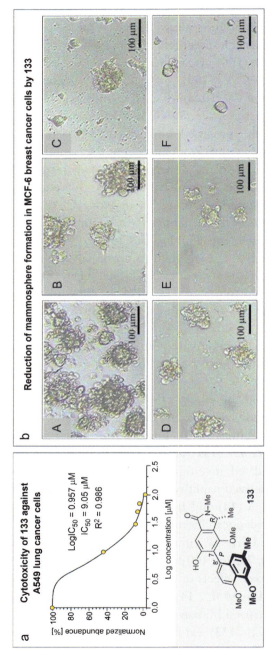

Fig. 114 Cytotoxic effects of ancistrobrevoline B (**133**) toward (a) A549 lung cancer cells and (b) significant reduction of the formation of highly metastatic spheroid clusters (mammospheres) in MCF-7 breast cancer cells. MCF-7 cells derived from mammospheres were treated with **133** at concentrations of 10 (**B**), 30 (**C**), 50 (**D**), 70 (**E**), and 100 µM (**F**) in comparison to untreated cells (**A**, control). The mammospheres were cultured for 5 h to form spheres on ultra-low attachment surface plates. The images of the mammospheres were taken using phase-contrast microscopy.

these non-tumorigenic cells, giving rise to a weak selectivity index of 1.4. Ancistrobrevoline A (**14**), by contrast, displayed no selectivity toward the lung cancer cells compared to the normal MCF-10A cells (SI ≈ 1).[59]

In a variety of tumors, cancer stem-like cells have been identified that possess the capacity of unlimited propagation and multipotent differentiation.[554–556] Such tumor-initiating cells are supposed to trigger tumor growth and metastasis, and to induce resistance to chemotherapeutic agents, thus causing failure in cancer treatment, leading to tumor recurrence.[557] MCF-7 human breast cancer cells have the potential to form discrete clusters of spheroids, termed mammospheres, which undergo asymmetric division and self-renewal. These MCF-7-derived stem-like cells are highly aggressive and metastatic.[558,559] Ancistrobrevolines A (**14**) and B (**133**) dose-dependently suppressed the formation of spheroids in MCF-7 cultures, compared to the production of large mammospheres in the non-treated control group (Fig. 114b).[59]

The results presented here again highlight the impressive structural diversity of the secondary metabolites produced by *A. abbreviatus*, and underline their potential as rewarding bioactive compounds for future investigations.

6.7 Highly oxygenated naphthoquinones from solid callus cultures of *Ancistrocladus abbreviatus*

Besides the living plants themselves, callus cultures of *A. abbreviatus* comprising entirely undifferentiated cells are versatile sources of new natural products, too.[151] Cultivation of *A. abbreviatus* as sterile callus cultures[187] generally imposes stress to the cells, eventually resulting in a largely increased formation of the naphthoquinones plumbagin (**16**) and droserone (**17**).[151] Solidification of the medium by addition of Gelrite® (also known as gellan gum) prevented the calli from sinking into the gel, hence exposing the cells to aerobic and thus oxidative conditions, which favored the formation of a series of structurally related, but now highly oxygenated naphthoquinones.[151]

Droserone (**17**) and plumbagin (**16**) are phytoalexins produced by intact plants as a reaction to all sorts of chemical, physical, or biotic stress.[5,268,560–566] Like the naphthylisoquinolines, also the two naphthoquinones are potent bioactive compounds, exerting strong cytotoxic, antioxidant, anti-inflammatory, and immunosuppressive properties, but also antibacterial, antifungal, antiviral, and antiprotozoal activities.[566–576] Resolution of a methanolic extract of dried powdered callus material of *A. abbreviatus* by preparative HPLC and repeated column chromatography

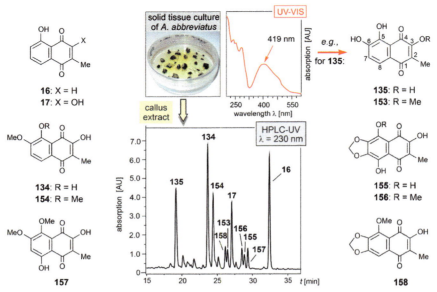

Fig. 115 Naphthoquinones with different oxygenation patterns produced by solid callus cultures of *A. abbreviatus*,[151] among them the five newly discovered ancistroquinones B (**134**), C (**154**), D (**157**), E (**155**), and F (**158**) and the anti-MM agent dioncoquinone B (**135**), which had, in parallel, also been isolated from likewise solidified callus cultures of *Triphyophyllum peltatum* (Dioncophyllaceae).[151] Furthermore, the already known[5,131,151,577,578] naphthoquinones plumbagin (**16**), droserone (**17**), malvone (**153**), and nepenthone A (**156**) were identified in the callus extract of *A. abbreviatus*.

afforded a total of 10 compounds, among them 5 new naphthoquinones, named ancistroquinones B–F (**134**, **154**, **157**, **155**, and **158**) and dioncoquinone B (**135**) (see Fig. 115).[151] The latter had, in parallel, also been isolated from the likewise solidified callus cultures of the Dioncophyllaceae liana *Triphyophyllum peltatum* from Coastal West Africa.[151,152]

Dioncoquinone B[151] (**135**) displayed a UV spectrum typical of a 3,5-dihydroxy-substituted naphthoquinone, with a UV maximum at $\lambda = 419$ nm (Fig. 115). ^{1}H NMR investigations on **135** corroborated the presence of a 3,5-dihydroxy-2-methyl-1,4-naphthoquinone derivative, as evidenced by the signal of a methyl group attached to the quinone ring (δ_H 2.10), two signals of aromatic protons, H-7 and H-8 (δ_H 7.20 and 7.65), and one chelated hydroxy function (δ_H 11.22) at C-5. The proton of a second hydroxy group resonated at δ_H 7.04; the position of this group at C-3 was deduced from HMBC interactions with C-2, C-3, and C-4. A third hydroxy group (δ_H 6.04), again not chelated, was assigned to be located at C-6 (Fig. 116a,b). The structure of a likewise imaginable 8-substituted analogue

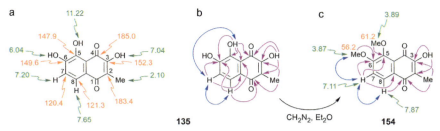

Fig. 116 Selected NMR data determining the constitution of dioncoquinone B (**135**): (a) ¹H (in green) and ¹³C NMR (in orange) shifts (methanol-d_4, δ in ppm); (b) HMBC (single purple arrows) and ¹H,¹H-COSY (double blue arrows) interactions; (c) selected NMR data of the dimethylether **154**, which was prepared from **135** by using CH$_2$N$_2$ in diethyl ether.

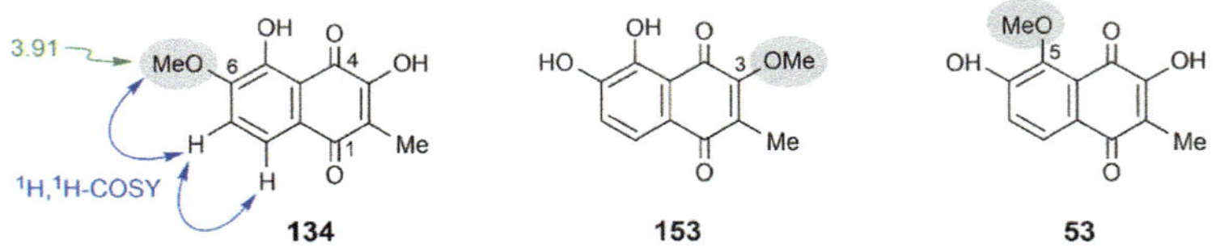

Fig. 117 Three regioisomeric 1,4-naphthoquinones, differing by the position of the O-methyl group: the newly discovered ancistroquinone B (**134**) and the known[577] phytoalexin malvone A (**153**), now for the first time identified in callus cultures of A. abbreviatus,[151] and ancistroquinone A (**53**) produced by the Indian liana A. heyneanus,[314] but not found to occur in cell cultures of A. abbreviatus.

was excluded by quantum-chemical ¹H and ¹³C NMR simulations, clearly confirming structure **135**.[151] Its 3,5,6-trioxygenation pattern was further corroborated by the selective conversion of this naphthoquinone into the dimethylether **154** by using diazomethane in diethyl ether. The presence of two additional O-methyl groups in **154** as compared to **135** was proven by the occurrence of two three-proton singlets (δ_H 3.87 and 3.89) in ¹H NMR and two corresponding signals in ¹³C NMR (δ_c 56.2 and 61.2) and by specific HMBC couplings to C-5 and C-6 (Fig. 116c).[151]

Four further compounds[151] were isolated from the callus cultures of A. abbreviatus showing a close structural relationship to dioncoquinone B (**135**) according to ¹H and ¹³C NMR analysis. In the first one (i.e., the second in total from the callus extract), the presence of an additional methoxy group as compared to **135** was indicated by a three-proton singlet resonating at δ_H 3.91 and by ¹H,¹H-COSY interactions in the series {OMe-6 ↔ H-7 ↔ H-8}, proving its location at C-6 (Fig. 117). From these data, the constitution of the naphthoquinone was established as

134. This compound had already been known as a side product in the total synthesis of ancistroquinone A (**53**) (Fig. 117), a metabolite from the leaves of the Indian liana *A. heyneanus*,[314] and as a degradation product of rubromycins.[579] Now isolated for the first time from a natural source and being a second example of a new highly oxygenated naphthoquinone isolated from Ancistrocladaceae plants, **134** was named ancistroquinone B.[151]

From its ^{1}H and ^{13}C NMR data, a further compound related to dioncoquinone B (**135**), i.e., the third naphthoquinone from *A. abbreviatus*, was easily identified as a regioisomer of ancistroquinone B (**134**), differing only by the position of the O-methyl group, now being located at C-3 (instead of C-6 as in **134**).[151] This compound, previously already isolated from *Malva sylvestris* (Malvaceae), named malvone (**153**) (Fig. 117), is a known phytoalexin; for its isolation, structural elucidation, X-ray structure analysis, and fungitoxicity, see Ref. 577.

The fourth naphthoquinone[151] from the callus cultures of *A. abbreviatus* displayed ^{1}H and ^{13}C NMR data revealing the presence of a compound possessing two more O-methyl groups than dioncoquinone B (**135**). Its structural data were identical to those obtained in the course of the derivatization of **135** (Fig. 116c), which had led to the dimethyl ether **154**, whose data had already been reported in the literature as a synthetic intermediate.[314] In view of the isolation of **154** as a metabolite of *A. abbreviatus*, it was named ancistroquinone C.[151]

A further, fifth naphthoquinone[151] showed NMR data evidencing the presence of a fourfold oxygenated 1,4-naphthoquinone, with signals of four quaternary atoms at δ_C 160.9, 160.0, 153.7, and 145.9 in ^{13}C NMR (Fig. 118a). Two three-proton signals, appearing at δ_H 3.94 and 3.89 in the ^{1}H NMR spectrum, indicated that two of the oxygen functions were

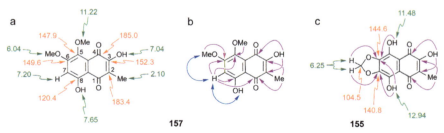

Fig. 118 Selected 1D and 2D NMR data relevant for the constitution of ancistroquinones D (**157**) and E (**155**): (a) ^{1}H (in green) and ^{13}C (in orange) NMR shifts (chloroform-*d*, δ in ppm) of **157**; (b) NOESY (double blue arrows) and HMBC (single purple arrows) interactions of **157**; (c) specific ^{1}H and ^{13}C NMR data and HMBC couplings of **155**.

methylated. The compound had two free hydroxy groups; one of them, whose proton signal resonated at δ_H 7.64, was located at C-3, the second OH function, appearing at δ_H 13.44, was downfield-shifted, and hence in the 8-position. The metabolite had only one aromatic proton, which was assigned to be located at C-7 due to NOESY interactions with 8-OH and 6-OMe, and from a long-range HMBC coupling to C-1 (Fig. 118b). The new metabolite possessed the structure **157** and was named ancistroquinone D.[151]

According to HRESIMS and ^{13}C NMR analysis, the sixth naphthoquinone[151] from *A. abbreviatus*, having a molecular formula of $C_{12}H_8O_7$, showed a high degree of oxygenation (seven of eight possible positions). The ^{1}H NMR spectrum displayed a two-proton singlet typical of a methylenedioxy group (δ_H 6.25); due to the achiral structure, these geminal protons were enantiotopic and, thus, isochronous. Three further singlets evidencing the presence of two chelated hydroxy functions in positions 5 and 8 (δ_H 11.48 and 12.94, respectively), and one methyl group on the quinoid ring (δ_H 2.07). HMBC experiments confirmed that the methylene group bridged the two oxygen atoms at C-6 and C-7 (Fig. 118c). The signal of the hydroxy group at C-3 was not detected, presumably due to a fast H-D exchange. This new 1,4-naphthoquinone, which possessed the structure **155**, was given the name ancistroquinone E.[151] It can also be addressed as the 5-O-demethyl analogue of the known[578] naphthoquinone nepenthone A (**156**) (Fig. 115), which had earlier been discovered in the roots of *Nepenthes rafflesiana* Jack (Nepenthaceae) from Southern Thailand. Nepenthone A (**156**) was found to be likewise produced by the solidified callus cultures of *A. abbreviatus*.[151] It was easily recognized due to the presence of an additional methoxy group and the missing signal of the chelated hydroxy group at C-5, when comparing its NMR data with those of ancistroquinone E (**155**).

The 1D and 2D NMR analysis of the seventh naphthoquinone isolated from *A. abbreviatus* again evidenced the presence of a highly oxygenated 1,4-naphthoquinone possessing a methylenedioxy group to bridge C-6 and C-7, differing from nepenthone A (**156**)[578] only by the disappearance of the chelated hydroxy group at C-8 (δ_H 13.21), showing the signal of an aromatic proton resonating at δ_H 7.39 instead (data not shown). Hence, the new metabolite had the constitution **158** (Fig. 115) and was named ancistroquinone F.[151]

In view of the reported strong antitumoral activities of some of the previously identified naturally occurring 1,4-naphthoquinones,[566–574] the highly oxygenated compounds isolated from the callus cultures of *A. abbreviatus* were tested for their inhibitory activities against human tumor cells of two different B-cell malignancies, B-cell lymphoma and multiple myeloma. B-cell lymphoma (also known as B non-Hodgkin lymphoma)[580,581] is a neoplasm of the immature B-cell, whereas multiple myeloma[464–466] is a neoplasm of the terminally differentiated B-cell (plasma cell). Excessive abnormal plasma cells are forming tumors in multiple locations in the bone marrow, thus leading to osteolytic bone destructions, impaired hematopoiesis, and renal failure.

Dioncoquinone B (**135**) exerted excellent cytotoxic activities against B-cell lymphoma (Fig. 119a) and multiple myeloma (MM) (Fig. 119b) cells, where it induced apoptosis.[151] Its activity, with effective concentrations (EC_{50} values) ranging from 7.6 to 18 µM, was comparable to that of melphalan[469] (Table 6), a DNA-alkylating agent used in the standard chemotherapy against B-cell malignancies. In contrast to melphalan, however, dioncoquinone B (**135**) did not display any significant toxicity toward normal peripheral mononuclear blood cells (PMBCs) (Fig. 119a,b). For this reason, dioncoquinone B (**135**) was classified as a promising lead compound for the development of novel anti-MM candidates. All of the other naphthoquinones from *A. abbreviatus*, ancistroquinones B (**134**), C (**154**), D (**157**), E (**155**), and F (**158**), malvone A (**153**), and nepenthone A (**156**), by contrast, were virtually inactive against the B-cell lymphoma and MM cell lines.[151]

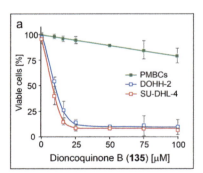 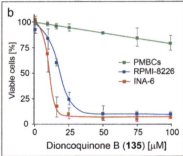

Fig. 119 Antiproliferative properties of dioncoquinone B (**135**) against human tumor cells of two different B-cell malignancies: (a) against the B-cell lymphoma cell lines DOHH-2 and SU-DHL-4, and (b) against the multiple myeloma cell lines INA-6 and RPMI-8226, in comparison to the non-cytotoxic effect of **135** toward normal peripheral mononuclear blood cells (PMBCs). The cells were treated with different concentrations of **135** for 3 days. The viable fractions of the treated cells were determined by annexin V-FITC/PI staining.[151] Error bars indicate the range of values derived from three independent experiments.

Table 6 EC$_{50}$ values (μM) of B-cell lymphoma and multiple myeloma cell lines and PMBCs treated with dioncoquinone B (**135**) and with the DNA-alkylating agent melphalan.[a]

	135	melphalan
B cell lymphoma cell line DOHH-2	9.4	1
B cell lymphoma cell line SU-DHL-4	7.6	4
Multiple myeloma cell line INA-6	11	2
Multiple myeloma cell line RPMI-8226	18	2
Peripheral mononuclear blood cells (PMBCs)	NR[b]	3

[a]Multiple myeloma cells were treated with different concentrations of dioncoquinone B (**135**), or with melphalan. Viability of cell fractions were determined by annexin V-FITC/PI staining.
[b]NR: not reached.

An efficient synthetic approach,[152] based on a directed *ortho*-metalation (DOM) reaction, was pursued to furnish dioncoquinone B (**135**) (Fig. 120) in sufficient quantities, e.g., for extended structure–activity relationship (SAR) studies to further optimize its structure.

The route started from the commercially available benzoic acid **159**, which was converted into the known[582,583] benzamide **160** using procedures described for related compounds.[584,585] Directed *ortho*-deprotonation of **160** using *sec*-butyllithium, followed by transmetalation of the intermediate aryllithium species with magnesium bromide diethyl etherate, afforded the Grignard reagent **161**. Its reaction with 2-methylallyl bromide (**162**) led to the formation of the *ortho*-allylbenzamide **163**.[152] This intermediate was converted into the naphthalene **164** by cyclization induced by methyl lithium, providing the targeted methylnaphthalene **164** in high yields (Fig. 120).[152] By oxidation of **164** with CuCl in the presence of air, the desired *para*-naphthoquinone **166** was obtained along with the respective *ortho*-naphthoquinone **165** (ratio 6.5:1). Introduction of the α-hydroxy functionality at C-3 in the target molecule was achieved by epoxidation of the 2,3-double bond in **166**, followed by ring opening of the resulting α,β-epoxydiketone **173** (for its structure, see Fig. 121), by treatment with sulfuric acid on silica gel.[152] O-Demethylation of the obtained ancistroquinone C (**154**) using boron tribromide finally provided the target molecule dioncoquinone B (**135**) in eight linear steps in an overall yield of 36%.[152]

This efficient synthetic route simultaneously gave access to a series of related naphthoquinones[152] by varying the substitution patterns of the starting material. Thus, a library of ca. 50 new analogues of dioncoquinone B were

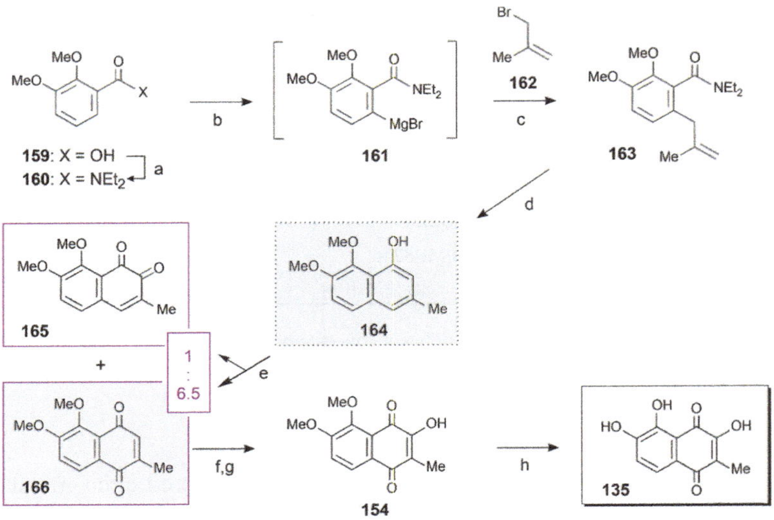

Fig. 120 Total synthesis of the anti-MM naphthoquinone dioncoquinone B (**135**) and the related ancistroquinone C (**154**),[152] based on a directed *ortho*-metalation (DOM) approach, by Grignard reaction of the benzamide **161** and 2-methylallyl bromide (**162**), to give the *ortho*-allyl benzamide **163**, followed by MeLi-induced cyclization to yield the methylnaphthalene **164**; oxidation with CuCl provided the naphthoquinones **165** and **166** (ratio: 1:6.5), and epoxidation with subsequent ring opening furnished **154**, and finally, after O-demethylation, the desired target molecule **135**. Reaction Conditions: (a) SOCl$_2$, reflux; then HNEt$_2$, THF, r.t., 99%; (b) sec-BuLi, TMEDA, THF, −90 °C, MgBr$_2$·2Et$_2$O; (c) **162**, THF, −78 °C → r.t., 60%; (d) MeLi, −78 °C → 0 °C, 99%; (e) CuCl, air, MeCN, **165** (8%) and **166** (66%) (ratio of 1:6.5, **165:166**); (f) H$_2$O$_2$, K$_2$CO$_3$, THF, 97%, providing the corresponding epoxide (not shown here), for its structure **173**, see Fig. 121, (g) conc. H$_2$SO$_4$, silica gel, THF, 95%; (h) BBr$_3$, CH$_2$Cl$_2$, r.t., 98%.

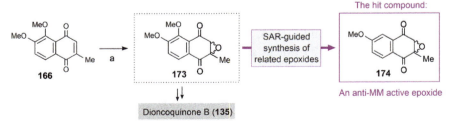

Fig. 121 Epoxidation of the 2,3-double bond of the naphthoquinone **166** en route to dioncoquinone B (**135**) leading to the epoxide **173**[152]; SAR-guided synthesis furnished its analogue **174** with strong inhibitory activity toward multiple myeloma (MM) cells.[586] Reaction Conditions: (a) aq. H$_2$O$_2$ (30%), 1N aq. Na$_2$CO$_3$, THF, r.t., 1 h, 97%.

Table 7 EC$_{50}$ values (µM) determined for the multiple myeloma cell line INA-6 and for peripheral blood mononuclear cells (PBMCs) treated with compounds **154**, **135**, and **167–172**, or with melphalan.[a]

	154	167	135	168	169	170	171	172	melphalan
X^1	OH	OH	OH	OH	OH	OH	OH	H	
X^2	OMe	OMe	OH	OH	OH	H	H	OH	
X^3	OMe	OMe	OH	OH	OH	OH	OH	OH	
X^4	H	OMe	H	OMe	OH	OH	H	H	
INA-6[a]	>100	>100	11	14	7	52	>100	75	2
PBMCs[b]	NR[c]	NR	NR	NR	70	NR	NR	100	3

[a]Multiple myeloma cells were treated with different concentrations of compounds **154**, **135**, and **167–172**, or with melphalan.
[b]Cytotoxicity against normal peripheral mononuclear blood cells (PMBCs).
Viability of cell fractions determined by annexin V-FITC/PI staining. [c]NR: not reached.

synthesized, comprising compounds with an additional oxygen function at C-7 as in **168** and **169** (see Table 7), or with only two oxygen functions on the aromatic ring in total (at C-5 and C-6 as in **172**). Biological evaluation of their antitumoral potential toward MM cell lines revealed the three hydroxy groups at C-3, C-5, and C-6 to be essential for anti-MM activity. The methyl group at C-2, however, was less important, it could even be omitted without loss of activity (data not shown). The test results showed that removal of one OH group in any of these positions led to an almost complete loss of inhibitory activity (see structures **154**, **167**, **171**, and **172** in Table 7).[152]

Epoxide **174** (Fig. 121), an analogue of the synthetic intermediate **173** en route to dioncoquinone B (**135**) (Fig. 120), was found to induce strong apoptosis in MM cells at a concentration (EC$_{50}$ = 3.5 µM) distinctly lower than that of **135** (EC$_{50}$ = 11 µM) and any of its related analogues,[152] without exerting significant toxicity against normal blood cells.[586]

Epoxidation of the respective naphthoquinone with hydrogen peroxide succeeded under different optimized conditions. Nearly all of the epoxides **174–181** (Table 8) were obtained in very good yields and high purity.[586] They were tested for their growth-retarding activities against MM cells, by determination of the viability of INA-6 cells 3 days after treatment. The results showed that all of the epoxides (except for compound **177**,

Table 8 EC$_{50}$ values (µM) of the active compounds **174–181** and melphalan[a] against multiple myeloma cell lines INA-6 and peripheral blood mononuclear cells (PBMCs).

	174	175	176	177	178	179	180	181	melphalan
X^1	H	OMe	H	H	H	H	OMe	OMe	
X^2	OMe	OMe	Me	OH	CO$_2$Me	OMe	OMe	Cl	
X^3	H	H	H	H	H	OMe	OMe	H	
INA-6[a]	3.5	2.8	2.0	13.0	0.6	5.0	2.2	3.0	2
PBMCs[b]	NR[c]	7.4	2.0	37.0	0.4	11.0	3.8	4.5	3

[a]Treatment of multiple myeloma cells with different concentrations of compounds **174–181** or with melphalan.
[b]Cytotoxicity against normal peripheral mononuclear blood cells (PMBCs).
Viability of cell fractions determined by annexin V-FITC/PI staining. [c]NR: not reached.

which was the only one with an EC$_{50}$ value of >10 µM) exhibited significant anti-MM activities in the low micromolar range, similar to melphalan as the standard. Within this series of epoxides, the most active compound was **174**, with an EC$_{50}$ value of 3.5 µM, while lacking any cytotoxic effects toward normal PMBCs.[586] The structure-activity relationship profile of the epoxides presented in Table 8 was clearly different from that of the naphthoquinones related to dioncoquinone B (**135**): For good anti-MM activities (and low cytotoxicities), C-5 and C-7 should be unsubstituted as in **176** and **178**, and C-6 should bear an alkoxy group, and there should be no free hydroxy functions. These two classes of agents, the epoxides[586] and the naphthoquinones,[151,152] thus should have different modes of action (and, consequently, divergent molecular targets).

The most active anti-MM epoxide **174**[586] was derivatized to a pulldown probe[587,588] to determine the cellular binding partners by a protein affinity isolation combined with mass-spectrometric analysis of the captured proteins. This was achieved by attachment of biotin to **174** through a triazol-containing linker, by applying a Cu(I)-catalyzed azide-alkyne cycloaddition (CuAAC, Huisgen click reaction). For the attachment of a linker-tethered biotin molecule to the agent, the oxygen function at C-6 of **174** was addressed, giving rise to the biotinylated compounds **182** and **183** (Fig. 122).[586]

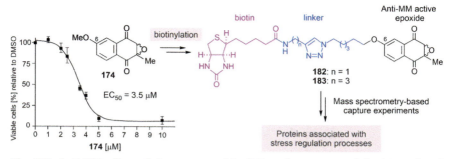

Fig. 122 Anti-MM effect of the new epoxide **174** and structures of the biotinylated compounds **182** and **183** subjected to mass spectrometry-based affinity capture experiments.[586]

MS-based affinity capture experiments with **182** and **183** showed a clear enrichment of several proteins,[586] among them peroxiredoxin 1 (PRDX1). This redox-regulating protein has been reported to eliminate reactive oxygen species (ROS) and to regulate cell growth, differentiation, and apoptosis.[589,590] Furthermore, thioredoxin domain-containing 5 (TXNDC5)[591,592] was found to be enriched; this more recently discovered endoplasmic reticulum (ER) resistant disulfide isomerase has been implicated in oxidative-stress related diseases including cancer. Likewise enriched were other ER-associated proteins such as thioredoxin-related transmembrane protein 1 (TMX1),[593] which catalyzes protein folding and thiol-disulfide interchange reactions. Among the proteins identified in the data set were SEC23B and SEC24C. These components of the coat protein complex II (COPII) promote formation of transport vesicles from the ER.[594] These enriched proteins indicate a functional line of the epoxide **174** to ER stress response mechanisms, which are considered to be critical for tumor cell survival in MM.[595,596]

It should be noted that many of the naturally occurring naphthoquinones and related, merely artificial compounds that were found to strongly affect the viability of cancer cells, simultaneously exerted high toxicity toward normal, non-malignant cells.[567,569,570,574,575] Dioncoquinone B (**135**), by contrast, did not display any significant toxicity toward normal peripheral mononuclear blood cells (PMBCs).[151] More recently, Khmelevskaya and Pelageev[597] reported on an alternative synthetic route to dioncoquinone B (**135**), aiming at an improved synthesis of this compound and of related highly oxygenated 1,4-naphthoquinones to make these compounds accessible in sufficient amounts for further modifications and more in-depth biological evaluations. Starting from simple and inexpensive reagents, the

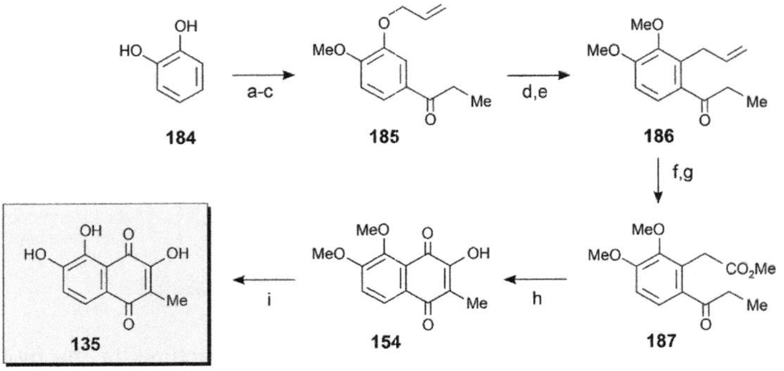

Fig. 123 Alternative synthetic approach to dioncoquinone B (**135**) starting from pyrocatechol (**184**), which was converted into the allyloxymethoxybenzene **185**, followed by Claisen rearrangement to give allylbenzene **186**; Dieckmann cyclization of ester **187** delivered ancistroquinone C (**154**), and, after O-demethylation, the target compound **135**. Reaction Conditions: (a) propionyl chloride, AlCl$_3$, (CH$_2$Cl)$_2$, 24 h, 0 °C, then r.t., 88%; (b) MeI, Li$_2$CO$_3$, DMF, 55 °C, 18 h, 81%; (c) allyl bromide, K$_2$CO$_3$, DMF, 18 h, r.t., 95%; (d) neat, 200 °C, 4 h, 98%; (e) MeI, K$_2$CO$_3$, 18 h, r.t., 98%; (f) NaIO$_4$, TBABr, 2.5 mol% Ru(OH)Cl$_3$, EtOAc/H$_2$O, 24 h, r.t., 85%; (g) MeOH, HCl, acetyl chloride, 3 h, reflux, 98%; (h) MeONa, MeOH, 4 h, reflux, then 5% aq. NaOH, air, 12 h, r.t., 82%; (i) AlCl$_3$, EtSH, CH$_2$Cl$_2$, 75%.

pathway provided dioncoquinone B (**135**) in a linear sequence of nine steps with an overall yield of 34%.[597]

As outlined in Fig. 123, Friedel–Crafts acylation of pyrocatechol (**184**) with propionyl chloride, followed by selective methylation at C-4 and allylation yielded the allyloxymethoxybenzene **185**, which underwent a Claisen rearrangement to give, after O-methylation, the allylbenzene **186**. Ruthenium-mediated oxidation of **186**, conversion to the respective methyl ester **187**, and Dieckmann cyclization with subsequent oxidation finally provided ancistroquinone C (**154**). Eventual twofold O-demethylation of **154** furnished dioncoquinone B (**135**) in good yields.[597]

Taken together, six new highly oxygenated 1,4-naphthoquinones were isolated from solid callus cultures of *A. abbreviatus*, among them dioncoquinone B (**135**),[151,152] which showed extraordinary antitumoral activities against B lymphoma and multiple myeloma cell lines. The related, yet merely artificial epoxide **174** induced strong apoptosis in MM cells at a concentration (EC$_{50}$ = 3.5 μM) even lower than that of dioncoquinone B (**135**, EC$_{50}$ = 11 μM), without exerting toxicity against normal blood cells.[586] Thus, dioncoquinone B (**135**) and the epoxide **174** constitute two promising basic structures for the development of new anti-MM candidates.

Pancreatic cancer is one of the most aggressive types of tumors with an extremely poor prognosis.[419–492] Standard medications suffer from impaired drug delivery pathways offering only limited survival benefit.[421–426] There is, consequently, an urgent need to develop new therapeutic approaches.

Pancreatic cancer cells are characterized by a severe tolerance to nutrient deficiency (i.e., austerity), allowing the cells to further proliferate and to infiltrate into nearby tissues and to migrate to distant organs.[420,424,432,434] Plumbagin (16)[598] and a series of shikimic-acid derived naphthoquinones[599,600] have recently emerged as promising anti-austerity agents efficiently inhibiting tumor progression by killing cells preferentially under nutrient-deprived conditions. The polyketidic naphthoquinone droserone (17) and related higher-oxygenated naphthoquinones like dioncoquinones B (135) and C (154) (Fig. 124), isolated from callus cultures of *A. abbreviatus*[151] and *Triphyophyllum peltatum*[151,152] (Dioncophyllaceae) were likewise found to target the capability of pancreatic cancer cells of tolerating a substantial lack of nutrients, by preferentially displaying significant cytotoxicity against PANC-1 cancer cells in nutrient-deprived medium (NDM), but without causing cell death in normal, nutrient-rich Dulbecco's modified Eagle's medium (DMEM).[601]

A library of natural and merely artificial 1,4-naphthoquinones related to droserone (17) (28 representatives) were evaluated for their anti-austerity activities, with the synthetically prepared compounds 188 and 189 exerting potent preferential cytotoxicity against PANC-1 cells.[601] The

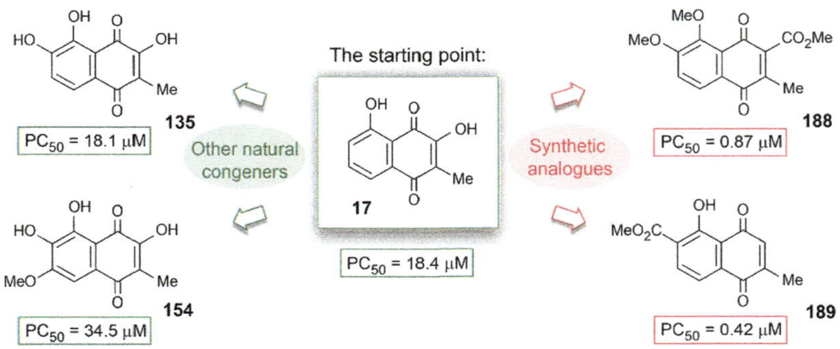

Fig. 124 Droserone (17) and dioncoquinones B (135) and C (154) from callus cultures of *A. abbreviatus* and *T. peltatum* and the two synthetically prepared hit compounds 188 and 189 displaying promising anti-austerity activities against PANC-1 pancreatic cancer cells.[601] PC$_{50}$ values represent concentration of the agent which causes the death of 50% of PANC-1 cells preferentially under nutrient-deprived conditions.

two naphthoquinones displayed PC$_{50}$ values in the submicromolar range, namely 0.87 µM (for **188**) and 0.42 µM (for **189**); these activities are, thus, in the same order of magnitude as arctigenin[428–430] (PC$_{50}$ = 0.83 µM), used as a positive control.

The naphthoquinones **188** and **189** were significantly more active than their parent compound droserone (**17**), by a factor of >20 for **188** and even of >40 in the case of **189**, but also more active than dioncoquinone B (**135**), which exhibited an antiproliferative effect similar to that of **17**. Dioncoquinone C (**154**) showed only weak growth-inhibitory activity, with a PC$_{50}$ value of 34.5 µM (see Fig. 124).[601]

The data hint at a significant contribution of the substituents and their positions on the 1,4-naphthoquinone molecular framework regarding the anti-austerity potential of the compounds. As outlined in Tables 9 and 10, the presence or absence of electron-donating or -withdrawing groups in the "western" (aromatic) or in the "eastern" (quinoid) part of the molecules had a substantial effect on the potency of the naphthoquinones. As an example, the presence of a strongly electron-donating group like OH at C-3 seems to diminish the activity drastically, by a factor of up to 37, as seen by comparison of two pairs of closely related analogues such as **189** (PC$_{50}$ = 0.42 µM) and **190** (PC$_{50}$ = 15.6 µM), and also in the case of **191** (PC$_{50}$ = 2.97 µM) and **192** (PC$_{50}$ = 98.8 µM) (see Table 9).[601] The presence of an electron-withdrawing group at C-3 like a methoxycarbonyl

Table 9 Influence of substituents on the "eastern" (quinoid) part of the naphthoquinone backbone on the anti-austerity activities of compounds **188–195**[a]

	189	190	191	192	193	188	194	195	arctigenin[b]
X^1	H	OH	H	OH	H	CO$_2$Me	Cl	Br	
X^2	OH	OH	H	H	OMe	OMe	OMe	OMe	
X^3	CO$_2$Me	CO$_2$Me	Me	Me	OMe	OMe	OMe	OMe	
PANC-1[a]	0.42	15.6	2.97	98.8	18.2	0.87	5.56	3.10	0.83

[a]PC$_{50}$; concentration at which 50% of the PANC-1 pancreatic cancer cells are killed preferentially in nutrient-deprived medium (NDM) without causing toxicity under nutrient-rich conditions in Dulbecco's modified Eagle's medium (DMEM).
[b]Used as a positive control.

Table 10 Influence of substituents on the "western" (aromatic) part of the naphthoquinone molecular framework, at C-5 and C-6, on the anti-austerity activities exerted by **189–191** and **196–201**.[a]

	196	191	197	189	198	190	199	200	201
X^1	H	H	H	H	H	OH	OH	H	H
X^2	OH	H	H	OH	OH	Oh	OH	OMe	OMe
X^3	Me	Me	CO_2Me	CO_2Me	CO_2H	CO_2Me	CO_2H	Cl	H
PANC-1[a]	1.21	2.97	3.36	0.42	12.4	15.6	25.4	2.06	7.80

[a]PC_{50}; concentration at which 50% of the PANC-1 pancreatic cancer cells are killed preferentially in nutrient-deprived medium (NDM) without causing toxicity under nutrient-rich conditions in Dulbecco's modified Eagle's medium (DMEM).

functionality (as in **188**) or of halogen atoms such as chlorine (in **194**) or bromine (in **195**) (see Table 9) resulted in an increase of the inhibitory activity compared to the parent compound **193**, which is unsubstituted at that position. In the case of **188** ($PC_{50} = 0.87\,\mu M$), the CO_2Me substituent led to an improvement of the PC_{50} value by a factor >20, compared to **193** ($PC_{50} = 18.2\,\mu M$).[601]

An activity-enhancing effect of electron-donating groups introduced at C-5 on the "western" (aromatic) part of the 1,4-naphthoquinone scaffold led a doubling of the potency for **196** ($PC_{50} = 1.21\,\mu M$) compared to **191** ($PC_{50} = 2.97\,\mu M$), and even to an increase of activity by the factor of 8 in the case of the hit compound **189** ($PC_{50} = 0.42\,\mu M$) compared to **197** ($PC_{50} = 3.36\,\mu M$) (see Table 10).[601] For plumbagin (**16**) and analogues, a likewise positive impact of an OH function at C-5 on the anti-austerity activity of the compounds was observed.[598]

Substituents at C-6, which act as H-bond acceptors possessing an electron-rich carbonyl oxygen of the ester group (while being electron-withdrawing to the ring) showed a distinctly stronger anti-austerity potency, as obvious from the test results of **189** ($PC_{50} = 0.42\,\mu M$) versus **198** ($PC_{50} = 12.4\,\mu M$) and **190** ($PC_{50} = 15.6\,\mu M$) versus **199** ($PC_{50} = 25.4\,\mu M$). Moreover, the electron-withdrawing chloro-substituent at C-6 in **200** ($PC_{50} = 2.06\,\mu M$) increased the activity by a factor of nearly 4 compared to **201** ($PC_{50} = 7.80\,\mu M$), which is unsubstituted at this position (see Table 10).[601]

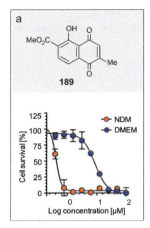 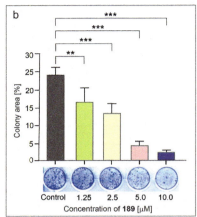

Fig. 125 (a) Preferential cytotoxic activity of the synthetically prepared 1,4-naphthoquinone **189** against PANC-1 human pancreatic cancer cells in nutrient-deprived medium (NDM) and in nutrient-rich Dulbecco's modified Eagle's medium (DMEM). (b) Effect of **189** on colony formation by PANC-1 cells: Graph (top) with mean values of the colony areas occupied by PANC-1 cells treated with different concentrations of **189** (three replicates), and representative wells (bottom) of the respective PANC-1 cell colonies; for statistical analysis, one-way ANOVA with Dunnett's multiple comparison test was used. ***$P < 0.001$, **$P < 0.005$, when compared with the untreated control group.

The two hit compounds, **188** and **189**, exerted pronounced cytotoxicity against PANC-1 cancer cells (as shown for **189** in Fig. 125a), with significant morphological and viability alterations and induction of apoptosis.[601] PANC-1 cells exposed to 1 μM of the naphthoquinones **188** and **189** in nutrient-deprived medium (NDM) for 24 h showed cellular damage with cell shrinkage and membrane blebbing, while untreated pancreatic cancer cells, serving as the control, maintained their normal morphology.[601]

Most of the patients suffering from pancreatic cancer develop liver and/or lung metastasis in the advanced stage of the disease.[602–604] Therefore, compounds with the ability to inhibit dissemination of cancer cells to remote organs have gained interest as promising candidates for targeting cancer metastasis. PANC-1 cells treated with different concentrations (1.25, 2.5, 5.0, and 10 μM) of **188** or **189** in DMEM for 24 h dose-dependently displayed a significant reduction of both the number and the size of colony. The cells in the control group (untreated) grew aggressively, occupying nearly about 25% of the total well area, while even the lowest concentration (1.25 μM) of **188** and **189** caused a drastic inhibition of colonization, by 80% (for **188**) and 84% (for **189**) (Fig. 125b).[601]

The serine/threonine kinase Akt (protein kinase B) is a signaling protein and key regulator of cell proliferation, survival, growth, and apoptosis.[605,606] Several elements of the Akt pathway are deregulated in many cancers, among them pancreatic cancer, resulting in its activation, in particular under conditions of nutritional stress.[607,608] Some of the anti-austerity agents, such as arctigenin[428–430] and the 5,8′-coupled naphthylisoquinoline alkaloid ancistrolikokine E$_3$ (**2**)[10] (for its structure, see Fig. 1), are strong inhibitors of the Akt cascade.

Western blot analysis (data not shown) of PANC-1 human pancreatic cancer cells treated with various concentrations of **188** and **189** under conditions of nutrient deprivation (NDM) and in nutrient-rich medium (DMEM) revealed that only **189** significantly inhibited the Akt signaling pathway, inducing a decrease in both Akt and p-Akt(S473) protein expression. Compound **188**, by contrast, did not show any downregulation of the Akt cascade,[601] thus suggesting different modes of actions for the two naphthoquinones, despite their—at first sight—structural similarity.

In conclusion, the results presented here on naturally occurring polyketidic naphthoquinones and structurally related, yet merely artificial congeners further confirm that *A. abbreviatus* is a promising source of potent lead compounds for the development of new agents against leukemia and pancreatic cancer.

7. Unprecedented biosynthesis of naphthylisoquinoline alkaloids—A novel acetate-polymalonate pathway to isoquinoline alkaloids in plants

Isoquinoline alkaloids are a fascinating class of natural products,[4,84–88] comprising about 2500 known, mostly pharmacologically important secondary metabolites, among them simple 1-alkylisoquinolines such as anhalonidine (**204**),[89] but also polycyclic representatives with complex molecular scaffolds like, e.g., the analgesic morphine.[91,92] Despite their great structural variety, all these compounds have in common that they originate from the biogenic amine dopamine (**203**) and, thus, ultimately from aromatic amino acids such as tyrosine (**202**), by a Pictet-Spengler-type condensation with aldehydes or α-keto acids, as a biosynthetic key step (Fig. 126, top).[84,85,87,89–94] The great structural diversity of isoquinoline alkaloids from

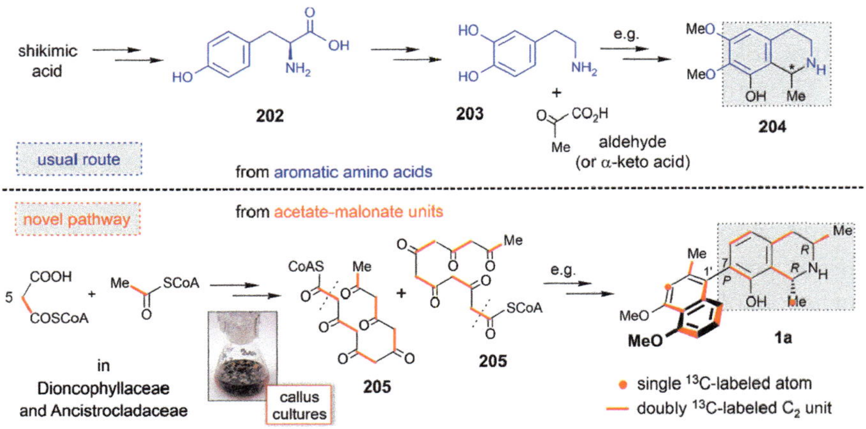

Fig. 126 Convergence in the biosynthesis of isoquinoline alkaloids in higher plants[83]—(a) the conventional biosynthetic pathway to tetrahydroisoquinoline alkaloids like anhalonidine (**204**) by Pictet-Spengler condensation starting from the biogenic amine dopamine (**203**), and, thus, ultimately from aromatic amino acids such as tyrosine (**202**), in comparison (b) to the novel acetate-polymalonate route to naphthylisoquinoline alkaloids such as dioncophylline A (**1a**), via a joint polyketide precursor **205**, giving rise to the entire carbon skeleton of **1a**.

higher plants results from the variation of the aldehyde precursor and from a series of follow-up reactions of the initially formed tetrahydroisoquinoline.

The naphthylisoquinoline alkaloids,[1–7] by contrast, with their unusual substitution patterns, in particular with the methyl group at C-3, the oxygen function at C-8, and the obviously acetogenic naphthalene substituent, hint at an unprecedented biosynthetic origin—not from aromatic amino acids as usual, but, rather, from acetate and malonate units, exclusively.[80–83] The entire carbon skeleton of naphthylisoquinolines, like dioncophylline A (**1a**), arises from a fundamentally different pathway to isoquinolines. Accordingly, both the naphthalene moiety and the isoquinoline half of these alkaloids are formed separately, yet via a joint open-chain hexaketide precursor **205** by stepwise cyclization steps—through aldol condensation reactions, and, in the case of the isoquinoline moiety, by nitrogen incorporation. The two molecular portions are then joined together convergently, by phenol-oxidative coupling, thus affording the complete final target molecule, here, e.g., dioncophylline A (**1a**),[80–82,268] as outlined in Fig. 126, bottom. Further metabolic follow-up reactions, such as reduction, oxidation, O- or N-methylation, or dimerization, then give rise to the plethora of structurally diverse naphthylisoquinoline alkaloids in Ancistrocladaceae and Dioncophyllaceae plants,[1–7] as described in Sections 3–6.

In conclusion, the naphthylisoquinolines are the as yet only known representatives of the large group of isoquinoline alkaloids that arise from polyketidic precursors,[80] and not, as usual, from aromatic amino acids.[84,91,92] The isoquinoline alkaloids are thus striking examples of biosynthetic convergence—structurally most similar compounds belonging to the same class of natural products formed via most different pathways.[83]

As presented in Fig. 127, key steps of the novel acetate-polymalonate biosynthetic route to naphthylisoquinoline alkaloids are the formation of the highly reactive β-pentaketone **205** from one acetate and five malonate units and its stepwise aldol condensation and aromatization, thus leading to the monocyclic intermediate **206**. From this diketone, the corresponding naphthalene portion **208** should arise by further aldol cyclization, while reductive amination of the more reactive aliphatic ketone **206** and intramolecular imine formation should lead to the dihydroisoquinoline **207**. This concept convincingly rationalizes the unusual substitution patterns both in the isoquinoline and in the naphthalene part. Oxidative phenolic cross-coupling of the two molecular moieties, **207** and **208**, will then directly give rise to the complete molecular skeleton of naphthylisoquinoline alkaloids, thus providing the target compounds, like e.g., dioncophylline A (**1a**), after some follow-up reactions (here in particular O-methylation reactions).[80–84,268]

The first clear hints at the chemical plausibility of the assumed acetogenic nature of naphthylisoquinoline alkaloids had been obtained, already early, from biomimetic cyclization[609] of the highly reactive polyketide chain **205** (R=H), leading to the monocyclic precursor **206** (R=H). This compound was smoothly transformed into the corresponding isoquinoline moiety **207** (dehydrogenated analogue of **207**) by reaction with ammonia as a nitrogen source, or into the naphthalene portion **208**, by a second aldol condensation, now with KOH/MeOH. The biomimetic imitation of the postulated biosynthetic pathway to naphthylisoquinoline alkaloids led to extremely short and efficient novel synthetic routes to isoquinolines and naphthalenes, via joint precursors, and, thus, to first, most rational total synthetic approaches to the molecular halves of the alkaloids.[5,609–611]

Another intriguing facet of the biosynthesis of naphthylisoquinoline alkaloids relates to the question at which stage and by which substrate properties the configuration at the centers and the axis is established, and which kinds of enzymes are involved in these reactions. Remarkably, all of the Dioncophyllaceae[5,6] and most of the Ancistrocladaceae[1–5] lianas (especially those from Asia[1]) strictly synthesize naphthylisoquinoline

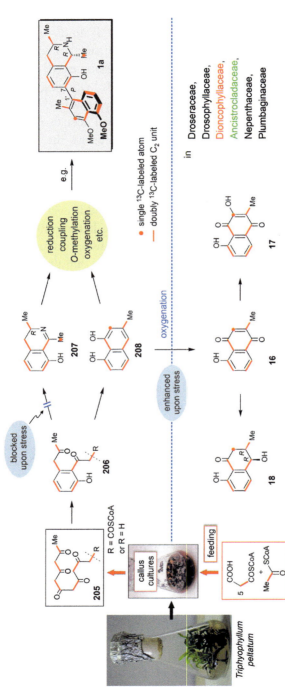

Fig. 127 Biosynthetic pathway to naphthylisoquinoline alkaloids like dioncophylline A (**1a**), from one acetate and five malonate units via joint polyketide and monocyclic diketo intermediates (e.g., **205** and **206**), providing the two molecular halves **207** and **208**, and stress-induced formation of related, yet N-free metabolites such as cis-isoshinanolone (**18**), plumbagin (**16**), and droserone (**17**). Feeding experiments with [$^{13}C_2$]-labeled acetate administered to callus cultures of *Triphyophyllum peltatum* (Dioncophyllaceae), proving the polyketidic origin of the isoquinoline part and the naphthalene moiety of **1a**.[80]

alkaloids in a highly regio- and stereoselective manner. Several West and Central African plants, by contrast, show a smaller degree of specificity and produce different categories[1,3,5] of alkaloids at the same time—typical Ancistrocladaceae-type alkaloids (as chemotaxonomically expected) and pure Dioncophyllaceae-type compounds and even mixed, hybrid-type structures with an oxygen function at C-6 and *R*-configuration at C-3.[3,5,11–16,28–35,53–65,175–182,288] Despite the formation of such complex mixtures of constitutionally and stereochemically complex diverse naphthylisoquinoline alkaloids produced by some of the African plants, the alkaloids are by no means built up in an uncontrolled or even chaotic manner. The appearance of the compounds in an enantiomerically pure form nearly exclusively, and the likewise mainly atroposelective construction of their biaryl linkage corroborates the assumption of a strictly enzyme-catalyzed pathway to naphthylisoquinolines, occurring with a high substrate specificity. This can be seen from the fact that never any of these alkaloids (except for fully dehydrogenated representatives) has been isolated as a mixture of all imaginable regio- and stereoisomeric forms.[1–3,5]

The unique biosynthetic cascade to acetogenic isoquinoline alkaloids is, however, extremely susceptible to chemical, physical, or biotic stress, which easily results in the alternative production of naphthoquinones like plumbagin (**16**) and droserone (**17**) and tetralones like isoshinanolone (**18**),[151–153,297] as a consequence of an impairment of the reductive incorporation of nitrogen into the assumed joint diketo precursor **206** to give the dihydroisoquinoline **207** (see Fig. 127). The co-occurrence of these provenly acetogenic natural products in Dioncophyllaceae and Ancistrocladaceae plants[5,6,68,151–153,268,269,297,560–562] further supports the formation of naphthylisoquinoline alkaloids from acetate and malonate units. The bicyclic compounds **16–18** are apparently formed from the free naphthalene part **108** of naphthylisoquinolines, by oxidation to give plumbagin (**16**), and by further oxygenation providing droserone (**17**) or by reduction, leading to isoshinanolone (**18**).[5,83,151,153,269]

1,4-Naphthoquinones like **16** and **17** and tetralones like **18** are well known metabolites of the closest phylogenetic neighbors to the Dioncophyllaceae and Ancistrocladaceae, such as the Plumbaginaceae, Droseraceae, Drosophyllaceae, and Nepenthaceae.[126,131,147,148] These related plant families, however, strictly produce nitrogen-free compounds, but no isoquinoline or even naphthylisoquinoline alkaloids. Thus, the capability to incorporate nitrogen into the postulated joint diketo precursor **206** to give the dihydroisoquinoline **207** is restricted to Dioncophyllaceae and Ancistrocladaceae plants, exclusively.[1–7]

7.1 Biosynthesis of isoshinanolone, plumbagin, and droserone in Ancistrocladaceae and Dioncophyllaceae

Particularly useful was the observation that the cultivation of suspended callus cultures of the Indian liana *Ancistrocladus heyneanus* (see Section 2.3.1) imposed stress to the cells, eventually resulting in the predominant formation of plumbagin (**16**) and isoshinanolone (**18**), while virtually no naphthylisoquinoline alkaloids were formed.[268,269] The occurrence of abundant amounts of these two compounds was used to study more closely their biosynthetic origin, by administration of differently [^{13}C]-labeled acetate to the cell cultures (established from sterile seeds, as described earlier in Section 2.3.1). An amount of 100 mg each of [2-^{13}C]-acetate or [^{13}C$_2$]-acetate was added to 100 mL of liquid medium with ca. 25 g of callus tissue, and the whole was incubated for 3 days. The calli (containing **16**) were then separated from the medium (containing **18**) by filtration.[269]

The ^{13}C NMR spectrum of isoshinanolone (**18**) obtained from the cell cultures after application of [2-^{13}C]-acetate expectedly confirmed the acetogenic nature of **18**, revealing a uniform incorporation rate of ca. 6% with enhanced signals at the carbon atoms 2-Me, C-1, C-8, C-6, C-10, and C-3 (Fig. 128a), and likewise showing the methylene positions alternating with the carbonyl-derived ones in the polyketide chain.[269] 1D and 2D INADEQUATE NMR experiments with material of **18** obtained from cell cultures incubated with [^{13}C$_2$]-acetate revealed the full folding mode of the polyketide chain, showing a pairwise coupling of the carbon atoms 2-Me/C-2, C-1/C-9, C-8/C-7, C-6/C-5, and C-10/C-4, while C-3 remained isolated, due to the loss of one carbon atom by decarboxylation (Fig. 128b).[269] This was the first proof of the so-called F folding mode[612] of an intermediate polyketide chain in higher plants, i.e., with two intact acetate-derived C$_2$ units in the first ring (referred to as folding mode F, as typical of fungal producers). Similar results (not shown) were obtained for plumbagin (**16**) revealing an incorporation rate of ca. 2–3%, and a labeling pattern that fully corresponded to that of isoshinanolone (**18**), thus proving the acetogenic nature of the two metabolites and confirming the folding of the polyketide chain of the intermediate **205** in the biosynthesis of **16** and **18** in Dioncophyllaceae and Ancistrocladaceae plants (see Fig. 127).[268,269]

Remarkably, droserone (**17**) was found to be the main metabolite in the medium of liquid cultures of *Triphyophyllum peltatum*.[153] These liquid cultures had been established from callus cultures[150] that had been initiated from axenically grown plants as described in Section 2.1. Besides droserone (**17**), the cell cultures of *T. peltatum* likewise produced plumbagin (**16**) and

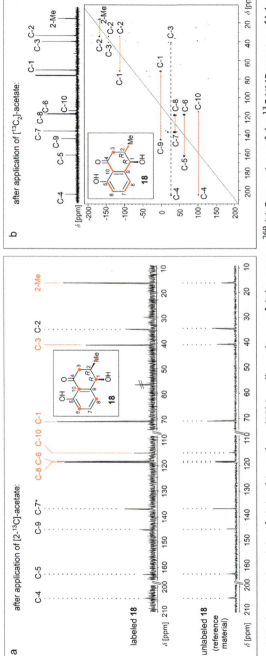

Fig. 128 Acetogenic origin of *cis*-isoshinanolone (**18**) in callus cultures of *A. heyneanus*.[269] (a) Comparison of the ^{13}C NMR spectra of labeled versus unlabeled **18** with the relative labeling intensities, after administration of [^{13}C$_1$]-acetate (relative peak areas standardized by the area of the peak at δ_C 136.9, corresponding to C-7). (b) 2D INADEQUATE spectrum of **18** from bond-labeling experiments with [^{13}C$_2$]-acetate, demonstrating an intact incorporation of C$_2$ units by the pairwise ^{13}C,^{13}C correlations; single ^{13}C-labeled carbon atoms are marked as red points, intact ^{13}C$_2$ units are indicated by a red bold bond line.

isoshinanolone (**18**), albeit in distinctly lower concentrations.[153] As already observed for the predominant formation of **16** and **18** in suspended callus cultures of *A. heyneanus*,[269] the release of droserone was a response to stress, too, caused by the transfer of the calli of *T. peltatum* from solid to liquid culture conditions.[153] The intact plants themselves, by contrast, did not produce significant amounts of this naphthoquinone.[5,6]

In all Dioncophyllaceae and Ancistrocladaceae plant species examined so far, droserone (**17**) had always been found to be accompanied by its oxygen-poorer analogue plumbagin (**16**), from which it might be derived biosynthetically.[1,5,151,153] The incorporation of intact [$^{13}C_2$]-labeled acetate units in droserone (**17**) isolated from liquid cultures of *T. peltatum* clearly revealed that **17** was, at least in part, synthesized de novo upon stress and was not e.g., the product of a degradation of naphthylisoquinoline alkaloids, which were likewise present in the calli of *T. peltatum*.[153] Neither did the formation of droserone (**17**) arise from a pre-formed depot precursor such as a leucoglycoside[613] or 5-O-glucosyldroserone,[614] as found in Droseraceae plants like e.g., *Dionaea muscipula*, *Drosera erythrorhiza* ssp. *magna*, or in cell cultures of *Drosera gigantea*.

7.2 Biosynthesis of dioncophylline A following an unprecedented acetate-polymalonate pathway to isoquinoline alkaloids

Given the high content of ancistrocladine (**20a**) (see Fig. 17 in Section 3) in the roots of *A. heyneanus*,[5,27,256–259] feeding experiments were performed by directly applying [^{14}C]- or [^{13}C]-labeled acetate and malonate via a syringe cannula to the roots of (unsterile) plants of *A. heyneanus*, which were grown in a thermostated and illuminated culture box. The incorporation rates of both acetate and malonate into **20a**, however, were too low for a localization of the labeled position by degradation or NMR spectroscopy.[5] Probable reasons for the unsatisfactory results were an inefficient administration of the acetate precursors to the roots, but possibly also the fact that general, primary metabolites are often rapidly catabolized by the plants or by co-occurring microorganisms, or that they are incorporated into numerous other primary and secondary metabolites. Another reason for the low incorporation rates may have been the slow growth of the plants.[5,224,225] Similar experiments using more advanced radiolabeled precursors[615] related to the diketo intermediate **206** likewise failed to yield significant incorporation rates.[5,615]

As reported in the previous paragraph, cell cultures of *Ancistrocladus heyneanus* produced large amounts of naphthoquinones and tetralones related to the naphthalene part of the alkaloids, but only very small quantities of naphthylisoquinoline alkaloids.[268,269] The breakthrough toward a reliable production of naphthylisoquinoline alkaloids finally came from callus cultures of *T. peltatum*,[150] which formed naphthoquinones, in particular droserone (**17**), too, but luckily, also larger amounts of dioncophylline A (**1a**).[80–82] For feeding experiments, the cells were cultivated on the same solid medium as used for general maintenance. Distributed over a period of 2 months, 250 mg of [1,2-$^{13}C_2$]-labeled sodium acetate were administered portion-wise to 832 mg (dry weight) of callus. From the methanolic extract of the lyophilized callus material, 1.7 mg of pure dioncophylline A (**1a**) were isolated and purified, and investigated by high-field NMR spectroscopy. The unique acetogenic origin of naphthylisoquinoline alkaloids was unambiguously proven by 2D INADEQUATE measurements for the identification of direct C—C connectivities, in combination with the NMR cryoprobe methodology to substantially improve the signal-to-noise ratio. As illustrated in Fig. 129, the NMR spectrum of [$^{13}C_2$]-labeled dioncophylline A (**1a**) thus obtained clearly revealed the presence of

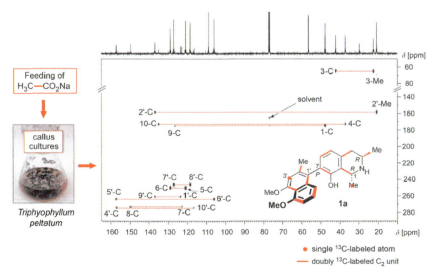

Fig. 129 2D INADEQUATE NMR spectrum of dioncophylline A (**1a**) from bond-labeling experiments with sodium [$^{13}C_2$]-acetate, proving the acetogenic nature of **1a** by visualizing the intact incorporation of [$^{13}C_2$]-units by their pairwise $^{13}C,^{13}C$-correlations[80]; likewise demonstrated was the formation of the two molecular units of **1a**, the naphthalene and the isoquinoline portion, according to the F-type folding mode,[612] here experimentally established for the first time in higher plants; single ^{13}C-labeled carbon atoms are marked as red points, intact [$^{13}C_2$]-units are indicated by a bold red bond line.

pairwise C—C correlations originating from the acetate-derived C_2-portions incorporated intact into **1a**, without bond cleavage.[80] The entire carbon skeletons of the two molecular halves of **1a** were derived from acetate units, displaying identical incorporation patterns for the naphthalene and the isoquinoline parts. In both molecular moieties of the alkaloid, decarboxylation had occurred at homologous positions, thus leaving the ultimate 1-methyl carbon atom and 3′-C as single ^{13}C atoms.[80] Furthermore, the full polyketide folding pattern was clearly and unambiguously identified. The naphthalene part and the isoquinoline portion were both found to arise from the F-type polyketide folding mode, as it had previously likewise been demonstrated for the related naphthoquinones plumbagin (**16**) and droserone (**17**) and for the tetralone isoshinanolone (**18**).[80,83,268]

7.3 Biosynthesis of dioncophylline A from late-stage dihydro- and tetrahydroisoquinoline precursors

One of the main features in the proposed biosynthetic route to naphthylisoquinoline alkaloids is the separate formation of the naphthalene and the isoquinoline portions, and their subsequent coupling to each other in a quite advanced, bicyclic form, not as a monocyclic intermediate or even as an open-chain precursor.[5,83] To investigate the biosynthesis of naphthylisoquinoline alkaloids from late-stage precursors in more detail, feeding experiments using cell cultures of *T. peltatum* were performed with the dihydroisoquinoline **207**, synthesized in a specifically $[1,1'$-$^{13}C_2]$-labeled form.[81]

Starting from the known[300,315] primary amine **208**, the labeled enantiomerically pure dihydroisoquinoline $[1,1'$-$^{13}C_2]$-**207** was prepared.[81] The synthesis basically followed previously established protocols,[315] which were further optimized in order to avoid loss of labeled material. For the introduction of the $[^{13}C_2]$-labeling, commercially available $[1,2'$-$^{13}C_2]$-acetyl chloride (**209**) was used and reacted with **208** to give the amide $[1,2'$-$^{13}C_2]$-**210**. Bischler-Napieralski cyclization gave the dihydroisoquinoline $[1,1'$-$^{13}C_2]$-**211**. Regioselective deoxygenation at C-6 afforded the monooxygenated compound $[1,1'$-$^{13}C_2]$-**212** in excellent yields. The desired target molecule, $[1,1'$-$^{13}C_2]$-**207**, was finally obtained by mild O-demethylation with sodium thiomethoxide in almost quantitative yield (see Fig. 130).[81]

The $[^{13}C_2]$-labeled putative precursor **207** was administered aseptically, in three portions (every 3 weeks), to the cell cultures of *T. peltatum* for a total period of 9 weeks.[81] The methanolic extract of the dried and ground callus

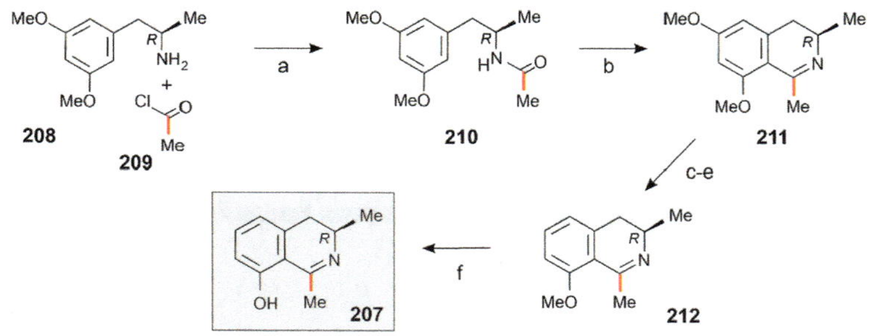

Fig. 130 Synthesis of the specifically [1,1'-^{13}C$_2$]-labeled, enantiomerically pure dihydroisoquinoline **207**[81] for administration to sterile callus cultures of *T. peltatum* as a putative precursor in the biosynthesis of naphthylisoquinoline alkaloids. Reaction Conditions: (a) NEt$_3$, CH$_2$Cl$_2$, r.t., 99%; (b) POCl$_3$, CH$_3$CN, 70 °C, 82%; (c) HBr, 110 °C, 69%; (d) NaH (dispersion in oil), triflic anhydride (Tf$_2$O), CH$_2$Cl$_2$, 0 °C, 92%; (e) ammonium formate, Pd/C, MeOH, 80 °C, 97%; (f) NaSMe, DMF, 160 °C, 97%.

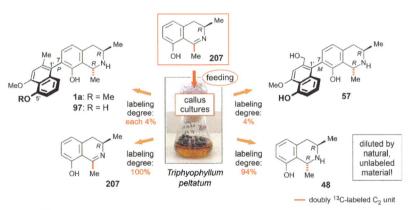

Fig. 131 [^{13}C$_2$]-Labeled metabolites identified in callus cultures of *T. peltatum* after administration of the [1,1'-^{13}C$_2$]-labeled dihydroisoquinoline **207**[81]—incorporation of the advanced precursor **207** into dioncophylline A (**1a**) and two minor naphthylisoquinoline alkaloids, 5'-O-demethyldioncophylline A (**97**) and habropetaline A (**57**); likewise observed was the labeled *trans*-configured tetrahydroisoquinoline **48**, isolated together with residual [1,1'-^{13}C$_2$]-**207**.

material was submitted to preparative HPLC, delivering dioncophylline A (**1a**) as the main metabolite, together with two minor alkaloids, 5'-O-demethyldioncophylline A (**97**) and habropetaline A (**57**), the *trans*-configured tetrahydroisoquinoline **48**, and residual dihydroisoquinoline **207** (Fig. 131).[81]

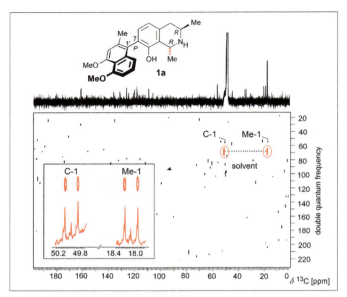

Fig. 132 INADEQUATE spectrum (in methanol-d_4) of dioncophylline A (**1a**) after feeding of [1,1'-$^{13}C_2$]-**207** to callus cultures of *T. peltatum*[81]; the expanded region (see included inset left side) indicates the two doublets of the [$^{13}C_2$]-unit of C-1 (δ_C 50.0) and Me-1 (δ_C 18.1), into which the [$^{13}C_2$]-label was incorporated.

^{13}C NMR investigations of the isolated compounds clearly revealed a significant incorporation of **207** into dioncophylline A (**1a**) and its two analogues **97** and **57** to an extent of ca. 4%, as evidenced from the two distinct doublets for the [$^{13}C_2$]-unit of C-1 (δ_C 50.0) and Me-1 (δ_C 18.1), into which the [$^{13}C_2$]-label was incorporated (see Fig. 132, here shown for **1a**).[81] These signals clearly exceeded the size of those originating from the non-labeled alkaloids (based on the natural ^{13}C abundance of ca. 1%). The results thus evidenced the first successful incorporation of a more differentiated precursor (compared to acetate) into a naphthylisoquinoline alkaloid, and clearly confirmed that the isoquinoline and naphthalene portions are formed separately and coupled to each other in a chemically differentiated form.[81]

The smooth conversion of the dihydroisoquinoline **207** to the *trans*-tetrahydroisoquinoline **48**, but not to the corresponding *cis*-diastereomer (not shown), which was not observed at all in the callus cultures of *T. peltatum* or the plants themselves,[81] clearly revealed that the reduction

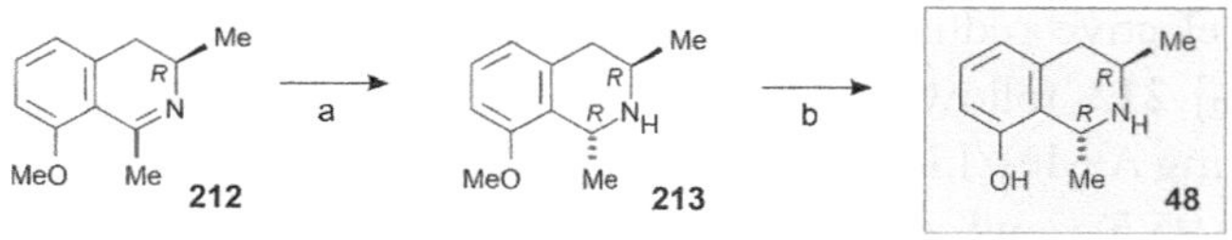

Fig. 133 Synthesis of the putative biosynthetic precursor phylline (**48**) in its [1,1'-^{13}C$_2$]-labeled form.[82] Reaction Conditions: (a) AlMe$_3$, LiAlH$_4$, THF, $-78\,°$C; then NaF, $-45\,°$C, warming up of the reaction slowly from $-45\,°$C to $0\,°$C stepwise within 4 h, 87%; (b) BBr$_3$, CH$_2$Cl$_2$, r.t., 99%.

proceeds highly diastereoselectively, thus indicating a specific, enzyme-mediated conversion of **207** to **48**. This is in accordance with the fact that a directed chemical synthesis of the *trans*-compound **48** in the lab is more demanding and requires particular reaction conditions[383] (see e.g., Fig. 133), because the use of common hydride transfer reagents such as NaBH$_4$ preferentially produces the *cis*-compound.[315] The newly formed *trans*-tetrahydroisoquinoline **48**, whose occurrence in the cell cultures had been triggered by the application of the [^{13}C$_2$]-labeled dihydroisoquinoline **207**, showed a very high labeling degree of 94%.[81] And yet, the fact that isolated **48** was not completely labeled, like 100%, as the administered compound, showed that [1,1'-^{13}C$_2$]-**48** was diluted by non-labeled—hence de novo formed—material.[81] This proved that **48** must be a genuine metabolite in the callus cultures of *T. peltatum*. The recovered dihydroisoquinoline **207**, by contrast, was found to be fully labeled.[81] It had so far been known as a merely synthetic compound, but had as yet never been detected in *T. peltatum* or related species—and still it might be formed in the callus cultures, present in very low steady-state concentrations, and might, upon formation, immediately be reduced to **48**, and then coupled to dioncophylline A (**1a**).

Remarkably, in *Habropetalum dawei*, a further Dioncophyllaceae liana closely related to *T. peltatum*, the *trans*-tetrahydroisoquinoline **48** had previously been identified as a genuine natural product, named phylline,[73,297] occurring in the plant only in trace amounts, thus hinting at its presence in low steady-state concentrations. This finding stimulated further feeding experiments, now with the natural product phylline (**48**) in a [^{13}C$_2$]-labeled form,[82] to address the question whether **207** or **48** is the authentic coupling substrate.

Synthetic access to the [1,1'-^{13}C$_2$]-labeled *trans*-configured tetrahydroisoquinoline **48**[82] was achieved in excellent chemical yields by a highly diastereoselective reduction of the monooxygenated dihydroisoquinoline [1,1'-^{13}C$_2$]-**212**, following a procedure[315] developed in the authors' group earlier, using AlMe$_3$/LiAlH$_4$ to give the tetrahydroisoquinoline [1,1'-^{13}C$_2$]-**213** (dr > 95:5), and subsequent O-demethylation of **213** with boron tribromide (see Fig. 133).

Following the protocol described above for the feeding experiments with the [^{13}C$_2$]-labeled dihydroisoquinoline **207**, administration of [1,1'-^{13}C$_2$]-**48** to solid callus cultures of *T. peltatum* likewise led to the isolation of dioncophylline A (**1a**) as the main metabolite, along with its 5'-O-demethyl analogue **97** and habropetaline A (**57**).[82] Unexpectedly, NMR experiments on a fraction containing **1a** and **97a** revealed the presence of another yet unknown naphthylisoquinoline alkaloid, displaying a mass peak of m/z 378.2063 ([M+H]$^+$), exactly matching that of dioncophylline A (**1a**).[82] This unknown metabolite was then also found to occur in callus material not treated with the [^{13}C$_2$]-labeled precursor phylline (**48**). While all attempts to resolve dioncophylline A (**1a**), 5'-O-demethyldioncophylline A (**56**), and the new alkaloid by HPLC or column chromatography failed, the separation succeeded by crystallization.[82] For this purpose, the mixture was dissolved in pure warm MeOH; after cooling down to ambient temperature, cold H$_2$O was added until a MeOH/H$_2$O ratio of 1:1 was reached. Storage of the solution at 4 °C for 24–48 h supported the precipitation of the new compound as confirmed by NMR measurements. This procedure had to be repeated at least once to obtain pure material of the new metabolite.[82]

^{1}H and ^{13}C NMR analysis revealed the new alkaloid to be a 1:1 mixture of two compounds. Their sets of signals were very similar to each other, with just slightly different chemical shifts, suggesting that they were diastereomers, probably atropo-diastereomers.[82]

The molecular formula was deduced to be C$_{24}$H$_{27}$NO$_3$, as evidenced from HRESIMS and from its ^{1}H NMR spectrum exhibiting the typical signals of a Dioncophyllaceae-type naphthyl-1,3-dimethyltetrahydroisoquinoline alkaloid with a spin system of six aromatic protons (two singlets and four doublets) (see Fig. 134a). A NOESY correlation sequence in the series {OMe-5'-H-6'-H-7'} excluded C-6' from being quaternary, so that it could not be the coupling position of the biaryl axis, whereas C-8' lacked a proton. The thus probable location of the biaryl axis at C-8' in the naphthalene moiety and at C-7 in the isoquinoline portion was firmly proven by HMBC interactions from H-1' and H-6 to C-8', and from H-7' to C-7, respectively.[82] This

Structural variety and pharmacological potential of naphthylisoquinoline alkaloids   245

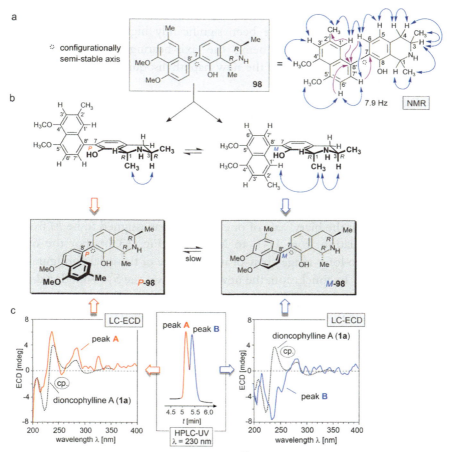

Fig. 134 5'-O-Methyldioncophylline D (**98**),[82] the first 7,8'-coupled naphthylisoquinoline alkaloid discovered in a Dioncophyllaceae plant: (a) ^{1}H,^{1}H-COSY and NOESY (double blue arrows) and HMBC (single pink arrows) interactions relevant for the constitution and the relative configuration at the stereogenic centers of **98**, and (c) assignment of the absolute axial configuration of the two configurationally semi-stable atropo-diastereomers, **98a** and **98b**, by LC-ECD coupling, by comparison of the LC-ECD spectra of peak **A** (left, in red) and peak **B** (right, in blue) with the ECD curve of the structurally related, P-configured dioncophylline A (**1a**).

assignment was further confirmed by an HMBC interaction of H-6' to C-8', and by examination of the coupling constant of the aromatic protons in the not methyl-substituted ring, i.e., of H-7' with H-6'. The new alkaloids showed a coupling constant of $J = 7.9$ Hz for these protons, typical of a 7,8'-coupling,[11,57,82] whereas in 7,6'-linked naphthylisoquinolines like e.g., dioncophyllinol B (**29**)[275] (see Fig. 20 in Section 3.1), the coupling constant of H-7' should have been significantly higher (ca. 8.4–8.6 Hz), thus again excluding C-6' being the axis-bearing carbon atom, and supporting the constitution of the new naphthylisoquinoline[82] as presented in Fig. 134a.

From a NOESY correlation from Me-1 to the likewise pseudo-axial proton at C-3 (see Fig. 134b), the relative configuration at C-1 versus C-3 was established to be *trans*. The absolute configuration at C-3 was determined as 3*R* by ruthenium-mediated oxidative degradation.[1,3,70,82] Thus, the new alkaloid was *R*-configured at C-3 and, given the relative *trans*-configuration assigned by NMR, it was *R*-configured at C-1, too. Although the two atropo-diastereomers could not be resolved, the absolute configuration at the biaryl axis of one of them was assigned to be *M* from a NOESY correlation from H-1' to Me-1 (see Fig. 133b). For the other atropisomer, no long-range cross-peak indicative of the axial configuration was detected.[82] In conclusion, the new alkaloid was the 5'-*O*-methyl derivative of the likewise 7,8'-coupled, yet merely synthetic naphthylisoquinoline dioncophylline D,[275] a previously postulated natural product, but not yet found in nature.

Online HPLC-ECD investigations giving rise to two very closely eluting, but not baseline separated peaks on a reversed-phase column provided two nearly mirror-imaged ECD spectra.[82] The absolute axial configurations of these two atropo-diastereomers were attributed by comparison of their ECD curves with that of the structurally closely related alkaloid dioncophylline A (**1a**), differing only by the (chiroptically less influential) position of the methyl group on the naphthalene portion. The more rapidly eluting peak **A** showed an ECD spectrum nearly identical to that of **1a**, while the more slowly eluting peak **B** displayed a ECD curve opposite to that of **1a**, thus evidencing the more rapidly eluting atropo-diastereomer as *P*-5'-*O*-methyldioncophylline D (**98a**), while the more slowly eluting isomer represented *M*-5'-*O*-methyldioncophylline D (**98b**) (see Fig. 134c).[82]

The significance of the natural tetrahydroisoquinoline phylline (**48**) for the biosynthetic route to naphthylisoquinoline alkaloids became apparent from its unambiguous incorporation into the final target molecules dioncophylline A (**1a**), its 5'-*O*-methyl derivative **97**, habropetaline A (**57**), and the newly discovered metabolite 5'-*O*-methyldioncophylline D (**98**).[82]

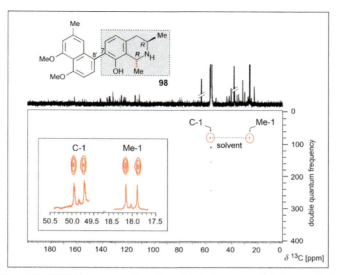

Fig. 135 2D INADEQUATE spectrum (methanol-d_4) of 5'-O-methyldioncophylline D (**98**) after feeding its molecular half, the natural product phylline (**48**), in a labeled [1,1'-$^{13}C_2$]-form (see the gray box), to solid callus cultures of *T. peltatum*[82]; the expanded region (see included window left side) indicates the two doublets of the [$^{13}C_2$]-unit of C-1 (δ_C 50.0) and Me-1 (δ_C 18.2), into which the [$^{13}C_2$]-label was incorporated.

^{13}C NMR and 2D INADEQUATE measurements on pure labeled 5'-O-methyldioncophylline D ([$^{13}C_2$]-**98**) revealed a significant incorporation of the applied precursor [$^{13}C_2$]-**48** into **98** to an extent of ca. 8% (Fig. 135), similar to that found for habropetaline A (**57**) and likewise also observed for the purified material of dioncophylline A (**1a**) and its 5'-O-methyl analogue **97** (not shown).[82] The two distinct doublets for C-1 and Me-1 (at δ_C 50.0 and 18.2, respectively) clearly exceeded the size of those of the corresponding peaks of the non-labeled alkaloids, i.e., based on the natural ^{13}C abundance of ca. 1% and, thus the double labeling degree of ca. 0.01%. Residual material of the applied [1,1'-$^{13}C_2$]-**48** showed a labeling degree that was significantly diluted down to 92%, thus clearly revealing that the labeled material of **48** had been diluted by natural **48** formed de novo, in the callus cultures of *T. peltatum*. The dihydroisoquinoline **207** (Fig. 136) was not detected in the callus extract.[82]

Feeding experiments with [1,1'-$^{13}C_2$]-**207** had likewise resulted in the formation of labeled naphthylisoquinolines, yet displaying a significantly lower labeling degree of ca. 4%.[81] With respect to the simultaneous isolation of the *trans*-configured tetrahydroisoquinoline **48**, showing a high labeling

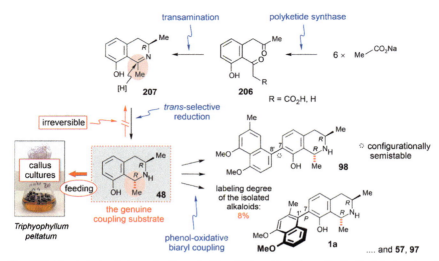

Fig. 136 The tetrahydroisoquinoline phylline (**48**) as the biosynthetically latest not yet coupled intermediate in the biosynthesis of Dioncophyllaceae-type naphthylisoquinolines like the new 7,8'-coupled alkaloid 5'-O-methyldioncophylline D (**98**), as well as the 7,1'-linked dioncophylline A (**1a**) and its 5'-O-demethyl derivative **97**, and habropetaline A (**57**).

degree of ca. 94 %, these results fully supported that **207** is biosynthetically "upstream," i.e., this intermediate is synthesized prior to the tetrahydroisoquinoline **48**, to which it is then immediately reduced *trans*-selectively and irreversibly.[81,83] These findings proved that **48**—and not the significantly electron-poorer dihydroisoquinoline **207**—was the authentic substrate (Fig. 135) for the enzymatic phenol-oxidative coupling reaction with the respective naphthalene unit, to give the final naphthylisoquinoline alkaloids.[82,83]

The nature of the enzymes that catalyze the reaction cascade of the acetate-polymalonate pathway leading to naphthylisoquinoline alkaloids is still unknown. Of particular interest is the most unusual transaminase that reductively incorporates nitrogen into the monocyclic precursor **206** to give the isoquinoline portion—maybe similar to the one involved in the biosynthesis of coniine. Likewise rewarding is the identification of the unusual polyketide synthase (PKS) and its biochemical and functional characterization in comparison to other, more "usual" plant PKSs (like e.g., chalcone synthases). The unique ability of the PKS in Ancistrocladaceae and Dioncophyllaceae plants to promote the production of the respective naphthalene and isoquinoline portions in parallel, in a 1:1 ratio, remains to be explored in more detail. The close organizational cooperation of the PKS with the transaminase and the coupling enzyme to give the final target molecules in a high regio- and stereoselective manner deserves particular interest. Also remarkable is the fact that, when exposed to stress, the production of corresponding *N*-free compounds prevails, such as naphthoquinones and tetralones, which is a further fascinating aspect in the biosynthesis of acetogenic natural products in Ancistrocladaceae and Dioncophyllaceae plants. All these unsolved questions make the biosynthesis of naphthylisoquinoline alkaloids a thrilling topic, worth further, in-depths investigations.

8. Chemo-ecological interactions and localization of naphthylisoquinoline alkaloids in intact plants

As reported in the previous sections, a great deal of information has meanwhile become available on naphthylisoquinoline alkaloids—on their structures,[1–6,54,55,58–60] their biosynthetic origin,[80–83] and their properties as potential chemotherapeutic agents for the treatment of tropical diseases such as malaria[1–4,7,36–41,58,180,182,250,254] and various malignancies, among them leukemia,[1–4,51–54,249,251,252,366] lung,[59,60] breast,[59,365] cervical,[3,4,31,55,56,183] and pancreatic cancer.[3,10,11,31,56,61,183] Furthermore, diverse synthetic pathways have been elaborated to provide access to these—in most cases rotationally hindered—biaryl natural products, independent from capricious plant material, including the development of efficient novel methods for the regio- and stereoselective construction of biaryl axes.[76–79,378,381,382,450,509,526,528,535,616–626] Far less, however, is known about the chemo-ecological interactions of the Dioncophyllaceae and Ancistrocladaceae with their natural environment and about the involvement of naphthylisoquinoline alkaloids and related compounds in defense-mediating processes, allowing the plants to cope with herbivores, with threatening microorganisms, and with neighboring plants. Another topic that is also little explored concerns the distribution, transport, and storage of naphthylisoquinoline alkaloids in distinct plant organs and tissues.

In this section, a short overview is given on investigations regarding the formation of naphthylisoquinoline alkaloids and related naphthoquinones as a result of abiotic or biotic stress,[5,80–83,151,264,268,560–562] along with first studies on the potential of these secondary metabolites to repel parasitic plants,[560] herbivorous insects,[98,99] and pathogenic fungi.[5,97] Fascinating insight into the anatomy of roots, stems, leaves, and fruits of *Ancistrocladus* species was obtained by the use of non-invasive analytical techniques such as FT-Raman microspectroscopy[108–110] or NMR microscopy,[104,107] the latter also in combination with NMR chemical-shift imaging (CSI) measurements.[105–107] In addition, remarkable results are described regarding the accumulation and storage of naphthylisoquinoline alkaloids and related compounds in special compartments of the roots, stems, and leaves of the plants.[104,105,108–110]

8.1 Chemo-ecological interactions

8.1.1 Ancistrocladus robertsoniorum *produces droserone as a "chemical weapon"*

Ancistrocladus robertsoniorum,[18–20,166–168] indigenous to the lowland moist evergreen coastal forests of Southeastern Kenya, produces a characteristic pattern of 5,8'-, 5,1'-, and 7,1'-coupled naphthylisoquinolines.[277,280] All of these compounds, among them the new ancistrobertsonines A–D (**214–216**, and **21**) (Fig. 137) belong to the subclass of Ancistrocladaceae-type[1,3,5] alkaloids (3S, 6-OR) exclusively. The most spectacular phytochemical peculiarity of *A. robertsoniorum*, however, concerns the discovery of orange-colored crystals in insect-damaged parts of the stems (Fig. 138). When mechanically separating bark and light wood, one could notice that the region around wounds was darker, and small cavities were filled with small orange-colored crystal needles.[561,562] Mass spectrometry in

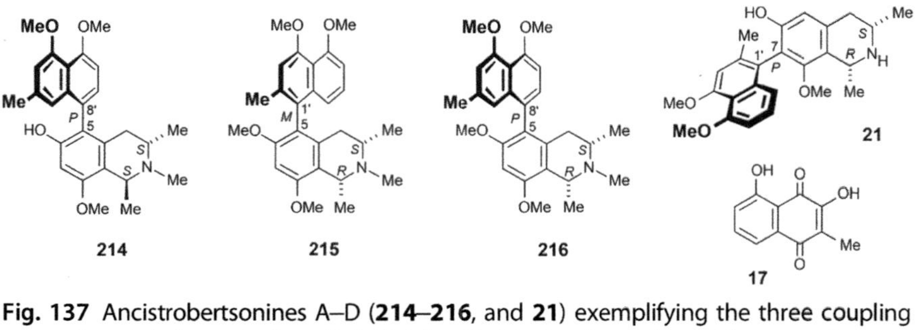

Fig. 137 Ancistrobertsonines A–D (**214–216**, and **21**) exemplifying the three coupling types of naphthylisoquinoline alkaloids identified in *A. robertoniorum*. They were isolated along with the naphthoquinone droserone (**17**), which was found in insect-wounded parts (see Fig. 138), presumably formed in the plant as a phytoalexin.

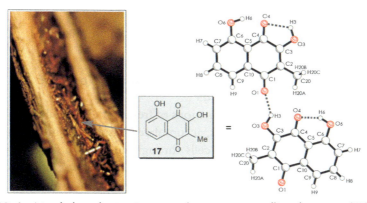

Fig. 138 *Ancistrocladus robertsoniorum* produces pure crystalline droserone (**17**) when wounded: X-ray structure of the "biogenic" crystals.

combination with NMR, IR, and UV spectroscopy revealed the presence of pure crystalline material of the naphthoquinone droserone (**17**). Apparently, when attacked by insects, *A. robertsoniorum* produces droserone (**17**) and locally concentrates it in the stems. Amazingly, the melting point (119 °C) of the crystals,[5,561,562] also highly pure, was lower than the value in the literature (181 °C).[131]

These "biogenic" crystals were of high purity and quality and size permitting an X-ray structure analysis performed directly on the natural material,[562] which confirmed the identity of the naphthoquinone droserone. As shown in Fig. 138, intra- and intermolecular hydrogen bonding between the carbonyl oxygen and the hydroxy groups of the molecules plays an important role in the crystals. Interestingly, recrystallization of the pure natural material from organic solvents provided crystals with a "correct," literature-conformous melting point (172–173 °C)—but these crystals were now no longer suitable for an X-ray structure analysis.[561,562] The low melting point of the pure "biogenic" crystals might be a hint at a new crystal modification of droserone, maybe occurring as a consequence of a slow sublimation process in the plant.

Droserone (**17**) as such had already been known for a long time,[131] it had also been isolated from other Ancistrocladaceae and Dioncophyllaceae plants.[1,5,6,151,153,311] The occurrence of **17** in *A. robertsoniorum*, moreover in a crystalline form, however, was completely unprecedented. As the crystals were found only in those parts that had been damaged either by insects or mechanically, and regarding the known antimicrobial activity of droserone (**17**) against a wide range of plant pathogens,[564–567] it can be assumed that this naphthoquinone is produced by *A. robertsoniorum* as a "chemical weapon" against herbivores and microorganisms.

8.1.2 Accumulation of ancistroheynine A in aging parts of Ancistrocladus heyneanus

During light microscopic investigations of fresh stem material of the Indian liana *A. heyneanus*, yellow solutes, accumulated in spherical inclusions, were detected in single parenchyma cells throughout the whole cross-section (Fig. 139a). A bright-yellow fluorescence (Fig. 139b) was observed upon irradiation with UV light ($\lambda = 365$ nm). This yellow material mainly occurred in the pith and bark regions of the shoots as obvious from fluorescence microscopy images (Fig. 139c).[104,108,264] Extraction with methanolic HCl of pith material prepared by peeling off the bark and the xylem/phloem regions provided a bright-yellow amorphous solid, which was chemically homogeneous. The accumulated compound turned out to be a new 7,8'-coupled naphthyldihydroisoquinoline alkaloid, which was named ancistroheynine A (**217**).[1,264] Its absolute configuration at C-3 was

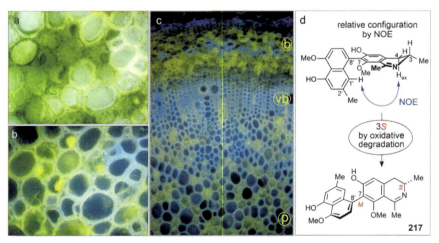

Fig. 139 Accumulation of ancistroheynine A (**217**) in spherical inclusions in the shoots of *A. heyneanus*.[264] (a) Bright-field microscopic image of the pith region with yellow cell inclusions (magnification ×472). (b) Corresponding image using fluorescence microscopy, showing a bright yellow fluorescence of the inclusions (irradiation, $\lambda = 365$ nm). (c) Fluorescence microscopy image of the shoot, showing a dense distribution of yellow material in the pith and bark regions (magnification ×215). The shown tissues are bark region (= b), vascular bundle region (= vb), and pith region (= p). (d) Elucidation of the full stereostructure of the yellow compound **217**, by ruthenium-mediated oxidative degradation and long-range NMR experiments. *Photos:* H. Schneider.

attributed to be S by ruthenium-mediated oxidative degradation. The axial configuration was established to be M by long-range NOE measurements covering the large distance between H_{ax}-4 and H-1' (Fig. 139d).[1,264]

A strong correlation of the yellow colored solutes as detected in aging parts of *A. heyneanus* by light and fluorescence microscopy to this new alkaloid ancistroheynine A (**217**) was also supported by comparison of localized Raman spectra monitored in situ in the shoot tip and the leaf midrib with the FT-Raman spectrum of the pure isolated compound (Section 8.3.1).[108] In addition, NMR microscopy and chemical shift imaging (CSI) experiments on shoot tissue also showed a precise correspondence of the yellow autofluorescence image arising from the presence of **217** with the NMR image obtained from the distribution of the aromatic signals (Section 8.2.2).[104,105] Owing to the enhanced accumulation of ancistroheynine A (**217**) in spherical inclusions, in particular in older cuttings of *A. heyneanus*, this naphthylisoquinoline alkaloid was assumed to originate as a stress metabolite, as already described for the naphthoquinones plumbagin (**16**) and droserone (**17**).

Plumbagin (**16**) was found to accumulate in *A. heyneanus*, too, mainly in stem bark cells (see Fig. 140) or directly at the site of invasion, e.g., when attacked by the holoparasitic plant *Cuscuta reflexa* (see next section), as observed by light-microscopic investigations.[5,560] Stress-induced formation of plumbagin (**16**) and droserone (**17**) was also found in *A. abbreviatus* and *Triphyophyllum peltatum*.[5,151–153] Also other, oxygen-richer quinones are produced as stress metabolites, such as ancistroquinone A (**53**), which was identified in the aerial parts of *A. heyneanus*.[314] As already described in Section 7.1, the cultivation of suspended callus cultures of *A. heyneanus* imposed stress to the cells, eventually resulting in the production of large

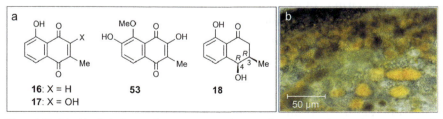

Fig. 140 (a) Naphthalene derivatives such as plumbagin (**16**), droserone (**17**), ancistroquinone A (**53**), and *cis*-isoshinanolone (**18**) isolated from *A. heyneanus*, presumably formed in the plant as phytoalexins. (b) Bright-field microscopic image of the root bark showing an accumulation of plumbagin (**16**), which is deeply orange-colored due to staining with ammonia solution. *Photo:* H. Schneider.

amounts of naphthoquinones and related tetralones.[151,268,269] These bicyclic compounds apparently arise from the naphthalene part of naphthylisoquinoline alkaloids, by oxidation to give plumbagin (16), and by further oxygenation providing droserone (17) and ancistroquinone A (53), or by reduction leading to *cis*-isoshinanolone (18).[268,269]

8.1.3 Interaction of Ancistrocladus heyneanus *with the herbal parasite* Cuscuta reflexa

Given the stress-induced variation of the metabolite pattern of *A. heyneanus*, stress exerted by plant parasites belonging to the genus *Cuscuta* might be a useful tool to study in more depth plant/plant interactions. *Cuscuta reflexa* Roxb. (Convolvulaceae) is a parasitic flowering plant with a low chlorophyll content growing on a wide range of herbaceous and woody host plants.[627–629] It is completely dependent on a host plant for support and food supply. While water and inorganic nutrients are absorbed through the xylem connections between a host and *C. reflexa*, organic compounds are transported from the phloem of the host to that of the parasite.[629–631] Although not yet concretely reported, an interaction between *A. heyneaus* and *C. reflexa* would be possible also in a natural environment, since their geographical distribution and habitats are overlapping.

Cuscuta reflexa was raised on the compatible host *Coleus blumei* Benth. (Labiatae). After coiling an unattached shoot of the parasite around young stems of *A. heyneanus*, *C. reflexa* immediately started to grow tightly around its new host and was found to insert a lot of close connections (so-called haustoria) between the phloem and xylem of *A. heyneanus* (Fig. 141a).[560] About 14 days after the first infection, an accumulation of a yellow-orange pigment surrounding the haustoria was observed by light microscopy (Fig. 141a,b).[560] Treatment of sections through the haustoria with ammonia resulted in a change of the color of the infected region to red-brown, thus hinting at the presence of plumbagin (16) and/or related other hydroxyquinones. HPLC analysis of the extracted pigment and comparison with authentic 16 proved that the compound accumulated at the sites of invasion to be plumbagin.[560] Hypersensitive reactions accompanied by lignification of the wound led to a degeneration of the haustoria, and to an expulsion of *C. reflexa* (Fig. 141c) from the stems of *A. heyneanus*.[5,560] The incompatible relationship between *Cuscuta* and *Ancistrocladus* was also demonstrated by the observation that *C. reflexa* rapidly began to turn black from the cut end and at the haustoria when the parasite was detached from its primary host *Coleus blumei* and left to grow on *A. heyneanus* exclusively.[560]

Structural variety and pharmacological potential of naphthylisoquinoline alkaloids 255

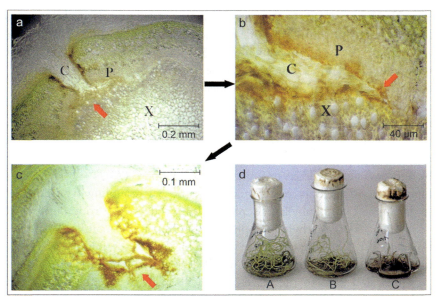

Fig. 141 Incompatible parasitic relationship between *Cuscuta reflexa* and *Ancistrocladus heyneanus*. (a,b) Accumulation of plumbagin (**16**) (see red arrow) in stems at the site of invasion by *C. reflexa* (C) between the phloem (P) and the xylem (X) of the host. (c) Dying of *C. reflexa* 3 days after experimental infection of *A. heyneanus* and detachment of the parasite from its compatible host, *Coleus blumei*. (d) Sterile shoot cultures of *C. reflexa* grown on Murashige & Skoog liquid media[271]: (A) as controls for 4 weeks, (B) 4 weeks after addition of *cis*-isoshinanolone (**18**), and (C) 4 weeks after addition of plumbagin (**16**); 100 µL of 25 mM solutions of **18** or **16** were added to a 100 mL Erlenmeyer flask. Photos: J. Schlauer.

Experiments with shoot cultures of *C. reflexa* (Fig. 141d) clearly revealed plumbagin (**16**) to display a strong toxicity toward the parasite leading to the complete death of the cultures 4 weeks after addition of **16**, while the controls still showed normal growth and proliferation. The effects caused by the structurally related tetralone *cis*-isoshinanolone (**18**) were less marked, but still led to an obvious growth inhibition 4 weeks after the treatment with **18**.[560]

8.1.4 Activities against herbivores: Insect-growth retarding and antifeedant activities of the alkaloids of Triphyophyllum peltatum

Triphyophyllum peltatum constitutes a rich source of naphthylisoquinoline alkaloids, of which so far about 20 compounds have been identified,[5,6,274,275,292–298,402] most of them occurring only in this peculiar liana, apart from the more wide-spread alkaloid dionocophylline A (**1a**) and its derivatives. Dioncophylline A (**1a**) has also been isolated as a main metabolite from the Dioncophyllaceae liana *Dioncophyllum thollonii*,[6,68,73] and from the African *Ancistrocladus* species *A. abbreviatus*,[5,63] *A. barteri*,[5,285] and *A. ileboensis*.[51] Dioncophylline B (**218**),[5,6,298] which possesses a configurationally unstable biaryl axis, is mainly present in the leaves of *T. peltatum*. This alkaloid does not form stable atropisomers, but undergoes a rapid rotation about its 7,6′-located axis, since its biaryl linkage is neighbored by only two small hydroxy groups. Dioncophylline C (**73**),[5,6,402] a main root metabolite, is the only representative of the 5,1′-coupling type detected so far in a Dioncophyllaceae plant. From callus and root cultures of *T. peltatum*, 5′-*O*-methyldioncophylline D (**98a**) and its epimer (**98b**) were isolated.[82,311] They are the only examples of the 7,8′-coupling type identified so far in organs of a Dioncophyllaceae plant. The lack of a methyl group located next to the biaryl axis of **98** leads to a reduced rotational barrier and hence to a slow interconversion of the two atropo-diastereomers at room temperature. All the compounds presented in Fig. 142—like all the other naphthylisoquinoline alkaloids isolated from *T. peltatum*—have in

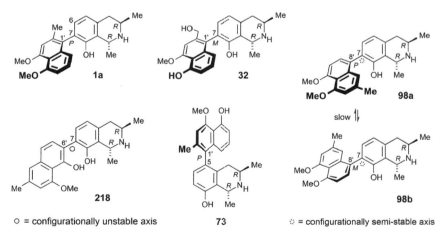

Fig. 142 *Triphyophyllum peltatum* produces alkaloids of four different coupling types, here some of the most prominent representatives: dioncophylline A (**1a**), and dioncopeltine A (**32**) (7,1′-coupling), dioncophylline B (**218**) (7,6′), dioncophylline C (**73**) (5,1′), and 5′-*O*-methyldioncophylline D (**98a**) and its atropo-diastereomer **98b** (7,8′); the latter were isolated exclusively from callus and root cultures, but not from the plant itself.

common the characteristic structural combination of "Dioncophyllaceae-type" alkaloids,[1,3,5] i.e., the *R*-configuration at C-3 and the absence of an oxygen function at C-6.

Dioncophylline A (**1a**) was shown to exert pronounced feeding deterrency and growth-retarding activity against larvae of the polyphagous herbivorous insect *Spodoptera littoralis* (Noctuidae), whereas compounds with additional hydroxy groups, such as dioncopeltine A (**32**) and the dioncophyllines B (**218**) and C (**73**), proved to be less active.[98,99] When neonate larvae were reared on artificial diet spiked with concentrations of **1a** as present in plant material of *T. peltatum*, high mortality of the larvae (e.g., 86% after 20 days at $1\,\text{mg}\,\text{g}^{-1}$ fresh weight of diet) and strong growth reduction (89% under the same conditions) was observed.[98] Feeding experiments to determine dose-response curves revealed dioncophylline A (**1a**) to strongly affect survival of the larvae in a concentration-dependent manner (Fig. 143a), with an EC_{50} value of $277\,\mu\text{g}/\text{g}$ fresh weight of diet, i.e., the concentration that inhibited larval growth by 50%.[98] Survival of the larvae on the diet treated with **1a** was monitored over a period of 28 days, by which time all of the control larvae (i.e., released to control diet without **1a**) had pupated. Surviving larvae kept on artificial diet spiked with dioncophylline A (**1a**), by contrast, were severely retarded with respect to growth and development (Fig. 143b), again in a dose-dependent manner.[98] None of the treated larvae had started to pupate, even after prolongation of the feeding on diet spiked with dioncophylline A (**1a**) for at least a further 7 days. The

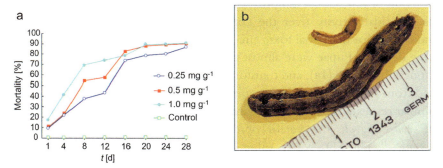

Fig. 143 (a) Mortality of larvae of *S. littoralis* on diet spiked with different concentrations of the insecticidal dioncophylline A (**1a**). (b) Growth retardation of larvae of *Spodoptera littoralis* after 25 days of treatment with dioncophylline A (**1a**, $1\,\text{mg}\,\text{g}^{-1}$ fresh weight of diet), in comparison to a control larvae, i.e., kept on the standard diet. *Photo:* B. Wiesen.

strong sublethal effects displayed by this naphthylisoquinoline were also demonstrated by the fact that the larvae needed another 20 days to pupate, after being transferred to control diet (i.e., without **1a**) on day 35 of the experiment.[98] This deleterious effect of **1a** was probably due to the strong antifeedant activity of **1a** since the larval net weight gain and the amount of diet consumed were drastically reduced compared to the controls when neonate larvae were released on normal diet with no dioncophylline A (**1a**). In addition, the feeding experiments gave hints at morphological changes induced by **1a**, since the larvae on the spiked diet significantly differed from control larvae on diet not containing **1a** in their size and also in their color.[98] Uptake of dioncophylline A (**1a**) resulted in a distinctly paler color of the larvae (Fig. 143b), thus indicating that **1a** may interfere with processes related to pigmentation and moulting of the insects.

8.1.5 Dioncophylline A as a larvicide against Anopheles stephensi and Aedes aegypti

Dioncophylline A (**1a**) was found to possess a potent activity against the larvae of *Anopheles stephensi*, one of the mosquito species responsible for the transmission of malaria.[100] The larvicidal activity of **1a** became apparent within the first 4 h of incubation in each instar larval stage. The alkaloid displayed a specific activity against the younger instar larval stages, with $LC_{50}/4h$ values (i.e., median lethal doses required to kill half the members of a tested population after a specified test duration, here after treatment for 4 h) below $1 \, mg \, L^{-1}$. Older larval stages were less affected ($LC_{50}/4h$, $4.1 \, mg \, L^{-1}$).[100] In each instar larval stage, the LC_{50} values decreased as a function of time, suggesting that dioncophylline A (**1a**) continued to exert its larvicidal activity over the complete time of incubation (0–48 h) or indicating that the initial toxic effect had an impact on larval mortality with time delay.[100] Pupae were almost completely insensitive to the presence of dioncophylline A (**1a**), and transformation processes from larvae into pupae and from pupae into adult mosquitoes were not influenced by **1a** at all.[101] Similar results were observed regarding the lethal effects of **1a** toward larvae of *Aedes aegypti*, the vector of dengue hemorrhagic fever and yellow fever.[5,101] The $LD_{100}/24h$ value of **1a** (dose having 100% probability of causing death in a tested population) was determined to be 5 ppm; for comparison, the phenylpropanoid β-asarone, used as a standard larvicide,[632] killed 100% of the larvae at a concentration of 16 ppm.[101] These findings showed that dioncophylline A (**1a**) can be regarded as a promising candidate as a phytolarvicide.

8.1.6 Molluscicidal activities of the alkaloids of Triphyophyllum peltatum against the snail Biomphalaria glabrata

Naphthylisoquinoline alkaloids containing extracts of *A. abbreviatus*, *A. barteri*, and *T. peltatum* were found to display significant molluscicidal activities against the tropical snail *Biomphalaria glabrata*, the intermediate host of the schistosomiasis parasite *Schistosoma mansoni*, with $LD_{100}/24h$ values of 100, 200, and 200 ppm, respectively.[102,103] Among the pure isolated alkaloids investigated, dioncophylline A (**1a**) proved to be the most potent agent, exhibiting an $LD_{100}/24h$ value of 20 ppm.[102,103] Other naphthylisoquinoline alkaloids, by contrast, like e.g., the dioncophyllines B (**218**) and C (**73**), were virtually inactive, showing $LD_{100}/24h$ values higher than 40 ppm.[102] Structure–activity relationship studies on dioncophylline A (**1a**) showed that 8-O-alkylation in nearly all cases significantly enhanced the efficacy (see Table 11).[103] The evaluation of 8-O-benzyldioncophylline A (**219**) ($LD_{100}/24h$, 10 ppm) and further related analogues, the substituted 8-O-benzyl derivatives **101** and **219–222**, showed that in particular 8-O-(p-bromobenzyl)dioncophylline A (**220**)

Table 11 $LD_{100}/24h$ values (ppm) of dioncophylline A (**1a**) and related 8-O-benzyl derivatives (**101**, and **219–222**) against *Biomphalaria glabrata*.

Compd.	R	$LD_{100}/24h$ [ppm]
1a	–	20
219	H	10
220	Br	3.13
221	Cl	6.25
101	NO$_2$	12.5
222	CF$_3$	6.25

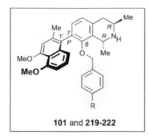

101 and 219-222

Trifenmorph was used at a concentration of 0.25 ppm as the standard, reducing parasitemia after 24 h completely.

(LD_{100}/24 h, 3.13 ppm) exerted a significantly enhanced molluscicidal activity as compared to **219**, whereas the *p*-nitrobenzyl derivative **101** (LD_{100}/24 h, 12.5 ppm) displayed a slightly weaker effect than **219**. For comparison, the well-known molluscicide trifenmorph,[633] killed all *B. glabrata* snails at a concentration of 0.25 ppm after 24 h.[103]

8.1.7 Fungicidal activities of the alkaloids of Triphyophyllum peltatum against plant-pathogenic fungi

Extracts of *Triphyophyllum peltatum* were shown to be active against grain-pathogen fungi such as *Leptosphaeria nodorum* and *Pyrenophora teres*.[97] The extracts exhibited particular activity against *Botrytis cinerea* and *Plasmopara viticola*, important plant-pathogenic species. Among the compounds tested, dioncophylline B (**218**) (Fig. 142) proved to be the most active one, exhibiting nearly no phytotoxicity.[97] Good activities were exhibited also by the related alkaloids dioncophylline A (**1a**), dioncopeltine A (**32**), and dioncophylline C (**73**) (Fig. 142). The molecular mode of action of these compounds is as yet unknown. Fungicidal activity is most likely to give the plant a specific advantage, especially in the humid rain forest.

8.2 NMR imaging of organs and seeds of Ancistrocladus species and localization of naphthylisoquinoline alkaloids in intact plants

Nuclear magnetic resonance (NMR) microscopy has become a valuable tool for non-invasive and, thus, non-destructive investigations on the anatomy, function, and metabolism of intact plants.[634–638] The method permits to investigate the plant tissue structure providing well-resolved NMR images with high spatial resolution. Based on the localized acquisition of the ^{1}H NMR signal of water in tissue, NMR images of the shoots, leaves, and roots of *A. heyneanus*[104] and the seeds of *A. abbreviatus*[107] with a spatial resolution of about 20 µm were monitored. The image contrast depended on the spin density, the NMR relaxation times T_1 and T_2, the diffusion coefficient of water, and the flow velocities in the vessels. In addition, the regional distribution of primary and secondary metabolites in *A. heyneanus* was observed using chemical-shift NMR imaging (CSI).[104,105] Using this method, which combines high-resolution NMR spectroscopy with imaging,[639–641] one complete NMR spectrum was monitored for each volume element. In order to support the findings from NMR microscopy and NMR micro-imaging, light and fluorescence microscopy images were recorded on the plant organs of *A. heyneanus* in parallel for comparison.[104,105,264]

8.2.1 Investigations on the anatomy of Ancistrocladus heyneanus by non-invasive NMR microscopy and micro-imaging

In the ^{1}H spin density image of the main root of *A. heyneanus*, three different tissue regions, *viz.*, cortical parenchyma, phloem, and xylem, were clearly recognizable (see Fig. 144a). Within the xylem region, radial alignments with slightly larger T_2 values were observed, presumably correlating with pith rays. The most striking region in the NMR image of the root, however, was the phloem, which appeared as a distinctly bright ring exhibiting T_2 values of 40–100 ms.[104]

^{1}H NMR images taken from the shoot of *A. heyneanus* revealed zones of different signal intensities corresponding to the plant tissue structures outer cortex, inner cortex, phloem, xylem, and pith. Fig. 144 shows the measured spin-density distribution according to a short echo time of 3.5 ms and a long

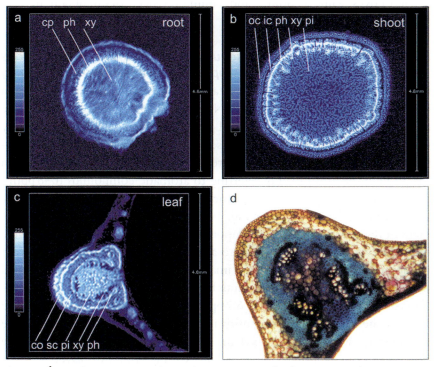

Fig. 144 ^{1}H NMR cross-sectional spin density images of *A. heyneanus* (echo time: 11 ms, rate of repetition: 3.1 s, spatial resolution: 17.9 µm). (a) Root, (b) shoot, and (c) leaf, for comparison: (d) the corresponding light micrograph of the leaf midrib stained with toluidine blue. The tissues are cortex (co), sclerenchyma (sc), pith (pi), xylem (xy), phloem (ph), outer cortex (oc), inner cortex (ic), and cortical parenchyma (cp).

repetition rate of 15 s. The low signal intensity observed in the xylem region was caused by the comparably thick cell walls, which did not contribute to the NMR signal. The pith region, by contrast, exhibited high T_1 values up to 1.8 s, whereas the xylem showed significantly diminished T_1 values of about 0.2–0.7 s. Moreover, bright spots of the protoxylem were visible.[104]

From a leaf of *A. heyneanus*, a ^{1}H NMR cross-sectional image was monitored by spin-echo NMR imaging providing a contrast NMR image based on the different T_2 NMR relaxation times of water in the tissue (Fig. 144c). This NMR microscopy image clearly permitted to distinguish different tissue regions such as cortex, sclerenchyma, phloem, xylem, and pith. The light micrograph of the leaf midrib stained with toluidine blue (Fig. 144d) confirmed the excellent agreement concerning the morphological information, obtained from the two different techniques. The NMR image, however, was recorded from intact tissue without the use of contrast agents or dyes, while sectioning and staining was necessary to obtain the light micrograph.[104]

8.2.2 Localization of naphthylisoquinoline alkaloids in Ancistrocladus heyneanus by NMR microscopy and radial spectroscopic imaging

In order to identify and localize certain metabolites in the shoot region, a combination of spatially resolved NMR spectroscopy, autofluorescence microscopy of the plant tissue, and high-resolution NMR spectroscopy of tissue extracts was applied. A slice-selective one-dimensional NMR spectrum of the whole shoot was monitored showing a number of resonances in this in vivo spectrum mainly referring to aromatic compounds (6–8 ppm), carbohydrates (3–4 ppm), and aliphatic molecules (0–3 ppm), besides a residual water signal at 4.8 ppm (see Fig. 145a).[104,105] Although an exact assignment of resonances to specific compounds was difficult due to the relatively broad linewidth of 20 Hz leading to an overlap of the lines from different metabolites, chemical-shift NMR imaging clearly gave rise to the regional distribution of individual compounds by integrating the NMR signal intensities of the aromatic region at 6.7 ppm (see Fig. 145b). This CSI image showed the highest signal intensity in the cortex and in the pith, while the intensity was low in the xylem. The fluorescence microscopy image of a shoot cross-section of *A. heyneanus* (see Fig. 145c) displayed a bright yellow autofluorescence (excitation at $\lambda = 450–490$ nm),[104] with a distribution similar to that of the resonances at δ_H 6.7 in the aromatic region,[104,105,107] thus revealing a strong resemblance of these two shoot images. Such a strong autofluorescence had been observed to be displayed

Structural variety and pharmacological potential of naphthylisoquinoline alkaloids   263

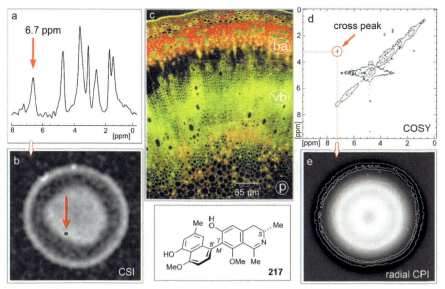

Fig. 145 ^{1}H NMR images of aromatic metabolites in the shoot of *A. heyneanus* using both chemical shift imaging (CSI) (left) and radial correlation peak imaging (CPI) (right).[104] (a) 300 MHz in vivo NMR spectrum with a distinct signal in the aromatic region at δ_H 6.7. (b) NMR CSI image obtained by integrating the NMR signal intensities of the CSI data in the aromatic region at δ_H 6.7. (d) COSY spectrum of the shoot showing a cross peak with resonances at δ_H 7.3 and 3.2 as typical of aromatic compounds. (e) Radial correlation peak image of this aromatic metabolite in the shoot with an overlayed contour plot of a conventional NMR shoot image similar to that presented in (b). For comparison, see: (c) Fluorescence microscopy image of the shoot showing a yellow autofluorescence at 450–490 nm, which was attributed to ancistroheynine A (**217**). The tissues are bark region (ba), vascular bundle region (vb), and pith region (p).

by ancistroheynine A (**217**), which had been discovered in spherical inclusions in the shoots of *A. heyneanus* and had unequivocally been identified after its isolation in a pure form.[264] Therefore, high-resolution NMR spectroscopy was applied to search for **217** and other metabolites in tissue extracts. The NMR spectrum of pure **217** revealed only a few resonance lines in the aromatic region (6–8 ppm),[264] which were also found to appear in the NMR spectrum of the extract.[104,105,107] In contrast to the resonances monitored in the aliphatic region, the aromatic resonances were overlayed in the spectrum of the extract only by few other lines. Thus, signals in the aromatic region were the most reliable markers to investigate the occurrence of naphthylisoquinoline alkaloids in the plant by NMR CSI.[105]

Similar NMR shoot images were obtained from radial spectroscopic ^{1}H NMR imaging and radial correlation peak imaging.[105,107] These two methods take advantage of the cylindrical symmetry of an object, as given, e.g., for plant stems, shoots, and roots, thus allowing to reduce the acquisition time dramatically, since only the radially dependent spatial information needs to be recorded. Fig. 145d shows a two-dimensional COSY spectrum of a shoot of *A. heyneanus*. A number of cross peaks were detected hinting at the presence of different metabolites, among them signals from 2.8 to 5 ppm referring to NMR resonances derived from carbohydrate molecules.[104] The distribution of the sucrose peak taken from a conventional CSI experiment (data not shown) revealed a high intensity in the pith region, while in the xylem region only very low intensities were detected. The corresponding radial chemical-shift image of the sucrose peak provided similar results, but more rapidly, with a dramatic acquisition time reduced from 11.5 h to 11 min with equivalent radial resolution (data not shown).[105]

With respect to the presence of naphthylisoquinoline alkaloids in the shoots of *A. heyneanus*, however, the by far most interesting cross-peak corresponded to a metabolite correlating with the resonances at 7.3 ppm in the aromatic region and 3.2 ppm in the aliphatic region (see Fig. 145d). The spatial distribution of this unknown metabolite was obtained from a radial correlation peak imaging data set (see Fig. 145e) referring to the presence of this metabolite only in the pith region. The acquisition time was 10 h, compared to 13 days for a comparable experiment using the correlation peak imaging technique.[105] Similar to the results described above for conventional CSI measurements, a strong resemblance of the NMR radial correlation peak image with the fluorescence micrograph of the shoot of *A. heyneanus* was likewise visible.[104,105,107]

8.2.3 NMR microscopy and chemical shift imaging (CSI) experiments during seed germination and early growth of Ancistrocladus abbreviatus

NMR microscopy and chemical shift imaging experiments furthermore gave valuable insight into the anatomy of the nut-like fruits of Ancistrocladaceae with their five enlarged sepals, here for the first time applied to monitor the viability of seeds of *Ancistrocladus abbreviatus* from Ivory Coast.[107] Zones of different signal intensities were clearly visible in the cross-sectional ^{1}H NMR spin-echo images of the seeds, providing an image of their typical shape

Structural variety and pharmacological potential of naphthylisoquinoline alkaloids 265

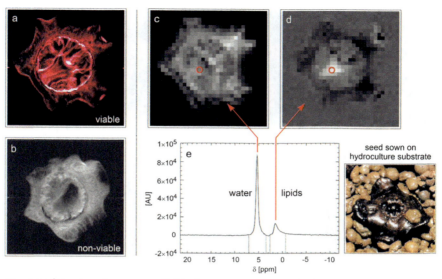

Fig. 146 ¹H spin-echo images (field of view: 2.0 cm; slice thickness: 1 mm) of seeds of *A. abbreviatus* showing the typical texture of the embryo, thus allowing to distinguish between (a) viable and (b) non-viable seeds.[107] ¹H CSI images of the (c) water and the (d) lipid peak. (e) Localized CSI spectrum originating from the encircled voxels in the two CSI images outlined in (c) and (d).

and allowing to clearly identify the contours of their five wings. Moreover, bright zones within the seeds referring to a high content of water were found to form a characteristic texture presumably representing the embryo (see Fig. 146a). This assumption was supported by CSI experiments providing images of the regional distribution of the water signal in the region of 3.5–7.0 ppm (see Fig. 146c) and of the lipid signal in the region of about 0–2.5 ppm (see Fig. 146d).[107] Remarkably, the CSI image referring to the lipid distribution in the seeds revealed a striking resemblance with the ¹H spin-echo image of the water signal likewise giving rise to the typical texture of the embryo. These two ¹H CSI images were obtained from the spatial distribution of the spectral regions marked in the localized CSI spectrum presented in Fig. 146e. About 40 seeds were investigated in a similar manner by ¹H NMR spin-echo imaging: in all of the viable seeds, a characteristic texture representing the embryo was clearly visible, while non-viable seeds were easily recognizable from the presence of deep black zones within the seeds hinting at a complete loss of water and/or severe damages caused by insects (see Fig. 146b).[107]

Fig. 147 *Ancistrocladus abbreviatus* germinating in the greenhouse of the Botanical Garden of the University of Würzburg. (a–d) ^{1}H NMR spin-echo images of germinating seeds producing roots and shoots, the seedlings are clearly visible as a bright zone left at the bottom (field of view: 2 cm; slice thickness: a,b—1 mm; c,d—6 mm). (e–g) For comparison: photos of seedlings just after germination with the hypocotyl growing and primary leaves sprouting, (h) young erect 2-months old sapling. *Photos: e–g*—K. Wolf; *h*—B. Wiesen.

The non-invasive character of magnetic resonance imaging permitted, for the first time, access to repeated specimen examinations, allowing—in contrast to conventional histological sections using light microscopy—a non-destructive monitoring of developmental plant anatomy and morphology of the same sample repeatedly over time.[107] The visualization of the germination process of seeds of *A. abbreviatus* illustrating the formation of roots and shoots on the newly formed seedlings (see Fig. 147) is a nice example highlighting the potential of magnetic resonance imaging to investigate anatomical features. All those seeds that were classified to be viable by NMR imaging due to the presence of the typical texture of the embryo[107] (see Fig. 146a), began to germinate about 4–8 weeks after sowing them in pure river sand (temperature 30–33 °C, relative humidity 90–100%), and keeping them always moist. As soon as the germination of the seeds started, ^{1}H NMR spin-echo images were recorded to visualize morphological changes in the embryo. Comparable to the measurements performed on the viable seeds (see Fig. 146a–c), NMR images were acquired from germinating seeds likewise with a cross-sectional view through their centers (see Fig. 147a and c).[107] Resolution in these experiments was sufficiently high to permit detection of the typical anatomy of the seeds, with the sprouting embryo in the center, giving rise to a distinct contrast due to the presence of water and lipids, and with the seedling occurring as a bright zone at the bottom left.[107] The NMR images also showed zones filled with air, therefore appearing as dark. From the longitudinal-sectional view (see Fig. 147b and d) the seedling was clearly recognizable with its root and shoot arising from the position where the former embryo had previously been localized.[107]

8.3 In vivo localization of naphthylisoquinoline alkaloids by non-invasive FT-Raman microspectroscopy

8.3.1 In situ localization of ancistroheynine A and ancistrocladine in plant organs of Ancistrocladus heyneanus

As described in the previous sections, light and fluorescence investigations,[1,104,264] but also NMR microscopy studies in combination with chemical-shift imaging (CSI) experiments[104,105,107] revealed the naphthyldihydroisoquinoline ancistroheynine A (**217**) to accumulate in distinctive parts of older shoots of *A. heyneanus*. This finding triggered further investigations on the anatomy of this Indian species and on the distribution of naphthylisoquinoline alkaloids in its tissues using FT-Raman microspectroscopy[642–645] as a non-invasive and, thus non-destructive technique. While NMR microscopy permits to study plant tissue structures with a good image contrast and a high spatial resolution based on the observation of the ^{1}H NMR signal of water in the tissue,[104–107] FT Raman spectroscopy even makes it possible to distinguish between certain secondary metabolites due to the high sensitivity of this analytical tool.[108–110]

The alkaloids of *A. heyneanus* showed very specific Raman spectra,[108] even though their chemical structures were, at first sight, most similar, differing only by the position of the linkage between their molecular moieties, their OH/OMe patterns, or the degree of hydrogenation in their isoquinoline portions. Micro-FT Raman studies clearly confirmed the presence of ancistroheynine A (**217**) in spherical yellowish cell inclusions mainly detected in the shoot pith regions.[108] Fig. 148a presents a fluorescence microscopic image of a cross section of fresh shoot material of *A. heyneanus*, showing the cell inclusions corresponding to ancistroheynine A (**217**) as yellow fluorescent spots.

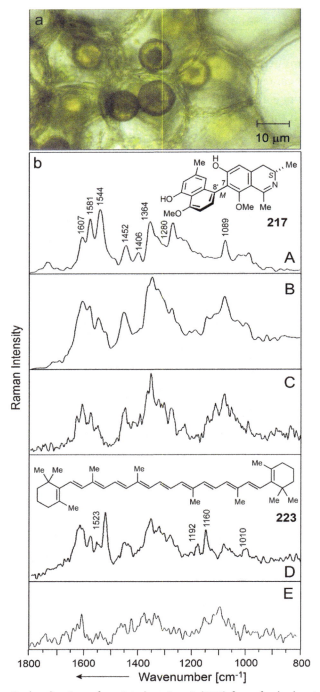

Fig. 148 In situ localization of ancistroheynine A (**217**) from fresh shoot material of *A. heyneanus*.[108] (a) Bright-field microscopic image showing the spherical yellowish cell inclusions (diameter ca. 20 μm) in the pith region corresponding to **217**. (b) FT-Raman investigations: spectrum of pure **217** (Section A), in situ spectrum of **217** monitored in the shoot tip (Section B), in situ spectrum recorded in the leaf midrib (Section C), in situ spectrum of **217** and β-carotene (**223**) taken in the leaf midrib (Section D), and in situ spectrum taken from cells without alkaloid inclusions (Section E).

The FT-Raman spectra (Fig. 148b) of these yellow-colored plant cells recorded from the shoot tip (see Section B) and the leaf midrib (see Section C) were found to be in good agreement with the spectrum obtained from pure ancistroheynine A (**217**) (see Section A).[108] Only small differences were observed between in vitro and in situ spectra of the cell inclusions caused by adjacent cell tissue and their components mainly giving rise to a broad shape of Raman signals in the regions between 1350–1450 and 1000–1200 cm^{-1}. Another compound co-occurring in the leaf midrib, together with ancistroheynine A (**217**), was identified as β-carotene (**223**) due to its characteristic signals as indicated in Fig. 148b (see Section D).[108]

Spatially resolved Raman studies of different parts of *A. heyneanus* led to the spectroscopic localization of another alkaloid in the outer root cortex.[108] Fig. 149 shows a microscopical root cross section, revealing spherical crystalline inclusions, which exhibited a weak fluorescence upon irradiation at 365 nm. Focussing the laser spot on a single crystal gave rise to the acquisition of in situ spectra with a good signal-to-noise ratio. The Raman signals thus obtained closely resembled those monitored from pure ancistrocladine (**20a**). The identification of **20a** was in agreement with previous isolation work which had shown that **20a** occurs in the roots of *A. heyneanus* in abundant amounts.

The fact that the Raman spectra obtained from ancistroheynine A (**217**) and ancistrocladine (**20a**) in a non-destructive manner, by measurements directly in fresh plant material, closely resembled those recorded from the pure isolated alkaloids suggested that the structures of **217** and **20a** were apparently not influenced by workup procedures during isolation.[108]

8.3.2 In vivo localization of dioncophylline A in inclusions of fresh stem material of Triphyophyllum peltatum by NIR-FT Raman microspectroscopy

The impressive results on the localization of ancistrocladine (**20a**) in inclusions of fresh shoot material of *A. heyneanus*[108] had already demonstrated the potential of FT-Raman microspectroscopy to gain detailed information about the structures of the biological molecules, and to characterize and identify them in a reliable and sensitive manner in their natural environment. For this reason, ultrasensitive near-infrared Fourier transform (NIR-FT) Raman spectroscopy was applied for the localization of dioncophylline A (**1a**) in different parts of *Triphyophyllum peltatum*.[109,110] Fluorescence

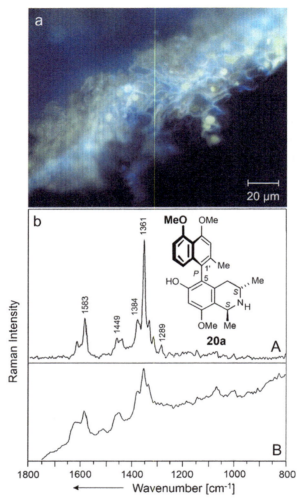

Fig. 149 Localization of spherical crystalline inclusions of ancistrocladine (**20a**) in the outer cortex of *A. heyneanus*.[108] (a) Fluorescence micrograph revealing crystals (size range: about 5 μm) with a weak fluorescence upon irradiation at 365 nm. (b) FT-Raman investigations: spectrum of pure crystalline **20a** (Section A), and in situ spectrum of **20a** from a root cross section (Section B).

microscopy images had previously given rise to the assumption that **1a** may be present in 10-μm sized inclusions located in the cortex of the stem or in the bark regions near the leaf petiole at the point of attachment (Fig. 150a).[109] Indeed, by means of NIR–FT Raman microspectroscopy

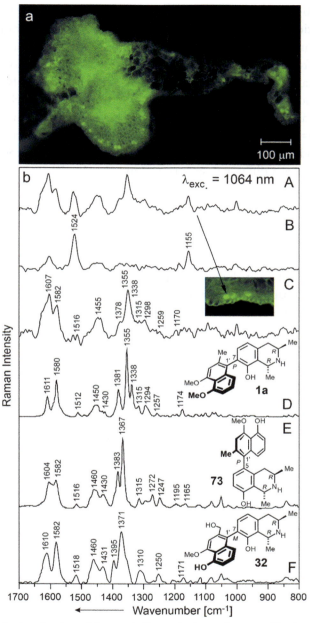

Fig. 150 In vivo localization of dioncophylline A (**1a**) in inclusions of fresh stem material of *T. peltatum*.[109] (a) Fluorescence microscopic image of a side stem near the leaf petiole at the point of attachment showing inclusions (ca. 10 µm). (b) FT Raman spectra of fresh plant material (Sections A and B) and spectra of pure dioncophylline A (**1a**) (Section D), dioncophylline C (**73**) (Section E), and dioncopeltine A (**32**) (Section F). In detail: FT Raman spectrum of an inclusion at the outer stem cortex, inset marked with an arrow is the fluorescence microscopic image of the stem fragment with three inclusions (Section A); in vivo Raman spectrum of an area of a plant cell far from an observable inclusion (Section B); difference spectrum "A minus B" (Section C).

($\lambda_{exc.} = 1064$ nm), dioncophylline A (**1a**) was clearly detected in high concentrations in these inclusions. The FT-Raman spectrum recorded in such inclusions was largely different from the Raman spectra obtained from outside these organelles, diverging by the loss of several Raman peaks (Fig. 150b, Lines A and B).[109] Highly intense Raman bands at 1155 and 1524 cm^{-1}, detected inside and outside the inclusions, were identified as the background signal of the plant material, mainly arising from νC=C and νC—C vibrations of β-carotene. Subtraction of the background signal from the FT-Raman spectrum monitored for the inclusions gave a difference spectrum (Line C = A − B, see Fig. 150b) very closely resembling the Raman spectrum obtained from pure dioncophylline A (**1a**) (Fig. 150b, Line D).[109] This in situ distribution of **1a** in *T. peltatum* was further confirmed by comparison with the Raman spectra of pure reference material of dioncophylline C (**73**) and dioncopeltine A (**32**), since the spectra of these two naphthylisoquinoline alkaloids can be clearly distinguished from that of dioncophylline A (**1a**).[109,110] In order to assign the observed Raman signals and interpret the differences in the experimental Raman spectra of **1a**, **73**, and **32**, NIR-Raman spectroscopy was combined with density functional theory (DFT) calculations.[110] The most prominent Raman modes were found to be those arising from C=C stretching and CH bending vibrations, i.e., νC=C naphthalene, νC=C isoquinoline, and νC—C bridge bonds. In particular, the two modes at 1356 and 1613 cm^{-1} turned out to permit a significant distinction between the three alkaloids **1a**, **73**, and **32**.

Outside the inclusions, however, dioncophylline A (**1a**) was not detectable even in traces, due to the low sensitivity of NIR Raman spectroscopy, and because the laser energy needed to obtain Raman spectra of the bright inclusions was too high for the plant material all around, which was destroyed under these conditions. Application of deep-UV resonance Raman spectroscopy, however, made it possible to localize **1a** very sensitively and selectively, even in low concentrations.[110] Owing to the fact that naphthylisoquinoline alkaloids were found to exhibit strong electronic absorptions in the deep-UV spectral range, the Raman excitation wavelengths of 244 and 257 nm were successfully applied to enhance the Raman signals of selective vibrations of **1a** in situ in *T. peltatum*. This resonance enhancement permitted to record signals with very high signal-to-noise ratios by using only a very low laser power, thus allowing to detect dioncophylline A (**1a**) in root and stem bark material, previously inaccessible by NIR-Raman spectroscopy.[110] Even very low concentrations of only ca. 60 ppm by weight of **1a** were located in the roots of *T. peltatum* by deep-UV resonance Raman microspectroscopy.[110] In this case, no fluorescence signal obscured the Raman spectrum, nor did any Raman signal from the surrounding plant material play a significant role because of the selective-resonance conditions.

Thus, in accordance with the studies on *A. heyneanus* (see previous paragraph), Raman microspectroscopy has proven to be a sensitive analytical tool for the localization of naphthylisoquinoline alkaloids also in plant organ tissues of *T. peltatum*. It permitted to unambiguously distinguish between different alkaloids by specific vibrational modes, even though their chemical structures were very similar. In agreement with the results obtained for ancistrocladine (**20a**) from *A. heyneanus*,[108] the spectra of dioncophylline A (**1a**) provided in situ, from *T. peltatum*, were nearly identical to the one monitored for the pure compound.[109,110]

8.4 Chemodiversity, storage, and ecological functions of naphthylisoquinoline alkaloids—Summary and outlook

For a long time, research on secondary plant metabolites was mainly driven by the search of highly bioactive compounds as promising leads for the development of new effective drugs for medical purposes.[646–649] However, it was assumed that they were of no importance for the plants themselves, and were thus considered to be useless waste products of the primary metabolism.[650,651] The essential role of these low-molecular weight compounds as efficient tools to protect plants against herbivores and microbes[564,566,571,652–654] or to function as signal molecules[652–654] has only been recognized a few decades ago. The multi-faceted tasks of secondary plant metabolites as mediators for all kinds of chemical communication regarding intra- and interspecific relationships and the importance of chemodiversity in this respect have stimulated intense research efforts on the involved biosynthetic pathways.[643,652] The fact that secondary metabolites can often be found in significant quantities in certain organs that do not produce them, stimulated investigations regarding their long-distant transport and their storage using non-invasive methods such as NMR chemical-shift imaging (CSI)[634–641] and FT-Raman microspectroscopy[642–645] to visualize their accumulation in various plant organs.

As summarized in this section, the significant antifungal,[97] antifeedant,[98,99] antiparasitic,[1,3,38,39,45,46,49,181] larvicidal,[100,101] and molluscicidal[103] effects displayed by some of the naphthylisoquinoline alkaloids and related compounds evidenced their role as decisive factors to protect Ancistrocladaceae and Dioncophyllaceae plants against herbivores, pathogens, or neighboring plants, and to cope with abiotic and biotic stress. The in vivo localization of accumulated naphthylisoquinoline alkaloids, for example in inclusions of fresh shoot material and in the outer cortex of *Ancistrocladus heyneanus* by use of non-destructive techniques such as NMR microscopy,[104,107] NMR radial spectroscopic imaging,[105,107] and FT-Raman microspectroscopy[108–110] further supported the involvement of these secondary metabolites in defense-mediating processes. Moreover, NMR microimaging is a valuable and efficient tool for plant science in general. Depending on differences in the ^{1}H signal, it provided noninvasively important information about the internal anatomy of the roots and leaves of *Ancistrocladus heyneanus* by visualizing the water distribution in the respective plant organs.[104,105] Particularly fascinating was the possibility to study morphological and structural details such as the structure of the embryo within the seeds of *Ancistrocladus abbreviatus*, and to track germination and early seedling growth.[107]

However, many aspects regarding the chemoecological functions of naphthylisoquinoline alkaloids and related compounds still remain unanswered and warrant further, more in-depth investigations for a better understanding of the intricate biochemistry, physiology, and chemodiversity of alkaloid formation to ensure the fitness of the plants. This will be an important task for future research activities on Dioncophyllaceae and Ancistrocladaceae plants and their secondary metabolite patterns.

9. Summary and outlook

This review reports on the tremendous advances in the field of naphthylisoquinoline alkaloids during the past decades. It summarizes the most important results from botanical research on their natural producers, the Dioncophyllaceae and Ancistrocladaceae, and covers the increasingly diverse molecular scaffolds of mono- and dimeric naphthylisoquinoline alkaloids,[1–6,55,59] their promising biological activities,[1–3,10,36–39,44,366,367] their polyketidic biosynthetic origin,[80–83] and describes first investigations hinting at essential chemoecological functions[98–103] of these secondary plant metabolites. Since the appearance of the last comprehensive overview on naphthylisoquinoline alkaloids in this series in 1995,[5] the number of known representatives has tremendously increased, accompanied by the discovery of many novel-type compounds with unprecedented molecular frameworks. Particularly noteworthy was the identification of the first N,C-coupled naphthylisoquinoline alkaloids[1,2] and of naphthylisoquinoline dimers[1,3] that possess three consecutive chiral biaryl axes such as jozimine A$_2$ (**4a**),[1,3,12] and the isolation of the first ring-opened (i.e., *seco*-type) alkaloids such as ancistrosecoline A (**13**)[55] and ring-contracted naphthylisoindolinones like e.g., ancistrobrevoline A (**14**).[59] The knowledge of the broad structural chemodiversity of naphthylisoquinoline alkaloids (Fig. 151), as acquired by today, has been decisively driven forward by extensive phytochemical studies on *Ancistrocladus* species from Central Africa,[3,11–16,29–35,51,52,175–183,287–290] but

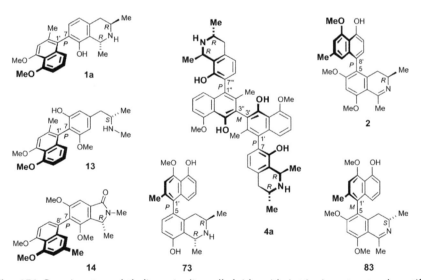

Fig. 151 Prominent naphthylisoquinoline alkaloids with intriguing structural motifs and/or promising biological activities: the antileukemic dioncophylline A (**1a**), the antiplasmodial compounds jozimine A$_2$ (**4a**) and dioncophylline C (**73**), the anticancer agents ancistrolikokine E$_3$ (**2**) and ancistrobrevidine C (**83**), and natural products with unprecedented molecular scaffolds, the ring-opened ancistrosecoline A (**13**), and the ring-contracted ancistrobrevoline A (**14**).

also, to a large extent, by detailed investigations on the outstanding metabolite pattern of *Ancistrocladus abbreviatus* from Coastal West Africa,[5,54–65] as described in this review.

Among the more than 280 naphthylisoquinoline alkaloids known so far,[1–7] dioncophylline A (**1a**) is one of the best-investigated representatives. It was among the very first known examples of this unique class of natural products discovered in nature.[5,6,8,9] Dioncophylline A (**1a**) has stimulated a plethora of research activities on its structure,[5,6,8,9,317,335,363,364,376,377] its regio- and stereoselective total synthesis,[5–9,76,364,378] its unusual polyketidic biosynthetic origin from acetate units,[80–83] its accumulation in distinct plant-organs,[109,110] and its chemoecological[98–103] and pharmacological[51,58,365–367,373] properties, including the mechanism of action.[365–367] It is still one of the most intensely studied monomeric naphthylisoquinoline alkaloids, and has only recently raised great interest because of its excellent cytotoxic activities against leukemia and multiple myeloma cells.[51,366] The outstanding position of dioncophylline A (**1a**) is highlighted in this review by accurately describing the advanced repertoire of methods applied (and, in part, specially developed) for the isolation, structural elucidation, and total synthesis of naphthylisoquinoline alkaloids using this alkaloid as a rewarding example. During the past decade, the tool box of modern chromatographic, spectroscopic, and chemical methods has strongly improved regarding effectiveness and sensitivity. Thus, it can clearly be excepted that future phytochemical investigations will permit to identify even minor and trace metabolites, and will thus provide information on the alkaloid patterns of the plants in a more precise manner. It may also be assumed that studies on so far less investigated or even unexplored *Ancistrocladus* species will furnish further new alkaloids with unprecedented structural motifs, and attractive bioactivities.

In view of the fact that there is still a controverse discussion about the species status of some of the *Ancistrocladus* plants,[18–20,24,25] phytochemical studies might also provide a more deepened understanding regarding the species-specific production of naphthylisoquinoline alkaloids and might thus contribute to a more reliable delineation of yet unidentified *Ancistrocladus* plant species from the already known taxa. This is, in many cases, still an unsolved task, because the taxonomy of the entire genus *Ancistrocladus* is only poorly evaluated due to the fact that some of the plants are fairly rare and, in addition, do not flower abundantly. Thus, a reliable species delineation, solely based on morphological criteria, is often difficult or even impossible.[20,22,24,25] This is one of the reasons why the hypothesis of a species differentiation within the Ancistrocladaceae plant family in Central Africa and Southeast Asia, as strongly supported by phylogenetic[24,25] and phytochemical[1–4] studies, still remains to be explored more thoroughly, in particular with respect to the (re)examination of morphological criteria. Thus, it can be assumed that an intensified collection of *Ancistrocladus* plants from different localities in Africa and Asia, further yet unknown taxa will come to light.

For a long-term eradication of severe tropical diseases such as malaria, trypanosomiasis, leishmaniasis, or schistosomiasis, proper diagnosis and for the development of novel drugs efficient approaches are urgently required.[405–417] Likewise dramatic is the situation in the fight against pancreatic cancer, which still is one of the deadliest forms of tumors, with a very low 5-year survival rate of less than 5%.[419–426] Some of the naphthylisoquinolines alkaloids, such as jozimine A_2 (**4a**)[1,3,12,58,183] or dioncophylline C (**73**),[36–39] exerted excellent antiparasitic activities, whereas other representatives, like e.g., ancistrolikokine E_3 (**2**)[10] or ancistrobrevidine C (**83**),[56] were highly effective against the aggressive and metastatic growth of pancreatic cancer cells. Chemical and biological investigations aiming at the development of novel therapeutic agents to combat malaria or pancreatic cancer will be high priority, challenging tasks for the future.

A further rewarding topic for future investigations deals with the unique acetogenic nature of the naphthylisoquinoline alkaloids.[80–83] While all as yet investigated isoquinoline alkaloids (more than 2500 known compounds) are formed from aromatic amino acids,[84,85,87,89–94] naphthylisoquinolines represent the as yet only known plant-derived polyketidic di- and tetrahydroisoquinolines originating from acetate and malonate units, exclusively. Of particular interest will be the identification of the polyketide synthase, which (unless inhibited by stress) always seems to produce a 1:1 ratio of the naphthalene and the isoquinoline building blocks, and the characterization of the remarkable coupling enzyme, which, for most of the alkaloids, joins the two molecular halves together with high regio- and stereoselectivity. Another fascinating aspect of this highly stress-sensitive pathway to plant isoquinoline alkaloids focusses on the unique reductive incorporation of nitrogen into the postulated diketo precursor (not into an α-keto acid or an aldehyde, as usual).

In view of the remarkable structural, pharmacological, and biosynthetic facets of naphthylisoquinoline alkaloids and the many open questions regarding the occurrence and taxonomy of their natural producers, the Dioncophyllaceae and Ancistrocladaceae, further promising developments in this rewarding field of research can be expected in the future.

10. Tables of the naphthylisoquinoline alkaloids and related compounds isolated from the West African liana *Ancistrocladus abbreviatus*

By the recent intense phytochemical investigations[54–61] on *Ancistrocladus abbreviatus* Airy Shaw,[18–20,159,160] as presented in this review, the liana, which is endemic to Coastal West Africa, has become a most impressive example of the tremendous development of the past years in the field of axially chiral naphthylisoquinoline alkaloids and related compounds.[1–3] This does not only apply to the considerable number of 80 natural products detected in this one single plant species,[5,54–65,151,404] but mainly refers to the unique structural diversity of the metabolites of *A. abbreviatus* and the occurrence of several novel-type alkaloids with unprecedented structural features.[5,54–56,58–60] In addition to their intriguing molecular framework, some of the alkaloids of *A. abbreviatus* display promising antiplasmodial[57,58] or antileukemic[51,52,54,366] activities, as well as pronounced cytotoxic effects against human pancreas,[56–58,61] colon,[58] lung,[59,60] breast,[59,365] or HeLa[55,56] cancer cells. The most important results on the secondary metabolites of *A. abbreviatus*, comprising 69 naphthylisoquinoline alkaloids and 11 related naphthoquinones,[5,54–65,151,404] are summarized in Tables 12.1–23.

As described in detail in Sections 5 and 6, *A. abbreviatus* is a rich source of naphthylisoquinoline alkaloids (see Fig. 152) belonging to five different coupling types (5,1′, 5,8′, 7,1′, 7,3′, and 7,8′).[5,54–65] Most of them possess a tetrahydroisoquinoline subunit like e.g., ancistrobrevine K (**80**),[58] which is a 5,1′-coupled typical Ancistrocladaceae-type[1,3,5] compound (i.e., with S-configuration at C-3 and an oxygen function at C-6). Many of the alkaloids with a 7,1′-biaryl linkage, however, are 3R-configured and have no oxygen group at C-6. These compounds, among them one of the main metabolites of *A. abbreviatus*, N-methyldioncophylline A (**94a**),[5,62] are categorized as Dioncophyllaceae-type[1,3,5] alkaloids, with respect to the fact that Dioncophyllaceae lianas, sharing a common habitat[20] with *A. abbreviatus*, exclusively produce naphthylisoquinolines with these structural characteristics.[5,6,68,73,274,275,292–298,402] In contrast to the monomeric naphthylisoquinoline alkaloids isolated from the Dioncophyllaceae, *A. abbreviatus* also contains Dioncophyllaceae-type dimers (Fig. 152) such

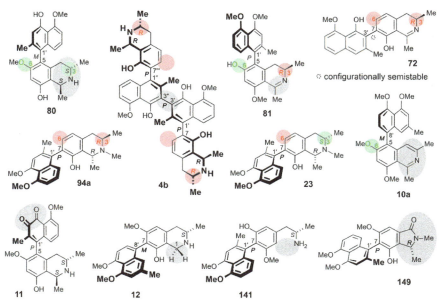

Fig. 152 Structures of selected naphthylisoquinoline alkaloids and related metabolites discovered in *Ancistrocladus abbreviatus* representing Ancistrocladaceae- (6-OR, 3S—"green/green"), Dioncophyllaceae- (6-H, 3R—"red/red"), hybrid- (6-OR, 3R—"green/red"), and inverse hybrid-type (6-H, 3S—"red/green") compounds. The structural units of **80**, **94a**, **4b**, **81**, **72**, **23**, and **10a** that are typical of the diverse subclasses of naphthylisoquinoline alkaloids, are labeled in light gray.

as jozibrevine A (**4b**),[58] which possesses three consecutive chiral biaryl axes. It is particularly noteworthy that a near-complete series of atropisomeric dimers (four out of six possible compounds) were identified only in *A. abbreviatus*, but so far never in any other *Ancistrocladus* species.

In addition, *A. abbreviatus* produces constituents like e.g., ancistrobrevidine A (**81**),[56] with an oxygen group at C-6 and *R*-configuration at C-3. Besides such hybrid-type compounds, the plant even synthesizes an inverse hybrid-type alkaloid, dioncoline A (**23**),[58] the as yet only known 3S-configured naphthylisoquinoline lacking an O-function at C-6 (Fig. 152). Thus, *A. abbreviatus* obviously does not form its metabolites in a highly regio- and stereoselective manner, but seems to pursue a nearly combinatorial approach. It is the only *Ancistrocladus* species known to produce naphthylisoquinolines of all of the mentioned four possible subclasses.[5,58]

Likewise remarkable was the isolation of some naphthyldihydroisoquinolines belonging to very rare subtypes (Fig. 152) of naphthylisoquinoline alkaloids, among them dioncophyllidine E (**72**), with its—likewise rare—configurationally semi-stable biaryl axis.[61] It is only the second example of a 7,3′-coupled Dioncophyllaceae-type compound found in nature. Furthermore, an unusually large series of seven alkaloids with a non-hydrogenated isoquinoline portion[56] like e.g., ancistrobreveine D (**10a**) were discovered in *A. abbreviatus*. So far, only few examples[1,5,6,53,176,252,273–275,291,403] of such naphthylisoquinolines with a rotationally hindered biaryl axis as the only element of chirality had been found in nature.

Even more intriguing was the discovery of novel-type alkaloids (Fig. 152) with unprecedented structural scaffolds[55,56,58–60] like ancistrobreviquinone A (**11**),[57] which possesses an *ortho*-quinone entity in the naphthalene portion, or 1-*nor*-8-*O*-demethylancistrobrevine H (**12**),[55] which lacks the otherwise generally present methyl group at C-1. Most remarkable was the isolation of a series of six *seco*-type (i.e., ring-opened) alkaloids like e.g., ancistrosecoline B (**141**),[55] and a further subfamily of likewise unique ring-contracted naphthylisoindolinones such as ancistrobrevoline C (**149**).[59]

Tables 12.1–22 comprise a complete list of all the monomeric[54,56–58,60–65] and dimeric[58] naphthylisoquinoline alkaloids, including *seco*-type[55] and ring-contracted[59] metabolites identified so far in the roots, stems, twigs, and leaves of *Ancistrocladus abbreviatus*, compiled according to their molecular scaffolds and coupling types. The alkaloids are categorized as Ancistrocladaceae-type, Dioncophyllaceae-type, hybrid-type, and inverse hybrid-type compounds. Table 23 summarizes naphthoquinones and tetralones produced by the plant itself or by callus cultures[151,187] established from leaf material. Fig. 153 shows the basic structures of all of the secondary metabolites isolated from *A. abbreviatus* and refers to the respective tables in which the representatives of the nine different subgroups are described in more detail.

The tables present a short overview on the structures, the natural occurrence, and the bioactivities of the presented secondary metabolites, including the spectroscopic, chemical, and computational methods used for their structure elucidation, with special emphasis on analytical procedures applied for the assignment of their complete stereostructures. Where appropriate, a summary is given on studies regarding the biosynthetic origin of naphthylisoquinoline alkaloids and related compounds. Moreover, the key steps of diverse approaches are briefly described that have been applied in the regio- and stereoselective total synthesis of selected representatives of mono- and dimeric naphthylisoquinoline alkaloids identified in *A. abbreviatus*.

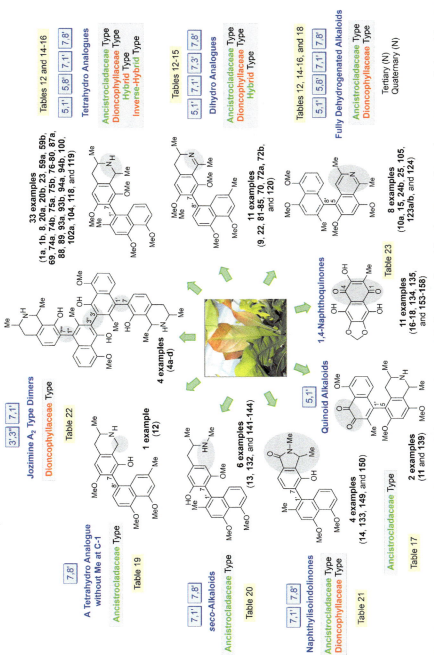

Fig. 153 Overview on the structural diversity of the secondary metabolites produced by the West African liana *Ancistrocladus abbreviatus* (configuration at stereogenic centers drawn flat for reasons of clarity).

Table 12.1 7,1′-Coupled Dioncophyllaceae-type naphthylisoquinoline alkaloids.
Name, formula (mol. wt.), and structure

Producing plants and their organs/physical data [mp]°C, $[\alpha]_D^T/c^2 g^{-1} 10^{-1}$ (solvent)]	Isolation, structural elucidation, partial and/or total synthesis, and further investigations	Biological activities
Dioncophylline A (1a) $C_{24}H_{27}NO_3$ (377.5)	Discovered for the first time in *Triphyophyllum peltatum* (Dioncophyllaceae) from Ivory Coast: Initially isolated (and named "triphyophylline") by Bruneton et al.[374,375], due to the incomplete and incorrect stereostructure of "triphyophylline,"[374,375] as obvious from later structural reinvestigations on its three stereoelements[5,6,8,66,317,335,363,376,377] and from total synthesis,[5,6,9,76,364,378] and because of wrong inter-relationships between the alkaloid and its congeners (e.g., "O-methyltriphyophylline," which is not the same as the O-methylation product of "triphyophylline"), the alkaloid "triphyophylline" was renamed into dioncophylline A (1a)[5,6,8,9]	Fungicidal activity against *Botrytis cinerea*[97] Molluscicidal activity against the tropical snail *Biomphalaria glabrata*, the vector of *Schistosoma mansoni*[102,103] Larvicidal activity against the vector of dengue and yellow fever, *Aedes aegypti*[101] Larvicidal activity against the malaria vector *Anopheles stephensi*[100]
A. abbreviatus: stem bark[63]	Isolation from *A. abbreviatus* along with its atropo-diastereomer 7-*epi*-dioncophylline A (1b); Constitution and complete stereostructure of 1a by HRESIMS, 1D and 2D NMR, oxidative degradation, and ECD investigations[63]	Feeding deterrency and growth-retarding activity against the herbivorous insect *Spodoptera littoralis*[98,99]
A. letestui: leaves and twigs[184]	Isolation from *A. letestui* by bioassay-guided fractionation of crude extracts, complete structural elucidation by NMR including NOE, HMQC, and HMBC experiments, by HRFABMS, by ECD spectral data, and by oxidative degradation[184]	Pronounced inhibitory activity in vitro against *T. brucei*[368]
A. ilebensis: root bark[51]	Isolation from *A. ilebensis* along with its 4′-O-demethyl analogue 93a; the two compounds were isolated together with their atropo-diastereomers, 1b and 93b[51]	Pronounced inhibitory activity in vitro against asexual erythrocytic forms of *Plasmodium falciparum* (strains: NF54, K1) and *P. berghei* (Anka)[369,370]
From a botanically yet undescribed Congolese *Ancistrocladus* species: root bark[12]	Isolation as the main metabolite of a botanically as yet undescribed *Ancistrocladus* species collected in the region of the town Ikela in the Democratic Republic of the Congo (Tshuapa Province) along with the related dimer jozimine A₂ (4a)[12]	Excellent inhibitory activity in vitro against the development of exoerythrocytic stages (liver forms) of *Plasmodium berghei* (Anka)[371]
T. peltatum: roots, stem bark,[5,6,8,9] cell cultures,[80–83] root cultures[311]		No suppression of parasitemia in mice infected with *Plasmodium berghei*[36]
H. dawei: stem bark[73,297]	Isolation from *Habropetalum dawei* (Dioncophyllaceae): Constitution and complete stereostructure of 1a established online, from crude extracts, by LC-MS/MS, LC-NMR, (^{1}H, 2D-TOCSY, 2D-ROESY), LC-ECD, and by oxidative degradation[73,297]	Strong inhibitory activity in vitro against the hematozoan parasite *Babesia canis*[373]
D. thollonii: roots[68,292]	Isolation from *Dioncophyllum thollonii* (Dioncophyllaceae): Co-occurrence of 1a and its 5′-O-demethyl analogue 97; identification online, by stop-flow two-dimensional HPLC-NMR-ROESY experiments,[68] followed by separation and isolation using high-speed countercurrent chromatography (HSCCC)[292]	Strong cytotoxic activities of 1a and its *N*-acetyl analogue against multiple myeloma INA-6 cells (even stronger than that of the standard drug **melphalan**)[51]
Colorless crystals (from CH₂Cl₂): mp 228–229; $[\alpha]_D^{25}$ −14.0 (c 0.64, CHCl₃)[63]		Strong cytotoxic activity of 1a against drug-sensitive acute lymphoblastic CCRF-CEM leukemia cells and their multidrug-resistant subline CEM/ADR5000[51,366]
White solid (from CH₂Cl₂/MeOH); $[\alpha]_D^{20}$ −13.8 (c 0.22, CHCl₃)[184]		
Yellowish amorphous powder (from MeOH): mp 214; $[\alpha]_D^{20}$ −16.0 (c 0.10, CHCl₃)[12]		
Colorless solid (from CH₂Cl₂/acetone); mp 214; $[\alpha]_D^{20}$ −14.9 (c 0.45, CHCl₃)[8]	*Structural Elucidation:*	
White microcrystalline solid (from CH₂Cl₂/MeOH): mp 214; $[\alpha]_D^{25}$ −13.8 (c 0.10, CHCl₃)[292]	Relative configuration at centers versus axis by long-range 1D and 2D NOESY and ROESY interactions[66]	

cf.: Compd. prepared synthetically: mp 215; $[\alpha]_D^{20}$ −14.9 (c 0.45, CH$_2$Cl$_2$)[9]	Absolute configuration established as R, both at C-1 and C-3, by ruthenium-mediated oxidative degradation[69,70]
	Assignment of the absolute axial configuration by experimental and computational ECD analysis of the dehydrogenated product of dioncophylline A (**1a**), dioncophylleine A (**24a**), possessing the biaryl axis as the only remaining chiral information,[333,334] and by MD-based ECD calculations directly for **1a**[335]
cf.: "Triphyophylline": Colorless crystals (from acetone): mp 215; $[\alpha]_D^{20}$ −14.0 (c 1.00, CHCl$_3$)[374]	Attribution of the absolute axial configuration of **1a** by a combination of solid-state ECD spectroscopy and quantum-chemical ECD calculations using three different theoretical approaches (CNDO/S-CI, TD-B3LYP, DFT/MRCI)[377]
	Assignment of axial chirality by an atropisomer-differentiating cyclization reaction, synthetically correlating the axis to the stereogenic centers for the elucidation of the relative configuration at the axis, accompanied by computational investigations[5,6,363]
	X-ray structure analysis of dioncophylline A (**1a**) (constitution and all relative configurations)[376]
	Full absolute stereostructure by anomalous X-ray dispersion of a bisbenzylated bromo derivative of dioncophylline A (**1a**)[317]

Total Synthesis and Further Synthetic Work:

Stereocontrolled total synthesis of dioncophylline A (**1a**) by applying the lactone method[76−79]; intramolecular biaryl coupling of the ester-type prefixed molecular moieties followed by atroposelective lactone cleavage, giving rise to **1a** or, optionally, to its atropo-diastereomer, 7-*epi*-dioncophylline A (**1b**)[5,6,9,76,364,378]

5,5-Dimerization of dioncophylline A (**1a**) to give the unnatural quateraryl jozimine A (**223**)[515] (for its structure, see left column) by a reductive Klar-Wittig coupling; the inner axis of **223** was stereochemically semi-stable, thus providing the two slowly interconverting atropo-diastereomers, **223a** and **223b**, in a ratio of ca. 1:1; the structure of **223** is unique, since its two halves are coupled via the isoquinoline moieties and not via the naphthalene parts, thus leading to an unprecedented connectivity with a sequence "naphthalene-isoquinoline-isoquinoline-naphthalene".[5,76,512,513,515]

	Significant antiproliferative activity of **1a** against two breast cancer cell lines, MCF-7 and MDA-MB-231, by inducing apoptotic cell death via the intrinsic pathway by causing deformation of the nuclear membrane, disruption of the mitochondrial membrane potential (MMP), and elevated production of reactive oxygen species (ROS)[365]

Continued

Table 12.1 7,1′-Coupled Dioncophyllaceae-type naphthylisoquinoline alkaloids.

Name, formula (mol. wt.), and structure

Producing plants and their organs/physical data [mp/°C, $[\alpha]_D^{T/°}$cm^2 g^{-1} 10^{-1} (solvent)]	Isolation, structural elucidation, partial and/or total synthesis, and further investigations	Biological activities
Jozimine A (223)[515]: Colorless powder (from CH$_2$Cl$_2$): **223a**: mp 223; $[\alpha]_D^{20}$ −130.8 (c 0.03, EtOH); **223b**: mp 225; $[\alpha]_D^{20}$ −79.8 (c 0.05, EtOH)	Treatment of dioncophylline A (**1a**) under phenol-oxidative conditions using lead tetraacetate to furnish two non-natural monomeric main products, namely the diastereomeric dioncotetralones A (**224a**) and B (**224b**) (for their structures, see left column); these possess an as yet undescribed Z-configured double bond, in contrast to the biaryl axis as present in the precursor **1a**, and an additional stereogenic center at C-2[520]	
Dioncotetralone A (**224a**)[520]: Colorless solid (from CH$_2$Cl$_2$/MeOH): mp >230; $[\alpha]_D^{20}$ +135 (c 021, MeOH)	*Further Investigations:* Dioncophylline A (**1a**) is formed via a novel biosynthetic pathway to isoquinoline alkaloids: the two molecular portions of dioncophylline A (**1a**)—the naphthalene and the isoquinoline part—are each built up from six acetate units, both through identical intermediate β-polycarbonyl precursors cyclized following the F-type folding mode[612]; proof by feeding experiments with ^{13}C-labeled precursors administered to callus cultures of *T. peltatum* and 2D INADEQUATE NMR analysis, applying the cryoprobe methodology[80–83,268] Incorporation of advanced ^{13}C$_2$-labeled isoquinoline precursors into dioncophylline A (**1a**) after administration of the precursors to callus cultures of *T. peltatum*[81–83] In vivo localization of dioncophylline A (**1a**) in *T. peltatum* by near-infrared Fourier transform (NIR FT) micro-Raman spectroscopy in combination with density functional theory (DFT) calculations: Fluorescence microscopy images revealing **1a** to be located in inclusions of 10μm size in the cortex of the stems or in outer stem parts, at the stalks of the leaves[109] Highly sensitive and selective in situ localization of dioncophylline A (**1a**) in low quantities by deep-UV resonance Raman microscopy in *T. peltatum*—a method capable of an in situ differentiation between **1a** and the structurally related alkaloids dioncophylline C (**73**) and dioncopeltine A (**32**) in the liana[110]	
Dioncotetralone B (**224b**)[520]: Colorless solid (from CH$_2$Cl$_2$/MeOH): mp 162–165; $[\alpha]_D^{20}$ −206 (c 0.21, MeOH)		

7-epi-Dioncophylline A (1b)

$C_{24}H_{27}NO_3$ (377.5)

Discovered for the first time in the West African liana *A. barteri*, apparently in an atropisomerically pure form, i.e., without its 7-epimer dioncophylline A (**1a**).[5,285]

Isolation from *A. abbreviatus* along with its 4′-O-demethyl and N-methyl analogues **93b** and **94b**; together with their atropo-diastereomers, **1a**, **93a**, and **94a**.[5,58,404]

Isolation from *A. ileboensis* along with its 4′-O-demethyl analogue **93b**; together with their atropo-diastereomers, **1a** and **93a**.[51]

Structural Elucidation:

Constitution and complete stereostructure by HRESIMS and by 1D and 2D NMR, oxidative degradation, and ECD—comparison of the data with those of the structurally closely related, yet P-configured dioncophylline A (**1a**).[5,51,58,285,404]

Relative configuration at centers versus axis by long-range 1D and 2D NOE and ROESY interactions.[66]

Total Synthesis:

Stereocontrolled total synthesis of 7-*epi*-dioncophylline A (**1b**) by applying the lactone method,[76–79] through intramolecular aryl coupling of the ester-type prefixed molecular moieties and atroposelective lactone cleavage, giving rise to **1b** or, optionally, to its likewise natural atropo-diastereomer, dioncophylline A (**1a**).[5,6,9,76,364,378]

Total synthesis of *ent*-7-*epi*-dioncophylline A (*ent*-**1b**), again by applying the lactone method.[9]

Yellow amorphous solid (from MeOH): $[\alpha]_D^{23}$ −36.5 (c 0.02, MeOH).[404]

White solid (from CH_2Cl_2/MeOH): mp 240–242; $[\alpha]_D^{20}$ −14.0 (c 1.00, $CHCl_3$).[5]

cf.: *ent*-7-*epi*-Dioncophylline A (*ent*-**1b**) prepared synthetically: mp 214; $[\alpha]_D^{20}$ +14.9 (c 0.45, CH_2Cl_2).[9]

A. abbreviatus: root bark.[5,58,404]

A. barteri: roots.[5,285]

A. ileboensis: root bark.[51]

Good to excellent growth-inhibiting activities in vitro against asexual blood stages of *Plasmodium falciparum* (NF54 strain)[369] and *P. berghei* (Anka).[370]

Inactive against exoerythrocytic stages of *Plasmodium berghei* (Anka) in vitro.[371]

Moderate inhibitory activity in vitro against the hematozoan parasite *Babesia canis*.[373]

N-Methyldioncophylline A (94a)

$C_{25}H_{29}NO_3$ (391.5)

Discovered for the first time in *A. abbreviatus*: separation of N-methyldioncophylline A (**94a**) and its atropo-diastereomer N-methyl-7-*epi*-dioncophylline A (**94b**), analytically on a chiral HPLC phase (Chiralcel OD®) and preparatively, after esterification with a chiral carboxylic acid.[62]

Isolation of **94a** from *A. barteri*: NMR, MS, and oxidative degradation of a mixture of the two atropo-diastereomers, **94a** and **94b**, resolution of **94a** and **94b** (natural occurrence: 1:1) on a chiral HPLC phase (Chiralcel OD®).[285]

Isolation of **94a** from *Habropetalum dawei* (Dioncophyllaceae) along with its atropo-diastereomer **94b**: Constitution and complete stereostructure of **94a** established online, from crude extracts, using the analytical triad LC-MS/MS,

Moderate feeding deterrency and growth-retarding activity against larvae of the herbivorous insect *Spodoptera littoralis*.[99]

Inactive against asexual erythrocytic forms of *Plasmodium falciparum* (strains: NF54, K1)[369] and *P. berghei* (Anka strain).[370]

Inactive against exoerythrocytic stages of *P. berghei* (Anka).[371]

Continued

Table 12.1 7,1′-Coupled Dioncophyllaceae-type naphthylisoquinoline alkaloids.

Name, formula (mol. wt.), and structure Producing plants and their organs/physical data [mp/°C, $[\alpha]_D^{T/°}$ cm^2 g^{-1} 10^{-1} (solvent)]	Isolation, structural elucidation, partial and/or total synthesis, and further investigations	Biological activities
A. abbreviatus: stem bark[62] *A. barteri*: roots[285] *H. dawei*: stem bark[297] *D. thollonii*: roots[292] Microcrystalline gray needles (from acetone): mp 93; $[\alpha]_D^{25}$ +15.0 (c 1.27, CHCl$_3$)[62] Microcrystalline white compd. (from MeOH): mp 190; $[\alpha]_D^{25}$ +13.9 (c 0.13, CHCl$_3$)[292] Microcrystalline white compd. (from acetone): atropo-diastereomeric mixture (ratio **94a**:**94b**, ca. 1:1): mp 208–212[285]	LC-NMR (^{1}H, 2D-TOCSY, 2D-ROESY), and LC-ECD, and, offline, by oxidative degradation[297] Isolation from *Dioncophyllum thollonii* (Dioncophyllaceae): Constitution and complete stereostructure after resolution of **94a** and its atropo-diastereomer **94b** (natural occurrence: *P*:*M* = 2:1) on a chiral HPLC phase (Chiralcel OD®), by NMR and ECD investigations, and by oxidative degradation[292] *Structural Elucidation:* Constitution and complete stereostructure by HRESIMS, 1D and 2D NMR, oxidative degradation, and ECD—comparison of the data of **94a** with those obtained from the structurally closely related, likewise *P*-configured dioncophylline A (**1a**)[6,9,62,285,292,297] *Partial Synthesis (Semisynthesis):* Partial synthesis (and thus total synthesis) from dioncophylline A (**1a**)[9,62,76]	Strong inhibitory activity in vitro against the hematozoan parasite *Babesia canis*[373]
N-Methyl-7-*epi*-dioncophylline A (94b) C$_{25}$H$_{29}$NO$_3$ (391.5)	Discovered for the first time in *A. abbreviatus*: separation of *N*-methyl-7-*epi*-dioncophylline A (**94b**) and its atropo-diastereomer *N*-methyldioncophylline A (**94a**), analytically on a chiral HPLC phase (Chiralcel OD®) and preparatively, after esterification with a chiral carboxylic acid[62] Isolation from *A. barteri*: NMR, MS, and oxidative degradation of a mixture of the two atropo-diastereomers, **94b** and **94a**, resolution of **94a** and **94b** (natural occurrence: 1:1) on a chiral HPLC phase (Chiralcel OD®)[285] Isolation from *Habropetalum dawei* (Dioncophyllaceae), along with its atropo-diastereomer **94a**: Constitution and complete stereostructure of **94b** established online from crude extracts, using the analytical triad LC-MS/MS,	Inactive against asexual erythrocytic forms of *Plasmodium falciparum* (NF54 strain)[369] and exoerythrocytic stages of *P. berghei* (Anka) in vitro[371] Strong inhibitory activity in vitro against the hematozoan parasite *Babesia canis*[373]

A. abbreviatus: stem bark[62] A. barteri: roots[285] H. dawei: stem bark[297] D. thollonii: roots[292] Microcrystalline gray needles (from acetone): mp 230; $[\alpha]_D^{25}$ −6.0 (c 1.12, CHCl$_3$)[62] Microcrystalline white compd. (from MeOH): mp 233; $[\alpha]_D^{25}$ −6.8 (c 0.11, CHCl$_3$)[292] Microcrystalline white compd. (from acetone): atropo-diastereomeric mixture (ratio **94a:94b**, ca. 1:1): mp 208–212[285]	LC-NMR (^{1}H, 2D-TOCSY, 2D-ROESY), and LC-ECD, and, offline, by oxidative degradation[297] Isolation from *Dioncophyllum thollonii* (Dioncophyllaceae): Constitution and complete stereostructure after resolution of **94b** and its atropo-diastereomer **94a** (natural occurrence: P:M = 1:2) on a chiral HPLC phase (Chiralcel OD®), by NMR and ECD investigations, and by oxidative degradation[292] *Structural Elucidation:* Constitution and complete stereostructure by HRESIMS, 1D and 2D NMR, oxidative degradation, and ECD—comparison of the chiroptical data of **94b** with those of the structurally closely related, likewise M-configured 7-*epi*-dioncophylline A (**1b**)[6,9,62,285,292,297] *Semisynthesis:* Semisynthesis of **94b** from 7-*epi*-dioncophylline A (**1b**) and, given the synthetic availability of **1b**,[9,62,76] its total synthesis[62]	
4′-O-Demethyldioncophylline A (93a) C$_{23}$H$_{25}$NO$_3$ (363.5) A. abbreviatus: roots, stem bark[5,58] A. iletoensis: root bark[51]	Isolation from the West African liana *A. abbreviatus*,[5,58] *A. iletoensis*,[51] along with its atropo-diastereomer, 4′-O-demethyl-7-*epi*-dioncophylline A (**93b**) *Structural Elucidation:* Constitution and complete stereostructure by MS, NMR, and ECD, and by oxidative degradation—comparison of the data of **93a** with those obtained from the P-configured parent compound, dioncophylline A (**1a**)[5,51,58] *Semisynthesis and Total Synthesis:* Semisynthesis of 4′-O-demethyldioncophylline A (**93a**)[12] by mono-O-demethylation of dioncophylline A (**1a**) using trimethylsilyl iodide in CHCl$_3$ giving rise to a 1:1	Pronounced—and specific—inhibitory activity in vitro against *Plasmodium falciparum*[58] Inactive against *Leishmania* and *Trypanosoma* parasites[58] Strong cytotoxic activity against multiple myeloma INA-6 cells, similar to the standard drug melphalan[51] Moderate cytotoxic activity against drug-sensitive acute lymphoblastic CCRF-CEM leukemia cells[51]

Continued

Table 12.1 7,1′-Coupled Dioncophyllaceae-type naphthylisoquinoline alkaloids. Name, formula (mol. wt.), and structure

Producing plants and their organs/physical data [mp/°C, $[\alpha]_D^\circ$/cm² g⁻¹ 10⁻¹ (solvent)]	Isolation, structural elucidation, partial and/or total synthesis, and further investigations	Biological activities
Microcrystalline white compd. (from CHCl₃/MeOH): mp 278; $[\alpha]_D^{20}$ −12.0 (c 0.85, EtOH)[5] cf.: Compd. prepared by semisynthesis: pale-yellow solid (from MeOH): mp 232–234; $[\alpha]_D^{20}$ −13.9 (c 0.10, MeOH)[12]	mixture of **93a** and its regioisomer, 5′-O-demethyldioncophylline A (**97**); subsequent resolution by preparative HPLC on a reversed-phase column to give pure **93a**	
4′-O-Demethyl-7-*epi*-dioncophylline A (93b) C₂₃H₂₅NO₃ (363.5)	Isolation from the West African liana *A. abbreviatus*,[5,58,404] and from the Congolese plant *A. ileboensis*,[51] along with its atropo-diastereomer, 4′-O-demethyldioncophylline A (**93a**)	Moderate to weak cytotoxic activity against multiple myeloma INA-6 cells and against drug-sensitive acute lymphoblastic CCRF-CEM leukemia cells[51]
	Isolation from the West African liana *A. barteri*, apparently in an atropisomerically pure form, i.e., without its 7-epimer, 4′-O-demethyldioncophylline A (**93a**)[5]	Pronounced—and specific—inhibitory activity in vitro against *Plasmodium falciparum*[58]
	Structural Elucidation:	Inactive against *Leishmania* and *Trypanosoma* parasites[58]
	Constitution and complete stereostructure by HRESIMS, 1D and 2D NMR, by oxidative degradation, and ECD—comparison of the data of **93b** with those obtained from the *P*-configured parent compound, 7-*epi*-dioncophylline A (**93a**)[5,51,58,404]	
A. abbreviatus: roots, stem bark[5,58,404]		
A. barteri: roots[5]		
A. ileboensis: root bark[51]		
Colorless crystals (from CHCl₃/MeOH): mp 244; $[\alpha]_D^{20}$ −22.0 (c 0.92, EtOH)[5]		
Yellow amorphous solid (from MeOH): $[\alpha]_D^{23}$ −120.0 (c 0.06, MeOH)[404]		
Microcrystalline white powder (from CHCl₃/MeOH): mp 240; $[\alpha]_D^{20}$ −24.4 (c 1.05, CHCl₃)[5]		

ent-Dioncophylleine A (24b)

$C_{24}H_{23}NO_3$ (373.5)

A. abbreviatus: roots[54,404]
A. benomensis: leaves[291]

Yellow amorphous solid (from MeOH): $[\alpha]_D^{23}$ −286.6 (*c* 0.03, MeOH)[404]
White solid (from MeOH): mp 265; $[\alpha]_D^{25}$ −29.7 (*c* 0.08, MeOH)[291]

cf.: *P*-Isomer dioncophylleine A (**24a**) prepared by semisynthesis, obtained as a white solid (from EtOH): mp 272[5,8,333]

Discovered for the first time in the Malaysian liana *A. benomensis*—categorized as a Dioncophyllaceae-type naphthylisoquinoline alkaloid, i.e., lacking an oxygen function at C-6, and thus, substantially differing from (nearly) all the other alkaloids found in Asian *Ancistrocladus* species with respect to the exclusive production of Ancistrocladaceae-type alkaloids by these plants[291]

Structural Elucidation:

Constitution established from HREIMS and by ^{1}H and ^{13}C NMR, indicating the presence of a non-hydrogenated naphthylisoquinoline alkaloid—assignment of the coupling position and the location of a Dioncophyllaceae-type specific proton at C-6 by NOESY and HMBC interactions[54,291,404]

Assignment of the absolute axial configuration as *M* due to the virtually opposite ECD spectrum of **24b** compared to that of its *P*-configured (naturally less frequently occurring, see below) enantiomer dioncophylleine A (**24a**),[54,291,404] as prepared semisynthetically, by dehydrogenation of dioncophylline A (**1a**)[5,8,333,334]

HPLC-ECD analysis on a chiral phase (Chiralcel OD®) revealing *A. benomensis* to produce *ent*-dioncophylleine A (**24b**) in a scalemic form, as a 93:7-mixture of **24b** and its *P*-configured enantiomer, dioncophylleine A (**24a**)[291]

Occurrence of **24b** in *A. abbreviatus* (from Ivory Coast) determined in an enantiomerically pure form, as deduced from HPLC-ECD analysis on a chiral phase (Lux Cellulose-1)[54]

Weak inhibitory activities against *Leishmania major*,[42] *Plasmodium falciparum* (K1 strain), and *Trypanosoma brucei rhodesiense*[291]

Inactive against *T. cruzi*[291]

Moderate antiproliferative effect against lymphoblastic CCRF-CEM leukemia cells and their multi-drug resistant subline CEM/ADR5000[54]

Table 12.2 7,1′-Coupled Ancistrocladaceae-type naphthylisoquinoline alkaloids.
Name, formula (mol. wt.), and structure

Producing plants and their organs/physical data [mp/°C, $[\alpha]_D^T$/°cm^2 g^{-1} 10^{-1} (solvent)]	Isolation, structural elucidation, partial and/or total synthesis, and further investigations	Biological activities
Ancistrobrevine D (102a) $C_{26}H_{31}NO_4$ (421.5) *A. abbreviatus*: stem bark,[5,65] roots[57,404] White solid (from MeOH): mp 172; $[\alpha]_D^{25}$ +24.9 (c 0.37, CHCl$_3$)[65] Yellow amorphous powder (from MeOH): $[\alpha]_D^{23}$ −115.7 (c 0.05, MeOH)[404] cf.: title compd. prepared by stereoselective total synthesis: white solid (from MeOH): mp 171; $[\alpha]_D^{20}$ +21.4 (c 1.00, CHCl$_3$)[65]	*Structural Elucidation:* Constitution and coupling position deduced from HRESIMS and from ^{1}H and ^{13}C NMR, including HMBC and NOESY correlations, indicating the presence of a 7,1′-linked *N*-methylated naphthyltetrahydroisoquinoline with three methoxy groups at C-4′, C-5′, and C-6 and a free hydroxy function at C-8[5,57,65,404] Relative *cis*-configuration at C-1 versus C-3 deduced from specific NOE interactions[5,57,65,404] Absolute configuration at C-3 established as *S* by ruthenium-mediated oxidative degradation[5,57,65,70,404] Assignment of the absolute axial configuration as *P* due to specific NOESY correlations in conjunction with the absolute 1*R*,3*S*-stereoarray in the tetrahydroisoquinoline part[5,57,65,404] *Total Synthesis:* Additional structural proof by stereoselective total synthesis of all four possible *cis*-configured stereoisomers and comparison of their physical properties, ^{1}H NMR data, and chromatographic behaviors on a chiral phase (Chiralcel OD) with those of the natural product **102a**[1,65]	Moderate to weak inhibitory activities in vitro against asexual erythrocytic stages of *Plasmodium falciparum* (strains: K1 and NF54/64)[57,369] and the rodent malaria parasite *P. berghei* (Anka)[370] No inhibitory activity against the development of exoerythrocytic stages (liver forms) of *P. berghei* (Anka) in vitro in human hepatoma cells (Hep G2)[371] No inhibitory activities against *Trypanosoma* and *Leishmania* parasites[57] Moderate preferential cytotoxicity against PANC-1 human pancreatic cancer cells under "austerity" (i.e., nutrient-deprived) conditions[57]

Ancistrocladisine B (103) $C_{25}H_{29}NO_4$ (407.2)	Isolated for the first time from *A. ileboensis* from the South-Central Congo Basin[51]	Biological activities of the pure compound not determined due to lack of isolated material
A. abbreviatus: roots[57,404]	*Structural Elucidation:*	
A. ileboensis: root bark[51]	Constitution and coupling position deduced from HRESIMS and from ^{1}H and ^{13}C NMR, including specific HMBC and long-range NOESY interactions, revealing **103** to differ from its parent compound ancistrobrevine D (**102a**) by the absence of an *N*-methyl group[51,404]	
Yellow amorphous solid (from MeOH): $[\alpha]_D^{23}$ −711 (*c* 0.01, MeOH)[404]	Absolute configuration at C-3 established as *S* by oxidative degradation[51]	
Brown amorphous powder (from MeOH): $[\alpha]_D^{20}$ +6.10 (*c* 0.10, MeOH)[51]	Absolute configuration at C-1 determined as *R* due to the relative *cis*-configuration at C-1 versus C-3 as deduced from NOE interactions in combination with the given absolute *S*-configuration at C-3[51,404]	
	Assignment of the absolute axial configuration as *P* due to specific NOESY correlations in conjunction with the known absolute 1*R*,3*S*-stereoarray in the tetrahydroisoquinoline part, confirmed by the nearly mirror-image like ECD spectrum of **103** compared to that of the structurally closely related and likewise 7,1′-coupled ancistrocladisine A (**99**), which is *M*-configured[51]	

Continued

Table 12.2 7,1′-Coupled Ancistrocladaceae-type naphthylisoquinoline alkaloids.—cont'd
Name, formula (mol. wt.), and structure

Producing plants and their organs/physical data [mp/°C, $[\alpha]_D^T$/°cm^2 g^{-1} 10^{-1} (solvent)]	Isolation, structural elucidation, partial and/or total synthesis, and further investigations	Biological activities
N-Methylancistrocladisine A (104) $C_{27}H_{33}NO_4$ (435.6)	*Structural Elucidation:* Constitution and complete stereostructure by HRESIMS, 1D and 2D NMR, oxidative degradation, and ECD—comparison of the data of **104** with those obtained from its parent compound ancistrocladisine A, evidencing the presence of an Ancistrocladaceae-type 7,1′-coupled *O,N*-permethylated naphthylisoquinoline alkaloid[404]	Biological activities of the pure compound not determined due to lack of isolated material
A. abbreviatus: leaves[404] Yellow amorphous solid (from MeOH): $[\alpha]_D^{25}$ + 30.4 (*c* 0.05, MeOH)[404]	Assignment of the absolute axial configuration as *P* due to the relative configuration of centers versus axis known from NMR in conjunction with the absolute 1*R*,3*S*-stereoarray in the tetrahydroisoquinoline part, corroborated by the nearly identical ECD spectrum of **104** compared to that of its parent compound, ancistrocladisine A (**99**)[404]	
Ancistrobreveine A (25) $C_{25}H_{25}NO_4$ (403.2)	*Structural Elucidation:* Constitution and coupling position deduced from HRESIMS and from ^{1}H and ^{13}C NMR, including HMBC and NOESY interactions, evidencing **25** to be a 7,1′-linked non-hydrogenated naphthylisoquinoline alkaloid with three methoxy groups at C-4′, C-5′, and C-6 and with a free hydroxy function at C-8[54]	Moderate antiproliferative activity in vitro against parental drug-sensitive CCRF-CEM leukemia cells[54]
	Assignment of the absolute axial configuration as *P* due the fact that **25** and the likewise 7,1′-coupled and	

A. abbreviatus: roots[54]	non-hydrogenated alkaloid *ent*-dioncophylleine A (**24b**) showed nearly identical ECD spectra; for formal reasons, however, the two alkaloids have opposite descriptors, according to the Cahn-Ingold-Prelog priority rules[54]
Yellow amorphous solid (from MeOH): $[\alpha]_D^{20}$ −11.1 (*c* 0.10, MeOH)[54]	HPLC-ECD analysis on a chiral phase (Lux Cellulose-1) revealed ancistrobreveine A (**25**) to occur enantiomerically pure in *A. abbreviatus*[54]

Ancistrobreveine B (105) $C_{26}H_{27}NO_4$ (417.2)	*Structural Elucidation:*	Moderate to weak cytotoxic effects against drug-sensitive CCRF-CEM leukemia cells and their multi-drug resistant subline CEM/ADR5000[54]
	Constitution and coupling position established by HRESIMS and by 1D and 2D NMR—comparison of the data of **105** with those obtained from ancistrobreveine A (**25**) revealed ancistrobreveine B (**105**) to be a 7,1′-linked non-hydrogenated naphthylisoquinoline alkaloid, but, in contrast to **25**, *O*-permethylated, equipped with an additional methoxy (instead of a hydroxy) group at C-8[54]	
A. abbreviatus: roots[54]	Assignment of the absolute axial configuration as *P* due the fact that **105** displayed a nearly mirror-image like ECD spectrum compared to that of ancistrobreveine A (**25**), thus revealing **105** to exhibit an opposite axial stereoarray; for formal reasons, however, the two alkaloids have the same *P*-descriptor, according to the Cahn-Ingold-Prelog priority rules[54]	
Yellow amorphous solid (from MeOH): $[\alpha]_D^{23}$ −180 (*c* 0.04, MeOH)[54]	HPLC-ECD analysis on a chiral phase (Lux Cellulose-1) revealed ancistrobreveine B (**105**) to occur enantiomerically pure in *A. abbreviatus*[54]	

Table 12.3 7,1′-Coupled hybrid-type naphthylisoquinoline alkaloids.

Name, formula (mol. wt.), and structure Producing plants and their organs/physical data [mp/°C, $[\alpha]_D^T$/°cm^2 g^{-1} 10^{-1} (solvent)]	Isolation, structural elucidation, partial and/or total synthesis, and further investigations	Biological activities
Ancistrobrevine C (22) $C_{25}H_{27}NO_4$ (405.5) *A. abbreviatus*: roots, stem bark[64] *A. ileboensis*: root bark[51] From a botanically yet undescribed Congolese *Ancistrocladus* species: root bark[12] Bright-yellow fluorescent amorphous solid (from MeOH/CH$_2$Cl$_2$): mp 180–183; $[\alpha]_D^{25}$ +13.0 (c 0.69, CHCl$_3$)[64] Yellow amorphous solid (from MeOH): $[\alpha]_D^{20}$ +26.0 (c 0.26, CHCl$_3$)[12]	Isolation from the West African liana *A. abbreviatus* as the first hybrid-type naphthylisoquinoline alkaloid discovered in nature, i.e., with the structural landmarks both of Ancistrocladaceae- (6-oxygenated) *and* Dioncophyllaceae-type (3*R*-configuration) representatives[64] *Structural Elucidation:* Constitution and coupling position established by HRESIMS, ^{1}H NMR and ^{13}C NMR, including ROESY, NOESY, and HMBC interactions, evidencing the presence of a 7,1′-linked naphthyldihydroisoquinoline with three *O*-methyl groups at C-4′, C-5′, and C-6, and a free hydroxy function at C-8[12,51,64] Absolute configuration at C-3 established as *R* by oxidative degradation[12,51,64,69,70] Attribution of the absolute axial configuration as *P* by comparison of the ECD spectrum with that of the likewise 7,1′-coupled, structurally closely related Ancistrocladaceae-type naphthyldihydroisoquinoline alkaloid ancistrocladisine, which is stereochemically well known from total synthesis[384]—nearly mirror-imaged ECD spectra for the two alkaloids; for formal reasons, however, they have the same *P*-descriptor according to the Cahn-Ingold-Prelog priority rules[64] *Semisynthesis:* Structural confirmation by 8-*O*-methylation of ancistrobrevine C (**22**) to give the enantiomer of the stereochemically known,[1,260,262,384,400] *O*-permethylated alkaloid ancistrocladisine, *ent*-**225**[64] Reaction Conditions: (*i*) CH$_2$N$_2$, Et$_2$O, MeOH, under N$_2$, 2 d, r.t.	Weak to very low inhibitory activities in vitro against the protozoan pathogens *Plasmodium falciparum*, *Trypanosoma brucei rhodesiense*, *T. cruzi*, and *Leishmania donovani*[57]

Table 12.4 A 7,1′-coupled inverse hybrid-type naphthylisoquinoline alkaloid.

Name, formula (mol. wt.), and structure	Producing plants and their organs/ physical data [mp/°C, $[\alpha]_D^T$/°cm^2 g^{-1} 10^{-1} (solvent)]	Isolation, structural elucidation, partial and/or total synthesis, and further investigations	Biological activities
Dioncoline A (23) C$_{25}$H$_{29}$NO$_3$ (391.5)	*A. abbreviatus*: roots, stem bark[58] Yellow amorphous solid (from MeOH): $[\alpha]_D^{23}$ −60.9 (c 0.08, MeOH)[58]	Dioncoline A (**23**) as the so far only known 3*S*-configured naphthylisoquinoline found in nature that lacks an *O*-functionality at C-6, thus being the first inverse hybrid-type naphthylisoquinoline alkaloid[58] *Structural Elucidation:* Constitution and complete stereostructure established by HRESIMS, 1D and 2D NMR, oxidative degradation, and ECD—comparison of the data of **23** with those obtained from the structurally closely related 7,1′-coupled, and likewise *P*-configured, alkaloid *N*-methyldioncophylline A (**94a**),[62] revealing **23** to be the C-3 epimer of **94a**, and thus, the first so-called "inverse hybrid-type" naphthylisoquinoline alkaloid[58] Relative *cis*-configuration of the two stereogenic centers at C-1 and C-3 assigned by specific NOESY interactions[58] Absolute configuration at C-3 established as *S* by oxidative degradation[58,70]	No substantial cytotoxic effects toward TH-29 human colon carcinoma cells[58]

Continued

Table 13 7,3′-Coupled Dioncophyllaceae-type naphthylisoquinoline alkaloids.

Producing plants and their organs/ Name, formula (mol. wt.), physical data [mp/°C, $[\alpha]_D^T$/°cm^2 g^{-1} 10^{-1} (solvent)]	Isolation, structural elucidation, partial and/or total synthesis, and further investigations	Biological activities
Dioncophyllidine E (72) $C_{23}H_{23}NO_3$ (361.5) 72a 72b ⁂ = configurationally semi-stable axis *A. abbreviatus*: leaves[61] Yellow amorphous solid (from MeOH):[61] $[\alpha]_D^{23}$ +28.2 (*c* 0.05, MeOH)[61]	*Structural Elucidation:* Constitution and complete stereostructure established by HRESIMS, 1D and 2D NMR, and oxidative degradation—comparison of the data of **72** with those of its parent compound dioncophylline E (**30**)[292] revealing **72** to be a 7,3′-coupled naphthyldihydroisoquinoline alkaloid with the *R*-configuration at C-3, lacking an oxygen function at C-6, thus being a Dioncophyllaceae-type alkaloid[61] As expected from the presence of only three (moreover relatively small) substituents next to the biaryl axis and as confirmed by LC-ECD experiments, slow rotation about the axis of dioncophylline E (**72**) occurs at room temperature, thus **72** exists as a pair of two slowly interconverting, since configurationally semi-stable atropo-diastereomers, **72a** and **72b**[61] After HPLC resolution of the two rotational isomers of dioncophyllidine E (**72**), online chiroptical assignment of the absolute configurations at the biaryl axes of **72a** and **72b**[61] by comparison of their LC-ECD spectra with the ECD curve of the likewise 7,3′-coupled, yet configurationally stable, Ancistrocladaceae-type naphthyldihydroisoquinoline ancistrocladidine (**116**),[1,261,262,265,279,388] with an *M*-configuration at the biaryl axis	Strong preferential cytotoxicity against PANC-1 human pancreatic cancer cells under "austerity" (i.e., nutrient-deprived) conditions[61] Moderate cytotoxic activity against human HeLa cervical cancer cells[61]

Table 14 7,8′-Coupled Ancistrocladaceae-type naphthylisoquinoline alkaloids.

Name, formula (mol. wt.), and structure	Producing plants and their organs/physical data [mp/°C, $[\alpha]_D^T$/°cm^2 g^{-1} 10^{-1} (solvent)]	Isolation, structural elucidation, partial and/or total synthesis, and further investigations	Biological activities
Ancistrobrevine A (100) $C_{27}H_{33}NO_4$ (435.6)	*A. abbreviatus*: stem bark,[5] roots[57,404] Yellow amorphous solid (from MeOH): $[\alpha]_D^{23}$ −31.1 (c 0.23, MeOH)[404]	*Structural Elucidation:* Constitution and coupling position established by HRESIMS and by ^{1}H and ^{13}C NMR, including specific NOESY and HMBC interactions, revealing the presence of a 7,8′-linked O,N-permethylated naphthyltetrahydroisoquinoline alkaloid[5,57,404] Absolute configuration at the two stereogenic centers established to be 1R,3S by oxidative degradation, in conjunction with the relative *cis*-configuration at C-1 versus C-3 assigned by specific NOESY interactions[5,57,404] Determination of the axial configuration as *P* by 1D and 2D NOE and ROE experiments, in combination with the results of the degradation procedure[66]	Weak inhibitory activity in vitro against *Plasmodium falciparum* (strain: NF54)[57,369] and *Trypanosoma* parasites[57] No inhibitory activity against *Leishmania donovani*[57] No inhibitory activity in vitro against the development of exoerythrocytic stages of *Plasmodium berghei* (Anka) in human hepatoma cells (Hep G2)[371] Moderate preferential activity against PANC-1 human pancreatic cancer cells under "austerity" (i.e., nutrient-deprived) conditions[57]

Continued

Table 14 7,8′-Coupled Ancistrocladaceae-type naphthylisoquinoline alkaloids.—cont'd

Name, formula (mol. wt.), and structure	Producing plants and their organs/physical data [mp/°C, $[\alpha]_D^T$/°cm^2 g^{-1} 10^{-1} (solvent)]	Isolation, structural elucidation, partial and/or total synthesis, and further investigations	Biological activities
6-O-Demethylancistrobrevine A (118) $C_{26}H_{31}NO_4$ (421.5) *A. abbreviatus*: root bark,[5] roots[57,404] *A. ealaensis*: twigs[10] From a botanically yet undescribed Congolese *Ancistrocladus* species: roots[29] Yellow amorphous solid (from MeOH): $[\alpha]_D^{23}$ −61.3 (c 0.07, MeOH)[404]	*Structural Elucidation:* Constitution and complete stereostructure established by HRESIMS, 1D and 2D NMR, oxidative degradation, and ECD—comparison of the data of **118** with those obtained from its parent compound ancistrobrevine A (**100**), revealing **118** to differ from **100** by the presence of a free hydroxy function at C-6 instead of having a methoxy group[5,29,57,404] Additional proof of the absolute axial configuration assigned as *P* by comparison of the ECD spectrum of **118**[5,29,57] with that of the structurally closely related 7,8′-coupled and likewise *P*-configured naphthylisoquinoline alkaloid ancistrogriffine A (**31**)[1,284]	Moderate inhibitory activity in vitro against *Plasmodium falciparum* (K1 and NF54 strains)[29,57] Weak inhibitory activities in vitro against *Trypanosoma cruzi* and *T. brucei rhodesiense*[29,57] No inhibitory activity against *Leishmania donovani*[29,57] Moderate to weak preferential cytotoxicity against PANC-1 human pancreatic cancer cells under "austerity" (i.e., nutrient-deprived) conditions[57]

Ancistrobrevine H (8)

$C_{26}H_{31}NO_4$ (421.2)

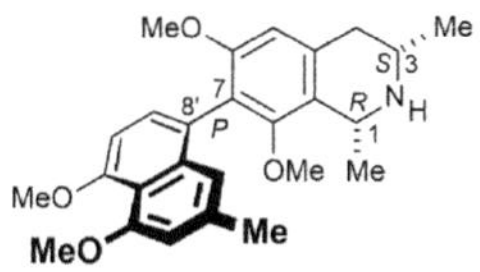

A. abbreviatus: roots[57]

Yellow amorphous solid (from MeOH): $[\alpha]_D^{20}$ −32.9 (c 0.05, MeOH)[57]

Structural Elucidation:

Constitution and relative configurations at the three stereogenic elements (two centers and one axis) deduced from HRESIMS and from ^{1}H and ^{13}C NMR, including HMBC and specific NOE interactions, revealing the presence of a 7,8′-coupled O-permethylated naphthyltetrahydroisoquinoline[57]

Absolute configuration at C-3 established as S by oxidative degradation and at C-1 as R due to the relative *cis*-configuration at C-1 versus C-3 (as known from NOE interactions)[57]

Assignment of the absolute configuration at the biaryl axis as P due to specific long-range NOESY interactions, in conjunction with the given absolute configuration at C-3 (as known from the degradation)[57,66]

Additional proof of the absolute axial configuration by the close resemblance of the ECD spectrum of **8** with that of the related N-methylated alkaloid ancistrobrevine A (**100**),[5,66] as likewise occurring in *A. abbreviatus*[57]

Moderate to weak inhibitory activity in vitro against *Plasmodium falciparum* (strain: NF54) and *Trypanosoma* parasites[57]

No inhibitory activities against *Leishmania donovani*[57]

Continued

Table 14 7,8′-Coupled Ancistrocladaceae-type naphthylisoquinoline alkaloids.—cont'd

Name, formula (mol. wt.), and structure	Producing plants and their organs/physical data [mp/°C, $[\alpha]_D^T/°cm^2 g^{-1} 10^{-1}$ (solvent)]	Isolation, structural elucidation, partial and/or total synthesis, and further investigations	Biological activities
6-O-Demethylancistrobrevine H (119) $C_{25}H_{29}NO_4$ (407.5)		*Structural Elucidation:* Constitution and complete stereostructure established by HRESIMS, 1D and 2D NMR, oxidative degradation, and ECD—comparison of the data of **119** with those obtained from its parent compound ancistrobrevine H (**8**), revealing **119** to differ from **8** by possessing a free hydroxy group (instead of having a methoxy group) at C-6[57]	Moderate preferential cytotoxicity against PANC-1 human pancreatic cancer cells under "austerity" (i.e., nutrient-deprived) conditions[57]
	A. abbreviatus: root bark[57] Yellow amorphous solid (from MeOH): $[\alpha]_D^{20}$ +183.6 (c 0.05, MeOH)[57]	Additional proof of the absolute axial configuration by the close resemblance of the ECD spectrum of **119** with that of the related N-methylated alkaloid 6-O-demethylancistrobrevine A (**118**),[5,10,29] as likewise occurring in *A. abbreviatus*[57]	
Ancistrobrevine I (120) $C_{25}H_{27}NO_4$ (405.2)		*Structural Elucidation:* Constitution deduced from HRESIMS, ^{1}H and ^{13}C NMR and from HMBC and NOESY interactions, revealing the presence of a 7,8′-coupled naphthyldihydroisoquinoline alkaloid with a free hydroxy function at C-6[57]	Biological activities of the pure compound not determined for lack of isolated material
		Absolute configuration at C-3 established to be *S* by oxidative degradation[57]	
		Assignment of the absolute axial configuration as *P*, by installing an additional stereogenic center at C-1, much closer to the	

A. abbreviatus: roots[57]

Yellow amorphous solid (from MeOH): $[\alpha]_D^{20}$ +183.6 (c 0.05, MeOH)[57]

axis than C-3. This was achieved by *cis*-selective reduction of **120** using NaBH$_4$ to give the respective tetrahydroisoquinoline analogue ancistrobrevine H (**8**); unambiguous NOE long-range interactions and, thus, a reliable attribution of the axial configuration of **120** had now become possible by the newly generated, stereochemically well-defined auxiliary center[57]

Final structural proof by the virtually identical ECD spectrum of the semisynthetic ancistrobrevine H (**8**) compared to the ECD curve of the authentic natural product **8** isolated from *A. abbreviatus*[57]

Ancistrobrevine J (70)

C$_{25}$H$_{27}$NO$_4$ (405.2)

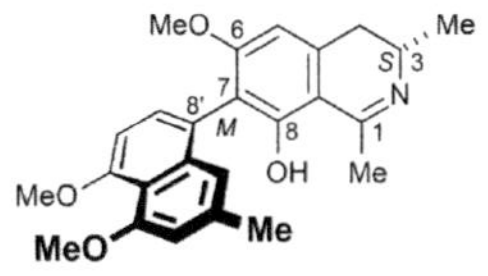

Compd. can also be addressed as 6-O-methyl-8-O-demethyl-ancistrobrevine I

Structural Elucidation:

Constitution deduced from HRESIMS and from ^{1}H and ^{13}C NMR, including HMBC and NOESY interactions, revealing the presence of a 7,8′-coupled naphthyldihydroisoquinoline alkaloid with a methoxy group at C-6 and a free hydroxy function at C-8[57]

Absolute configuration at C-3 established to be *S* by oxidative degradation[57]

Assignment of the absolute axial configuration as *M* due the nearly identical ECD spectra of **70** and the likewise 7,8′-coupled naphthyldihydroisoquinoline ancistrobrevine I (**120**); for formal reasons, however, the two alkaloids have

Biological activities of the pure compound not determined for lack of isolated material

Continued

Table 14 7,8′-Coupled Ancistrocladaceae-type naphthylisoquinoline alkaloids.—cont'd
Name, formula (mol. wt.), and structure

Producing plants and their organs/physical data [mp/°C, $[\alpha]_D^T$/°cm^2 g^{-1} 10^{-1} (solvent)]	Isolation, structural elucidation, partial and/or total synthesis, and further investigations	Biological activities
A. abbreviatus: roots[57] Yellow amorphous solid (from MeOH): $[\alpha]_D^{20}$ −77.5 (*c* 0.05, MeOH)[57]	opposite descriptors, according to the Cahn-Ingold-Prelog priority rules[57] Confirmation of the assigned absolute axial configuration by generation of an auxiliary stereogenic center at C-1 by stereoselective reduction of **70** using NaBH$_4$, to give the (as yet merely semisynthetic) tetrahydroisoquinoline **121**[57] (see Fig. 73); detection of a diagnostically significant long-range NOESY cross-peak[57] Additional proof of the absolute axial configuration by comparison of the ECD spectrum of the semisynthetic tetrahydroisoquinoline **121**[57] (see Fig. 73) with the, expectedly, virtually opposite ECD curve of the known, structurally closely related and likewise 7,8′-coupled, but *P*-configured naphthyltetrahydroisoquinoline ancistrogriffine C (**122**)[1,284]	

Ancistrobreveine C (123a) and ent-Ancistrobreveine C (123b)

$C_{26}H_{27}NO_4$ (417.2)

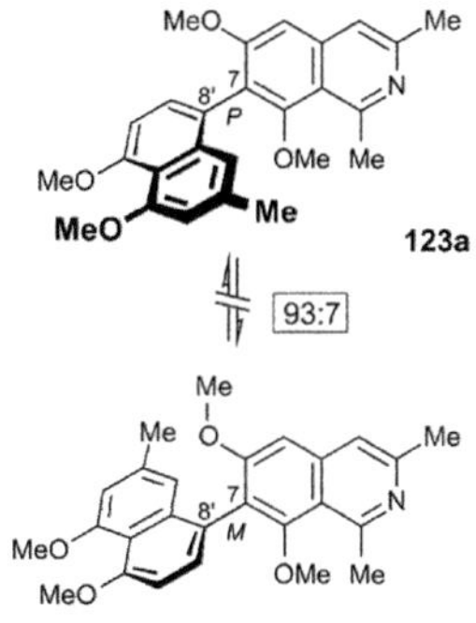

A. abbreviatus: roots[54]

Yellow amorphous solid (from MeOH): $[\alpha]_D^{20}$ +5.4 (c 0.10, MeOH)[54]

Structural Elucidation:

Constitution and complete stereostructure established by HRESIMS, 1D and 2D NMR, and by ECD—comparison of the data of **123** with those of the 7,1′-coupled non-hydrogenated, O-permethylated alkaloid ancistrobreveine B (**105**)[54] revealing **123** to differ from **105** only by the position of the methyl group on the naphthalene moiety, and, thus, being a representative of 7,8′-coupled naphthylisoquinolines[54]

Absolute axial configuration established to be *P* by comparison of the ECD spectrum of **123** with those of *ent*-dioncophylleine A (**24b**)[1,291] and ancistrobreveine B (**105**)[54]—the ECD curve of **123** being nearly identical to that of the likewise *P*-configured ancistrobreveine B (**105**), and virtually opposite to that of the *M*-configured *ent*-dioncophylleine A (**24b**)[54]

HPLC-ECD analysis on a chiral phase (Lux Cellulose-1) showing ancistrobreveine C (**123**) to occur in *A. abbreviatus* in a scalemic form, as a 93:7-mixture of **123a** and its *M*-configured enantiomer, *ent*-ancistrobreveine C (**123b**)[54]

Pronounced antiproliferative activities against drug-sensitive CCRF-CEM leukemia cells and their multi-drug resistant subline CEM/ADR5000, no cross-resistance observed[54]

Table 15.1 5,1′-Coupled Ancistrocladaceae-type naphthylisoquinoline alkaloids. Name, formula (mol. wt.), and structure

Producing plants and their organs/physical data [mp/°C, $[\alpha]_D^T$/°cm^2 g^{-1} 10^{-1} (solvent)]	Isolation, structural elucidation, partial and/or total synthesis, and further investigations	Biological activities
Ancistrocladine (20a) C$_{25}$H$_{29}$NO$_4$ (407.5) *A. abbreviatus*: stem bark[63] *A. barteri*: roots[5,285] *A. ileboensis*: root bark[51] From two botanically yet undescribed Congolese *Ancistrocladus* species: root bark,[12] roots[29] *A. robertsoniorum*: stem bark[280] *A. heyneanus*: roots, aerial parts[27,28,256,257,259,655] *A. hamatus*: roots[28,281,404,405]	Isolated for the first time from the Indian liana *A. heyneanus*[27,28,256,257,259] Occurrence in *Ancistrocladus* plants along with its atropo-diastereomer hamatine (20b),[1,12,28,29,65,243,249,250,280,281,284,285] except for *A. heyneanus*[27,28,256,257] and *A. ileboensis*.[51] From *A. griffithii*: identification online by LC-MS, and by LC-UV-ECD, by comparison of the structural data of the isolated compound with those of reference material[1,284] *Structural Elucidation:* Constitution and complete stereostructure by MS, NMR, ECD, and X-ray analysis—evidence of the structure and substitution pattern of ancistrocladine (20a) became evident mainly from the spectral data, from degradation experiments, and from spectroscopic studies on related derivatives[27,28,256,257,259] Assignment of the relative *trans*-configuration at the stereocenters at C-1 and C-3 by NMR investigations of 20a and related derivatives[27,259] Elucidation of the absolute configuration at C-3 as S by exhaustive ozonolysis to give the stereochemically known compound (*S*)-3-aminobutyric acid[27,259] Determination of the axial configuration by ECD and application of the Exciton Chirality method to the corresponding fully	Moderate to weak inhibitory activities in vitro against asexual erythrocytic forms of *Plasmodium falciparum* (strains: NF54/64 and K1)[369,418] No inhibitory activity against asexual erythrocytic forms of the rodent malaria parasite *P. berghei* (Anka strain)[370] No inhibitory activity against exoerythrocytic stages of *P. berghei* (Anka) in vitro in human hepatoma cells (Hep G2)[371] Pronounced antitumoral activity of the crude extract of *A. tectorius* from China (Hainan Island) in vivo[243] Ancistrocladine (20a), a main component of the alkaloidal fraction of *A. tectorius*, was inactive against the human

A. tectorius (from Laos): roots[238]	dehydrogenated, merely axially chiral naphthylisoquinoline ancistrocladine (125), prepared from 6-O-methylancistrocladine (74a) by dehydrogenation[27,259]	tumor cell line INA-6 (derived from a patient with multiple myeloma)[250] and showed only very weak activity against drug-sensitive lymphoblastic CCRF-CEM leukemia cells[251]
A. tectorius (from China): stems, twigs, branches[243,246,250,251,656]	X-ray diffraction analysis of ancistrocladine hydrobromide, evidencing the constitution and the relative configuration[259]	
A. griffithii: twigs[284]	Determination of the relative axial configuration by NOE and ROE experiments[66]	
	Attribution of the axial configuration as P by semiempirical ECD calculations using the CNDO/2S method[333,334,657]	
	Ab initio calculations of the optical-rotatory power (ORP) by means of the Hartree–Fock (HF) and the density functional theory (DFT) methods[658]	
White powder (from EtOH): mp 263–265; $[\alpha]_D^{25}$ −20.0 (c 0.15, CHCl$_3$)[63]	Total Synthesis and Related Synthetic Work:	
Amorphous solid (from MeOH): mp 260–263; $[\alpha]_D^{20}$ −16.0 (c 0.09, MeOH)[12]	Stereocontrolled total synthesis of ancistrocladine (20a) by applying the lactone method[76–79]: intramolecular mixed aryl coupling of the ester-type prefixed molecular moieties with subsequent reductive lactone cleavage giving rise to 20a, or, optionally, to its atropo-diastereomer, hamatine (20b)[1,76,78,383], even more stereo-economical: the directed, atropo-divergent synthesis of 20a and 20b via the respective cyclic biaryl ethers[659]	
Colorless amorphous solid (from CH$_2$Cl$_2$): mp 260; $[\alpha]_D^{20}$ −19.0 (c 0.10, MeOH)[29]		
Crystalline powder (from MeOH/H$_2$O): mp 261–263; $[\alpha]_D^{25}$ −21.0 (c 0.07, CHCl$_3$)[280]		
Amorphous solid,[256] white heavy cubes (from MeOH): mp 265–267 (dec)[27]; $[\alpha]_D^{25}$ −32.4 (pyridine)[257]	Oxidative dimerization of 20a to give the unnatural, highly unsymmetrically coupled quateraryl jozimine B (226)[76,512,513,516]	
As its hydrochloride salt, white needles (from MeOH/acetone): mp 220–224 (dec); $[\alpha]_D^{25}$ −25.5 (c 2.29, MeOH)[27,256]		
As its hydrobromide salt, white needles (from MeOH): mp 229–231 (dec); $[\alpha]_D^{25}$ −25.51 (c 2.29, MeOH)[256]		

Continued

Table 15.1 5,1′-Coupled Ancistrocladaceae-type naphthylisoquinoline alkaloids.—cont'd

Name, formula (mol. wt.), and structure	Producing plants and their organs/physical data [mp/°C, $[\alpha]_D^T/°\text{cm}^2\,\text{g}^{-1}\,10^{-1}$ (solvent)]	Isolation, structural elucidation, partial and/or total synthesis, and further investigations	Biological activities
	Amorphous solid: mp 265[655]		
	Amorphous solid: mp 262–263; $[\alpha]_D^{23}$ −20.5 (c 014, CHCl$_3$)[243]		
	Colorless amorphous solid (from MeOH): mp 260–263; $[\alpha]_D^{20}$ −15.4 (c 0.08, MeOH)[251]		
	cf.: Compd. prepared by stereoselective total synthesis: mp 261; $[\alpha]_D^{20}$ −23.0 (c 0.16, MeOH)[383,659]	Jozimine B (**226**)[516]: Amorphous solid (from MeOH): mp ≥229 (dec); $[\alpha]_D^{25}$ −14.7 (c 0.107, CHCl$_3$) *Further Investigations:* Localization and identification of ancistrocladine (**20a**) in the outer cortex of the roots of *A. heyneanus* by non-invasive FT-Raman microspectroscopy[108]	

Hamatine (20b) $C_{25}H_{29}NO_4$ (407.5)	For the first time discovered in *A. hamatus* from Sri Lanka[28,281,282]
	Occurrence in *Ancistrocladus* plants along with its atropo-diastereomer ancistrocladine (**20a**)[1,5,12,28,29,63,243,249,250,280,281,284,285]
	From *A. griffithii*: identification online by LC-MS, and by LC-UV-ECD, by comparison of the structural data of the detected compound with those of reference material[1,284]
	Structural Elucidation:
	Constitution and complete stereostructure of hamatine (**20b**) by MS, NMR, and ECD[28,281,282]—comparison of the structural data with those obtained from ancistrocladine (**20a**),[28] which is stereochemically well known from partial and total synthesis.[1,383,659]
	Absolute configuration at C-3 established as *S* by chemical degradation[69,70,281]
	Attribution of the absolute configuration at C-1 as *S* by NMR spectroscopy[281] in combination with chemical modification and degradation reactions[28,282]
	Determination of the relative axial configuration by long-range NOESY experiments[66]
	Assignment of the absolute axial configuration by ECD measurements and by application of the Exciton Chirality method[318–323,326,329] to the corresponding non-hydrogenated, merely axially chiral naphthylisoquinoline 6-O-methylhamateine (**124**), as obtained by dehydrogenation of 6-O-methylhamatine (**74b**)[28,281,282]
A. abbreviatus: stem bark[57,63]	Moderate inhibitory activity against asexual erythrocytic forms of *Plasmodium falciparum* (NF54 strain)[369,418]
A. barteri: roots[5,285]	
A. congolensis: root bark[178]	No inhibitory activity of an atropo-diastereomeric mixture of ancistrocladine (**20a**) and hamatine (**20b**) against the development of exoerythrocytic stages (liver forms) of *P. berghei* (Anka) in vitro in human hepatoma cells (Hep G2)[371]
From two botanically yet undescribed Congolese *Ancistrocladus* species: root bark,[12] roots[29]	
A. robertsoniorum: stem bark[280]	Moderate inhibitory activity against the human tumor cell line INA-6 (derived from a patient with multiple myeloma)[250]
A. hamatus: roots[28,281,282]	
A. tectorius (from China): stem bark, twigs[243,249–251,656]	Weak inhibitory activity against drug-sensitive acute lymphoblastic CCRF-CEM leukemia cells[251]
A. griffithii: twigs[284]	Moderate preferential cytotoxicity against PANC-1 human pancreatic cancer cells under

Continued

Table 15.1 5,1′-Coupled Ancistrocladaceae-type naphthylisoquinoline alkaloids.—cont'd

Name, formula (mol. wt.), and structure	Producing plants and their organs/physical data [mp/°C, $[\alpha]_D^T/°\text{cm}^2\text{g}^{-1}\,10^{-1}$ (solvent)]	Isolation, structural elucidation, partial and/or total synthesis, and further investigations	Biological activities
	Amorphous solid (from MeOH/acetone): mp 247–248; $[\alpha]_D^{25}$ +64.0 (c 0.24, CHCl$_3$)[63]	Determination of the axial configuration as M by semiempirical ECD calculations using the CNDO/2S method[657]	"austerity" (i.e., nutrient-deprived) conditions[57]
	Amorphous solid (from MeOH): mp 235–237; $[\alpha]_D^{20}$ +46.0 (c 0,24, CHCl$_3$)[12]	Ab initio calculations of the optical rotatory power (ORP) by means of the Hartree-Fock (HF) and the density functional theory (DFT) methods[658]	Moderate to weak growth inhibitory effects against human leukemia HL-60 and myelogenous leukemia K562 cells, and against human lung lymphoblast U937 cell lines[249]
	Colorless amorphous solid (from CH$_2$Cl$_2$): $[\alpha]_D^{20}$ +10.51 (c 0.10, MeOH)[29]	Additional confirmation of the structural assignment by semisynthetic conversion of hamatine (20b) into its known 6-O-methyl- and N-methyl analogues: comparison of the data (MS, NMR, ECD) with those obtained from the respective derivatives of ancistrocladine (20a)[28,281,282]	
	Microcrystalline powder (from MeOH/H$_2$O): mp 240–242; $[\alpha]_D^{25}$ +16.0 (c 0.32, CHCl$_3$)[280]		
	White cubes (from MeOH): mp 250–252; $[\alpha]_D^{20}$ +77.4 (c 0.86, CHCl$_3$)[28,281]	Total Synthesis:	
	Amorphous solid: mp 250–252; $[\alpha]_D^{20}$ +66.1 (c 1.07, CHCl$_3$)[243]	Stereocontrolled total synthesis of hamatine (20b) by applying the lactone method[76–79]: intramolecular mixed aryl coupling of the ester-type prefixed molecular moieties with subsequent reductive lactone cleavage, giving rise to 20b, or, optionally, to its atropo-diastereomer, ancistrocladine (20a)[1,76,78,383]; even more stereo-economical: the directed, atropo-divergent synthesis of 20a and 20b via the respective cyclic biaryl ethers[659]	
	Colorless amorphous powder (from MeOH): mp 235–237; $[\alpha]_D^{20}$ +47 (c 0.09, MeOH)[250]		
	cf.: Compd. prepared by stereoselective total synthesis: mp 250; $[\alpha]_D^{20}$ +68.0 (c 0.04, CHCl$_3$)[383,659]		

Ancistrocline (77)

$C_{26}H_{31}NO_4$ (421.5)

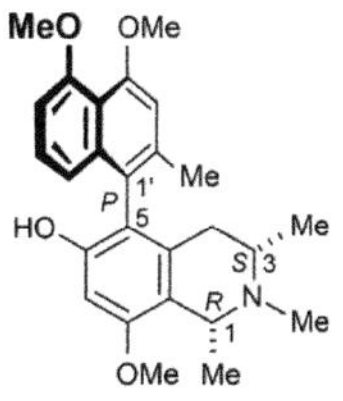

MeO OMe

Me
1'
P 5
HO
Si 3
Me
R 1
N—Me
OMe Me

A. abbreviattus: leaves[366]

A. tectorius (from China): stems, stem bark, twigs, branches[243,244,247,250–252,656]

Amorphous solid (from MeOH): mp 227–228; $[\alpha]_D^{25}$ +61.7 (c 2.11, CHCl$_3$)[243]

Colorless amorphous powder (from MeOH); mp 220–222; $[\alpha]_D^{20}$ +58.6 (c 1.30, CHCl$_3$)[250]

Amorphous solid (from MeOH): mp 223–224; $[\alpha]_D^{25}$ +62.0 (c 0.15, CHCl$_3$)[244]

Amorphous solid (from MeOH): 229–233[247]

Isolated for the first time from *A. tectorius* from the Chinese Hainan Island[243,244]

Structural Elucidation:

Constitution, coupling position, and relative *cis*-configuration at C-1 versus C-3 evidenced by HRESIMS and by ^{1}H and ^{13}C NMR, including specific HMBC and NOE correlations[76,243,244,247,250]

Absolute configurations at C-1 and C-3 established as *S* by oxidative degradation[69,70,244,250]

Assignment of the absolute axial configuration as *P* due to specific NOE interactions in conjunction with the absolute 1*S*,3*S*-stereoarray in the tetrahydroisoquinoline portion,[243,244,247,250] and by the nearly identical ECD spectrum of **77** compared to that of the known, likewise 5,1′-coupled alkaloid ancistrocladine (**20a**)[5,76,244]

Semisynthesis and Total Synthesis:

Semisynthesis of **77** by *cis*-selective reduction of the naphthyldihydroisoquinoline ancistrocladinine (**19a**) using NaBH$_4$, followed by reductive *N*-methylation[244]

Convergent total synthesis of the (unnatural) 6-*O*-methyl derivative of ancistrocline (**77**) by stereoselective biaryl coupling between a naphthalene precursor activated by a chiral oxazoline moiety and a Grignard reagent derived from a suitably substituted bromoisoquinoline—confirmation of the absolute stereostructure of the synthetic target molecule at the stereocenters by ruthenium-mediated oxidative degradation and at the axis by ECD calculations[660]

Moderate inhibitory activity against the human tumor cell line INA-6 (derived from a patient with multiple myeloma)[250]

Weak inhibitory activity against drug-sensitive lymphoblastic CCRF-CEM leukemia cells[251]

Pronounced inhibitory effect against *Staphylococcus aureus*[656]

Continued

Table 15.1 5,1′-Coupled Ancistrocladaceae-type naphthylisoquinoline alkaloids.—cont'd

Name, formula (mol. wt.), and structure	Producing plants and their organs/physical data [mp/°C, $[\alpha]_D^T$/°cm^2 g^{-1} 10^{-1} (solvent)]	Isolation, structural elucidation, partial and/or total synthesis, and further investigations	Biological activities
	cf.: Compd. prepared by semisynthesis[244] and by total synthesis[1,76,244]; mp 223–224; $[\alpha]_D^{25}$ +59.1 (c 0.23, CHCl$_3$)	Total synthesis of ancistrocline (**77**) by applying the lactone methodology[76–79]: stereocontrolled construction of the biaryl axis via the respective biaryl lactone intermediate, by intramolecular aryl coupling of the ester-type prefixed molecular moieties to give the helicene-like distorted lactone atropisomers, with the P-configured diastereomer prevailing, followed by ring cleavage giving rise to **77**[1,76,244]	
5-*epi*-Ancistectorine A$_2$ (76) C$_{25}$H$_{29}$NO$_4$ (407.5) Compd. can likewise be addressed as 6-*O*-methyl-8-*O*-demethylhamatine	*A. abbreviatus*: roots[57] *A. tectorius* (from China): twigs[250]	Isolated for the first time from *A. tectorius* from the Chinese Hainan Island[250] Occurrence in *A. tectorius*, along with its *P*-configured atropo-diastereomer, ancistectorine A$_2$ (not shown)[250] *Structural Elucidation:* Constitution, coupling position, and relative *trans*-configuration at C-1 versus C-3 evidenced by HRESIMS and by ^{1}H and ^{13}C NMR, including specific HMBC and NOE correlations[57,250] Absolute configurations at C-1 and C-3 established as *S* by oxidative degradation[250] Assignment of the absolute axial configuration as *M* due to specific NOE interactions in combination with the absolute 1*S*,3*S*-stereoarray in the tetrahydroisoquinoline portion,[250] and by the nearly identical ECD spectrum of **76** compared to that of the structurally closely related known[28,281] 5,1′-coupled, and likewise *M*-configured, alkaloid hamatine (**20b**)[57,250]	Strong—and selective—inhibitory activity against the K1[250] and the NF54 strain[57] of *Plasmodium falciparum*. Given its extremely weak cytotoxicity, **76** has a high selectivity index, and may thus be considered as a lead compound according to the TDR-WHO guidelines[310] Inactive against *Trypanosoma cruzi*, *T. brucei rhodesiense*, and *Leishmania donovani*[250] **Weak inhibitory activity** in vitro against the hematozoan parasite *Babesia canis*[373]

Yellow amorphous solid (from MeOH): $[\alpha]_D^{23}$ −14.4 (c 0.16, MeOH)[404] Colorless amorphous powder (from MeOH): mp 230; $[\alpha]_D^{20}$ +24.4 (c 0.10, MeOH)[250]	Strong preferential cytotoxicity against PANC-1 human pancreatic cancer cells under "austerity" (i.e., nutrient-deprived) conditions[57]
Ancistrobrevine E (71a) $C_{24}H_{27}NO_4$ (393.5)	Occurrence in *A. abbreviatus* along with its atropo-diastereomer, 5-*epi*-ancistrobrevine E (**71b**)[57]
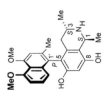 Compd. can likewise be addressed as 8-*O*-demethylancistrocladine *A. abbreviatus*: roots[57] Yellow amorphous solid (from MeOH): $[\alpha]_D^{20}$ −30.3 (c 0.05, MeOH)[57]	Moderate preferential cytotoxicity against PANC-1 human pancreatic cancer cells under "austerity" (i.e., nutrient-deprived) conditions[57] Pronounced—and selective—inhibitory activity in vitro against *Plasmodium falciparum* (strain: NF54)[57] Inactive against *Trypanosoma cruzi*, *T. brucei rhodesiense*, and *Leishmania donovani*[57]
Structural Elucidation: Constitution, coupling position, and relative *trans*-configuration at C-1 versus C-3 evidenced by HRESIMS and by ^{1}H and ^{13}C NMR, including specific HMBC and NOE correlations[57] Absolute configurations at C-1 and C-3 established as *S* by oxidative degradation[57] Assignment of the absolute axial configuration as *P* due to specific NOE interactions in combination with the absolute 1*S*,3*S*-stereoarray in the tetrahydroisoquinoline portion,[57] and by the virtually mirror-imaged ECD spectrum of **71a** compared to that of the structurally related, likewise 5,1′-coupled, but *M*-configured alkaloid 5-*epi*-ancistectorine A$_2$ (**76**)[57,250]	

Continued

Table 15.1 5,1′-Coupled Ancistrocladaceae-type naphthylisoquinoline alkaloids.—cont'd

Name, formula (mol. wt.), and structure	Producing plants and their organs/physical data [mp/°C, $[\alpha]_D^T$/°cm^2 g^{-1} 10^{-1} (solvent)]	Isolation, structural elucidation, partial and/or total synthesis, and further investigations	Biological activities
5-*epi*-Ancistrobrevine E (**71b**) $C_{24}H_{27}NO_4$ (393.5) [structure: MeO, OMe, Me, HO, OH, Me, NH, with positions 1′, 5, M, 3, 8, 1 labeled] Compd. can likewise be addressed as 8-*O*-demethylhamatine	*A. abbreviatus*: roots[57] Yellow amorphous solid (from MeOH): $[\alpha]_D^{20}$ + 21.8 (*c* 0.10, MeOH)[57]	Occurrence in *A. abbreviatus* along with its atropo-diastereomer ancistrobrevine E (**71a**)[57] *Structural Elucidation:* Constitution, coupling position, and relative *trans*-configuration at C-1 versus C-3 evidenced by HRESIMS and by ^{1}H and ^{13}C NMR, including specific HMBC and NOE correlations[57] Absolute configurations at C-1 and C-3 established as *S* by oxidative degradation[57] Assignment of the absolute axial configuration as *M* due to specific NOE interactions in combination with the absolute 1*S*,3*S*-stereoarray in the tetrahydroisoquinoline portion,[57] and by the virtually identical ECD spectrum of **71b** compared to that of the structurally related, likewise 5,1′-coupled and *M*-configured 5-*epi*-ancistectorine A$_2$ (**76**)[57,250]	Moderate preferential cytotoxicity against PANC-1 human pancreatic cancer cells under "austerity" (i.e., nutrient-deprived) conditions[57] Pronounced—and selective—inhibitory activity in vitro against *Plasmodium falciparum* (strain: NF54)[57] Inactive against *Trypanosoma cruzi*, *T. brucei rhodesiense*, and *Leishmania donovani*[57]

Ancistrobrevine L (79)

$C_{25}H_{29}NO_4$ (407.2)

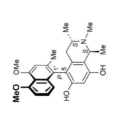

Compd. can likewise be addressed as
N-methylancistrobrevine E or as
N-methyl-8-*O*-demethylancistrocladine

A. abbreviatus: root bark[58]

Yellow amorphous solid (from MeOH):
$[\alpha]_D^{25}$ −27.9 (*c* 0.08, MeOH)[58]

Structural Elucidation:

Constitution and complete stereostructure by comparison of the data of **79** obtained from HRESIMS, 1D and 2D NMR, oxidative degradation, and ECD investigations[58] with those obtained from the structurally closely related, likewise 5,1′-coupled alkaloids ancistrobrevines E (**71a**)[57] and K (**80**)[58]

Assignment of the absolute axial configuration as *P* by the virtually opposite ECD spectrum of **79** compared to that of the likewise 5,1′-coupled, but *M*-configured alkaloid ancistrobrevine K (**80**)[58]

Biological activities of the pure compound not determined for lack of isolated material

Continued

Table 15.1 5,1′-Coupled Ancistrocladaceae-type naphthylisoquinoline alkaloids.—cont'd

Name, formula (mol. wt.), and structure Producing plants and their organs/physical data [mp/°C, $[\alpha]_D^T$/°cm^2 g^{-1} 10^{-1} (solvent)]	Isolation, structural elucidation, partial and/or total synthesis, and further investigations	Biological activities
6-O-Methylancistrocladine (74a) $C_{26}H_{31}NO_4$ (421.5) *A. abbreviatus*: roots, stem bark[5] *A. iloboensis*: root bark[661] *A. heyneanus*: roots[28,662] *A. hamatus*: stem bark[5,663] *A. tectorius* (from China): twigs[252]	For the first time isolated from the Indian liana *A. heyneanus*[28,662] Metabolic screening of root bark extracts of the newly described[23] Central African plant *A. iloboensis*: detection of **74a** by capillary electrophoresis coupled to ion trap mass spectrometry (CE-MS/MS)[661] *Structural Elucidation:* Constitution and complete stereostructure by comparison of the MS, NMR, and ECD data of **74a** with those of the structurally closely related, likewise *P*-configured ancistrocladine (**20a**)[28,661,662] *Semisynthesis and Total Synthesis:* Semisynthesis of **74a** from *N*-formyl-6-*O*-methylancistrocladine by hydrolysis under drastic conditions (KOH in refluxing ethylene glycol)[27] Total synthesis of **74a** by asymmetric construction of the biaryl linkage between a naphthalene precursor activated by a chiral oxazoline moiety and a Grignard reagent derived from a protected isoquinoline precursor[664,665]	Biological activities of the pure compound not determined for lack of isolated material

Amorphous solid (from MeOH): $[\alpha]_D^{20}$ +3.3 (c 0.12, CH_2Cl_2)[51,661]

cf.: Compd. prepared synthetically:

White fluffy needles (from MeOH): mp 200–202[27]

As its hydrochloride salt: white needles (from acetone): mp 315–317 (dec); $[\alpha]_D^{25}$ −56.1 (c 1.9, $CHCl_3$)[27]

As its hydrochloride salt: colorless prisms (from acetone/n-hexane): mp 270–274 (dec); $[\alpha]_D^{25}$ −38.5 (c 0.76, $CHCl_3$)[664,665]

Formal total synthesis of **74a** by preparation of a suitably substituted chiral naphthyl-3,5-dimethoxybenzaldehyde precursor by asymmetric construction of the biaryl axis between a chiral naphthyloxazoline and an aryl Grignard reagent[666]

6-O-Methylhamatine (74b)

$C_{26}H_{31}NO_4$ (421.5)

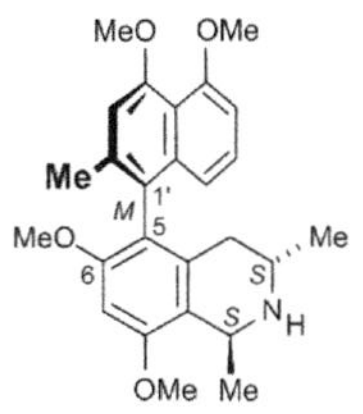

Isolated for the first time from the Sri Lankan liana *A. hamatus*[5,663]

Metabolic screening of root bark extracts of the newly described[23] Central African plant *A. ileboensis*: detection of **74b** by capillary electrophoresis coupled to ion trap mass spectrometry (CE-MS/MS)[661]

Structural Elucidation:

Constitution and complete stereostructure by comparison of the MS, NMR, and ECD data of **74b** with those of the structurally closely related, likewise *M*-configured alkaloid hamatine (**20b**)[5,28–30,57,663]

Moderate preferential cytotoxicity against PANC-1 human pancreatic cancer cells under "austerity" (i.e., nutrient-deprived) conditions[30,57]

Pronounced—and selective—inhibitory activity in vitro against *Plasmodium falciparum* (strain: NF54)[57]

Inactive against *Trypanosoma cruzi, T. brucei rhodesiense,* and *Leishmania donovani*[57]

A. abbreviatus: stem bark,[5] roots[57]

A. congolensis: root bark[178]

A. ileboensis: root bark[51,661]

Semisynthesis and Total Synthesis:

Semisynthesis of **74b** from hamatine (**20b**) by 6-O-methylation using diazomethane[281]

Continued

Table 15.1 5,1′-Coupled Ancistrocladaceae-type naphthylisoquinoline alkaloids.—cont'd

Name, formula (mol. wt.), and structure	Producing plants and their organs/physical data [mp/°C, $[\alpha]_D^T$/°cm^2 g^{-1} 10^{-1} (solvent)]	Isolation, structural elucidation, partial and/or total synthesis, and further investigations	Biological activities
	From two botanically yet undescribed Congolese *Ancistrocladus* species: roots,[29] root bark[30] *A. hamatus*: stem bark[5,663] *A. tectorius* (from China): twigs[252] Yellow amorphous solid (from MeOH): $[\alpha]_D^{23}$ −150 (c 0.05, MeOH)[404] Amorphous solid (from MeOH): $[\alpha]_D^{20}$ +26.2 (c 0.15, CH$_2$Cl$_2$)[51,661] cf.: Compd. prepared synthetically: As its hydrochloride salt: white needles (from MeOH/acetone): mp 318–322 (dec); $[\alpha]_D^{20}$ −20.53* (c 0.76, CHCl$_3$)[281] As its hydrochloride salt: colorless prisms (from CH$_2$Cl$_2$/n-hexane): mp 271–273 (dec); $[\alpha]_D^{25}$ +28.0 (c 0.40, MeOH)[665]	Total synthesis of **74b** by asymmetric construction of the biaryl axis between a naphthalene precursor activated by a chiral naphthyloxazoline moiety and a Grignard reagent derived from a protected isoquinoline precursor[664,665]	

*The published sign of this optical rotation may be a misprint

Ancistrobrevine K (80)

$C_{24}H_{27}NO_4$ (393.2)

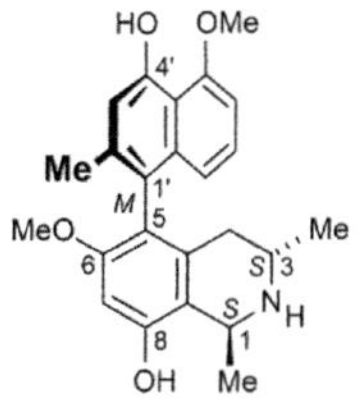

Me
S|3
HO OMe   N
   4' 1'   H
MeO 5   S 1
   6   8 Me
Me   OH

Compd. can likewise be addressed as 8,4'-O,O-didemethy-6-O-methyl-hamatine

A. abbreviatus: root bark[58]

Yellow amorphous solid (from MeOH): $[\alpha]_D^{25}$ −13.6 (c 0.10, MeOH)[58]

Ancistrobrevine K (80)[58] is the atropo-diastereomer of 4'-O-demethylancistectorine A₂,[252] which is a minor constituent occurring in the twigs of the Chinese species *A. tectorius*

Structural Elucidation:

Constitution and complete stereostructure by comparison of the data of **80** obtained from HRESIMS, 1D and 2D NMR, oxidative degradation, and ECD investigations[58] with those of the structurally related alkaloids ancistrobrevines E (71a)[57] and G (78)[57]

Assignment of the absolute axial configuration as *M* on the basis of specific NOESY correlations in conjunction with the relative *trans*-configuration at C-1 versus C-3, and by the virtually opposite ECD spectrum of **80** compared to that of the likewise 5,1'-coupled, but *P*-configured alcaloid ancistrobrevine G (78)[57,58]

No substantial cytotoxic effects on HT-29 human colon carcinoma cells[58]

Continued

Table 15.1 5,1′-Coupled Ancistrocladaceae-type naphthylisoquinoline alkaloids.—cont'd

Name, formula (mol. wt.), and structure Producing plants and their organs/physical data [mp/°C, $[\alpha]_D^T$/°cm^2 g^{-1} 10^{-1} (solvent)]	Isolation, structural elucidation, partial and/or total synthesis, and further investigations	Biological activities
Ancistrobrevine F (59a) C$_{23}$H$_{25}$NO$_4$ (379.5) [structure] Compd. can likewise be addressed as 4′-O-demethylancistrobrevine E or as 4′,8-O,O-didemethylancistrocladine *A. abbreviatus*: roots[57] Yellow amorphous solid (from MeOH): $[\alpha]_D^{23}$ −25.2 (c 0.02, MeOH)[57]	Occurrence in *A. abbreviatus* along with its atropo-diastereomer, 5-*epi*-ancistrobrevine F (**59b**)[57] *Structural Elucidation:* Constitution, coupling position, and relative *trans*-orientation of the methyl groups at C-1 versus C-3 evidenced by HRESIMS and by ^{1}H and ^{13}C NMR, including specific HMBC and NOE correlations[57] Absolute configurations at C-1 and C-3 established as *S* by oxidative degradation[57] Assignment of the absolute axial configuration as *P* due to specific NOE interactions in combination with the absolute 1*S*,3*S*-stereoarray in the tetrahydroisoquinoline portion, and by the virtually mirror-imaged ECD spectrum of **59a** compared to that of the structurally related, but *M*-configured alkaloid 5-*epi*-ancistectorine A$_2$ (**76**)[57,250]	Pronounced—and selective—inhibitory activity in vitro against *Plasmodium falciparum* (strain: NF54)[57] Inactive against *Trypanosoma cruzi*, *T. brucei rhodesiense*, and *Leishmania donovani*[57]

5-*epi*-Ancistrobrevine F (59b) $C_{23}H_{25}NO_4$ (379.5) Compd. can likewise be addressed as 4′-*O*-demethyl-5-*epi*-ancistrobrevine E or as 4′,8-*O*,*O*-didemethylhamatine[57] *A. abbreviatus*: roots[57] Yellow amorphous solid (from MeOH): $[\alpha]_D^{20} + 9.2$ (*c* 0.10, MeOH)[57]	Occurrence in *A. abbreviatus* along with its atropo-diastereomer, ancistrobrevine F (**59a**)[57] *Structural Elucidation:* Constitution, coupling position, and relative *trans*-configuration at C-1 versus C-3 evidenced by HRESIMS and by ^{1}H and ^{13}C NMR, including specific HMBC and NOE correlations[57] Absolute configurations at C-1 and C-3 established as *S* by oxidative degradation[57] Assignment of the absolute axial configuration as *M* due to specific NOE interactions in combination with the absolute 1*S*,3*S*-stereoarray in the tetrahydroisoquinoline portion, and by the virtually identical ECD spectrum of **59b** compared to that of the likewise *M*-configured alkaloid 5-*epi*-ancistectorine A₂ (**76**)[57,250]	Moderate preferential cytotoxicity against PANC-1 human pancreatic cancer cells under "austerity" (i.e., nutrient-deprived) conditions[57] Pronounced—and selective—inhibitory activity in vitro against *Plasmodium falciparum* (strain: NF54)[57] Inactive against *Trypanosoma cruzi*, *T. brucei rhodesiense*, and *Leishmania donovani*[57]

Continued

Table 15.1 5,1′-Coupled Ancistrocladaceae-type naphthylisoquinoline alkaloids.—cont'd

Name, formula (mol. wt.), and structure	Producing plants and their organs/physical data [mp/°C, $[\alpha]_D^T$/cm^2 g^{-1} 10^{-1} (solvent)]	Isolation, structural elucidation, partial and/or total synthesis, and further investigations	Biological activities
6-O-Methyl-4′-O-demethyl-ancistrocladine (75a) $C_{25}H_{29}NO_4$ (407.5) *A. abbreviatus*: roots[373] *A. tectorius* (from China): stem bark, twigs, leaves[245]	Yellow amorphous solid (from MeOH): $[\alpha]_D^{23}$ −16.7 (*c* 0.07 MeOH)[373] Light-yellow powder (from CHCl$_3$): $[\alpha]_D^{25}$ −3.14 (*c* 0.38, MeOH)[245]	Isolated for the first time from *A. tectorius* from the Chinese Hainan Island[245] *Structural Elucidation*: Constitution, coupling position, and relative *trans*-configuration at C-1 versus C-3 by HRESIMS and by ^{1}H and ^{13}C NMR, including HMBC and NOESY correlations, indicative of the presence of a 5,1′-linked naphthyltetrahydroisoquinoline with a free hydroxy group at C-4′[245,373] Absolute configuration at C-3 established as *S* by oxidative degradation[373] Assignment of the axial configuration as *P* due to specific NOESY correlations in combination with the absolute 1*S*,3*S*-stereoarray in the tetrahydroisoquinoline portion, and by the nearly identical ECD spectrum of **75a** compared to that of the likewise *P*-configured alkaloid ancistrocladine (**20a**)[245,373]	Moderate inhibitory activity in vitro against the hematozoan parasite *Babesia canis*[373]

6-O-Methyl-4′-O-demethyl-hamatine (75b)	Isolated for the first time from *A. tectorius* from the Chinese Hainan Island[245]	Strong preferential cytotoxicity against PANC-1 human pancreatic cancer cells under "austerity" (i.e., nutrient-deprived) conditions[57]
$C_{25}H_{29}NO_4$ (407.5)		Weak inhibitory activity in vitro against the hematozoan parasite *Babesia canis*[373]
	Structural Elucidation:	
	Constitution, coupling position, and relative *trans*-configuration at C-1 versus C-3 by HRESIMS and by ^1H and ^{13}C NMR, including HMBC and NOESY correlations, indicative of the presence of a 5,1′-linked naphthyltetrahydroisoquinoline with a free hydroxy group at C-4′[57,245,404]	
	Absolute configuration at C-3 established as *S* by oxidative degradation[57,404]	
	Assignment of the axial configuration as *M* due to specific NOESY correlations in combination with the absolute 1*S*,3*S*-stereoarray in the tetrahydroisoquinoline portion, and by the nearly identical ECD spectrum of **75b** compared to that of the likewise *M*-configured alkaloid hamatine (**20b**)[57,245,404]	
A. abbreviatus: roots[57]		
A. tectorius (from China): stem bark, leaves[245]		
Yellow amorphous solid (from MeOH): $[\alpha]_D^{23}$ −46.4 (*c* 0.18, MeOH)[404]		
Light-yellow powder (from CHCl$_3$): $[\alpha]_D^{25}$ +12.4 (*c* 0.36, MeOH)[245]		

Continued

Table 15.1 5,1′-Coupled Ancistrocladaceae-type naphthylisoquinoline alkaloids.—cont'd Name, formula (mol. wt.), and structure

Producing plants and their organs/physical data [mp/°C, $[\alpha]_D^T$/°cm^2 g^{-1} 10^{-1} (solvent)]	Isolation, structural elucidation, partial and/or total synthesis, and further investigations	Biological activities
Ancistrobrevine G (78) $C_{23}H_{25}NO_4$ (379.5) Compd. can likewise be addressed as 5′-O-demethylancistrobrevine E or as 5′,8-O-didemethylancistrocladine	*Structural Elucidation:* Constitution, coupling position, and relative *trans*-orientation of the methyl groups at C-1 and C-3 evidenced by HRESIMS and by ^{1}H and ^{13}C NMR, including specific HMBC and NOESY correlations[57] Absolute configurations at C-1 and C-3 established as *S* by oxidative degradation[57] Assignment of the absolute axial configuration as *P* due to specific NOE interactions in combination with the relative *trans*-configuration at C-1 versus C-3, and by the virtually mirror-imaged ECD spectrum of **78** compared to that of the structurally related, but *M*-configured alkaloid 5-*epi*-ancistectorine A$_2$ (**76**)[57,250]	Pronounced—and selective—inhibitory activity in vitro against *Plasmodium falciparum* (strain: NF54)[57] Inactive against *Trypanosoma cruzi*, *T. brucei rhodesiense*, and *Leishmania donovani*[57]

A. abbreviatus: roots[57]

Yellow amorphous solid (from MeOH): $[\alpha]_D^{20}$ −36.8 (*c* 0.05, MeOH)[57]

| 6-O-Methylhamatinine (84) $C_{26}H_{29}NO_4$ (419.5) | Isolated for the first time from the Vietnamese liana *A. cochinchinensis* by column chromatography (CC) on silica gel[403]

Isolation from a botanically so far undescribed Congolese *Ancistrocladus* species by fast centrifugal partition chromatography (FCPC) in combination with preparative HPLC[29]

Structural Elucidation:

Constitution and coupling position by HRESIMS and by ^{1}H and ^{13}C NMR, including HMBC, HMQC, H,H-COSY, and NOE interactions, indicative of the presence of an O-permethylated 5,1′-linked naphthyldihydroisoquinoline alkaloid[29,56,57,252,403]

Absolute configuration at C-3 by oxidative degradation[29,56,57,252,403]

Assignment of the absolute axial configuration as *M* due to specific long-range NOE interactions, in combination with the absolute configuration at C-3 determined as *S*[29,56,252,403]

Additional proof of the absolute axial configuration by comparison of the ECD spectrum of **84** with that of the *P*-configured naphthyldihydroisoquinoline ancistrocladinine (**19a**), which was found to be opposite[252,403] | Weak to moderate inhibitory activities in vitro against *Plasmodium falciparum* (K1 strain), *Leishmania donovani*, *Trypanosoma cruzi*, and *T. brucei rhodesiense*[29]

Pronounced inhibitory activity in vitro against the hematozoan parasite *Babesia canis*[373]

Pronounced preferential cytotoxicity against PANC-1 human pancreatic cancer cells under "austerity" (i.e., nutrient-deprived) conditions[57] |
| *A. abbreviatus*: roots,[57] leaves[56]

From a botanically yet undescribed Congolese *Ancistrocladus* species: roots[29]

A. tectorius (from China): twigs[252]

A. cochinchinensis: leaves[403]

Amorphous solid (from acetone/MeOH): $[\alpha]_D^{23} +34.1$ (*c* 1.00, CHCl$_3$)[403]

Colorless amorphous powder (from MeOH): $[\alpha]_D^{20} +15.6$ (*c* 0.10, MeOH)[252] | | |

Continued

Table 15.1 5,1′-Coupled Ancistrocladaceae-type naphthylisoquinoline alkaloids.—cont'd

Name, formula (mol. wt.), and structure	Producing plants and their organs/physical data [mp/°C, $[\alpha]_D/°cm^2 g^{-1} 10^{-1}$ (solvent)]	Isolation, structural elucidation, partial and/or total synthesis, and further investigations	Biological activities
6-O-Methylancistectorine A₃ (85) $C_{25}H_{27}NO_4$ (405.5) Compd. can likewise be considered as 4′-O-demethyl-6-O-methylancistrocladinine	*A. abbreviatus*: leaves[56] *A. tectorius* (from China): twigs[252] Yellow amorphous solid (from MeOH): $[\alpha]_D^{20}$ −35.3 (c 0.05, MeOH)[56] Colorless amorphous powder (from MeOH): $[\alpha]_D^{20}$ −40.5 (c 0.10, MeOH)[252]	Isolated for the first time from *A. tectorius* from the Chinese Hainan Island[252] *Structural Elucidation:* Constitution and coupling position by HR-ESIMS and by ^1H and ^{13}C NMR, including specific HMBC and NOE correlations, indicative of the presence of a 5,1′-linked naphthyldihydroisoquinoline with a free hydroxy group at C-4′[56,252] Absolute configuration at C-3 established as *S*, by oxidative degradation[56,252] Assignment of the absolute axial configuration as *P* due to specific NOESY interactions and the given absolute configuration at C-3[56,252] Additional proof of the absolute axial configuration by comparison of the ECD spectrum of **85**[56,252] with that of the likewise *P*-configured parent compound ancistectorine A_3^{250} (not shown)	Biological activities of the pure compound not determined for lack of isolated material

Ancistrobrevidine C (83)

$C_{25}H_{27}NO_4$ (405.5)

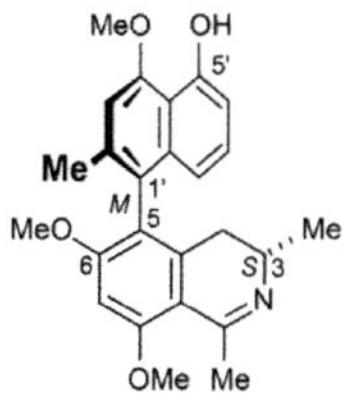

Compd. can likewise be considered as 5′-O-demethyl-6-O-methylhamatinine

A. abbreviatus: leaves[56]

Yellow amorphous solid (from MeOH);
$[\alpha]_D^{23}$ +12.7 (*c* 0.13, MeOH)[56]

Structural Elucidation:

Constitution and coupling position evidenced by HRESIMS and by ^{1}H and ^{13}C NMR, including specific HMBC and NOESY correlations indicative of a 5,1′-linked naphthyldihydroisoquinoline with a free hydroxy function at C-5′[56]

Absolute configuration at C-3 established as *S*, by oxidative degradation[56]

Assignment of the absolute axial configuration as *M* by the virtually identical ECD spectrum of 83[56] compared to that of the likewise 5,1′-linked and also *M*-configured naphthyldihydroisoquinoline 6-O-methylhamatinine (84)[56,252,403]

Pronounced preferential cytotoxicity against PANC-1 human pancreatic cancer cells under "austerity" (i.e., nutrient-deprived) conditions[56]

Significant inhibition of PANC-1 cell migration and colony formation, and induction of dramatic alteration in cell morphology, leading to cell death of pancreatic cancer cells[56]

Strong cytotoxic activity against human HeLa cervical cancer cells[56]

Continued

Table 15.1 5,1′-Coupled Ancistrocladaceae-type naphthylisoquinoline alkaloids.—cont'd

Name, formula (mol. wt.), and structure Producing plants and their organs/physical data [mp/°C, $[\alpha]_D^T$/°cm^2 g^{-1} 10^{-1} (solvent)]	Isolation, structural elucidation, partial and/or total synthesis, and further investigations	Biological activities
6-O-Methylhamateine (124) $C_{26}H_{27}NO_4$ (417.5) *A. abbreviatus*: roots[54] *A. cochinchinensis*: leaves[403] Crystalline solid (from CHCl$_3$/n-hexane): mp 233–236; $[\alpha]_D^{23}$ −41.4 (c 0.46, CHCl$_3$)[403] cf.: Compd. prepared semisynthetically: Needles (from diethyl ether): mp 240–241; $[\alpha]_D^{23}$ −50.5 (c 1.6, CHCl$_3$)[281]	Isolated for the first time from the Vietnamese liana *A. cochinchinensis*, but the enantiomeric purity of **124** was not analyzed[403] Occurrence of **124** isolated from the roots of *A. abbreviatus* in an enantiomerically pure form in this West African liana, by HPLC-ECD analysis on a chiral phase (Lux Cellulose-1)[54] *Structural Elucidation:* Constitution and coupling position established by HRESIMS and by 1D and 2D NMR, revealing the presence of a 5,1′-linked non-hydrogenated and O-permethylated naphthylisoquinoline alkaloid[54,403] Absolute axial configuration assigned as *M* by ECD investigations using the Exciton Chirality method[281,403] Additional proof of the axial configuration of **124** by comparison of its ECD spectrum with those of the likewise 5,1′-linked, but *P*-configured fully dehydrogenated naphthylisoquinolines 6-O-methylancistrocladeine[27,259] and ancistrocladeine (**125**)[54,333]—the ECD spectrum of **124** being virtually opposite to those of **125** and its 6-O-methyl analogue	Strong cytotoxic activity against parental drug-sensitive CCRF-CEM leukemia cells and their multi-drug resistant subline CEM/ADR5000 in the low micromolar range[54]

P-Isomer: Unnatural

6-O-methylancistrocladeine, prepared synthetically[27,664,665]:

Amorphous solid (from diethylether): mp 240–242; $[\alpha]_D^{25}$ +58.9 (*c* 1.6, CHCl$_3$)[27]

Prisms (from diethylether/*n*-hexane): mp 238–240; $[\alpha]_D^{23}$ +51.0 (*c* 0.30, CHCl$_3$)[665]

Semisynthesis and Total Synthesis:

Semisynthesis of **124** from 6-O-methylhamatine (**74b**) by catalytic dehydrogenation[281]

Total synthesis of the *P*-isomer of **124** (i.e., 6-O-methylancistrocladeine), by asymmetric construction of the biaryl linkage between a naphthalene precursor activated by a chiral oxazoline moiety and a suitably substituted aryl Grignard reagent as the precursor of the isoquinoline portion[664,665]

Table 15.2 5,1′-Coupled Dioncophyllaceae-type naphthylisoquinoline alkaloids.

Name, formula (mol. wt.), and structure	Producing plants and their organs/physical data [mp/°C, $[\alpha]_D^T$/°cm^2 g^{-1} 10^{-1} (solvent)]	Isolation, structural elucidation, partial and/or total synthesis, and further investigations	Biological activities
5-*epi*-Dioncophyllidine C$_2$ (9) C$_{24}$H$_{25}$NO$_3$ (375.2)	*A. abbreviatus*: leaves[56] Yellow amorphous solid (from MeOH): $[\alpha]_D^{23}$ −4.2 (*c* 0.10, MeOH)[56]	5-*epi*-Dioncophyllidine C$_2$ (**9**) as the only third 5,1′-coupled Dioncophyllaceae-type alkaloid found in nature and the first one of this subgroup possessing a dihydroisoquinoline subunit and displaying the *M*-configuration at the biaryl axis[56] *Structural Elucidation:* Constitution and coupling position evidenced by HRESIMS and by ^{1}H and ^{13}C NMR, along with specific HMBC and NOESY interactions, revealing the presence of a naphthyldihydroisoquinoline alkaloid with two methoxy groups at C-4′ and C-5′, with a free hydroxy function at C-8, and without an oxygen function at C-6[56] Absolute configuration at C-3 established as *R*, by oxidative degradation[56] Assignment of the absolute configuration as *M* due to the given *R*-configuration at C-3 in conjunction with specific NOESY correlations revealing the methyl-free ring system of the naphthalene	Biological activities of the pure compound not determined for lack of isolated material

Table 15.3 5,1′-Coupled hybrid-type naphthylisoquinoline alkaloids.

Name, formula (mol. wt.), and structure	Producing plants and their organs/physical data [mp/°C, $[\alpha]_D^T$/°cm^2 g^{-1} 10^{-1} (solvent)]	Isolation, structural elucidation, partial and/or total synthesis, and further investigations	Biological activities
Ancistrobrevine M (69) $C_{25}H_{29}NO_4$ (407.2)	*A. abbreviatus*: root bark[58] Yellow amorphous solid (from MeOH): $[\alpha]_D^{25}$ −235 (*c* 0.03, MeOH)[58]	*Structural Elucidation:* Constitution and coupling position evidenced by HRESIMS and by 1D and 2D NMR, revealing the presence of a 5,1′-linked naphthyl-1,3-dimethyltetrahydroisoquinoline alkaloid equipped with a 4′,5′-dimethoxy-3′-methylnaphthalene portion[58] Relative configuration at C-1 versus C-3 established as *cis* due to specific NOESY interactions[58] Absolute configuration at C-3 attributed as *R*, by oxidative degradation[58] Assignment of the absolute axial configuration as *M* by long-range NOESY correlations in combination with the absolute 1*S*,3*R*-stereoarray in the tetrahydroisoquinoline portion, and by the nearly identical ECD spectrum of **69** compared to that of the likewise 5,1′-coupled and *M*-configured ancistrobrevine K (**80**)[58]	No substantial cytotoxic effects against HT-29 human colon carcinoma cells[58]
Compd. can likewise be addressed as 3-*epi*-hamatine			

Continued

Table 15.3 5,1′-Coupled hybrid-type naphthylisoquinoline alkaloids.—cont'd

Name, formula (mol. wt.), and structure	Producing plants and their organs/physical data [mp/°C, $[\alpha]_D^T$/°cm^2 g^{-1} 10^{-1} (solvent)]	Isolation, structural elucidation, partial and/or total synthesis, and further investigations	Biological activities
Ancistrobrevidine A (81) C$_{25}$H$_{27}$NO$_4$ (405.5) Compd. can likewise be addressed as 3-*epi*-ancistrocladinine	*A. abbreviatus*: leaves[56] Yellow amorphous solid (from MeOH): $[\alpha]_D^{23}$ −56.6 (*c* 0.10, MeOH)[56]	*Structural Elucidation:* Constitution and coupling position established by HRESIMS and by ^{1}H and ^{13}C NMR, including specific HMBC and NOESY interactions, indicating the presence of a 5,1′-linked naphthyldihydroisoquinoline with a free hydroxy function at C-6[56] Absolute configuration at C-3 elucidated as *R*, by oxidative degradation[56] Assignment of the absolute axial configuration as *P* due to long-range NOESY correlations and the given *R*-configuration at C-3, and by the virtually opposite ECD spectrum of **81**[56] compared to that of the likewise 5,1′-coupled, but *M*-configured 6-*O*-methylhamatinine (**84**)[29,56,57,403]	Moderate to weak preferential cytotoxicity against PANC-1 human pancreatic cancer cells under "austerity" (i.e., nutrient-deprived) conditions[56] Inactive against human HeLa cervical cancer cells[56]

Ancistrobrevidine B (82)

$C_{26}H_{29}NO_4$ (419.5)

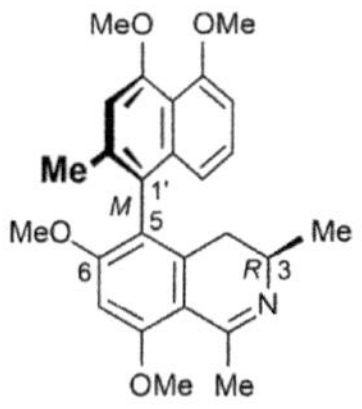

Compd. can likewise be addressed as 3-*epi*-6-*O*-methylhamatinine

A. abbreviatus: leaves[56]

Yellow amorphous solid (from MeOH): $[\alpha]_D^{23}$ −82.1 (*c* 0.10, MeOH)[56]

Structural Elucidation:

Constitution and complete stereostructure assigned by HRESIMS, 1D and 2D NMR, and oxidative degradation—comparison of the data of **82** with those obtained from the likewise 5,1'-coupled ancistrobrevidine A (**81**), indicating the presence of an *O*-permethylated naphthyldihydroisoquinoline alkaloid with *R*-configuration at C-3[56]

Absolute axial configuration established as *M* due to long-range NOESY correlations, in combination with the given *R*-configuration at C-3[56]; confirmed by the virtually opposite ECD spectrum of **82** compared to that of the *P*-configured ancistrobrevidine A (**81**)[56], and also corroborated by the fact that **82** and its likewise *M*-configured 3-epimer, 6-*O*-methylhamatinine (**84**),[29,56,57,403] displayed nearly identical ECD spectra[56]

Weak preferential cytotoxicity against PANC-1 human pancreatic cancer cells under "austerity" (i.e., nutrient-deprived) conditions[56]

Moderate cytotoxic activity against human HeLa cervical cancer cells[56]

Table 16 5,8'-Coupled Ancistrocladaceae-type naphthylisoquinoline alkaloids.

Name, formula (mol. wt.), and structure	Producing plants and their organs/physical data [mp/°C, $[\alpha]_D^T$/°cm^2 g^{-1} 10^{-1} (solvent)]	Isolation, structural elucidation, partial and/or total synthesis, and further investigations	Biological activities
Ancistrobrevine B (87a) $C_{25}H_{29}NO_4$ (407.5)	*A. abbreviatus*: stem bark[57,63] *A. congolensis*: root bark[178] From a botanically yet undescribed Congolese *Ancistrocladus* species: root bark[30] *A. robertsoniorum*: stem bark, twigs[280]	Isolated for the first time from *A. abbreviatus* by column chromatography (CC) and preparative TLC on silica gel[63] Isolation from *A. robertsoniorum* by high-speed countercurrent chromatography (HSCCC) and preparative TLC on silica gel[280] *Structural Elucidation:* Constitution and relative *trans*-configuration at C-1 versus C-3 deduced from HRESIMS and from ^{1}H and ^{13}C NMR, including specific NOE interactions[30,63,178,280] Absolute configuration determined by oxidative degradation to be *S*, both for C-1 and C-3[63,178,280] Absolute axial configuration established as *M* due to specific long-range NOE interactions and the given absolute configuration at C-3 (as known from the degradation)[63,178,280] Additional proof of the absolute axial configuration by comparison of the ECD spectrum of **87a** with that of	Moderate inhibitory activity in vitro against *Plasmodium falciparum* (strains: K1, NF54)[57,277,418] Weak inhibitory activities in vitro against *Trypanosoma* parasites[57] No inhibitory activity in vitro against *Leishmania donovani*[57] No inhibitory activity against the development of exoerythrocytic stages of *P. berghei* (Anka) in vitro in human hepatoma cells (Hep G2)[371] Pronounced preferential cytotoxicity against PANC-1 human pancreatic cancer cells under "austerity" (i.e., nutrient-deprived) conditions[30]

Amorphous powder (from MeOH/acetone): mp 122–124; $[\alpha]_D^{25}$ +68.0 (c 0.81, CHCl$_3$)[63]

Amorphous fine powder (from MeOH/H$_2$O): mp 150–151; $[\alpha]_D^{25}$ +84.0 (c 0.13, CHCl$_3$)[280]

cf.: Ancistrobrevine B (**87a**) and its atropisomer **87b** prepared synthetically:

White solid, structural assignment by FAB-MS and by ^{1}H and ^{13}C NMR[78,449,450]

the *P*-configured alkaloid ancistrocladine (**20a**),[27,28,256,257,259] which was found to be opposite, and with the ECD curve of the *M*-configured hamatine (**20b**),[28,281,282] which was found to be nearly identical to that of **87a**[30,63,178,280]

Total Synthesis:

Total synthesis of ancistrobrevine B (**87a**) and its (unnatural) atropo-diastereomer **87b** by
(a) Watanabe-type benzyne-diene cycloaddition reaction to form the naphthalene portion activated by a boronic acid entity, (b) synthesis of a specifically protected iodo-activated tetrahydroisoquinoline, (c) formation of the sterically hindered biaryl axis by Pd-catalyzed Suzuki-Miyaura cross-coupling, and (d) separation of **87a** and its atropo-diastereomer by HPLC on an amino-bonded column[78,449,450]

Continued

Table 16 5,8′-Coupled Ancistrocladaceae-type naphthylisoquinoline alkaloids.—cont'd

Name, formula (mol. wt.), and structure	Producing plants and their organs/physical data [mp/°C, $[\alpha]_D^T/°\text{cm}^2\,\text{g}^{-1}\,10^{-1}$ (solvent)]	Isolation, structural elucidation, partial and/or total synthesis, and further investigations	Biological activities
5′-O-Demethylancistrobrevine B (88) $C_{24}H_{27}NO_4$ (393.5) Compd. can also be addressed as 5-*epi*-ancistroguineine A (for the structure of ancistroguineine A, see Ref. 67)	*A. abbreviatus*: roots[57] Yellow amorphous solid (from MeOH): $[\alpha]_D^{20}$ −27.8 (*c* 0.05, MeOH)[57]	*Structural Elucidation:* Constitution and complete stereostructure by HRESIMS, 1D and 2D NMR, oxidative degradation, and ECD—comparison of the data of **88** with those obtained from its parent compound ancistrobrevine B (**87a**)[57] Assignment of the absolute configuration at the biaryl axis as *M* due to specific long-range NOESY interactions, in combination with the given absolute configuration at C-3 (known from the degradation)[57] Additional proof of the absolute axial configuration by the close resemblance of the ECD spectrum of **88**[57] with that of its parent compound ancistrobrevine B (**87a**)[57,63]	Moderate inhibitory activity in vitro against *Plasmodium falciparum* (strains: K1, NF54)[57] Weak inhibitory activities against *Trypanosoma* parasites[57] No inhibitory activity against *Leishmania donovani*[57]

Ancistroealaine D (89)

$C_{25}H_{29}NO_4$ (407.5)

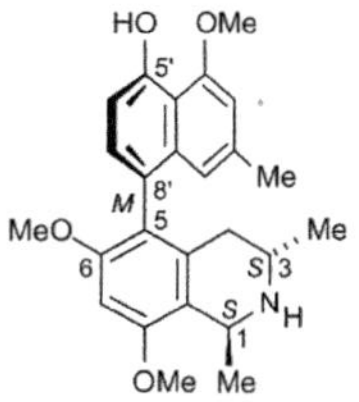

[Structure diagram with labels: HO, OMe, Me, MeO, OMe, Me, N, H, M, S, S, and positions 5', 8', 3, 5, 6, 1]

Compd. can likewise be addressed as 5'-O-demethyl-6-O-methylancistrobrevine B

A. abbreviatus: root bark[404]

A. ealaensis: leaves, twigs[181]

Yellow amorphous solid (from MeOH): $[\alpha]_D^{23}$ −8.7 (c 0.07, MeOH)[404]

Yellow amorphous solid (from MeOH): $[\alpha]_D^{23}$ +17.0 (c 0.05, MeOH)[181]

Isolated for the first time from the Central African liana *A. ealaensis*, along with its atropo-diastereomer ancistroealaine B (not shown)[179,181]

Structural Elucidation:

Constitution and relative *trans*-configuration at C-1 versus C-3 deduced from HRESIMS and from ^{1}H and ^{13}C NMR, including HMBC, ROESY, NOESY, and COSY interactions[181,404]

Absolute configuration at the stereogenic centers determined by oxidative degradation to be S, both for C-1 and C-3[181]

Assignment of the axial configuration as M due to specific long-range NOESY correlations and the given absolute configuration at C-3 (from degradation), and based on the comparison of the ECD spectrum of **89**[181] with that of the related, yet P-configured alkaloid korupensamine A (**27**),[5,287] which was virtually opposite[181]

Pronounced activity against erythrocytic forms of *Plasmodium falciparum* (strains: K1 and NF54) in vitro[181]

Moderate inhibitory activities in vitro against *Trypanosoma brucei rhodesiense*, *T. cruzi*,[181] and *Leishmania donovani*[181]

Continued

Table 16 5,8′-Coupled Ancistrocladaceae-type naphthylisoquinoline alkaloids.—cont'd Name, formula (mol. wt.), and structure

Producing plants and their organs/physical data [mp/°C, $[\alpha]_D^T$/°cm^2 g^{-1} 10^{-1} (solvent)]	Isolation, structural elucidation, partial and/or total synthesis, and further investigations	Biological activities
Ancistrobreveine D (10a) C$_{26}$H$_{27}$NO$_4$ (417.2) Compd. can likewise be addressed as *ent*-ancistrolikokine J$_3$ *A. abbreviatus*: root bark[54] Yellow amorphous solid (from MeOH): $[\alpha]_D^{20}$ −11.4 (*c* 0.10, MeOH)[54]	*Structural Elucidation:* Constitution and coupling position established by HRESIMS and by 1D and 2D NMR, revealing **10a** to be a 5,8′-linked non-hydrogenated and O-permethylated naphthylisoquinoline alkaloid[54] Assignment of the absolute axial configuration as *M* due to the fact that **10a** displayed a virtually mirror-imaged ECD spectrum compared to that of the constitutionally identical, yet enantiomeric, naphthylisoquinoline ancistrolikokine J$_3$ (**10b** = *ent*-**10a**) from the Congolese liana *A. likoko*,[53] which is *P*-configured at the biaryl axis[54] HPLC-ECD analysis on a chiral phase (Lux Cellulose-1) revealed ancistrobreveine D (**10a**) to occur enantiomerically pure in *A. abbreviatus*[54]	Moderate growth-retarding activities against drug-sensitive CCRF-CEM leukemia cells and their multi-drug resistant subline CEM/ADR5000[54]

Table 17 *ortho*-Quinoid naphthylisoquinoline alkaloids. Name, formula (mol. wt.), and structure

Producing plants and their organs/physical data [mp/°C, $[\alpha]_D^T$/°cm^2 g^{-1} 10^{-1} (solvent)]	Isolation, structural elucidation, partial and/or total synthesis, and further investigations	Biological activities
Ancistrobreviquinone A (11) $C_{24}H_{25}NO_5$ (407.2)	*Structural Elucidation:* Constitution and coupling position deduced from HRESIMS and by ^{1}H and ^{13}C NMR, including HMBC and NOESY correlations, indicating the presence of a 5,1′-linked naphthyltetrahydroisoquinoline alkaloid possessing a diketo entity in the naphthalene portion at C-3′ and C-4′ located in a close proximity to the methyl group at C-2′[56]	Low cytotoxic activity against human HeLa cervical cancer cells[56]
	Absolute configuration at C-3 established to be *S*, by oxidative degradation[56]	Moderate preferential cytotoxicity against PANC-1 human pancreatic cancer cells under "austerity" (i.e., nutrient-deprived) conditions[56]
A. abbreviatus: root bark[56]	Absolute configuration at C-1 determined as *S* due to the relative *trans*-configuration at C-1 versus C-3 as assigned by specific NOESY interactions[56]	
Dark orange-yellow amorphous solid (from MeOH): $[\alpha]_D^{23}$ −169.7 (*c* 0.04, MeOH)[56]	Assignment of the absolute axial configuration as *P* due to specific long-range NOESY interactions evidencing the large naphthalene part to be directed down relative to the "plane" of the isoquinoline half, and in conjunction with the established 1*S*,3*S*-stereoarray in the tetrahydroisoquinoline subunit[56]	
	Additional proof of the absolute configuration at the biaryl axis due to a nearly identical ECD spectrum of **11** compared to that of the likewise 5,1′-coupled, but *M*-configured alkaloid 5-*epi*-ancistrobrevine E (**71b**)[56,57]; application of the Cahn-Ingold-Prelog priority rules, however, leading to opposite *M/P*-descriptors	

Continued

Table 17 *ortho*-Quinoid naphthylisoquinoline alkaloids.—cont'd Name, formula (mol. wt.), and structure

Producing plants and their organs/physical data [mp/°C, $[\alpha]_D^T/°cm^2\ g^{-1}\ 10^{-1}$ (solvent)]	Isolation, structural elucidation, partial and/or total synthesis, and further investigations	Biological activities
Ancistrobreviquinone B (131) $C_{25}H_{27}NO_5$ (421.2)	*Structural Elucidation:* Constitution and complete stereostructure assigned by HRESIMS, 1D and 2D NMR, oxidative degradation, and ECD—comparison of the data of **131** with those obtained from its parent compound ancistrobreviquinone A (**11**), revealing **131** to be the 8-O-methyl analogue of **11**[56]	Biological activities of the pure compound not determined for lack of isolated material
	Absolute configuration at C-3 established to be *S*, by oxidative degradation[56]	
Compd. can also be addressed as 8-O-methylancistrobreviquinone A	Assignment of the absolute axial configuration as *P* due to specific long-range NOESY interactions, evidencing the large naphthalene part to be directed down relative to the isoquinoline "plane," and in conjunction with the 1*S*,3*S*-stereoarray in the tetrahydroisoquinoline subunit established by the oxidative degradation procedure[56]	
A. abbreviatus: root bark[56]		
Dark orange-yellow amorphous solid (from MeOH); $[\alpha]_D^{23}$ −18.3 (*c* 0.05, MeOH)[56]	Additional proof of the absolute configuration at the biaryl axis due to a nearly identical ECD spectrum of **131** compared to that of its likewise *P*-configured parent compound ancistrobreviquinone A (**11**)[56]	

Table 18 A cationic naphthylisoquinolinium alkaloid.

Name, formula (mol. wt.), and structure	Producing plants and their organs/physical data [mp/°C, $[\alpha]_D^T$/°cm^2 g^{-1} 10^{-1} (solvent)]	Isolation, structural elucidation, partial and/or total synthesis, and further investigations	Biological activities
Ancistrobrevinium A (15) C$_{27}$H$_{30}$NO$_4$ (432.2)	*A. abbreviatus*: root bark[60] Yellow amorphous powder (from MeOH): $[\alpha]_D^{23}$ +7.2 (c 0.03, MeOH)[60]	Ancistrobrevinium A (15) as the first (and so far only!) C,C-coupled cationic naphthylisoquinolinium alkaloid discovered in nature; no information yet available on the natural counterion[60] *Structural Elucidation:* Constitution and complete stereostructure established by HRESIMS, 1D and 2D NMR, and ECD—comparison of the data of 15 with those obtained from its parent compound, the non-hydrogenated 7,8′-coupled alkaloid ancistrobreveine C (123), revealing 15 to differ from 123 by an additional methyl group on the—hence cationic—nitrogen atom[60]	Weak inhibitory activities against the non-small cell lung cancer (NSCLC) A549 cell line[60]

Table 19 Naphthylisoquinoline alkaloids lacking a methyl substituent at C-1. Name, formula (mol. wt.), and structure

Producing plants and their organs/physical data [mp/°C, $[\alpha]_D^T$/°cm^2 g^{-1} 10^{-1} (solvent)]	Isolation, structural elucidation, partial and/or total synthesis, and further investigations	Biological activities
1-*nor*-8-O-Demethylancistrobrevine H (12) C$_{24}$H$_{27}$NO$_4$ (393.2) *A. abbreviatus*: root bark[55] Yellow amorphous powder (from MeOH): $[\alpha]_D^{23}$ −7.7 (c 0.06, MeOH)[55]	*Structural Elucidation:* Constitution and coupling position established by HRESIMS and by 1D and 2D NMR, revealing the presence of a 7,8′-coupled naphthyltetrahydroisoquinoline lacking the otherwise generally present methyl group at C-1[55] Determination of the constitution of **12** as the 1-*nor* derivative of the (yet unknown) 8-O-demethyl analogue of ancistrobrevine H (**8**) by ^{1}H NMR and specific NOESY interactions, evidencing the loss of the usual Me group at C-1, with this C-atom thus bearing two diastereotopic protons, as confirmed by a respective signal in the DEPT-135 spectrum of **12**[55] Absolute configuration at C-3 established to be *S*, by oxidative degradation[55] Assignment of the absolute axial configuration as *M* by comparison of the ECD spectrum of **12** with that of the known, structurally related and likewise 7,8′-coupled alkaloid ancistrobrevine H (**8**)—the two alkaloids display nearly identical ECD curves; for formal reasons, however, **12** and **8** have opposite descriptors, according to the Cahn-Ingold-Prelog priority rules[55]	Biological activities of the pure compound not determined for lack of isolated material

Table 20 *seco*-Type naphthylisoquinoline alkaloids.
Name, formula (mol. wt.), and structure

Producing plants and their organs/physical data [mp/°C, $[\alpha]_D^T$/°cm^2 g^{-1} 10^{-1} (solvent)]	Isolation, structural elucidation, partial and/or total synthesis, and further investigations	Biological activities
Ancistrosecoline A (13) $C_{24}H_{29}NO_4$ (395.2) *A. abbreviatus*: root bark[55] Yellow amorphous solid (from MeOH): $[\alpha]_D^{23}$ −149.5 (*c* 0.05, MeOH)[55]	Ancistrosecoline A (13) is the first *seco*-type naphthylisoquinoline alkaloid discovered in nature, the tetrahydroisoquinoline ring being cleaved, with loss of its entire Me–C–1–H unit[55] *Structural Elucidation:* Constitution and 7,1′-coupling site established by HRESIMS and by ^{1}H and ^{13}C NMR, including specific HMBC and NOESY interactions, revealing the presence of a *seco*-type naphthylisoquinoline alkaloid, consisting of a naphthalene portion and a *meta*-dioxygenated phenyl ring system equipped with a flexible *N*-methyl-2-aminopropyl side chain[55] Absolute configuration at "C-3'' (numbering adjusted to that of normal naphthylisoquinoline alkaloids) established to be *S*, by oxidative degradation[55] Assignment of the absolute axial configuration as *P* by comparison of the ECD spectrum of 13 with that of a configurationally known, though simplified unnatural phenylnaphthalene 145 (available from previous synthetic work[532,533]), with an axially chiral chromophore similar to that of 13—the almost mirror-image like ECD curves of the two compounds, evidencing opposite axial	Moderate cytotoxic activity against human HeLa cervical cancer cells[55]

Continued

Table 20 seco-Type naphthylisoquinoline alkaloids.—cont'd

Name, formula (mol. wt.), and structure	Producing plants and their organs/physical data [mp/°C, $[\alpha]_D^T/°$cm^2 g^{-1} 10^{-1} (solvent)]	Isolation, structural elucidation, partial and/or total synthesis, and further investigations	Biological activities
		configurations, yet to be addressed with the same descriptor P, for formal reasons, according to the Cahn-Ingold-Prelog priority rules[55]	
		Additional proof of the assignment of the absolute axial configuration of ancistrosecoline A (13) by quantum-chemical ECD calculations at the TDωB97XD3/def2-TZVP//B3LYP-D3/def2-TZVP level, the ECD spectrum calculated for the P-configured structure 13 showing a good agreement with the experimental one[55]	
Ancistrosecoline B (141) $C_{23}H_{27}NO_4$ (381.2)		*Structural Elucidation:* Constitution and complete stereostructure assigned by HRESIMS, 1D and 2D NMR, oxidative degradation, and ECD—by comparing the data of 141 with those obtained from its parent compound ancistrosecoline A (13), revealing 141 to be the N-demethylated analogue of 13[55]	Low cytotoxic activity against human HeLa cervical cancer cells[55]
		Absolute configuration at C-3 established to be S, by oxidative degradation[55]	
A. abbreviatus: root bark[55]		Absolute configuration at the biaryl axis established as P due to the nearly identical ECD spectrum of 141 compared to that of the likewise P-configured parent compound ancistrosecoline A (13)[55]	
Yellow amorphous solid (from MeOH): $[\alpha]_D^{23}$ −163.2 (c 0.05, MeOH)[55]			

Ancistrosecoline C (132)

C$_{24}$H$_{29}$NO$_4$ (395.2)

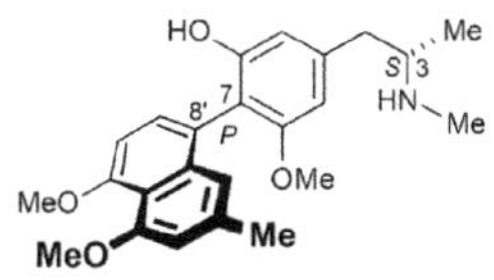

A. abbreviatus: root bark[55]

Yellow amorphous powder (from MeOH): $[\alpha]_D^{23}$ −163.5 (c 0.05, MeOH)[55]

Structural Elucidation:

Constitution and complete stereostructure established by HRESIMS, 1D and 2D NMR, and by ECD—comparison of the data of **132** with those of the 7,1′-coupled *seco*-type alkaloid ancistrosecoline A (**13**), revealing **132** to differ from **13** only by the position of the methyl group on the naphthalene moiety, thus displaying a 7,8′-coupling site[55]

Absolute configuration at C-3 established to be S, by oxidative degradation[55]

Absolute configuration at the biaryl axis established as P due to the nearly identical ECD spectrum of **132** compared to that of the structurally closely related, likewise P-configured, but 7,1′-coupled *seco*-type alkaloid ancistrosecoline A (**13**)[55]

Biological activities of the pure compound not determined for lack of isolated material

---

Ancistrosecoline D (142)

C$_{23}$H$_{27}$NO$_4$ (381.2)

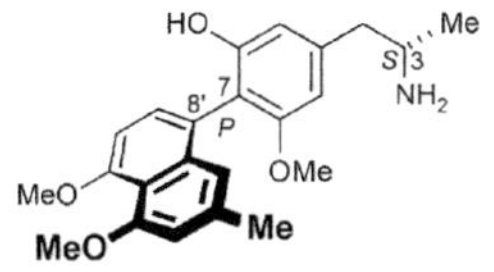

A. abbreviatus: root bark[55]

Structural Elucidation:

Constitution and complete stereostructure assigned by HRESIMS, 1D and 2D NMR, oxidative degradation, and ECD—comparison of the data of **142** with those obtained from its parent compound ancistrosecoline C (**132**), revealing **142** to be the N-demethylated analogue of **132**[55]

Absolute configuration at C-3 established to be S, by oxidative degradation[55]

Absolute configuration at the biaryl axis established as P due to the nearly identical ECD spectrum of **142** compared to those of the likewise P-configured parent compound ancistrosecoline C (**132**) and the

Strong cytotoxic activity against HeLa cervical cancer cells, displaying massive nuclei fragmentation, leading to apoptotic cell death[55]—the cytotoxicity of **142** being the same magnitude of activity compared to that of 5-fluoro-uracil,[446–448] a well-established drug for cancer treatment

Continued

Table 20 *seco*-Type naphthylisoquinoline alkaloids.—cont'd
Name, formula (mol. wt.), and structure

Producing plants and their organs/physical data [mp/°C, $[\alpha]_D^T$/°cm^2 g^{-1} 10^{-1} (solvent)]	Isolation, structural elucidation, partial and/or total synthesis, and further investigations	Biological activities
Yellow amorphous powder (from MeOH): $[\alpha]_D^{23}$ −143.0 (c 0.05, MeOH)[55]	N-demethylated, but 7,1′-coupled *seco*-type compound ancistrosecoline B (**141**)[55]	
Ancistrosecoline E (143)	*Structural Elucidation:*	Biological activities of the pure compound not determined for lack of isolated material
C$_{25}$H$_{31}$NO$_4$ (409.2)	Constitution and coupling position assigned by HRESIMS and by ^{1}H and ^{13}C NMR, including specific HMBC and NOESY interactions—comparison of the data of **143** with those obtained from the 7,1′-coupled *seco*-type alkaloid ancistrosecoline A (**13**) revealing **143** to differ from **13** by an additional O-methyl group at C-6[55]	
■ = axis rotationally hindered, but not chiral	Absolute configuration at C-3 established to be S, by oxidative degradation[55]	
△ = diastereotopic methoxy groups	Ancistrosecoline E (**143**) as the very first naphthylisoquinoline-derived alkaloid found in nature whose biaryl axis, although rotationally hindered, is non-stereogenic, due to its symmetric substitution pattern, with two methoxy functions at C-6 and C-8 in the phenyl subunit[55]—unprecedented as compared to all other naphthylisoquinoline alkaloids[1–7]—is the diastereotopic character of these two methoxy groups to each other, not constitutionally heterotopic[55]	
A. abbreviatus: root bark[55]		
Yellow amorphous powder (from MeOH): $[\alpha]_D^{23}$ −151.1 (c 0.05, MeOH)[55]	Virtually no ECD effect due to the lack of axial chirality[55]	

Ancistrosecoline F (144)

C$_{24}$H$_{29}$NO$_4$ (395.2)

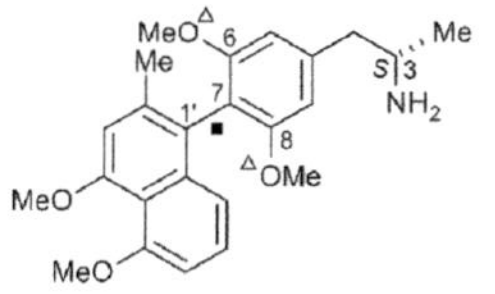

■ = axis rotationally hindered, but not chiral
△ = diastereotopic methoxy groups

A. abbreviatus: root bark[55]

Yellow amorphous powder (from MeOH): $[\alpha]_D^{23}$ −5.1 (*c* 0.10, MeOH)[55]

Structural Elucidation:

Constitution and coupling position assigned by HRESIMS and by ^{1}H and ^{13}C NMR, including specific HMBC and NOESY interactions—comparison of the data of 144 with those obtained from its 7,1′-coupled parent *seco*-type alkaloid ancistrosecoline E (143) revealing 144 to differ from 143 by a free, primary unsubstituted amino group (instead of an *N*-methyl function as in 143)[55]

Absolute configuration at C-3 established to be *S*, by oxidative degradation[55]

In addition to its parent compound ancistrosecoline E (143, see entry above), ancistrosecoline F (144) is a further example of a naphthylisoquinoline with a rotationally hindered biaryl axis that does not constitute an element of chirality[55]

Virtually no ECD effect due to the lack of axial chirality[55]

Biological activities of the pure compound not determined for lack of isolated material

Table 21 Naphthylisoquinoline-derived isoindolinones.

Producing plants and their organs/physical data [mp/°C, $[\alpha]_D^T$/°cm^2 g^{-1} 10^{-1} (solvent)]	Isolation, structural elucidation, partial and/or total synthesis, and further investigations	Biological activities
Ancistrobrevoline A (14) $C_{25}H_{27}NO_5$ (421.2) *A. abbreviatus*: root bark[59] Yellow amorphous powder (from MeOH): $[\alpha]_D^{23}$ + 11.5 (c 0.05, MeOH)[59]	Ancistrobrevoline A (14) is the first ring-contracted naphthylisoquinoline-related compound discovered in nature, consisting of a naphthalene portion and a 6,8-dioxygenated isoindolinone subunit, i.e., the six-membered heterocyclic part of the isoquinoline molecular half of a usual naphthylisoquinoline being contracted to give a five-membered ring system[59] *Structural Elucidation:* Constitution and 7,8'-coupling site established by HRESIMS and by ^{1}H and ^{13}C NMR, with specific HMBC and NOESY interactions, revealing the presence of a naphthylisoquinoline-related compound that consists of a 4',5'-dimethoxy-2'-methylnaphthalene portion and a *meta*-dioxygenated six-membered nitrogen-containing ring system fused to a five-membered benzene ring system (= isoindolinone entity), with the carbonyl function at C-4 (atom numbers adjusted to that of normal naphthylisoquinoline alkaloids)[59] Absolute configuration at C-1 established as *R*, by the ruthenium-mediated oxidative degradation procedure, as usually applied for di- and tetrahydroisoquinolines, still, despite the strong *N,C*-lactam bond, here leading to its cleavage, too, thus affording (*R*)-methylalanine as the predominant enantiomer of the degradation product[59]	Low growth-inhibitory activities against MCF-7 breast adenocarcinoma cells[59] Significant reduction of the formation of MCF-7-derived highly aggressive and metastatic breast cancer stem-like cells (mammospheres)[59] Moderate cytotoxic effects against the non-small cell lung cancer (NSCLC) cell line A549[59]

Assignment of the absolute axial configuration as P by comparison of the experimental ECD curve with the spectra obtained by quantum-chemical ECD calculations at the TDωB97XD3/def2-TZVP//B3LYP-D3/def2-TZVP level for the P- and M-atropodiastereomers—comparison of the experimental ECD spectrum with those quantum-chemically simulated showing a good agreement with the one calculated for the P-configured atropisomer[59]

Ancistrobrevoline B (133)	*Structural Elucidation:*	Low growth-inhibitory activities against MCF-7 breast adenocarcinoma cells[59]
$C_{24}H_{25}NO_5$ (407.2)	Constitution and coupling position assigned by HRESIMS and by ^{1}H and ^{13}C NMR, including specific HMBC and NOESY interactions—comparison of the data of **133** with those of its likewise 7,8'-coupled ring-contracted parent alkaloid ancistrobrevoline A (**14**) revealing **133** to differ from **14** by a free hydroxy function at C-6 (instead of methoxy group as in **14**)[59]	
		Significant reduction of the formation of MCF-7-derived highly aggressive and metastatic breast cancer stem-like cells (mammospheres)[59]
	Absolute configuration at C-1 established to be R, by oxidative degradation[59]	
	Absolute configuration at the biaryl axis established as P due to the virtually identical ECD spectrum of **133** compared to that of the likewise P-configured parent compound ancistrobrevoline A (**14**)[59]	Pronounced cytotoxic effects against the non-small cell lung cancer (NSCLC) cell line A549[59]
A. abbreviatus: root bark[59]		
Yellow amorphous solid (from MeOH): $[\alpha]_D^{23}$ −8.3 (c 0.07, MeOH)[59]		

Continued

Table 21 Naphthylisoquinoline-derived isoindolinones.—cont'd
Name, formula (mol. wt.), and structure

Producing plants and their organs/physical data [mp/°C, $[\alpha]_D^T$ 10^{-1} (solvent)]	Isolation, structural elucidation, partial and/or total synthesis, and further investigations	Biological activities
Ancistrobrevoline C (149) $C_{24}H_{25}NO_5$ (407.2) *A. abbreviatus*: root bark[59] Yellow amorphous powder (from MeOH): $[\alpha]_D^{23}$ −18.9 (c 0.05, MeOH)[59]	*Structural Elucidation:* Constitution and complete stereostructure established by HRESIMS, 1D and 2D NMR, and by ECD—comparison of the data of **149** with those of the 7,8′-coupled naphthylisoquinoline-related isoindolinone ancistrobrevoline B (**133**), revealing **149** to differ from **133** by the position of the methyl group on the naphthalene moiety, thus displaying a 7,1′-coupling site, and by the substitution pattern of the benzene ring system of the isoindolinone unit being 6-OMe/8-OH in **149** instead of 6-OH/8-OMe as in **133**[59] Absolute configuration at C-1 established to be *R*, by oxidative degradation[59] Assignment of the absolute axial configuration as *P* due the nearly mirror-image like ECD spectrum of **149** as compared to those of the structurally related, but 7,8′-coupled naphthylisoindolinones ancistrobrevolines A (**14**) and B (**133**), thus revealing **149** to exhibit an opposite axial stereoarray, for formal reasons, however, leading to the same *P*-descriptor for all the three alkaloids, according to the Cahn-Ingold-Prelog priority rules[59]	Biological activities of the pure compound not determined for lack of isolated material

Ancistrobrevoline D (150)

$C_{23}H_{23}NO_4$ (377.2)

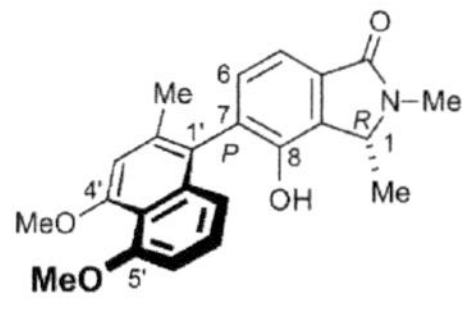

[structure drawing with labels: N-Me, R, Me, O, 8, OH, Me, 6, 7, 1', P, 4', 5', MeO, MeO]

A. abbreviatus: root bark[59]

Yellow amorphous powder (from MeOH); $[\alpha]_D^{23}$ −24.0 (c 0.04, MeOH)[59]

Structural Elucidation:

Constitution and 7,1'-coupling type assigned by HRESIMS and by ^{1}H and ^{13}C NMR, including specific HMBC and NOESY interactions—comparison of the data of **150** with those of the structurally related naphthylisoindolinone ancistrobrevoline C (**149**) showing **150** to be similar to **149**, equipped with two methoxy groups at C-4' and C-5', and a free hydroxy function at C-8, but lacking an O-functionality at C-6, thus tentatively categorized as a Dioncophyllaceae-type compound (despite the non-fulfillment of the second criterion, viz., the required presence of an *R*-configuration at C-3)[59]

Absolute configuration at C-1 established to be *R*, by oxidative degradation[59]

Absolute configuration at the biaryl axis assigned as *P* due to the virtually opposite ECD spectrum of **150** compared to that of its parent alkaloid ancistrobrevoline C (**149**), thus revealing **150** to exhibit an opposite axial stereoarray, for formal reasons, however, leading to the same *P*-descriptor for the two alkaloids, according to the Cahn-Ingold-Prelog priority rules[59]

Biological activities of the pure compound not determined for lack of isolated material

Table 22 3′,3″-Coupled Dioncophyllaceae-type dimeric naphthylisoquinoline alkaloids.

Name, formula (mol. wt.), and structure Producing plants and their organs/physical data [mp/°C, $[\alpha]_D^T$/cm^2 g^{-1} 10^{-1} (solvent)]	Isolation, structural elucidation, partial and/or total synthesis, and further investigations	Biological activities
Jozimine A$_2$ (4a) C$_{46}$H$_{48}$N$_2$O$_6$ (724.9)	Discovered for the first time from the roots of an as yet botanically undescribed *Anistrocladus* liana endemic to the Central Congo Basin[12]; jozimine A$_2$ (4a) as the first Dioncophyllaceae-type naphthylisoquinoline dimer found in nature, i.e., lacking oxygen functions at C-6 and bearing R-configurations at C-3 in both of its two isoquinoline portions[1,2,12]	Excellent—and highly specific—inhibitory activity in vitro against the malaria parasite *Plasmodium falciparum* (NF54 and K1 strains), in the low-nanomolar range[12,180]
	Isolation of jozimine A$_2$ (4a) from the root bark of *A. abbreviatus*, along with a series of three further Dioncophyllaceae-type dimers, the jozibrevines A–C (4b–d)[58]—all four dimers consisting of two 7,1′-coupled naphthylisoquinoline monomers linked via the sterically constrained 3′,3″-positions of their two naphthalene units, and all having the same constitution and identical absolute configurations at the four stereogenic centers, but differing by their axial chirality[58]	Due to its excellent antiplasmodial effects, combined with an extremely weak cytotoxicity, and, thus a high selectivity index (>11.400)[12,180]; validation of 4a as a lead compound according to the guidelines of TDR-WHO[310]
	Isolation of jozimine A$_2$ (4a) from the roots of the Congolese liana *A. ileoboensis*, along with two further Dioncophyllaceae-type dimers, the jozilebomines A (44a) and B (44b), consisting of two 7,1′-coupled naphthylisoquinoline monomers linked through an unprecedented 3′,6″-coupling in the binaphthalene core[2,183]	Moderate to weak inhibitory activities in vitro against *Trypanosoma brucei rhodesiense*, *T. cruzi*, and *Leishmania donovani*[12]
		Moderate to low inhibitory activity in vitro against the hematozoan parasite *Babesia canis*[373]
	Structural Elucidation:	Strong cytotoxicity with half-maximum inhibition values in the low micromolar range against the HeLa human cervical cancer cell line[183]
A. abbreviatus: root bark[58]		
A. ileoboensis: root bark[183]	Constitution deduced from HRESIMS, MALDI-TOF-MS, and from ^{1}H and ^{13}C NMR, including HMBC, HMQC, and ROESY interactions, indicative of a C$_2$-symmetric dimeric naphthylisoquinoline alkaloid, i.e., with two homomorphous halves, both lacking oxygen functions at C-6 in the two isoquinoline portions[2,12,58,183]	
From a botanically yet undescribed Congolese *Anistrocladus* species: root bark[12]		

Yellow amorphous solid (from MeOH): $[\alpha]_D^{23}$ −68.1 (c 0.02, MeOH)[404]	Coupling positions established from HMBC and ROESY interactions indicative of a dimer consisting of two 7,1′-coupled naphthylisoquinoline monomers, identified as 4′-O-demethyldioncophylline A (**93a**),[2,12,58] linked via the sterically constrained C-3′ and C-3″ positions of the central binaphthalene core, thus evidencing jozimine A$_2$ (**4a**) to possess three consecutive stereogenic axes[1,2,12]	Highly potent preferential cytotoxicity against PANC-1 human pancreatic cancer cells under "austerity" (i.e., nutrient-deprived) conditions[183]
Colorless crystals (from CH$_2$Cl$_2$/MeOH): mp 238–240; $[\alpha]_D^{20}$ −29.8 (c 0.20, MeOH)[12]	Relative configuration of the stereocenters at C-1 versus C-3 and at C-1‴ versus C-3‴ determined to be *trans* by specific long-range ROESY correlations[12]	Strong inhibition of colony formation and induction of dramatic morphological alterations such as membrane rupture and disintegration of cellular organelles, leading to total cell death of PANC-1 cancer cells[183]
	Absolute *R*-configuration both at C-1 and C-3, and at C-1‴ and C-3‴, determined by oxidative degradation[12]	Strong cytotoxic activity in the low micromolar range against drug-sensitive acute lymphoblastic CCRF-CEM leukemia cells and their multidrug-resistant subline CEM/ADR5000[52]
	Absolute axial configuration of the two outer axes deduced as *P* from specific long-range ROESY correlations in combination with the 1*R*,3*R*-configurations in the isoquinoline moieties as established by degradation[2,12]	
	Assignment of the absolute configuration at the central axis as *M* by comparison of the ECD spectrum of **4a**[2,12,58] with that of the structurally closely related, likewise 3′,3″-coupled dimer shuangancistrotectorine A (**43**),[254] consisting of two 7,1′-coupled Ancistrocladaceae-type naphthylisoquinoline monomers—the ECD spectra of the two dimers **4a** and **43** being virtually identical[2,12]	Pronounced cytotoxicity against human colon carcinoma (HT-29), fibrosarcoma (HT-1080), and multiple myeloma (MM.1S) cells[58]
	Further proof and validation of the constitution and the relative configuration of **4a** by X-ray diffraction analysis[12]	
	Semisynthesis:	
	Semisynthesis of jozimine A$_2$ (**4a**) and its 3′-atropo-diastereomer (= jozibrevine A, **4b**), by biomimetic phenol-oxidative coupling of the O,N-dibenzyl derivative of 4′-O-demethyldioncophylline A (**93a**) using lead tetraacetate—deprotection and resolution on a preparative reversed-phase HPLC column, yielding the two atropo-diastereomers **4a** and **4b**[12]	

Continued

Table 22 3′,3‴-Coupled Dioncophyllaceae-type dimeric naphthylisoquinoline alkaloids.—cont'd

Name, formula (mol. wt.), and structure Producing plants and their organs/physical data [mp/°C, $[\alpha]_D^T$/°cm^2 g^{-1} 10^{-1} (solvent)]	Isolation, structural elucidation, partial and/or total synthesis, and further investigations	Biological activities
Jozibrevine A (4b) C$_{46}$H$_{48}$N$_2$O$_6$ (724.9) Compd. can also be addressed as 3′-*epi*-jozimine A$_2$ *A. abbreviatus*: root bark[58] Yellow amorphous solid (from MeOH): $[\alpha]_D^{23}$ −23.0 (*c* 0.03, MeOH)[58]	Isolation of jozibrevine A (**4b**) as a minor metabolite from the roots of *A. abbreviatus*, along with three further Dioncophyllaceae-type naphthylisoquinoline dimers, jozimine A$_2$ (**4a**), jozibrevine B (**4c**), and jozibrevine C (**4d**)[58] *Structural Elucidation:* Constitution deduced from HRESIMS and from ^{1}H and ^{13}C NMR, including HMBC, HMQC, and ROESY interactions, indicative of a C$_2$-symmetric naphthylisoquinoline dimer,[58] consisting of two homomorphous halves and possessing the same constitution as jozimine A$_2$ (**4a**)[2,12,58] Relative configuration of the stereocenters at C-1 versus C-3 and at C-1‴ versus C-3‴ determined to be *trans* by specific long-range ROESY correlations[58] Absolute configuration at C-1 and C-3 and at C-1‴ and C-3‴ established as all being *R* by oxidative degradation[58] Absolute axial configuration of the two outer axes established as *P* from specific long-range ROESY correlations in combination with the 1*R*,3*R*-configurations in the isoquinoline moieties as established by degradation—hence, showing **4b** to consist of two identical 4′-*O*-demethyldioncophylline A (**93a**) halves,[58] all like in the case of jozimine A$_2$ (**4a**)[1,2,12] Absolute configuration at the central biaryl axis assigned as *P* due to a virtually opposite ECD spectrum of **4b** compared to that of jozimine A$_2$ (**4a**)[58]	Strong—and highly specific—inhibitory activity in vitro against the malaria parasite *Plasmodium falciparum* (NF54 strain), but less pronounced than that of jozimine A$_2$ (**4a**)[58] Moderate to weak inhibitory activities in vitro against *Trypanosoma brucei rhodesiense* and *T. cruzi*[58] No inhibitory activity in vitro against *Leishmania donovani*[58] No significant cytotoxic effect against TH-29 human colon carcinoma cells[58]

Semisynthesis:

Semisynthesis of jozibrevine A (**4b**), along with its 3′-atropo-diastereomer (= jozimine A₂, **4a**), by biomimetic phenol-oxidative coupling of the O,N-dibenzyl derivative of 4′-O-demethyldioncophylline A (**93a**) using lead tetraacetate—deprotection and resolution on a preparative reversed-phase HPLC column, to give the two atropo-diastereomers **4b** and **4a**.[12]

Jozibrevine B (4c) $C_{46}H_{48}N_2O_6$ (724.9) Compd. can also be addressed as 7-*epi*-jozibrevine A *A. abbreviatus*: root bark[58]	Isolation of jozibrevine B (**4c**) as a minor metabolite from the roots of *A. abbreviatus*, along with three further Dioncophyllaceae-type naphthylisoquinoline dimers, jozimine A₂ (**4a**), jozibrevine A (**4b**), and jozibrevine C (**4d**)[58] *Structural Elucidation:* Constitution deduced from HRESIMS and from ¹H and ¹³C NMR, including HMBC, HMQC, and ROESY interactions, evidencing the presence of an unsymmetric naphthylisoquinoline dimer, with a 3′,3″-linkage and spectral features similar to those of jozimine A₂ (**4a**)[2,12] and jozibrevine A (**4b**),[58] consisting of two constitutionally identical, yet configurationally different monomeric halves[58] Relative configuration of the stereocenters at C-1 versus C-3 and at C-1‴ versus C-3‴ established to be *trans* by specific long-range ROESY correlations[58] Absolute configuration at C-1 and C-3 and at C-1‴ and C-3‴ all established as *R* by oxidative degradation[58] Assignment of the absolute configuration at the two outer biaryl axes by long-range ROESY correlations for the relative configuration at centers versus axes in combination with the 1*R*,3*R*-configurations in the isoquinoline portions as established by degradation, proving jozibrevine B (**4c**) to possess two differently configured outer axes,	Strong—and highly specific—inhibitory activity in vitro against the malaria parasite *Plasmodium falciparum* (NF54 strain), but less pronounced than that of jozimine A₂ (**4a**)[58] Moderate to weak inhibitory activities in vitro against *Trypanosoma brucei rhodesiense* and *T. cruzi*[58] No inhibitory activity in vitro against *Leishmania donovani*[58] No significant cytotoxic effect against TH-29 human colon carcinoma cells[58]

Continued

Table 22 3′,3‴-Coupled Dioncophyllaceae-type dimeric naphthylisoquinoline alkaloids.—cont'd

Name, formula (mol. wt.), and structure	Producing plants and their organs/physical data [mp/°C, [α]$_D^T$/°cm^2 g^{-1} 10^{-1} (solvent)]	Isolation, structural elucidation, partial and/or total synthesis, and further investigations	Biological activities
	Yellow amorphous solid (from MeOH): [α]$_D^{23}$ −81.5 (c 0.04, MeOH)[58]	therefore showing a full set of signals in both the ^{1}H and the ^{13}C spectra, thus being the cross-coupling product of the two atropo-diastereomeric alkaloids 4′-O-demethyldioncophylline A (**93a**) and 4′-O-demethyl-7-*epi*-dioncophylline A (**93b**)[58]	
		Absolute configuration at the central biaryl axis assigned as *P* due to a virtually opposite ECD spectrum of **4c** compared to that of the related dimer jozimine A$_2$ (**4a**), which is *M*-configured at the central axis,[2,12] and due to a nearly identical ECD curve of **4c** compared to that of the *P*-configured jozibrevine A (**4b**)[58]	
Jozibrevine C (4d) C$_{46}$H$_{48}$N$_2$O$_6$ (724.9)		Isolation of jozibrevine B (**4c**) as a minor metabolite from the roots of *A. abbreviatus*, along with three further Dioncophyllaceae-type naphthylisoquinoline dimers, jozimine A$_2$ (**4a**), jozibrevine A (**4b**), and jozibrevine C (**4d**)[58]	Strong—and highly specific—inhibitory activity in vitro against the malaria parasite *Plasmodium falciparum* (NF54 strain), but less pronounced than that of jozimine A$_2$ (**4a**)[58]
			Moderate to weak inhibitory activities in vitro against *Trypanosoma brucei rhodesiense* and *T. cruzi*.[58]
		Structural Elucidation:	No inhibitory activity in vitro against *Leishmania donovani*[58]
		Constitution deduced from HRESIMS, ^{1}H and ^{13}C NMR, including HMBC, HMQC, and ROESY interactions, evidencing the presence of an unsymmetric naphthylisoquinoline dimer, with a 3′,3‴-linkage and spectral features similar to those of jozimine A$_2$ (**4a**),[2,12] jozibrevine A (**4b**),[58] and jozibrevine B (**4c**),[58] consisting of two constitutionally identical, yet configurationally different monomeric halves[58]	
		Relative configuration of the stereocenters at C-1 versus C-3 and at C-1‴ versus C-3‴ established to be *trans* by specific long-range ROESY correlations[58]	
		Absolute configuration at C-1 and C-3 and at C-1‴ and C-3‴ all established as *R* by oxidative degradation[58]	
		Assignment of the absolute configuration at the two outer biaryl axes by long-range ROESY correlations for the relative configuration at centers	

Table 23 Naphthoquinones and tetralones related to naphthylisoquinoline alkaloids.

Name, formula (mol. wt.), and structure	Producing plants and their organs/physical data [mp/°C, $[\alpha]_D^T$/°cm^2 g^{-1} 10^{-1} (solvent)]	Isolation, structural elucidation, partial and/or total synthesis, and further investigations	Biological activities
Plumbagin (16) $C_{11}H_8O_3$ (188.2) Compd. can also be addressed as 5-hydroxy-2-methyl-1,4-naphthoquinone *A. abbreviatus*: stem bark, twigs,[5] solid callus cultures[151] *A. barteri*: roots[5,285] *T. peltatum*: roots,[5,6,374] cell cultures,[151,153,268] root cultures[311] *D. thollonii*: roots, stem bark[5,6,375] *H. dawei*: stem bark, leaves[6,297] *A. heyneanus*: roots, stem bark, cell cultures[269,560] *A. cochinchinensis*: leaves[667,668] Orange powder (from CH$_2$Cl$_2$): mp 78[5]	Plumbagin (**16**) as a polyketide-derived 1,4-naphthoquinone[566,567] of widespread occurrence in various plant families, among them the Dioncophyllaceae,[5,6,151,153,268,375] and the Ancistrocladaceae,[1,5,151,269,285,560,667,668] but also closely related families such as the Plumbaginaceae,[670–673] the Droseraceae,[674] the Droseraceae,[613,675–677] and the Nepenthaceae[563,578] *Further Investigations:* Formation of **16** from six acetate units via a β-pentaketo precursor according to the F-type folding mode[612]—proof of the acetogenic origin of **16** by feeding experiments with $^{13}C_2$-labeled acetate administered to callus cultures of *A. heyneanus*[269] or *T. peltatum*,[80–83,268] followed by 2D INADEQUATE NMR analysis, assisted by the cryoprobe methodology[80–83]—after earlier hints at a biosynthesis of **16** from acetate units by feeding experiments on cell cultures of *Plumbago europaea* using $^{14}C_2$-labeled acetate[672] Abundant amounts of **16**, apparently derived from the naphthalene precursor of naphthylisoquinoline alkaloids, produced in Dioncophyllaceae and Ancistrocladaceae plants, and, in particular, in cell cultures of *Triphyophyllum peltatum*,[5,6,151,153,268,311] *Ancistrocladus heyneanus*,[5,268,269,560] and *A. abbreviatus*,[151] when exposed to chemical, physical, or biotic stress Plumbagin (**16**) evidenced as the major antipathogenic factor involved in the incompatible relationship between the parasitic plant *Cuscuta reflexa* (Convolvulaceae) and *A. heyneanus*, leading to a damage of the haustoria of the parasite and eventually to a degeneration of the parasitic tissues[560]	Strong antifeedant activity against *Spodoptera littoralis*[676] Pronounced inhibitory activity against the fungus *Penicillium citrinum*, strong inhibition of seed germination (test system: lettuce seeds), and strong ichthyotoxicity (fish toxicity) against male guppies of the species *Poecilia reticulata*[678] Highly potent inhibitory activity in vitro against *Trypanosoma brucei brucei*[368] Pronounced inhibitory activities against extracellular promastigotes of *Leishmania donovani*, *L. major*, and other strains, and against intracellular amastigotes of *L. donovani*,[679,680] as well as strong to good activities against lesion development in cutaneous leishmaniasis[669,681] Strong cytotoxic activities against multiple myeloma cells[152] Anti-proliferation and anti-metastasis effects of **16**

Continued

Table 23 Naphthoquinones and tetralones related to naphthylisoquinoline alkaloids.—cont'd

Name, formula (mol. wt.), and structure	Producing plants and their organs/physical data [mp/°C, $[\alpha]_D^T$/°cm^2 g^{-1} 10^{-1} (solvent)]	Isolation, structural elucidation, partial and/or total synthesis, and further investigations	Biological activities
	Orange crystals (from petroleum ether): mp 74[375]		against various cancer types both in vitro and in vivo; for detailed reviews on the anti-cancer properties of **16**: see Refs. 566–575
	Orange crystals (from n-hexane/CHCl$_3$): mp 73–74[667]		High antibacterial activities against drug-sensitive and -resistant strains of *Mycobacterium tuberculosis*.[682,683]
	cf. Ref. 669: Isolation from *Pera benensis* as orange needles (from CHCl$_3$): mp 78–79		Pronounced inhibitory activities against sensitive and multidrug-resistant Gram-negative and Gram-positive bacteria, among them *Escherichia coli*, *Klebsiella pneumonia*, and *Staphylococcus aureus*[569,571,684,685]
	cf. Ref. 131: Orange needles: mp 77		Inhibition of biofilm-assisted infections caused by *Candida albicans*[569,571,684,686]
			Neuroprotective[571,687] and anti-inflammatory[571,688] properties combined with the ability to improve memory performance[569,688,689]

Droserone (17)

$C_{11}H_8O_4$ (204.2)

Compd. can also be addressed as 3,5-dihydroxy-2-methyl-1,4-naphthoquinone

A. abbreviatus: solid callus cultures[151]

T. peltatum: (solid) callus cultures,[5,6,151–153] root cultures[311]

D. thollonii: roots, stem bark, twigs[5,6]

H. dawei: twigs, stem bark[5,6,297]

A. robertsoniorum: stem bark[561,562]

A. heyneanus[655]

Yellow solid (from MeOH): mp 179–181[153]

Yellow prisms (from MeOH): mp 178–180[151]

Orange-red needles (from diethyl ether): mp 172–173[561,655]

Needle-shaped red-orange "biogenic" crystals found in insect-dug holes in plant stems: mp 119[561,562]

cf. Ref. 131: Yellow needles: mp 181

Droserone (17) as a polyketide-derived 1,4-naphthoquinone[566,567] of widespread occurrence in various plant families, among them the Dioncophyllaceae,[5,6,151,153,268] and the Ancistrocladaceae,[5,561,562,655] but also closely related families such as the Plumbaginaceae,[673] the Droseraceae,[613,677,690] and the Nepenthaceae[578]

Further Investigations:

Pure orange-colored "biogenic" crystals of 17 produced by the Kenyan liana *A. robertsoniorum*, well-suited for X-ray crystallography in their genuine form, found in small insect-dug holes in twigs and stems in the otherwise nearly colorless wood[561,562]

Formation of 17 from six acetate units via a β-pentaketo precursor according to the F-type folding mode[612]—proof of the acetogenic origin of 17 by feeding experiments with $^{13}C_2$-labeled acetate administered to callus cultures of *T. peltatum*,[153] analyzed by the 2D INADEQUATE NMR method, assisted by the cryoprobe methodology[80–83,268]

Upon exposure to chemical, physical, or biotic stress, formation of abundant amounts of droserone (17), apparently originating from 1,8-dihydroxy-3-methylnaphthalene, which is the isocyclic precursor of naphthylisoquinoline alkaloids in Dioncophyllaceae and Ancistrocladaceae plants, and, in particular, in cell cultures of *T. peltatum*,[5,6,151,153,268,311] *A. heyneanus*,[5,268,269] and *A. abbreviatus*[151]

Moderate inhibition of seed germination (test system: lettuce seeds), and strong ichthyotoxicity (fish toxicity) against male guppies of the fish species *Poecilia reticulata*[678]

No antitumoral activity in vitro against the murine leukemia cell line P388[677]

No antimicrobial activity against *Candida albicans*, *Bacillus subtilis*, *Cladosporium resinae*, *Trichophyton mentagrophytes*, *Escherichia coli*, or *Pseudomonas aeruginosa*[677]

Moderate antiviral activity against Polio virus type 1, but no inhibition of *Herpes simplex* type 1[677]

Considerable reduction of measles virus infection induced by droserone (17), but only upon addition of 17 during virus entry, not when added to the cells prior to virus uptake or after virus uptake, thus indicating interaction of 17 with the viral particles to reduce infectivity[691]

Weak inhibitory activity in vitro against *Plasmodium falciparum*[692]

Continued

Table 23 Naphthoquinones and tetralones related to naphthylisoquinoline alkaloids.—cont'd

Name, formula (mol. wt.), and structure Producing plants and their organs/physical data [mp/°C, $[\alpha]_D^T$/°C cm² g⁻¹ 10⁻¹ (solvent)]	Isolation, structural elucidation, partial and/or total synthesis, and further investigations	Biological activities
Dioncoquinone B (135) $C_{11}H_8O_5$ (220.2) Compd. can also be addressed as 3,5,6-trihydroxy-2-methyl-1,4-naphthoquinone *A. abbreviatus*: solid callus cultures[151] *T. peltatum*: solid callus cultures[151,152] Red needles (from CHCl₃): mp 218[151] cf.: Title compd. prepared by total synthesis: red solid (from CH₂Cl₂): mp 216–217[152]	*Structure Elucidation:* Constitution deduced from HRESIMS and from ¹H and ¹³C NMR, including HMBC and H,H-COSY interactions, indicative of a 1,4-naphthoquinone structurally closely related to droserone (17), yet equipped with an additional hydroxy function at C-6[151] Confirmation of the assigned constitution by quantum-chemical ¹H and ¹³C NMR simulations, unambiguously excluding the presence of a likewise imaginable 8-substituted droserone analogue (instead of a 6-substituted congener)[151] *Semisynthesis and Further Synthetic Work:* Further proof of the structure of 135 by its selective conversion into ancistroquinone C (154) using an excess of diazomethane, thus clearly confirming the 3,5,6-trihydroxysubstitution pattern of 135[151] Development of three synthetic approaches to dioncoquinone B (135) and related analogues—the most efficient one based on a directed *ortho*-metalation (DOM) reaction starting from appropriately substituted benzamides and providing the targeted key naphthalene intermediates in a very short manner and in high yields[151]	Strong cytotoxic activities in vitro against human tumor cells derived from two different B cell malignancies, B cell lymphoma and multiple myeloma—135 showed activity in concentration ranges similar to those of the therapeutically used melphalan,[469] a DNA-alkylating agent[151] Highly selective cytotoxic activity—no effect on normal peripheral mononuclear blood cells (PMBCs) isolated from healthy donors[151] Pronounced activities against *Leishmania donovani*, but virtually no activity against *Plasmodium falciparum*,

	Structure-activity relationship (SAR) studies using a small library of analogues (~50 compounds) of dioncoquinone B (135), showing each of the three hydroxy groups, at C-3, C-5, and C-6, to be required for improved antitumoral activities and decreased cytotoxicities, in contrast to the only little influence of position 7 on the antiproliferative activities of the derivatives of 135[151]	*Trypanosoma cruzi*, and *T. brucei rhodesiense*[151]
Malvone A (153) $C_{12}H_{10}O_5$ (234.2)	Previously discovered in *Malva sylvestris* (Malvaceae)—malvone A (153) as a known phytoalexin; for its isolation, structural elucidation, X-ray structure analysis, and investigations on its fungitoxicity, see Ref. 577 *Structural Elucidation:* Constitution deduced from HRESIMS and from ^{1}H and ^{13}C NMR, including HMBC, HMQC, and NOESY interactions, indicating the presence of a droserone derivative with an additional hydroxy function at C-6, equipped with an O-methyl group (instead of a hydroxy function) at C-3[151] Malvone A (153) as a regioisomer of ancistroquinone B (134), differing from 134 by the position of the methoxy group—in the case of 153 located at C-3, in contrast to 134, where it is at C-6[151]	No cytotoxic activity in vitro against human tumor cells derived from two different B cell malignancies, B cell lymphoma and multiple myeloma[151] Low inhibitory activities against *Leishmania donovani*, but virtually no inhibitory activity against *Plasmodium falciparum*, *Trypanosoma cruzi*, and *T. brucei rhodesiense*[151]
Compd. can likewise be addressed as 5,6-dihydroxy-3-methoxy-2-methyl-1,4-naphthoquinone A. *abbreviatus*: solid callus cultures[151] Red amorphous solid (from CH_2Cl_2): mp 168[151] cf. Ref. 577: Isolation from *Malva sylvestris* as red crystals (from $CHCl_3$): mp 169–172		

Continued

Table 23 Naphthoquinones and tetralones related to naphthylisoquinoline alkaloids.—cont'd

Producing plants and their organs/physical data [mp/°C, $[\alpha]_D^T$/cm^2 g^{-1} 10^{-1} (solvent)]	Isolation, structural elucidation, partial and/or total synthesis, and further investigations	Biological activities
Ancistroquinone B (134) $C_{12}H_{10}O_5$ (234.2) Compd. can likewise be addressed as 3,5-dihydroxy-6-methoxy-2-methyl-1,4-naphthoquinone *A. abbreviatus*: solid callus cultures[151] Orange needles (from H$_2$O/MeCN): mp 218–220[151] cf.: Title compd. prepared as a semisynthetic intermediate en route to ancistroquinone A (53): yellow needles (from MeOH): mp 215–217[314] cf.: Title compd. prepared semisynthetically, as a degradation product of rubromycins: red-yellow needles (from CHCl$_3$): mp 168–170[579]	*Structural Elucidation:* Constitution deduced from HRESIMS and from ^{1}H and ^{13}C NMR, including HMBC and H,H-COSY interactions indicative of a 1,4-naphthoquinone structurally closely related to dioncoquinone B (135), yet with a methoxy group (instead of a hydroxy group) at C-6[151] Ancistroquinone B (134) as a regioisomer of the known natural product malvone A (153),[577] differing from 153 only by the position of the O-methyl group (at C-6, and not at C-3, as in 147)[151] *Total Synthesis:* Total synthesis of 134 starting from 2,3-dimethoxyphenacetyl chloride: condensation with diethyl methylmalonate in the presence of magnesium ethoxide, and cyclization of the resulting ketoester in polyphosphoric acid affording ancistroquinone C (154); selective O-demethylation of 154 using anhydrous AlCl$_3$ in nitrobenzene eventually providing ancistroquinone B (134)[314]	No cytotoxic activity in vitro against human tumor cells derived from two different B cell malignancies, B cell lymphoma and multiple myeloma[151] Moderate inhibitory activities against *Leishmania donovani*, but virtually inactive against *Plasmodium falciparum*, *Trypanosoma cruzi*, and *T. brucei rhodesiense*.[151]

Ancistroquinone C (154)

$C_{13}H_{12}O_5$ (248.2)

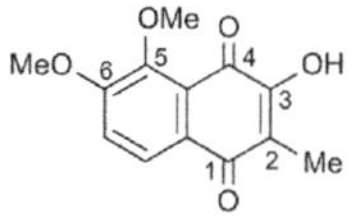

Compd. can likewise be addressed as 3-hydroxy-5,6-dimethoxy-2-methyl-1,4-naphthoquinone

A. abbreviatus: solid callus cultures[151]

Yellow needles (from MeOH): mp 231[151]

cf.: Title compd. prepared as a synthetic intermediate en route to ancistroquinone A (53): orange solid: mp 235–238[314]

cf.: Title compd. prepared by O-methylation of dioncoquinone B (135): yellow solid (from MeCN): mp 217[151]

Structural Elucidation:

Constitution deduced from HRESIMS and from ^{1}H and ^{13}C NMR, including HMBC and H,H-COSY interactions indicative of a 1,4-naphthoquinone structurally closely related to ancistroquinone B (134), yet with a methoxy group (instead of an hydroxy group) at C-5[151]

Semisynthesis and Total Synthesis:

Semisynthesis of 154 by O-methylation of dioncoquinone B (131) with diazomethane[151]

Total synthesis of 154 starting from 2,3-dimethoxyphenacetyl chloride: condensation with diethyl methylmalonate in the presence of magnesium ethoxide, and cyclization of the resulting ketoester in polyphosphoric acid to give 154[314]

No cytotoxic activity in vitro against human tumor cells derived from two different B cell malignancies, B cell lymphoma and multiple myeloma[151]

Virtually inactive against *Plasmodium falciparum*, *Leishmania donovani*, *Trypanosoma cruzi*, and *T. brucei rhodesiense*[151]

Continued

Table 23 Naphthoquinones and tetralones related to naphthylisoquinoline alkaloids.—cont'd

Name, formula (mol. wt.), and structure	Producing plants and their organs/physical data [mp/°C, $[\alpha]_D^T$/°cm^2 g^{-1} 10^{-1} (solvent)]	Isolation, structural elucidation, partial and/or total synthesis, and further investigations	Biological activities
Ancistroquinone D (157) $C_{13}H_{12}O_6$ (264.2) Compd. can likewise be addressed as 3,8-dihydroxy-5,6-dimethoxy-2-methyl-1,4-naphthoquinone	*A. abbreviatus*: solid callus cultures[151] Red needles (from MeOH): mp 218–220[151]	*Structural Elucidation:* Constitution deduced from HRESIMS and from ^{1}H and ^{13}C NMR, including HMBC and NOESY interactions indicative of a fourfold oxygenated 2-methyl-1,4-naphthoquinone, equipped with two methoxy groups at C-5 and C-6, and two free hydroxy functions at C-3 and C-8[151]	No cytotoxic activity in vitro against human tumor cells derived from two different B cell malignancies, B cell lymphoma and multiple myeloma[151] Low inhibitory activities against *Leishmania donovani*[151] Virtually no inhibitory activities against *Plasmodium falciparum*, *Trypanosoma cruzi*, and *T. brucei rhodesiense*[151]

Ancistroquinone E (155)

$C_{12}H_8O_7$ (264.2)

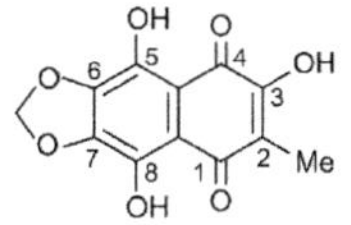

Compd. can also be addressed as 3,5,8-trihydroxy-2-methyl-6,7-methylenedioxy-1,4-naphthoquinone or as 5-O-demethylnepenthone A

A. abbreviatus: solid callus cultures[151]

Red amorphous solid (from MeOH): mp 242 (subl)[151]

Structural Elucidation:

Constitution deduced from HRESIMS and from ^{1}H and ^{13}C NMR, including HMBC interactions indicative of a 3-hydroxy-2-methyl-1,4-naphthoquinone with two chelated hydroxy functions in positions 5 and 8, and a methylenedioxy group bridging the two oxygens at C-6 and C-7—thus, ancistroquinone E (155) being the 5-O-demethyl analogue of the known natural product nepenthone A (156),[578] likewise found to occur in cell cultures of *A. abbreviatus*.[151]

No cytotoxic activity in vitro against human tumor cells derived from two different B cell malignancies, B cell lymphoma and multiple myeloma[151]

Moderate inhibitory activities against *Leishmania donovani*[151]

Virtually no inhibitory activity against *Plasmodium falciparum*, *Trypanosoma cruzi*, and *T. brucei rhodesiense*[151]

Continued

Table 23 Naphthoquinones and tetralones related to naphthylisoquinoline alkaloids.—cont'd

Name, formula (mol. wt.), and structure	Producing plants and their organs/physical data [mp/°C, $[\alpha]_D^T$/°cm^2 g^{-1} 10^{-1} (solvent)]	Isolation, structural elucidation, partial and/or total synthesis, and further investigations	Biological activities
Ancistroquinone F (158) $C_{13}H_{10}O_6$ (262.2) Compd. can also be addressed as 3-hydroxy-5-methoxy-2-methyl-6,7-methylenedioxy-1,4-naphthoquinone	*A. abbreviatus*: solid callus cultures[151] Yellow amorphous solid (from MeOH): mp 220 (dec)[151]	*Structural Elucidation:* Constitution deduced from HRESIMS and from ^{1}H and ^{13}C NMR, including HMBC interactions indicative of a 3-hydroxy-2-methyl-1,4-naphthoquinone possessing a methoxy function at C-5 and a methylenedioxy group to bridge C-6 and C-7[151]	No cytotoxic activity in vitro against human tumor cells derived from two different B cell malignancies, B cell lymphoma and multiple myeloma[151] Virtually no inhibitory activity against *Plasmodium falciparum*, *Leishmania donovani*, *Trypanosoma cruzi*, and *T. brucei rhodesiense*[151]

Nepenthone A (156)

$C_{13}H_{10}O_7$ (278.2)

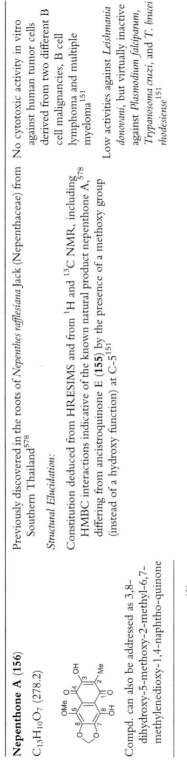

Compd. can also be addressed as 3,8-dihydroxy-5-methoxy-2-methyl-6,7-methylenedioxy-1,4-naphtho-quinone

A. abbreviatus: solid callus cultures[151]

Red amorphous solid (from CH$_2$Cl$_2$): mp 242 (dec)[151]

cf. Ref. 578: Isolation from *Nepenthes rafflesiana* as feathery orange needles (from acetone): mp 255–260 (dec with shrinkage from 245)

Previously discovered in the roots of *Nepenthes rafflesiana* Jack (Nepenthaceae) from Southern Thailand[578]

Structural Elucidation:

Constitution deduced from HRESIMS and from ^{1}H and ^{13}C NMR, including HMBC interactions indicative of the known natural product nepenthone A,[578] differing from ancistroquinone E (**155**) by the presence of a methoxy group (instead of a hydroxy function) at C-5[151]

No cytotoxic activity in vitro against human tumor cells derived from two different B cell malignancies, B cell lymphoma and multiple myeloma[151]

Low activities against *Leishmania donovani*, but virtually inactive against *Plasmodium falciparum*, *Trypanosoma cruzi*, and *T. brucei rhodesiense*[151]

Continued

Table 23 Naphthoquinones and tetralones related to naphthylisoquinoline alkaloids.—cont'd

Name, formula (mol. wt.), and structure	Producing plants and their organs/physical data [mp/°C, $[\alpha]_D^T$/°cm^2 g^{-1} 10^{-1} (solvent)]	Isolation, structural elucidation, partial and/or total synthesis, and further investigations	Biological activities
cis-Isoshinanolone (18) $C_{11}H_{12}O_3$ (192.2) Compd. can also be addressed as (3R,4R)-4,8-dihydroxy-3-methyl-3,4-dihydro-2H-naphthalen-1-one A. abbreviatus: root[57] A. barteri: roots[5] A. likoko: roots[176] A. ileboensis: roots[406] T. peltatum: cell cultures[153,268] D. thollonii: roots[5,6,68,693–695] H. dawei: stem bark[5,6,297,696]	cis-Isoshinanolone (**18**) as a widely distributed known natural tetralone isolated from various Dioncophyllaceae[5,6,153,268,297,693–696] and Ancistrocladaceae[5,176,406,667] plant species cis-Isoshinanolone (**18**) with 3R,4R-configuration and its three further possible stereoisomeric forms, i.e., with 3R,4S- (trans), 3S,4R- (trans), and 3S,4S-configuration (cis), also known from other plant families, such as Plumbaginaceae,[698–700] Nepenthaceae,[692,701] Iridaceae,[702] and Ebenaceae[697,703–706] Identification in *Dioncophyllum thollonii* (Dioncophyllaceae) by HPLC-NMR online investigations on a root extract, applying time-slice experiments to distinguish between the two co-eluting diastereomers, cis- and trans-isoshinanolone (cis-**18** and trans-**18**)[68] Identification in *Habropetalum dawei* (Dioncophyllaceae): Constitution and complete stereostructure of cis-**18** established online, directly from crude extracts, by LC-MS/MS, LC-NMR, and LC-ECD, and by comparison of the data with those obtained from synthetic reference material[297] *Structural Elucidation and Stereoanalysis:* Attribution of the absolute configuration of the cis- and trans-diastereomers of isoshinanolone (and their enantiomers), from the known relative configuration at C-3 versus C-4 (through NMR), and from a newly developed	Fish-stunning activity occurring at a concentration of 25 ppm of **18**, stunning of the test fish *Barbus liberiensis* in a few minutes and killing within 45 min[696] Potent larvicidal activity against the yellow fever mosquito *Aedes aegypti*[699] Slight inhibitory activity in vitro against *Plasmodium falciparum*[692]

A. heyneanus: roots,[5] cell cultures[269,560]	ruthenium-catalyzed oxidative degradation to give chiral 2-methylsuccinic acid as a stereochemically well known and easily analyzable degradation product[693]
A. cochinchinensis: leaves[667,668]	Ultimate proof of the absolute configuration of (+)-isoshinanolone as 3R,4R, by X-ray diffraction analysis of its 5,7-dibromo derivative[695]—in agreement with previous ECD investigations on a dibenzoate derivative[697] and with the results from the above-described oxidative degradation[693]
Oil: $[\alpha]_D^{20} +20.0$ (c 1.00, CHCl$_3$)[695]	
Oil: $[\alpha]_D^{25} +22.2$ (c 1.00, CHCl$_3$)[693]	
Oil: bp 130 °C at 0.8 mm Hg; $[\alpha]_D^{20} +33.0$ (c 0.97, CHCl$_3$)[696]	Development of an HPLC-based stereoanalytical device for the rapid discrimination between the four possible stereoisomeric forms of isoshinanolone, i.e., with 3R,4R(cis)-, 3R,4S(trans)-, 3S,4R(cis)-, and 3S,4S(trans)-configuration, by chromatography on a chiral stationary phase[693]
Oil: $[\alpha]_D^{20} +20.0$ (c 1.00, CHCl$_3$)[5,374]	
Oil: $[\alpha]_D^{25} +46.7$ (c 0.60, CHCl$_3$)[53]	
Oil: $[\alpha]_D^{25} +25.4$ (c 0.78, CHCl$_3$)[667]	Development of a chiroptical on-line stereoanalytical device for the discrimination of all four stereoisomeric forms of isoshinanolone by HPLC on a chiral phase coupled to ECD spectroscopy, providing the ECD spectra of the four stereoisomers directly from stereoisomeric mixtures (even applicable to the analysis in crude plant extracts)[695]
Semisynthesis of the four possible stereoisomers of isoshinanolone by reduction of plumbagin (16) with LiAlH$_4$, giving rise to the cis- and trans-isomers (ratio ca. 1:4), both as racemates[693,697]	
rac-cis-Isoshinanolone: yellow semi-solid (from MeOH/water)[693]	

Continued

Table 23 Naphthoquinones and tetralones related to naphthylisoquinoline alkaloids.—cont'd

Name, formula (mol. wt.), and structure		
Producing plants and their organs/physical data [mp/°C, $[\alpha]_D^T$/cm^2 g^{-1} 10^{-1} (solvent)]	Isolation, structural elucidation, partial and/or total synthesis, and further investigations	Biological activities

Further Investigations:

Formation of *cis*-**18** (3R,4R) from six acetate units via a β-pentaketo precursor according to the F-type folding mode[612]—proof of the acetogenic origin of *cis*-**18** by feeding experiments with $^{13}C_2$-labeled acetate administered to callus cultures of *A. heyneanus*, using 2D INADEQUATE NMR analysis, assisted by the cryoprobe methodology[80–83,268,269]

Abundant amounts of *cis*-**18**—apparently arising from the naphthalene precursor of naphthylisoquinoline alkaloids, by reduction of the naphthoquinone plumbagin (**16**)—produced by Dioncophyllaceae and Ancistrocladaceae plants, and, in particular, in cell cultures of *T. peltatum*,[5,6,151,268,311] *A. heyneanus*,[5,268,269,560] and *A. abbreviatus*,[151] when exposed to chemical, physical, or biotic stress

cis-Isoshinanolone (*cis*-**18**) as an inducible pathogen repellent involved in the incompatible relationship between the parasitic plant *Cuscuta reflexa* (Convolvulaceae) and *A. heyneanus*, leading to a damage of the haustoria of the parasites and eventually to the degeneration of the parasitic tissues[560]

Acknowledgments

Cordial thanks are due to those numerous enthusiastic students and co-workers of our group, who have strongly promoted the work presented here with great commitment and skill, by the isolation, structure elucidation, and total or partial synthesis of numerous new structurally intriguing mono- and dimeric naphthylisoquinoline alkaloids and related compounds. The research activities of these young academics also included biosynthetic investigations and, in particular, efforts to adapt or even further develop powerful analytical and computational methods like e.g., HPLC-ECD coupling and quantum-chemical ECD calculations. Our particular thank goes to Dr. Shaimaa Fayez for her outstanding phytochemical work on *Ancistrocladus abbreviatus* presented in this review in Sections 5 and 6. For their valuable contributions, we also thank Dr. Stefan Busemann, Dr. Tobias Büttner, Alessia Cacciatore, Dr. Michael Dreyer, Dr. Johan H. Faber, Dr. Torsten Geuder, Prof. Dr. Ralf God, Dr. Roland Götz, Dr. Klaus-Peter Gulden, Prof. Dr. Tanja Gulder, Dr. Christian Günther, Dr. Anastasia Hager, Dr. Andreas Hamm, Yasmin Hemberger, Dr. Barbara Hertlein-Amslinger, Dr. Jörg Holenz, Dr. Andreas Irmer, Dr. Johannes R. Jansen, Dr. Inga Kajahn, Dr. Dagmar Koppler, Dr. Jun Li, Dr. Dietmar Lisch, Dr. Blaise Kimbadi Lombe, Dr. Katja Maksimenka, Dr. Kim Messer, Michael Moos, Prof. Dr. Jean-Pierre Mufusama, Dr. Jörg Mühlbacher, Dr. Michael Ochse, Dr. Matthias Reichert, Dr. Helmut Reuscher, Dr. Heiko Rischer, Dr. Martin Rübenacker, Dr. Markus Rückert, Dr. Stefan Rüdenauer, Dr. Wael Saeb, Dr. Raina Seupel, Dr. Anu Schaumlöffel, PD Dr. Jan Schlauer, Dr. Claudia Steinert, Dr. Ralf Stowasser, Dr. Friedrich Teltschik, Prof. Dr. Dieudonné Tshitenge Tshitenge, Dr. Ralf Weirich, Dr. Matthias Wenzel, Barbara Wiesen, Dr. Michael Wohlfarth, Dr. Kristina Wolf, Dr. Minjuan Xu, Rainer Zagst, Dr. Guoliang Zhang, and, in particular, Dr. Torsten Bruhn—the names of all involved co-workers can also be seen from the literature cited. Furthermore, we are indebted to Manuela Michel, Michaela Schraut, Stefanie Schmitt, and Frank Meyer for their valuable technical assistance. The phytochemical investigations presented here would not have been possible without the engaged help of our competent scientific partners from tropical countries, late Dr. René Haller, Dr. S. A. Robertson (Kenya), Prof. A. S. Sankara Narayanan, Prof. Dr. M. R. Almeida, Prof. Dr. N. J. De Souza (India), Prof. Dr. Hamid A. Hadi (Malaysia), Prof. Dr. Luu Hoang Ngoc (Vietnam), Prof. Dr. Jun Wu (PR China), and, in particular late Prof. Dr. Laurent Aké Assi (Ivory Coast) and Prof. Dr. Virima Mudogo (Democratic Republic of the Congo). Moreover, we are especially grateful to Prof. Dr. Günther Heubl (Germany) and Prof. Dr. Harald Meimberg (Austria) for their long-term cooperation focussing on the molecular phylogeny, chemotaxonomy, and botanical classification of *Ancistrocladus* species. Our special thanks also go to Prof. Dr. Markus Riederer und Dr. Gerd Vogg for their generous hospitality in allowing us to cultivate and study more closely *Triphyophyllum peltatum* and numerous species of *Ancistrocladus* in the Botanical Garden of the University of Würzburg. Furthermore, we gratefully acknowledge the excellent horticultural expertise and continuous support and commitment of all involved gardeners (in particular Andreas Kreiner and Friedrich Thiele) regarding the cultivation of these fascinating lianas from seeds to adult plants. Moreover, we want to thank Prof. Dr. Reto Brun, Dr. Marcel Kaiser, Prof. Dr. Kurt Hostettmann (Switzerland), Prof. Dr. Thomas Efferth, Dr. Mona Dawood, Prof. Dr. Heiko Ihmels, Prof. Harald Wajant, Prof. Dr. Peter Proksch, late Prof. Dr. Franz-Christian Czygan (Germany), Prof. Dr. Shashank Kumar (India), Dr. Łukasz Adaszek (Poland), Prof. Dr. Michael R. Boyd (USA), and Prof. Dr. Suresh Awale (Japan) for fruitful and intense cooperations regarding the evaluation of the bioactivities of naphthylisoquinoline alkaloids. Furthermore, we are grateful to Prof. Reiko Kuroda (Japan) for experimental solid-state ECD measurements, late Dr. Karl Peters and Dr. Christian Burschka (Germany) for X-ray diffraction analyses, Prof. Dr. Wolfgang Kiefer and Prof. Dr. Jürgen Popp (Germany) for FT-Raman investigations, Prof. Dr. Axel Haase and Prof. Dr. Peter M. Jakob (Germany) for nuclear magnetic microscopy and NMR chemical-shift imaging studies, and Prof. Dr. Ulrich Zimmermann (Germany) for light and fluorescence microscopy investigations. Finally, we gratefully acknowledge generous financial support by the Deutsche Forschungsgemeinschaft (projects Br 699/14-1 and 14-2; "Molecular Phylogeny and Chemotaxonomy of the Ancistrocladaceae Plant Family," Collaborative Research Center SFB 630 "Recognition, Preparation, and Functional Analysis of Agents against Infectious Diseases," Collaborative Research Center SFB 251 "Ecology, Physiology, and Biochemistry of Plants and Animals under Stress," Clinical Research Unit KFO 216 "Characterization of the Oncogenic Signaling Network in Multiple Myeloma: Development of Targeted Therapies," Priority Program SPP 1152 "Evolution of Metabolic Diversity," and Research Training Group "NMR in vivo and in vitro for Biological Basic Research"). This work was also funded by grants from the Bundesministerium für Bildung und Forschung (BMBF), project no. 0310722, the BASF AG, the Fonds der Chemischen Industrie (FCI), and the Max Buchner Foundation.

References

1. Feineis, D.; Bringmann, G. Asian *Ancistrocladus Lianas* As Creative Producers of Naphthylisoquinoline Alkaloids. In *Progress in the Chemistry of Organic Natural Products*; Kinghorn, A. D., Falk, H., Gibbons, S., Asakawa, Y., Liu, Y.-K., Dirsch, V. M., Eds.; Vol. 119; Springer Nature Switzerland AG: Cham, Switzerland, 2023; pp. 1–335.
2. Tajuddeen, N.; Bringmann, G. N,C-Coupled Naphthylisoquinoline Alkaloids: A Versatile New Class of Axially Chiral Natural Products. *Nat. Prod. Rep.* **2021**, *38*, 2154–2186.
3. Lombe, B. K.; Feineis, D.; Bringmann, G. Dimeric Naphthylisoquinoline Alkaloids: Polyketide-Derived Axially Chiral Bioactive Quateraryls. *Nat. Prod. Rep.* **2019**, *36*, 1513–1545.
4. Shang, X. F.; Yang, C. J.; Morris-Natschke, S. L.; Li, J. C.; Yin, X. D.; Liu, Y. Q.; Peng, J. W.; Goto, M.; Zhang, J. Y.; Lee, K. H. Biologically Active Isoquinoline Alkaloids Covering 2014-2018. *Med. Res. Rev.* **2020**, *40*, 2212–2289.
5. Bringmann, G.; Pokorny, F. The Naphthylisoquinoline Alkaloids. In *The Alkaloids. Chemistry and Pharmacology*; Cordell, G. A., Ed.; Vol. 46; Academic Press: San Diego, 1995; pp. 127–271.
6. Bringmann, G.; François, G.; Aké Assi, L.; Schlauer, J. The Alkaloids of *Triphyophyllum peltatum* (Dioncophyllaceae). *Chimia* **1998**, *52*, 18–28.
7. Ibrahim, S. R. M.; Mohamed, G. A. Naphthylisoquinoline Alkaloids Potential Drug Leads. *Fitoterapia* **2015**, *106*, 194–225.
8. Bringmann, G.; Rübenacker, M.; Jansen, J. R.; Scheutzow, D.; Aké Assi, L. On the Structure of the Dioncophyllaceae Alkaloids Dioncophylline A ("Triphyophylline") and "O-Methyl-triphyophylline". *Tetrahedron Lett.* **1990**, *31*, 639–642.
9. Bringmann, G.; Jansen, J. R.; Reuscher, H.; Rübenacker, M.; Peters, K.; von Schnering, H. G. First Total Synthesis of (-)-Dioncophylline A ("Triphyophylline") and of Selected Stereoisomers: Complete (Revised) Stereostructure. *Tetrahedron Lett.* **1990**, *31*, 643–646.
10. Awale, S.; Dibwe, D. F.; Balachandran, C.; Fayez, S.; Feineis, D.; Lombe, B. K.; Bringmann, G. Ancistrolikokine E$_3$, a 5,8'-Coupled Naphthylisoquinoline Alkaloid, Eliminates the Tolerance of Cancer Cells to Nutrition Starvation by Inhibition of the Akt/mTOR/Autophagy Signaling Pathway. *J. Nat. Prod.* **2018**, *81*, 2282–2291.

11. Tshitenge, D. T.; Bruhn, T.; Feineis, D.; Schmidt, D.; Mudogo, V.; Kaiser, M.; Brun, R.; Würthner, F.; Awale, S.; Bringmann, G. Ealamines A-H, a Series of Naphthylisoquinolines With the Rare 7,8′-Coupling Site, from the Congolese Liana *Ancistrocladus ealaensis*, Targeting Pancreatic Cancer Cells. *J. Nat. Prod.* **2019**, *82*, 3150–3164.
12. Bringmann, G.; Zhang, G.; Büttner, T.; Bauckmann, G.; Kupfer, T.; Braunschweig, H.; Brun, R.; Mudogo, V. Jozimine A$_2$: The First Dimeric Dioncophyllaceae-Type Naphthylisoquinoline Alkaloid, With Three Chiral Axes and High Antiplasmodial Activity. *Chem. -Eur. J.* **2013**, *19*, 916–923.
13. Manfredi, K. P.; Blunt, J. W.; Cardellina, J. H., II; McMahon, J. B.; Pannell, L. L.; Cragg, G. M.; Boyd, M. R. Novel Alkaloids From the Tropical Plant *Ancistrocladus abbreviatus* Inhibit Cell Killing by HIV-1 and HIV-2. *J. Med. Chem.* **1991**, *34*, 3402–3405.
14. Boyd, M. R.; Hallock, Y. F.; Cardellina, J. H., II; Manfredi, K. P.; Blunt, J. W.; McMahon, J. B.; Buckheit, R. W., Jr.; Bringmann, G.; Schäffer, M.; Cragg, G. M.; Thomas, D. W.; Jato, J. G. Anti-HIV Michellamines From *Ancistrocladus korupensis*. *J. Med. Chem.* **1994**, *37*, 1740–1745.
15. Lombe, B. K.; Bruhn, T.; Feineis, D.; Mudogo, V.; Brun, R.; Bringmann, G. Antiprotozoal Spirombandakamines A$_1$ and A$_2$, Fused Naphthylisoquinoline Dimers From a Congolese *Ancistrocladus* Plant. *Org. Lett.* **2017**, *19*, 6740–6743.
16. Bringmann, G.; Kajahn, I.; Reichert, M.; Pedersen, S. E. H.; Faber, J. H.; Gulder, T.; Brun, R.; Christensen, S. B.; Ponte-Sucre, A.; Moll, H.; Heubl, G.; Mudogo, V. Ancistrocladinium A and B, the First *N,C*-Coupled Naphthyldihydroisoquinoline Alkaloids, From a Congolese *Ancistrocladus* Species. *J. Org. Chem.* **2006**, *71*, 9348–9356.
17. Bringmann, G.; Günther, C.; Ochse, M.; Schupp, O.; Tasler, S. Biaryls in Nature: A Multi-Facetted Class of Stereochemically, Biosynthetically, and Pharmacologically Intriguing Secondary Metabolites. In *Progress in the Chemistry of Organic Natural Products*; Herz, W., Falk, H., Kirby, G. W., Moore, R. E., Eds.; Vol. 82; Springer-Verlag: Wien, NY, 2001.
18. Gereau, R. E. Typification of Names of *Ancistrocladus* Wallich (Ancistrocladaceae). *Novon* **1997**, *7*, 242–245.
19. Cheek, M. A Synoptic Revision of *Ancistrocladus* (Ancistrocladaceae) in Africa, With a NEW species From Western Cameroon. *Kew Bull.* **2000**, *55*, 871–882.
20. Taylor, C. M.; Gereau, R. E.; Walters, G. M. Revision of *Ancistrocladus* Wall. Ancistrocladaceae. *Ann. MO Bot. Gard.* **2005**, *92*, 360–399.
21. Thomas, D. W.; Gereau, R. E. *Ancistrocladus korupensis* (Ancistrocladaceae): A New Species of Liana From Cameroon. *Novon* **1993**, *3*, 494–498.
22. Rischer, H.; Heubl, G.; Meimberg, H.; Dreyer, M.; Hadi, H. A.; Bringmann, G. *Ancistrocladus benomensis* (Ancistrocladaceae): A New Species From Peninsular Malaysia. *Blumea* **2005**, *50*, 357–365.
23. Heubl, G.; Turini, F.; Mudogo, V.; Kajahn, I.; Bringmann, G. *Ancistrocladus ileboensis* (DR Congo), A New Liana With Unique Alkaloids. *Plant Ecol. Evol.* **2010**, *143*, 63–69.
24. Meimberg, H.; Rischer, H.; Turini, F. G.; Chamchumroon, V.; Dreyer, M.; Sommaro, M.; Bringmann, G.; Heubl, G. Evidence for Species Differentiation Within the *Ancistrocladus tectorius* Complex (Ancistrocladaceae) in Southeast Asia: A Molecular Approach. *Plant Syst. Evol.* **2010**, *284*, 77–98.
25. Turini, F. G.; Steinert, C.; Heubl, G.; Bringmann, G.; Lombe, B. K.; Mudogo, V.; Meimberg, H. Microsatellites Facilitate Species Delimitation in Congolese *Ancistrocladus* (Ancistrocladaceae), a Genus With Pharmacologically Potent Naphthylisoquinoline Alkaloids. *Taxon* **2014**, *63*, 329–341.

26. Airy Shaw, H. K. On the Dioncophyllaceae, a Remarkable New Family of Flowering Plants. *Kew Bull.* **1951**, *6*, 327–347.
27. Govindachari, T. R.; Parthasarathy, P. C. Ancistrocladine, A New Type of Isoquinoline Alkaloid From *Ancistrocladus heyneanus*. *Tetrahedron* **1971**, *27*, 1013–1026.
28. Govindachari, T. R.; Parthasarathy, P. C. Alkaloids of Ancistrocladaceae. *Heterocycles* **1977**, *7*, 661–684.
29. Bringmann, G.; Spuziak, J.; Faber, J. H.; Gulder, T.; Kajahn, I.; Dreyer, M.; Heubl, G.; Brun, R.; Mudogo, V. Six Naphthylisoquinoline Alkaloids and a Related Benzopyranone From a Congolese *Ancistrocladus* Species Related to *Ancistrocladus congolensis*. *Phytochemistry* **2008**, *69*, 1065–1075.
30. Kavatsurwa, S. M.; Lombe, B. K.; Feineis, D.; Dibwe, D. F.; Maharaj, V.; Awale, S.; Bringmann, G. Ancistroyafungines A-D, 5,8'- and 5,1'-Coupled Naphthylisoquinoline Alkaloids From a Congolese *Ancistrocladus* Species, With Antiausterity Activities Against Human PANC-1 Pancreatic Cancer Cells. *Fitoterapia* **2018**, *130*, 6–16.
31. Lombe, B. K.; Feineis, D.; Mudogo, V.; Brun, R.; Awale, S.; Bringmann, G. Michellamines A$_6$ and A$_7$, and Further Mono- and Dimeric Naphthylisoquinoline Alkaloids From a Congolese *Ancistrocladus* Liana and Their Antiausterity Activities Against Pancreatic Cancer Cells. *RSC Adv.* **2018**, *8*, 5243–5254.
32. Bringmann, G.; Lombe, B. K.; Steinert, C.; Ndjoko Ioset, K.; Brun, R.; Turini, F.; Heubl, G.; Mudogo, V. Mbandakamines A and B, Unsymmetrically Coupled Dimeric Naphthylisoquinoline Alkaloids, From a Congolese *Ancistrocladus Species*. *Org. Lett.* **2013**, *15*, 2590–2593.
33. Lombe, B. K.; Bruhn, T.; Feineis, D.; Mudogo, V.; Brun, R.; Bringmann, G. Cyclombandakamines A$_1$ and A$_2$, Oxygen-Bridged Naphthylisoquinoline Dimers From a Congolese *Ancistrocladus* Liana. *Org. Lett.* **2017**, *19*, 1342–1345.
34. Lombe, B. K.; Feineis, D.; Mudogo, V.; Kaiser, M.; Bringmann, G. Spirombandakamine A$_3$ and Cyclombandakamines A$_8$ and A$_9$, Polycyclic Naphthylisoquinoline Dimers, With Antiprotozoal Activity, From a Congolese *Ancistrocladus* Plant. *J. Nat. Prod.* **2021**, *84*, 1335–1344.
35. Mufusama, J. P.; Feineis, D.; Mudogo, V.; Kaiser, M.; Brun, R.; Bringmann, G. Antiprotozoal Dimeric Naphthylisoquinolines, Mbandakamines B$_3$ and B$_4$, and Related 5,8'-Coupled Monomeric Alkaloids, Ikelacongolines A-D, From a Congolese *Ancistrocladus* Liana. *RSC Adv.* **2019**, *9*, 12034–12046.
36. François, G.; Timperman, G.; Eling, W.; Aké Assi, L.; Holenz, J.; Bringmann, G. Naphthylisoquinoline Alkaloids Against Malaria: Evaluation of the Curative Potentials of Dioncophylline C and Dioncopeltine A Against *Plasmodium berghei* In Vivo. *Antimicrob. Agents Chemother.* **1997**, *41*, 2533–2539.
37. Moyo, P.; Shamburger, W.; van der Watt, M. E.; Reader, J.; de Sousa, A. C. C.; Egan, T. J.; Maharaj, V. J.; Bringmann, G.; Birkholtz, L. M. Naphthylisoquinoline Alkaloids, Validated as Hit Multistage Antiplasmodial Natural Products. *Int. J. Parasitol.: Drugs Drug Resist.* **2020**, *13*, 51–58.
38. Tajuddeen, N.; Van Heerden, F. R. Antiplasmodial Natural Products: An Update. *Malar. J.* **2019**, *18*, 404.
39. Kingston, D. G. I.; Cassera, M. B. Antimalarial Natural Products. In *Progress in the Chemistry of Organic Natural Products*; Kinghorn, A. D., Falk, H., Gibbons, S., Asakawa, Y., Liu, Y.-K., Dirsch, V. M., Eds.; Vol. 117; Springer Nature Switzerland AG: Cham, Switzerland, 2022; pp. 1–106.
40. Kumar, V.; Mahajan, A.; Chibale, K. Synthetic Medicinal Chemistry of Selected Antimalarial Natural Products. *Bioorg. Med. Chem.* **2009**, *17*, 2236–2275.
41. Onguéné, P. A.; Ntie-Kang, F.; Lifongo, L. L.; Ndom, J. C.; Sippl, W.; Meva'a Mbaza, L. The Potential of Anti-Malarial Compounds Derived From African Medicinal Plants. Part I: A Pharmacological Evaluation of Alkaloids and Terpenoids. *Malar. J.* **2013**, *12*, 449.

42. Ponte-Sucre, A.; Faber, J. H.; Gulder, T.; Kajahn, I.; Pedersen, S. E. H.; Schultheis, M.; Bringmann, G.; Moll, H. Activities of Naphthylisoquinoline Alkaloids and Synthetic Analogs Against *Leishmania major*. *Antimicrob. Agents. Chemother.* **2007**, *51*, 188–194.
43. Ponte-Sucre, A.; Gulder, T.; Wegehaupt, A.; Albert, C.; Rikanović, C.; Schaeflein, L.; Frank, A.; Schultheis, M.; Unger, M.; Holzgrabe, U.; Bringmann, G.; Moll, H. Structure-Activity Relationship and Studies on the Molecular Mechanism of Leishmanicidal *N,C*-Coupled Arylisoquinolinium Salts. *J. Med. Chem.* **2009**, *52*, 626–636.
44. Ponte-Sucre, A.; Gulder, T.; Gulder, T. A. M.; Vollmers, G.; Bringmann, G.; Moll, H. Alterations to the Structure of *Leishmania major* Induced by *N*-Arylisoquinolines Correlate With Compound Accumulation and Disposition. *J. Med. Microbiol.* **2010**, *59*, 69–75.
45. Simoben, C. V.; Ntie-Kang, F.; Akone, S. H.; Sippl, W. Compounds From African Medicinal Plants With Activities Against Selected Parasitic Diseases: Schistosomiasis, Trypanosomiasis and Leishmaniasis. *Nat. Prod. Bioprospect.* **2018**, *8*, 151–169.
46. Scotti, M. T.; Scotti, L.; Ishiki, H.; Ribeiro, F. F.; Duarte da Cruz, R. M.; de Oliveira, M. P.; Mendonça, F. J. B., Jr. Natural Products as a Source for Antileishmanial and Antitrypanosomal Agents. *Comb. Chem. High Throughput Screening* **2016**, *19*, 537–553.
47. Singh, N.; Mishra, B. B.; Bajpai, S.; Singh, R. K.; Tiwari, V. K. Natural Product Based Leads to Fight Against Leishmaniasis. *Bioorg. Med. Chem.* **2014**, *22*, 18–45.
48. Izumi, E.; Ueda-Nakamura, T.; Dias Filho, B. P.; Veiga, V. F., Jr.; Nakamura, C. V. Natural Products and Chagas' Disease: A Review of Plant Compounds Studied for Activity Against *Trypanosoma cruzi*. *Nat. Prod. Rep.* **2011**, *28*, 809–823.
49. Zofou, D.; Ntie-Kang, F.; Sippl, W.; Efange, S. M. N. Bioactive Natural Products Derived From the Central African Flora Against Neglected Tropical Diseases and HIV. *Nat. Prod. Rep.* **2013**, *30*, 1098–1120.
50. Ntie-Kang, F.; Lifongo, L. L.; Simoben, C. V.; Babiaka, S. B.; Sippl, W.; Meva'a Mbaze, L. The Uniqueness and Therapeutic Value of Natural Products from West African Medicinal Plants. Part I: Uniqueness and Chemotaxonomy. *RSC Adv.* **2014**, *4*, 28728–28755.
51. Li, J.; Seupel, R.; Feineis, D.; Mudogo, V.; Kaiser, M.; Brun, R.; Brünnert, D.; Chatterjee, M.; Seo, E. J.; Efferth, T.; Bringmann, G. Dioncophyllines C$_2$, D$_2$, and F, and Related Naphthylisoquinoline Alkaloids From the Congolese Liana *Ancistrocladus ileboensis* With Potent Activities Against *Plasmodium falciparum* and Against Multiple Myeloma and Leukemia Cell Lines. *J. Nat. Prod.* **2017**, *80*, 443–458.
52. Li, J.; Tajuddeen, N.; Feineis, D.; Mudogo, V.; Kaiser, M.; Seo, E. J.; Efferth, T.; Bringmann, G. Jozibrevine D from *Ancistrocladus ileboensis*, the Fifth Alkaloid in a Series of Six Possible Atropo-Diastereomeric Naphthylisoquinoline Dimers, Showing Antiparasitic and Antileukemic Activities. *Bioorg. Med. Chem. Lett.* **2023**, *86*, 129258.
53. Fayez, S.; Feineis, D.; Mudogo, V.; Seo, E. J.; Efferth, T.; Bringmann, G. Ancistrolikokine I and Further 5,8′-Coupled Naphthylisoquinoline Alkaloids From the Congolese Liana *Ancistrocladus likoko* and Their Cytotoxic Activities Against Drug-Sensitive and Multi-Drug Resistant Human Leukemia Cells. *Fitoterapia* **2018**, *129*, 114–125.
54. Fayez, S.; Feineis, D.; Aké Assi, L.; Seo, E. J.; Efferth, T.; Bringmann, G. Ancistrobreveines A-D and Related Dehydrogenated Naphthylisoquinoline Alkaloids With Antiproliferative Activities Against Leukemia Cells, From the West African Liana *Ancistrocladus abbreviatus*. *RSC Adv.* **2019**, *9*, 15738–15748.
55. Fayez, S.; Bruhn, T.; Feineis, D.; Aké Assi, L.; Awale, S.; Bringmann, G. Ancistrosecolines A-F, Unprecedented *Seco*-Naphthylisoquinoline Alkaloids From the Roots of *Ancistrocladus abbreviatus*, With Apoptosis-Inducing Potential Against HeLa Cancer Cells. *J. Nat. Prod.* **2020**, *83*, 1139–1151.

56. Fayez, S.; Cacciatore, A.; Sun, S.; Kim, M.; Aké Assi, L.; Feineis, D.; Awale, S.; Bringmann, G. Ancistrobrevidines A-C and Related Naphthylisoquinoline Alkaloids With Cytotoxic Activities Against HeLa and Pancreatic Cancer Cells, From the Liana *Ancistrocladus abbreviatus*. *Bioorg. Med. Chem.* **2021**, *30*, 115950.
57. Fayez, S.; Feineis, D.; Aké Assi, L.; Kaiser, M.; Brun, R.; Awale, S.; Bringmann, G. Ancistrobrevines E-J and Related Naphthylisoquinoline Alkaloids From the West African Liana *Ancistrocladus abbreviatus* With Inhibitory Activities Against *Plasmodium falciparum* and PANC-1 Human Pancreatic Cancer Cells. *Fitoterapia* **2018**, *131*, 245–259.
58. Fayez, S.; Li, J.; Feineis, D.; Aké Assi, L.; Kaiser, M.; Brun, R.; Anany, M. A.; Wajant, H.; Bringmann, G. A Near-Complete Series of Four Atropisomeric Jozimine-A$_2$ Type Naphthylisoquinoline Dimers With Antiplasmodial and Cytotoxic Activities and Related Alkaloids From *Ancistrocladus abbreviatus*. *J. Nat. Prod.* **2019**, *82*, 3033–3046.
59. Fayez, S.; Bruhn, T.; Feineis, D.; Akè Assi, L.; Kushwaha, P. P.; Kumar, S.; Bringmann, G. Naphthylisoindolinone Alkaloids: the First Ring-Contracted Naphthylisoquinolines, From the Tropical Liana *Ancistrocladus abbreviatus*, With Cytotoxic Activity. *RSC Adv.* **2022**, *12*, 28916–28928.
60. Tajuddeen, N.; Fayez, S.; Kushwaha, P. P.; Feineis, D.; Aké Assi, L.; Kumar, S.; Bringmann, G. Ancistrobrevinium A, the First N-Methylated, Cationic Naphthylisoquinoline Alkaloid, From the Tropical Liana *Ancistrocladus abbreviatus* (Ancistrocladaceae). *Nat. Prod. Res.* **2023**. https://doi.org/10.1080/14786419.2023.2194648.
61. Fayez, S.; Cacciatore, A.; Maneenet, J.; Nguyen, H. H.; Tajuddeen, N.; Feineis, D.; Aké Assi, L.; Awale, S.; Bringmann, G. Dioncophyllidine E: the First Configurationally Semi-Stable, 7,3′-Coupled Naphthyldihydroisoquinoline Alkaloid, From *Ancistrocladus abbreviatus*, With Antiausterity Activity Against PANC-1 Human Pancreatic Cancer Cells. *Bioorg. Med. Chem. Lett.* **2023**, *86*, 129234.
62. Bringmann, G.; Lisch, D.; Reuscher, H.; Aké Assi, L.; Günther, K. Atrop-Diastereomer Separation by Racemate Resolution Techniques: N-Methyldioncophylline A and Its 7-epimer From *Ancistrocladus abbreviatus*. *Phytochemistry* **1991**, *30*, 1307–1310.
63. Bringmann, G.; Zagst, R.; Reuscher, H.; Aké Assi, L. Ancistrobrevine B, the First Naphthylisoquinoline Alkaloid With 5,8′-Coupling Site, and Related Compounds From *Ancistrocladus abbreviatus*. *Phytochemistry* **1992**, *31*, 4011–4014.
64. Bringmann, G.; Pokorny, F.; Stäblein, M.; Schäffer, M.; Aké Assi, L. Ancistrobrevine C From *Ancistrocladus abbreviatus*: the First Mixed 'Ancistrocladaceae/Dioncophyllaceae-Type' Naphthylisoquinoline Alkaloid. *Phytochemistry* **1993**, *33*, 1511–1515.
65. Bringmann, G.; Weirich, R.; Lisch, D.; Aké Assi, L. Ancistrobrevine D: An Unusual Alkaloid From *Ancistrocladus abbreviatus*. *Planta Med.* **1992**, *58*, A703–A704.
66. Bringmann, G.; Koppler, D.; Scheutzow, D.; Porzel, A. Determination of Configuration at the Biaryl Axes of Naphthylisoquinoline Alkaloids by Long-Range NOE Effects. *Magn. Reson. Chem.* **1997**, *35*, 297–301.
67. Bringmann, G.; Günther, C.; Schlauer, J.; Rückert, M. HPLC-NMR Online Coupling Including the ROESY Technique: Direct Characterization of Naphthylisoquinoline Alkaloids in Crude Plant Extracts. *Anal. Chem.* **1998**, *70*, 2805–2811.
68. Bringmann, G.; Rückert, M.; Messer, K.; Schupp, O.; Louis, A. M. Use of On-Line High-Performance Liquid Chromatography—Nuclear Magnetic Resonance Spectrometry Coupling in Phytochemical Screening Studies: Rapid Identification of Metabolites in *Dioncophyllum thollonii*. *J. Chromatogr. A* **1999**, *837*, 267–272.
69. Bringmann, G.; Geuder, T.; Rübenacker, M.; Zagst, R. A Facile Degradation Procedure for Determination of Absolute Configuration in 1,3-Dimethyltetra- and Dihydroisoquinolines. *Phytochemistry* **1991**, *30*, 2067–2070.

70. Bringmann, G.; God, R.; Schäffer, M. An Improved Degradation Procedure for Determination of the Absolute Configuration in Chiral Isoquinoline and β-Carboline Derivatives. *Phytochemistry* **1996**, *43*, 1393–1403.
71. Bringmann, G.; Bruhn, T.; Maksimenka, K.; Hemberger, Y. The Assignment of Absolute Stereostructures Through Quantum Chemical Circular Dichroism Calculations. *Eur. J. Org. Chem.* **2009**, 2717–2727.
72. Bruhn, T.; Schaumlöffel, A.; Hemberger, Y.; Bringmann, G. SpecDis: Quantifying the Comparison of Calculated and Experimental Electronic Circular Dichroism Spectra. *Chirality* **2013**, *25*, 243–249.
73. Bringmann, G.; Lang, G. Full Absolute Stereostructures of Natural Products Directly From Crude Extracts: The HPLC-MS/MS-NMR-CD 'Triad'. In *Marine Molecular Biotechnology*; Müller, W. E. G., Ed.; Springer Verlag: Berlin, Heidelberg, 2003; pp. 89–116.
74. Bringmann, G.; Gulder, T. A. M.; Reichert, M.; Gulder, T. The Online Assignment of the Absolute Configuration of Natural Products: HPLC-CD in Combination With Quantum Chemical CD Calculations. *Chirality* **2008**, *20*, 628–642.
75. Bringmann, G.; Götz, D.; Bruhn, T. The Online Stereochemical Analysis of Chiral Compounds by HPLC-ECD Coupling in Combination With Quantum Chemical Calculations. In *Comprehensive Chiroptical Spectroscopy—Applications in Stereochemical Analysis of Synthetic Compounds, Natural Products, and Biomolecules*; Berova, N., Polavarapu, P. L., Nakanishi, K., Woody, R. W., Eds.; Vol. 2; John Wiley & Sons, Inc., 2012; pp. 353–386.
76. Bringmann, G.; Breuning, M.; Tasler, S. The Lactone Concept: An Efficient Pathway to Axially Chiral Natural Products and Useful Reagents. *Synthesis* **1999**, 525–558.
77. Bringmann, G.; Menche, D. Stereoselective Total Synthesis of Axially Chiral Natural Products via Biaryl Lactones. *Acc. Chem. Res.* **2001**, *34*, 615–624.
78. Bringmann, G.; Gulder, T.; Gulder, T. A. M.; Breuning, M. Atroposelective Total Synthesis of Axially Chiral Biaryl Natural Products. *Chem. Rev.* **2011**, *111*, 563–639.
79. Tajuddeen, N.; Feineis, D.; Ihmels, H.; Bringmann, G. The Stereoselective Total Synthesis of Axially Chiral Naphthylisoquinoline Alkaloids. *Acc. Chem. Res.* **2022**, *55*, 2370–2383.
80. Bringmann, G.; Wohlfarth, M.; Rischer, H.; Grüne, M.; Schlauer, J. A New Biosynthetic Pathway to Alkaloids in Plants: Acetogenic Isoquinolines. *Angew. Chem.* **2000**, *112*, 1523–1525. *Angew. Chem., Int. Ed.* **2000**, *39*, 1464–1466.
81. Bringmann, G.; Mutanyatta-Comar, J.; Greb, M.; Rüdenauer, S.; Noll, T. F.; Irmer, A. Biosynthesis of Naphthylisoquinoline Alkaloids: Synthesis and Incorporation of an Advanced $^{13}C_2$-Labeled Isoquinoline Precursor. *Tetrahedron* **2007**, *63*, 1755–1761.
82. Bringmann, G.; Irmer, A.; Rüdenauer, S.; Mutanyatta-Comar, J.; Seupel, R.; Feineis, D. 5′-O-Methyldioncophylline D, a 7,8′-Coupled Naphthylisoquinoline Alkaloid From Callus Cultures of *Triphyophyllum peltatum*, and Its Biosynthesis From a Late-Stage Tetrahydroisoquinoline Precursor. *Tetrahedron* **2016**, *72*, 2906–2912.
83. Bringmann, G.; Irmer, A.; Feineis, D.; Gulder, T. A. M.; Fiedler, H. P. Convergence in the Biosynthesis of Acetogenic Natural Products From Plants, Fungi, and Bacteria. *Phytochemistry* **2009**, *70*, 1776–1786.
84. Bentley, K. W. The Isoquinoline Alkaloids. In *Chemistry and Biochemistry of Organic Natural Products*; Ravindranath, B., Ed.; Harwood Academic Publishers: Amsterdam, 1998.
85. Bentley, K. W. β-Phenylethylamines and the Isoquinoline Alkaloids. *Nat. Prod. Rep.* **2006**, *23*, 444–463.
86. Iranshahy, M.; Quinn, R. J.; Iranshahi, M. Biologically Active Isoquinoline Alkaloids With Drug-Like Properties From the Genus *Corydalis*. *RSC Adv.* **2014**, *4*, 15900–15913.

87. Weber, C.; Opatz, T. Bisbenzylisoquinoline Alkaloids. In *The Alkaloids. Chemistry and Biology*; Knölker, H. J., Ed.; Vol. 81; Academic Press: San Diego, 2019; pp. 1–114.
88. Chrzanowska, M.; Grajewska, A.; Rozwadowska, M. D. Asymmetric Synthesis of Isoquinoline Alkaloids: 2004–2015. *Chem. Rev.* **2016**, *116*, 12369–12465.
89. Staunton, J. Biosynthesis of Isoquinoline Alkaloids. *Planta Med.* **1979**, *36*, 1–20.
90. Kutchan, T. M.; Dittrich, H.; Bracher, D.; Zenk, M. H. Enzymology and Molecular Biology of Alkaloid Biosynthesis. *Tetrahedron* **1991**, *47*, 5945–5954.
91. O'Connor, S. E. Alkaloids. In *Comprehensive Natural Products II—Chemistry and Biology*; Mander, E. L., Lui, H. W., Eds.; Vol. 1; Elsevier: Oxford, 2010; pp. 977–1007.
92. O'Connor, S. E. Alkaloids. In *Natural Products in Chemical Biology*; Civjan, N., Ed.; John Wiley & Sons, Inc.: Hoboken, NJ, 2012; pp. 209–238.
93. Stöckigt, J.; Chen, Z.; Ruppert, M. Enzymatic and Chemo-Enzymatic Approaches Towards Natural and Non-Natural Alkaloids: Indoles, Isoquinolines, and Others. *Top. Curr. Chem.* **2010**, *297*, 67–103.
94. Hagel, J. M.; Facchini, P. J. Benzylisoquinoline Alkaloid Metabolism: A Century of Discovery and a Brave New World. *Plant Cell Physiol.* **2013**, *54*, 647–672.
95. Pal, T.; Pal, A. Oxidative Phenol-Coupling: A Key Step for the Biomimetic Synthesis of Many Important Natural Products. *Curr. Sci.* **1996**, *71*, 106–108.
96. Quideau, S.; Deffieux, D.; Pouysegu, L. Oxidative Coupling of Phenols and Phenol Ethers. In *Comprehensive Organic Synthesis*; Knochel, P., Molander, G. A., Eds, 2nd ed.; Vol. 3; Elsevier: Amsterdam, The Netherlands, 2014; pp. 656–740.
97. Bringmann, G.; Rübenacker, M.; Ammermann, E.; Lorenz, G.; Aké Assi, L. Use of Dioncophyllines as Fungicides. US Patent No. 5,260,315, 1993b.
98. Bringmann, G.; Gramatzki, S.; Grimm, C.; Proksch, P. Feeding Deterrency and Growth Retarding Activity of the Naphthylisoquinoline Alkaloid Dioncophylline A Against *Spodoptera littoralis*. *Phytochemistry* **1992**, *31*, 3821–3825.
99. Bringmann, G.; Holenz, J.; Wiesen, B.; Nugroho, B. W.; Proksch, P. Dioncophylline A as a Growth-Retarding Agent Against the Herbivorous Insect *Spodoptera littoralis*: Structure-Activity Relationships. *J. Nat. Prod.* **1997**, *60*, 342–347.
100. François, G.; Van Looveren, M.; Timperman, G.; Chimanuka, B.; Aké Assi, L.; Holenz, J.; Bringmann, G. Larvicidal Activity of the Naphthylisoquinoline Alkaloid Dioncophylline A Against the Malaria Vector *Anopheles stephensi*. *J. Ethnopharmacol.* **1996**, *54*, 125–130.
101. Bringmann, G.; Holenz, J.; Saeb, W.; Aké Assi, L.; Hostettmann, K. Dioncophylline A as a Larvicide Against *Aedes aegypti*. *Pharm. Pharmacol. Lett.* **1999**, *9*, 24–25.
102. Bringmann, G.; Holenz, J.; Aké Assi, L.; Zhao, C.; Hostettmann, K. Molluscicidal Activity of Naphthylisoquinoline Alkaloids From *Triphyophyllum peltatum* and *Ancistrocladus species*. *Planta Med.* **1996**, *62*, 556–557.
103. Bringmann, G.; Holenz, J.; Aké Assi, L.; Hostettmann, K. Molluscicidal Activity (*Biomphalaria glabrata*) of Dioncophylline A: Structure-Activity Investigations. *Planta Med.* **1998**, *64*, 485–486.
104. Meininger, M.; Stowasser, R.; Jakob, P. M.; Schneider, H.; Koppler, D.; Bringmann, G.; Zimmermann, U.; Haase, A. Nuclear Magnetic Microscopy of *Ancistrocladus heyneanus*. *Protoplasma* **1997**, *198*, 210–217.
105. Meininger, M.; Jakob, P. M.; von Kienlin, M.; Koppler, D.; Bringmann, G.; Haase, A. Radial Spectroscopic Imaging. *J. Magn. Reson.* **1997**, *125*, 325–331.
106. Bringmann, G.; Wolf, K.; Meininger, M.; Rokitta, M.; Haase, A. In vivo ^{19}F NMR Chemical-Shift Imaging of *Ancistrocladus Species*. *Protoplasma* **2001**, *218*, 134–143.
107. Wolf, K. In-Vivo-NMR zur Metabolitendetektion an Pflanzen und Quantenchemische Berechnungen. PhD Thesis, University of Würzburg, 2000.

108. Urlaub, E.; Popp, J.; Kiefer, W.; Bringmann, G.; Koppler, D.; Schneider, H.; Zimmermann, U.; Schrader, B. FT-Raman Investigation of Alkaloids in the Liana *Ancistrocladus heyneanus*. *Biospectroscopy* **1998**, *4*, 113–120.
109. Frosch, T.; Schmitt, M.; Schenzel, K.; Faber, J. H.; Bringmann, G.; Kiefer, W.; Popp, J. In vivo Localization and Identification of the Antiplasmodial Alkaloid Dioncophylline A in the Tropical Liana *Triphyophyllum peltatum* by a Combination of Fluorescence, Near Infrared Fourier Transform Raman Microscopy, and Density Functional Theory Calculations. *Biopolymers* **2006**, *82*, 295–300.
110. Frosch, T.; Schmitt, M.; Noll, T. F.; Bringmann, G.; Schenzel, K.; Popp, J. Ultrasensitive In situ Tracing of the Alkaloid Dioncophylline A in the Tropical Liana *Triphyophyllum peltatum* by Applying Deep-UV Resonance Raman Microscopy. *Anal. Chem.* **2007**, *79*, 986–993.
111. Hutchinson, J.; Dalziel, J. M. Dioncophyllaceae. In *Flora of West Tropical Africa*; Hutchinson, J., Dalziel, J. M., Eds, 2nd ed.; Crown Agents Overseas Governments and Administration: London, 1954; pp. 191–194 (revised by Keay, R. W. J.).
112. Porembski, S.; Barthlott, W. Dioncophyllaceae. In *The Families and Genera of Vascular Plants. V. Flowering Plants—Dicotyledons, Malvales, Capparales and Non-betalain Caryophyllales*; Kubitzki, K., Bayer, C., Eds.; Vol. 5; Springer: Heidelberg, 2002; pp. 178–181.
113. Porembski, S. Ancistrocladaceae. In *The Families and Genera of Vascular Plants. V. Flowering Plants—Dicotyledons, Malvales, Capparales and Non-betalain Caryophyllales*; Kubitzki, K., Bayer, C., Eds.; Vol. 5; Springer: Heidelberg, 2002; pp. 25–27.
114. Schmid, R. Die Systematische Stellung der Dioncophylleen. *Bot. Jahrb. Syst.* **1964**, *83*, 1–56.
115. Metcalfe, C. R. The Anatomical Structure of the Dioncophyllaceae in Relation to the Taxonomic Affinities of the Family. *Kew Bull.* **1951**, *6*, 351–368.
116. McPherson, S. *Triphyophyllum peltatum*. In *Glistening Carnivores: The Sticky-Leaved Insect-Eating Plants*; Mc Pherson, S., Ed.; Redfern Natural History Productions Ltd.: Poole, Dorset, England, 2008; pp. 68–94.
117. Bringmann, G.; Rischer, H.; Schlauer, J.; Wolf, K.; Kreiner, A.; Duschek, M.; Aké Assi, L. The Tropical Liana *Triphyophyllum peltatum* (Dioncophyllaceae): Formation of Carnivorous Organs Is Only a Prerequisite for Shoot Elongation. *Carniv. Pl. Newslett.* **2002**, *31*, 44–52.
118. Rembold, K.; Irmer, A.; Poppinga, S.; Rischer, H.; Bringmann, G. Propagation of *Triphyophyllum peltatum* (Dioncophyllaceae) and Observations on Its Carnivory. *Carniv. Pl. Newslett.* **2010**, *39*, 71–77.
119. Winkelmann, T.; Bringmann, G.; Herwig, A.; Hedrich, R. Carnivory on Demand: Phosphorus Deficiency Induces Glandular Leaves in the African Liana *Triphyophyllum peltatum*. *New Phytol.* **2023**, *239*, 1140–1152.
120. Marburger, J. E. Glandular Leaf Structure of *Triphyophyllum peltatum* (Dioncophyllaceae): A "Fly-Paper" Insect Trapper. *Am. J. Bot.* **1979**, *66*, 404–411.
121. Green, S.; Green, T. L.; Heslop-Harrison, Y. Seasonal Heterophylly and Leaf Gland Features in *Triphyophyllum* (Dioncophyllaceae), a New Carnivorous Plant Genus. *Bot. J. Linn. Soc.* **1979**, *78*, 99–116.
122. Baillon, H. Observations Sur Quelques Nouveaux Types du Congo. *Bull. Mens. Soc. Linn. Paris* **1890**, *109*, 868–872.
123. Warburg, O. Flacourtiaceae. In *Die natürlichen Pflanzenfamilien nebst ihren Gattungen und wichtigeren Arten insbesondere den Nutzpflanzen III*; Engler, A., Prantl, K., Eds.; Vol. 6a; Verlag Wilhelm Engelmann: Leipzig, 1895; pp. 28–30.
124. Gilg, E. Flacourtiaceae Africanae. *Engl. Bot. Jahrb.* **1908**, *40*, 486–487.
125. Sprague, T. A. Dioncophyllum. *Bull. Misc. Inform. Kew* **1916**, *4*, 89–92.

126. Kubitzki, K. Drosophyllaceae. Drosophyllaceae. In *The Families and Genera of Vascular Plants. V. Flowering Plants—Dicotyledons, Malvales, Capparales and Non-betalain Caryophyllales*; Kubitzki, K., Bayer, C., Eds.; Vol. 5; Springer: Heidelberg, 2002; pp. 203–205.
127. Meimberg, H.; Dittrich, P.; Bringmann, G.; Schlauer, J.; Heubl, G. Molecular Phylogeny of Caryophyllidae s.l. Based on *matK* Sequences With Special Emphasis on Carnivorous Taxa. *Plant Biol.* **2000**, *2*, 218–228.
128. Cuénoud, P.; Savolainen, V.; Chatrou, L. W.; Powell, M.; Grayer, R. J.; Chase, M. W. Molecular Phylogenetics of Caryophyllales Based on Nuclear 18S rDNA and Plastid *rbc*L, *atp*B, and *matK* DNA Sequences. *Am. J. Bot.* **2002**, *89*, 132–144.
129. Hilu, K. W.; Liang, H. The *matK* Gene: Sequence Variation and Application in Plant Systematics. *Am. J. Bot.* **1997**, *84*, 1735–1741.
130. Hegnauer, R. Comparative Phytochemistry and plant Taxonomy. *Giorn. Bot. Ital.* **1986**, *120*, 15–26.
131. Thomson, R. H. Naphthoquinones. *Naturally Occurring Quinones IV. Recent Advances*; Chapman and Hall: London, 1997; pp. 112–308.
132. Renner, T.; Specht, C. D. A Sticky Situation: Assessing Adaptations for Plant Carnivory in the Caryophyllales by Means of Stochastic Character Mapping. *Int. J. Plant Sci.* **2011**, *172*, 889–901.
133. Heubl, G.; Bringmann, G.; Meimberg, H. Molecular Phylogeny and Character Evolution of Carnivorous Plant Families in Caryophyllales—Revisited. *Plant Biol.* **2006**, *8*, 821–830.
134. Cameron, K. M.; Chase, M. W.; Swensen, S. M. Molecular Evidence for the Relationships of *Triphyophyllum* (Dioncophyllaceae) and *Ancistrocladus* (Ancistrocladaceae). *Am. J. Bot.* **1995**, *82*, 117–118.
135. Fay, M. F.; Cameron, K. M.; Prance, G. T.; Lledo, M. D.; Chase, M. W. Familial Relationships of Rhabdodendron (Rhabdodendraceae): Plastid *rbcL* Sequences Indicate a Caryophyllid Placement. *Kew Bull.* **1997**, *52*, 923–932.
136. Nandi, O.; Chase, M. W.; Endress, P. K. A Combined Cladistic Analysis of Angiosperms Using *rbcL* and Non-Molecular Data Sets. *Ann. MO Bot. Gard.* **1998**, *79*, 249–265.
137. Lledó, M. D.; Crespo, M. B.; Cameron, K. M.; Fay, M. F.; Chase, M. W. Systematics of Plumbaginaceae Based Upon Cladistic Analysis of *rbcL* Sequence Data. *Syst. Bot.* **1998**, *23*, 21–29.
138. Gottwald, H.; Parameswaran, N. Das Sekundäre Xylem und DIE systematische Stellung der Ancistrocladaceae und Dioncophyllaceae. *Bot. Jahrb. Syst.* **1968**, *88*, 49–69.
139. Albert, V. A.; Stevenson, D. W. M. Morphological Cladistics of the Nepenthales. *Am. J. Bot.* **1996**, *83*, 135.
140. Schlauer, J. "New" Data relating to the Evolution and Phylogeny of Some Carnivorous Plant Families. *Carniv. Pl. Newslett.* **1997**, *26*, 34–38.
141. Erdtman, G. A Note on the Pollen Morphology in the Ancistrocladaceae and Dioncophyllaceae. *Veröff. Geobot. Inst. Rübel Zürich* **1958**, *33*, 47–49.
142. Ellis, R. Carnivores on Stamps and Currency. *Carniv. Pl. Newslett.* **2000**, *29*, 90–93.
143. Hallier, H. Beiträge zur Kenntnis der Linaceae. *Beih. Bot. Centralbl.* **1923**, *39*, 1–178.
144. Bringmann, G.; Wenzel, M.; Bringmann, H.; Schlauer, J.; Aké Assi, L. Die "Teilzeitfleischfressende" Pflanze *Triphyophyllum peltatum* (Dioncophyllaceae): Nutzung der Fangorgane zur Erforschung der Alkaloidbildung. *Der Palmengarten* **1996**, *60* (2), 32–37.
145. Bringmann, G.; Wenzel, M.; Bringmann, H. P.; Schlauer, J.; Aké Assi, L.; Haas, F. Uptake of the Amino Acid Alanine by Digestive Leaves: Proof of Carnivory in the Tropical Liana *Triphyophyllum peltatum* (Dioncophyllaceae). *Carniv. Pl. Newslett.* **2001**, *30*, 15–21.

146. Bringmann, G.; Schlauer, J.; Wolf, K.; Rischer, H.; Buschbom, U.; Kreiner, A.; Thiele, F.; Duschek, M.; Aké Assi, L. Cultivation of *Triphyophyllum peltatum* (Dioncophyllaceae), the Part-Time Carnivorous Plant. *Carniv. Pl. Newslett.* **1999**, *28*, 7–13.
147. Kubitzki, K. Droseraceae. In *The Families and Genera of Vascular Plants. V. Flowering Plants—Dicotyledons, Malvales, Capparales and Non-betalain Caryophyllales*; Kubitzki, K., Bayer, C., Eds.; Vol. 5; Springer: Heidelberg, 2002; pp. 198–202.
148. Kubitzki, K. Nepenthaceae. In *The Families and Genera of Vascular Plants. V. Flowering Plants—Dicotyledons, Malvales, Capparales and Non-betalain Caryophyllales*; Kubitzki, K., Bayer, C., Eds.; Vol. 5; Springer: Heidelberg, 2002; pp. 320–324.
149. Renner, T.; Specht, C. D. Molecular and Functional Evolution of Class I Chitinases for Plant Carnivory in the Caryophyllales. *Mol. Biol. Evol.* **2012**, *29*, 2971–2985.
150. Bringmann, G.; Rischer, H. In Vitro Propagation of the Alkaloid-Producing Rare African Liana *Triphyophyllum peltatum* (Dioncophyllaceae). *Plant Cell Rep.* **2001**, *20*, 591–595.
151. Bringmann, G.; Rüdenauer, S.; Irmer, A.; Bruhn, T.; Brun, R.; Heimberger, T.; Stühmer, T.; Bargou, R.; Chatterjee, M. Antitumoral and Antileishmanial Dioncoquinones and Ancistroquinones From Cell Cultures of *Triphyophyllum peltatum* (Dioncophyllaceae) and *Ancistrocladus abbreviatus* (Ancistrocladaceae). *Phytochemistry* **2008**, *69*, 2501–2509.
152. Bringmann, G.; Zhang, G.; Hager, A.; Moos, M.; Irmer, A.; Bargou, R.; Chatterjee, M. Antitumoral Activities of Dioncoquinones B and C and Related Naphthoquinones Gained From Total Synthesis or Isolation From Plants. *Eur. J. Med. Chem.* **2011**, *46*, 5778–5789.
153. Bringmann, G.; Rischer, H.; Wohlfarth, M.; Schlauer, J.; Aké Assi, L. Droserone From Cell Cultures of *Triphyophyllum peltatum* (Dioncophyllaceae) and Its Biosynthetic Origin. *Phytochemistry* **2000**, *53*, 339–343.
154. Anderson, W. C. Tissue Culture Propagation of Red and Black Raspberries, *Rubus idaeus* and *R. occidentalis*. *Acta Hortic.* **1980**, *112*, 13–20.
155. Wallich, N. *A Numerical List of Dried Specimens of Plants in the East India Company's Museum: Collected Under the Superintendence of Dr. Wallich of the Company's Botanic Garden at Calcutta, nos. 1-2153*; Lithographed manuscript, London, 1829.
156. Planchon, M. J. E. Essai Monographique d'une Nouvelle Famille de Plantes Proposée Sous le nom d'Ancistrocladacées. *Ann. Sci. Nat. Bot. 3 Ser. Bot.* **1849**, *13*, 316–320.
157. Oliver, D. Dipterocarpaceae. In *Flora of Tropical Africa*; Oliver, D., Ed.; Lovell Reeve & Co.: London, 1868; pp. 172–175.
158. Freson, R. Note sur la Distribution Africaine du Genre *Ancistrocladus* Wall (Ancistrocladaceae). *Bull. Jard. Bot. Nat. Belg.* **1967**, *37*, 73–76.
159. Airy Shaw, H. K. *Ancistrocladus barteri*. *Kew Bull.* **1949**, *4*, 67–68.
160. Airy Shaw, H. K. Further Notes on West African *Ancistrocladus*. *Kew Bull.* **1950**, *5*, 147–150.
161. Hutchinson, J.; Dalziel, J. M. Ancistrocladaceae. In *Flora of West Tropical Africa*; Hutchinson, J., Dalziel, J. M., Eds.; Vol. 1, Part 1; The Crown Agents for the Colonies: London, 1927; pp. 195–196.
162. Hutchinson, J.; Dalziel, J. M. Ancistrocladaceae. In *Flora of West Tropical Africa*; Hutchinson, J., Dalziel, J. M., Eds, 2nd ed.; Crown Agents Overseas Governments and Administration: London, 1954; pp. 233–234 (revised by Keay, R. W. J.).
163. Scott-Elliot, G. F. On the Botanical Results of the Sierra Leone Boundary Commission. *Bot. J. Linn. Soc.* **1894**, *30*, 64–100.
164. Léonard, J. Une Nouvelle et Curieuse famille Pour la Flore Phanérogamique du Congo Belge: Les Ancistrocladaceae. *Bull. Soc. R. Bot. Belg.* **1949**, *82*, 27–40.
165. Léonard, J. Ancistrocladaceae. In *Flore D'Afrique Centrale (Zaïre-Rwanda-Burundi)*; Bamps, P., Ed.; Jardin Botanique National de Belgique: Meise, 1982; pp. 1–9.

166. Léonard, J. *Ancistrocladus robertsoniorum* J. Léonard (Ancistrocladaceae), Espèce Nouvelle du Kenya. *Bull. Jard. Bot. Nat. Belg.* **1984**, *54*, 465–470.
167. Léonard, J. Ancistrocladaceae. In *Flora of Tropical East Africa*; Polhill, R. M., Ed.; A. A. Balkema: Rotterdam, Boston, 1986; pp. 1–4.
168. Bringmann, G.; Haller, R. D.; Bär, S.; Isahakia, M. A.; Robertson, S. A. *Ancistrocladus robertsoniorum* J. Léonard: Eine Erst Spät Entdeckte *Ancistrocladus*-Art. *Der Palmengarten* **1994**, *58*, 148–153.
169. Cheek, M.; Frimodt-Møller, C.; Hørlyck, V. A New Submontane Species of *Ancistrocladus* From Tanzania. *Kew Bull.* **2000**, *55*, 207–212.
170. Gilg, E. Ancistrocladaceae. In *Die natürlichen Pflanzenfamilien nebst ihren Gattungen und wichtigeren Arten insbesondere den Nutzpflanzen*; Engler, A., Ed, 2nd ed.; Vol. 21; Duncker & Humblot: Berlin, 1925; pp. 589–592.
171. Van Steenis, C. G. G. J. Ancistrocladaceae. In *Flora Malesiana*; Van Steenis, C. G. G. J., Ed.; Vol. 4; Noordhoff-Kolff N.V.: Djakarta, 1948; pp. 8–10.
172. Baillon, H. *Histoire des Plantes*, Vol. 4; Librairie Hachette & C. Boulevard Saint-Germain: Paris, 1873; pp. 206–207.
173. Hutchinson, J.; Dalziel, J. M. Ancistrocladaceae. *Bull. Misc. Inform. Kew* **1928**, *6*, 218–219.
174. Pellegrin, F. Deux Plantes Congolaises à Affinités Asiatique. *Bull. Soc. Bot. France* **1951**, *98*, 17–19.
175. Bringmann, G.; Günther, C.; Saeb, W.; Mies, J.; Wickramasinghe, A.; Mudogo, V.; Brun, R. Ancistrolikokines A-C: New 5,8′-Coupled Naphthylisoquinoline Alkaloids From *Ancistrocladus likoko*. *J. Nat. Prod.* **2000**, *63*, 1333–1337.
176. Fayez, S.; Feineis, D.; Mudogo, V.; Awale, S.; Bringmann, G. Ancistrolikokines E-H and Related 5,8′-Coupled Naphthylisoquinoline Alkaloids From the Congolese Liana *Ancistrocladus likoko* With Antiausterity Activities Against PANC-1 Human Pancreatic Cancer Cells. *RSC Adv.* **2017**, *7*, 53740–53751.
177. Bringmann, G.; Messer, K.; Brun, R.; Mudogo, V. Ancistrocongolines A-D, New Naphthylisoquinoline Alkaloids From *Ancistrocladus congolensis*. *J. Nat. Prod.* **2002**, *65*, 1096–1101.
178. Bringmann, G.; Steinert, C.; Feineis, D.; Mudogo, V.; Betzin, J.; Scheller, C. HIV-Inhibitory Michellamine-Type Dimeric Naphthylisoquinoline Alkaloids From the Central African Liana *Ancistrocladus congolensis*. *Phytochemistry* **2016**, *128*, 71–81.
179. Bringmann, G.; Hamm, A.; Günther, C.; Michel, M.; Brun, R.; Mudogo, V. Ancistroealaine A and B, Two New Bioactive Naphthylisoquinolines, and Related Naphthoic Acids From *Ancistrocladus ealaensis*. *J. Nat. Prod.* **2000**, *63*, 1465–1470.
180. Tshitenge, D. T.; Feineis, D.; Mudogo, V.; Kaiser, M.; Brun, R.; Bringmann, G. Antiplasmodial Ealapasamines A-C, 'Mixed' Naphthylisoquinoline Dimers From the Central African Liana *Ancistrocladus ealaensis*. *Sci. Rep.* **2017**, *7*, 5767.
181. Tshitenge, D. T.; Feineis, D.; Mudogo, V.; Kaiser, M.; Brun, R.; Seo, E. J.; Efferth, T.; Bringmann, G. Mbandakamine-Type Naphthylisoquinoline Dimers and Related Alkaloids From the Central African Liana *Ancistrocladus ealaensis* With Antiparasitic and Antileukemic Activities. *J. Nat. Prod.* **2018**, *81*, 918–933.
182. Tshitenge, D. T.; Bruhn, T.; Feineis, D.; Mudogo, V.; Kaiser, M.; Brun, R.; Bringmann, G. An Unusually Broad Series of Seven Cyclombandakamines, Bridged Dimeric Naphthylisoquinoline Alkaloids From the Congolese Liana *Ancistrocladus ealaensis*. *Sci. Rep.* **2019**, *9*, 9812.
183. Li, J.; Seupel, R.; Bruhn, T.; Feineis, D.; Kaiser, M.; Brun, R.; Mudogo, V.; Awale, S.; Bringmann, G. Jozilebomines A and B, Naphthylisoquinoline Dimers From the Congolese Liana *Ancistrocladus ileboensis*, With Antiausterity Activities Against the PANC-1 Human Pancreatic Cancer Cell Line. *J. Nat. Prod.* **2017**, *80*, 2807–2817.
184. Hallock, Y. F.; Hughes, C. B.; Cardellina, J. H., II; Schäffer, M.; Gulden, K. P.; Bringmann, G.; Boyd, M. R. Dioncophylline A, the Principal Cytotoxin From *Ancistrocladus letestui*. *Nat. Prod. Lett.* **1995**, *6*, 315–320.

185. Fasina, F. O.; Olaokun, O. O.; Oladipo, O. O.; Fasina, M. M.; Makinde, A. A.; Heath, L.; Bastos, A. D. S. Phytochemical Analysis and In-Vitro Anti-African Swine Fever Virus Activity of Extracts and Fractions of *Ancistrocladus uncinatus*, Hutch and Dalziel (Ancistrocladaceae). *BMC Vet. Res.* **2013**, *9*, 120.
186. Cremers, G. Architecture de Quelques Lianes d'Afrique Tropicale. *Candollea* **1974**, *29*, 57–110.
187. Bringmann, G.; Rischer, H.; Schlauer, J.; Aké Assi, L. In Vitro Propagation of *Ancistrocladus abbreviatus* Airy Shaw (Ancistrocladaceae). *Plant Cell, Tissue Org. Cult.* **1999**, *57*, 71–73.
188. Hallé, N. Crochets de Lianes du Gabon: *Ancistrocladus* et *Anacolosa* (Ancistrocladacées et Olacacées). *Adansonia* **1973**, *13*, 299–306.
189. Foster, P. F.; Sork, V. L. Population and Genetic Structure of the West African Rain Forest Liana *Ancistrocladus korupensis* (Ancistrocladaceae). *Am. J. Bot.* **1997**, *84*, 1078–1091.
190. Thomas, A. L.; Carpenter-Boggs, L. A. Mycorrhizal Colonization of *Ancistrocladus korupensis*, a New Tropical Forest Species With Anti-HIV Activity. *J. Herbs, Spices Med. Plants* **1995**, *3*, 51–54.
191. Thiselton Dyer, W. T. XXV. Dipterocarpaceae. In *Flora of British India*; Hooker, J. D., Ed.; Vol. 1; Lovell Reeve & Co. Ltd.: Kent, 1874; pp. 299–300.
192. Craib, W. G. Contributions to the Flora of Siam—Additamentum XV. *Bull. Misc. Inform. Kew* **1925**, 19–20.
193. Gagnepain, F. Plantes Nouvelles d'Indo-Chine. In *Notulae Systematicae*; Lecomte, H., Ed.; Vol. 4; J. Dumoulin: Paris, 1909; pp. 114–116.
194. Hayata, B. Dipterocarpaceae. *Ic. Pl. Formos* **1913**, *3*, 46–47.
195. Walker-Arnott, G. Malphigiaceae. *Nova Acta Acad. Caes. Leop.-Carol. German. Nat. Cur.* **1836**, *18*, 325–326.
196. Van Tieghem, M. P. Sur les Ancistrocladacées. *J. Bot.* **1903**, *17*, 151–168.
197. Vahl, M. Beskrivelse Over Nye Planteslaegter—Wormia. *Skr. Naturhist. Selsk.* **1810**, *6*, 104–107.
198. Graham, J. 206. *Ancistrocladus heyneanus*. In *A Catalogue of the Plants Growing in Bombay and Its Vicinity*; Graham, J., Ed.; Government Press: Bombay, 1839; p. 28.
199. Ridley, H. N. Ancistrocladaceae. In *The Flora of the Malay Peninsula (With Illustrations by J. Hutchinson)*; Ridley, H. N., Ed.; Vol. 1; Lovell Reeve & Co.: London, 1922; pp. 250–252.
200. Loureiro, D. Genus XXVII. Bembix. In *Flora Cochinchinensis. Typis*, Vol. 1; Et Expensis Academics: Ulyssipone, 1790; pp. 282–283.
201. Merrill, E. D. A Second Supplementary List of Hainan Plants. *Lingnan Sci. J.* **1928**, *6*, 323–332.
202. De Candolle, A. Ancistrocladaceae. In *Prodromus Systematis Naturalis Regni Vegetabilis*; De Candolle, A., Ed.; Vol. 16; Victoris Masson et Filii: Paris, 1864; pp. 601–603.
203. Warburg, O. Beiträge zur Kenntnis der Papuanischen FLORA. *Bot. Jahrb. Syst.* **1891**, *13*, 383.
204. Dagar, J. C.; Singh, T. Ancistrocladaceae. In *Plant Resources of the Andaman & Nicobar Islands (Enumeration and Utilisation of Vascular Plants)*, Vol. 2; Bishen Singh Mahendra Pal Singh: Dehradun, 1999; p. 308.
205. Kundu, S. R. A Synopsis of Ancistrocladaceae in Indian Subcontinent: Its Distribution and Endemism. *Geobios* **2005**, *32*, 221–308.
206. Parkinson, C. E. Ancistrocladaceae. In *A Forest Flora of the Andaman Islands*; Bishen Singh Mahendra Pal Singh: Dehradun, 1972; pp. 94–95.
207. Hajra, P. K.; Rao, P. S. N.; Mudgal, X. Ancistrocladaceae. In *Flora of Andaman and Nicobar Islands*, Vol. 1; Botanical Survey of India: Calcutta, 1999; p. 153.

208. Pandey, R. P.; Diwakar, P. C. An Integrated Check-List Flora of Andaman and Nicobar Islands. *J. Econ. Taxon. Bot.* **2008**, *32*, 403–500.
209. Brandis, D. Order XVIII. Ancistrocladaceae. In *Indian Trees—An Account of Trees, Shrubs, Woody Climbers, Bamboos and Palms Indigenous or Commonly Cultivated in the British Indian Empire*; Brandis, D., Ed.; Bishen Singh Mahendra Pal Singh: Dehradun, 1906; pp. 73–74.
210. Das, S. Ancistrocladaceae. In *Flora of India, Portulacaceae-Ixonanthaceae*; Sharma, B. D., Sanjappa, M., Eds.; Vol. 3; Botanical Survey of India: Calcutta, 1993; pp. 252–256.
211. Gilg, E. Ancistrocladaceae. In *Die natürlichen Pflanzenfamilien nebst ihren Gattungen und wichtigeren Arten insbesondere den Nutzpflanzen*; Engler, A., Ed.; Teil 3, Abteilung 6; Verlag Wilhelm Engelmann: Leipzig, 1895; pp. 274–276.
212. Craib, W. G. Ancistrocladaceae. In *Florae Siamensis Enumeratio—A List of Plants Known From Siam With Records of Their Occurrence*, Vol. 1; Siam Society: Bangkok, 1931; pp. 148–149.
213. Moore, S.; Tandy, G. Notes on Two Species of Loureiro's Flora Cochinchinensis. *J. Bot.* **1927**, *65*, 279–281.
214. Pham-Hoang, H. Dipterocarpaceae. In *Câycô Việtnam—An Illustrated Flora of Vietnam*, Vol. 1; Mekong Printing: Santa Ana, Canada, 1991; p. 555.
215. Savajol, N.; Toun, V.; Sam, J. Ancistrocladus cochinchinensis. In *Traditional Therapeutic Knowledge of the Bunong People in Southeastern Cambodia*; Savajol, N., Toun, V., Sam, J., Eds.; Nomad RSI Cambodia Edition: Phnom Penh, Cambodia, 2011; pp. 58–59.
216. Prasad, P. R. C.; Reddy, C. S.; Iakshmi, R. K. V.; Kumari, P. V.; Raza, S. H. Angiosperms of North Andaman, Andaman and Nicobar Islands, India. *Check List* **2009**, *5*, 254–269.
217. Hu, H. H. Ancistrocladaceae. In *Prodromus Florae Sinensis—Contributions From the Biological Laboratory of the Science Society of China*, Vol. 5; The Science Society of China: Nanking, China, 1929; pp. 43–45.
218. Trimen, H. Ancistrocladaceae. In *A Handbook to the Flora of Ceylon (Containing Descriptions of All the Species of Flowering Plants Indigenous to the Island, and Notes on Their History, Distribution, and Uses), With an Atlas of Plates (Illustrating Some of the More Interesting Species)*, Part 1; Dulau & Co.: London, 1893; pp. 138–139 (Plate XVI).
219. Thwaites, G. H. K. Note on the Genus *Ancistrocladus* Wallich. *Trans. Linn. Soc. London* **1849**, *21*, 225–226 (Figs. 1–18).
220. Harriman, N. A. Ancistrocladaceae. In *A Revised Handbook to the Flora of Ceylon*; Dassanayake, M. D., Ed.; Vol. 6; Amerind Publishing: New Delhi, 1987; pp. 1–2.
221. Boorsma, W. G. Ancistrocladaceae. In *Onderzoek naar de Plantenstoffen van Nederlandsch-Indie. Mededeelingen uit 'S Lands Plantentuin*, Vol. 31; G. Kolff & Co.: Batavia, 1899; pp. 4–5.
222. Abeywickrama, B. A. A Provisional CHECK list of the Flowering Plants of Ceylon. *Ceylon J. Sci. (Biol. Sci.)* **1959**, *2*, 119–120.
223. Ramamoorthy, T. P. Ancistrocladaceae. In *Flora of Hassan District, Karnataka, India*; Saldanha, C. J., Nicolson, D. H., Eds.; Amerind Publishing: New Delhi, 1976; pp. 171–172.
224. Bringmann, G.; Pokorny, F.; Zinsmeister, H. D. *Ancistrocladus*, Eine Botanisch und Chemisch Bemerkenswerte Gattung. *Der Palmengarten* **1991**, *55* (3), 13–18.
225. Bringmann, G.; Schneider, C.; Pokorny, F.; Lorenz, H. J.; Fleischmann, H.; Sankaranarayanan, T. R.; Aké Assi, L. The Cultivation of the Genus *Ancistrocladus*. *Planta Med.* **1993**, *59*, A623–A624.
226. Van Rhede tot Draekestein, H. *Ancistrocladus heyneanus* Wallich Ex J. Graham (as *Modira valli*). In *Hortus Indicus Malabaricus*, Vol. 7; Van Someren & Dyck: Amstelodami, 1686; p. 87 (tab 46).

227. Wight, R. *Ancistrocladus heyneanus* Wallich Ex J. Graham. In *Icones Plantarum Indiae Orientalis (or Figures of Indian Plants)*, Vol. 6; American Mission Press: Madras, 1853; p. 12 (no. 1987–88), including plate (no. 1987–88).
228. Hutchinson, J. 149. Ancistrocladaceae. In *The Families of Flowering Plants—Arranged According to a New System Based on Their Probable Phylogeny*; Hutchinson, J., Ed, 3rd ed.; Vol. 1; Clarendon Press: Oxford, 1973; pp. 351–352.
229. Santapau, H. The Flora of Khandala on the Western Ghats of India. In *Records of the Botanical Survey of India*, 3rd ed, Vol. 16; The Manager of Publication, Civil Lines, Calcutta, Ed.; Loyal Art Press Private Limited: Calcutta, 1967; p. 15.
230. Rao, V. S. The Floral Anatomy of *Ancistrocladus*. *Proc. - Indian Acad. Sci., Sect. B* **1969**, *70*, 215–222.
231. Panday, S.; Roy, D. K.; Odyuo, N.; Sinha, B. K. Notes on a Little Known Genus *Ancistrocladus* Wallich ex Wight & Arnott [Ancistrocladaceae] From Northeast India. *Pleione* **2012**, *6*, 336–340.
232. Datar, M. N.; Lakshminarasimhan, P. Check List of Wild Angiosperms of Bhagwan Mahavir (Molem) National Park, Goa, India. *Check List* **2013**, *9*, 186–207.
233. Yunoh, S. M. M. *Ancistrocladus tectorius* (Lour.) Merill. In *Flora of Malaysia; MyCHM i-Newsletter Part 3*; Forest Research Institute Malaysia (FRIM), 2007; pp. 46–50.
234. Keng, H. Observations on *Ancistrocladus tectorius*. *Gard. Bull. Singapore* **1967**, *22*, 113–121.
235. Keng, H. Further Observations on *Ancistrocladus tectorius* (Ancistrocladaceae). *Gard. Bull. Singapore* **1970**, *25*, 235–237.
236. Mo, S. L.; Zhong, Y. C. Plantae Novae Guangxiensis. *Guihaia* **1987**, *7*, 287–291.
237. Mabberley, D. J. Ancistrocladaceae. In *Mabberley's Plant Book—A Portable Dictionary of Plants, Their Classification and Uses*; Mabberley, D. J., Ed, 3rd ed.; Cambridge University Press: United Kingdom, 2008; pp. 43–44.
238. Foucher, J. P.; Pousset, J. L.; Cavé, A.; Paris, R. R. Chimiotaxonomie des Ancistrocladacées II.—Sur les Alcaloïdes de l'*Ancistrocladus tectorius* Loureiro, du Laos. *Plantes Méd. Phytothér.* **1975**, *9*, 26–31.
239. Ruangrungsi, N.; Wongpanich, V.; Tantivatana, P.; Cowe, H. J.; Cox, P. J.; Funayama, S.; Cordell, G. A. Traditional Medicinal Plants of Thailand. V. Ancistrotectorine, a New Naphthalene-Isoquinoline Alkaloid From *Ancistrocladus tectorius*. *J. Nat. Prod.* **1985**, *48*, 529–535.
240. Meksuriyen, D.; Ruangrungsi, N.; Tantivatana, P.; Cordell, G. A. NMR Spectroscopic Analysis of Ancistrocladidine. *Phytochemistry* **1990**, *29*, 2750–2752.
241. Manfredi, K. P.; Britton, M.; Vissieche, V.; Pannell, L. L. Three New Naphthylisoquinoline Alkaloids from *Ancistrocladus tectorius*. *J. Nat. Prod.* **1996**, *59*, 854–859.
242. Montagnac, A.; Hadi, H. A.; Remy, F.; Païs, M. Isoquinoline Alkaloids from *Ancistrocladus tectorius*. *Phytochemistry* **1995**, *39*, 701–704.
243. Chen, Z.; Wang, B.; Qin, K.; Zhang, B.; Su, Q.; Lin, Q. Isolation and Identification of the Alkaloids From *Ancistrocladus tectorius*. *Acta Pharm. Sin.* **1981**, *16*, 519–523.
244. Bringmann, G.; Kinzinger, L. (+)-Ancistrocline, a Naphthylisoquinoline Alkaloid From *Ancistrocladus tectorius*. *Phytochemistry* **1992**, *31*, 3297–3299.
245. Tang, C. P.; Yang, Y. P.; Zhong, Y.; Zhong, Q. X.; Wu, H. M.; Ye, Y. Four New Naphthylisoquinoline Alkaloids From *Ancistrocladus tectorius*. *J. Nat. Prod.* **2000**, *63*, 1384–1387.
246. Su, Z.; Liu, M.; Li, Z.; Kang, S.; Hua, H. Chemical Constituents of *Ancistrocladus tectorius*. *Chin. J. Med. Chem.* **2007**, *17*, 383–385.
247. Liu, M.; Su, Z.; Zhang, J.; Li, Z.; Hua, H. Studies on Constituents of the Stems and Branches of *Ancistrocladus tectorius*. *Chin. Pharm. J.* **2008**, *43*, 1060–1063.
248. Tang, C. P.; Xin, Z. Q.; Li, X. Q.; Ye, Y. Two New Naphthylisoquinoline Alkaloids From Stems and Leaves of *Ancistrocladus tectorius*. *Nat. Prod. Res.* **2010**, *24*, 989–994.

249. Jiang, C.; Li, Z. L.; Gong, P.; Kang, S. L.; Liu, M. S.; Pei, Y. H.; Jing, Y. K.; Hua, H. M. Five Novel Naphthylisoquinoline Alkaloids With Growth Inhibitory Activities Against Human Leukemia Cells HL-60, K562 and U937 From Stems and Leaves of *Ancistrocladus tectorius*. *Fitoterapia* **2013**, *91*, 305–312.
250. Bringmann, G.; Zhang, G.; Ölschläger, T.; Stich, A.; Wu, J.; Chatterjee, M.; Brun, R. Highly Selective Antiplasmodial Naphthylisoquinoline Alkaloids From *Ancistrocladus tectorius*. *Phytochemistry* **2013**, *91*, 220–228.
251. Bringmann, G.; Seupel, R.; Feineis, D.; Zhang, G.; Xu, M.; Wu, J.; Kaiser, M.; Brun, R.; Seo, E. J.; Efferth, T. Ancistectorine D, a Naphthylisoquinoline Alkaloid With Antiprotozoal and Antileukemic Activities, and Further 5,8′- and 7,1′-Linked Metabolites From the Chinese Liana *Ancistrocladus tectorius*. *Fitoterapia* **2016**, *115*, 1–8.
252. Bringmann, G.; Seupel, R.; Feineis, D.; Xu, M.; Zhang, G.; Kaiser, M.; Brun, R.; Seo, E. J.; Efferth, T. Antileukemic Ancistrobenomine B and Related 5,1′-Coupled Naphthylisoquinoline Alkaloids From the Chinese Liana *Ancistrocladus tectorius*. *Fitoterapia* **2017**, *121*, 76–85.
253. Bringmann, G.; Xu, M.; Seupel, R.; Feineis, D.; Wu, J. Ancistrotectoquinones A and B, the First Quinoid Naphthylisoquinoline Alkaloids, From the Chinese Liana *Ancistrocladus tectorius*. *Nat. Prod. Commun.* **2016**, *11*, 971–976.
254. Xu, M.; Bruhn, T.; Hertlein, B.; Brun, R.; Stich, A.; Wu, J.; Bringmann, G. Shuangancistrotectorines A-E, Dimeric Naphthylisoquinoline Alkaloids With Three Chiral Biaryl Axes, From the Chinese Plant *Ancistrocladus tectorius*. *Chem. -Eur. J.* **2010**, *16*, 4206–4216.
255. Seupel, R.; Hemberger, Y.; Feineis, D.; Xu, M.; Seo, E. J.; Efferth, T.; Bringmann, G. Ancistrocyclinones A and B, Unprecedented N,C-Coupled Naphthylisoquinoline Alkaloids, From the Chinese Liana *Ancistrocladus tectorius*. *Org. Biomol. Chem.* **2018**, *16*, 1581–1590.
256. Govindachari, T. R.; Parthasarathy, P. C. Ancistrocladine, a Novel Isoquinoline Alkaloid FROM *Ancistrocladus heyneanus* Wall. *Indian J. Chem.* **1970**, *8*, 567–569.
257. Govindachari, T. R.; Parthasarathy, P. C.; Desai, H. K. Chemical investigation of *Ancistrocladus heyneanus* Wall: Part III—Further Studies on Ancistrocladine. *Indian J. Chem.* **1971**, *9*, 931–935.
258. Govindachari, T. R.; Parthasarathy, P. C.; Desai, H. K. Ancistrocladinine, a Minor Alkaloid From *Ancistrocladus heyneanus* Wall. *Indian J. Chem.* **1971**, *9*, 1421–1422.
259. Govindachari, T. R.; Nagarajan, K.; Parthasarathy, P. C.; Rajagopalan, T. G.; Desai, H. K.; Kartha, G.; Chen, S. M. L.; Nakanishi, K. Absolute Stereochemistry of Ancistrocladine and Ancistrocladinine. *J. Chem. Soc., Perkin Trans. 1* **1974**, 1413–1417.
260. Govindachari, T. R.; Parthasarathy, P. C.; Desai, H. K. Chemical Investigation of *Ancistrocladus heyneanus* Wall: Part VI—Isolation & Structure of Ancistrocladisine, a Novel Alkaloid. *Indian J. Chem.* **1972**, *10*, 1117–1119.
261. Govindachari, T. R.; Parthasarathy, P. C.; Desai, H. K. Chemical Investigation of *Ancistrocladus heyneanus* Wall.—Ancistrocladidine, a New Isoquinoline Alkaloid. *Indian J. Chem.* **1973**, *11*, 1190–1191.
262. Govindachari, T. R.; Parthasarathy, P. C.; Rajagopalan, T. G.; Desai, H. K.; Ramachandran, K. S.; Lee, E. Absolute Configuration of Ancistrocladisine and Ancistrocladidine. *J. Chem. Soc., Perkin Trans. 1* **1975**, 2134–2136.
263. Bringmann, G.; Kinzinger, L.; Ortmann, T.; De Souza, N. J. Isoancistrocladine from *Ancistrocladus heyneanus*: the First Naturally Occurring N-Unsubstituted *cis*-Configurated Naphthyltetrahydroisoquinoline Alkaloid. *Phytochemistry* **1994**, *35*, 259–261.
264. Bringmann, G.; Koppler, D.; Wiesen, B.; François, G.; Sankara Narayanan, A. S.; Almeida, M. R.; Schneider, H.; Zimmermann, U. Ancistroheynine A, the First 7,8′-Coupled Naphthylisoquinoline Alkaloid from *Ancistrocladus heyneanus*. *Phytochemistry* **1996**, *43*, 1405–1410.

265. Bringmann, G.; Dreyer, M.; Michel, M.; Tayman, F. S. K.; Brun, R. Ancistroheynine B and Two Further 7,3′-Coupled Naphthylisoquinoline Alkaloids from *Ancistrocladus heyneanus* Wall. *Phytochemistry* **2004**, *65*, 2903–2907.
266. Yang, L. K.; Glover, R. P.; Yoganathan, K.; Sarnaik, J. P.; Godbole, A. J.; Soejarto, D. D.; Buss, A. D.; Butler, M. S. Ancisheynine, a Novel Naphthylisoquinoline Alkaloid From *Ancistrocladus heyneanus*. *Tetrahedron Lett.* **2003**, *44*, 5827–5829.
267. Bringmann, G.; Gulder, T.; Reichert, M.; Meyer, F. Ancisheynine, the first *N,C*-Coupled Naphthylisoquinoline Alkaloid: Total Synthesis and Stereochemical Analysis. *Org. Lett.* **2006**, *2006* (8), 1037–1040.
268. Bringmann, G.; Feineis, D. Stress-Related Polyketide Metabolism of Dioncophyllaceae and Ancistrocladaceae. *J. Exp. Bot.* **2001**, *52*, 2015–2022.
269. Bringmann, G.; Wohlfarth, M.; Rischer, H.; Rückert, M.; Schlauer, J. The Polyketide Folding Mode in the Biogenesis of Isoshinanolone and Plumbagin From *Ancistrocladus heyneanus* (Ancistrocladaceae). *Tetrahedron Lett.* **1998**, *39*, 8445–8448.
270. Rischer, H. Acetogenine Sekundärmetabolite und Ihre Produzenten: Physiologie und Botanik ausgewählter Vertreter der Ancistrocladaceae, Dioncophyllaceae und Nepenthaceae Sowie Von *Antidesma* (Euphorbiaceae). PhD Thesis, University of Würzburg, 2002.
271. Murashige, T.; Skoog, F. Revised Medium for Rapid Growth and Bioassays With Tobacco Tissue Cultures. *Physiol. Plant.* **1962**, *15*, 473–497.
272. Irmer, A. Naturstoffe aus Zell- und Wurzelkulturen Von *Triphyophyllum peltatum*, Experimente zur Biosynthese der Naphthylisochinolin-Alkaloide Sowie Kalluskulturen und Sekundärmetabolite aus *Aloe saponaria*. PhD Thesis, University of Würzburg, 2012.
273. Bringmann, G.; Dreyer, M.; Rischer, H.; Wolf, K.; Hadi, H. A.; Brun, R.; Meimberg, H.; Heubl, G. Ancistrobenomine A, the First Naphthylisoquinoline Oxygenated at Me-3, and Related 5,1′-Coupled Alkaloids, from the "New" Plant Species *Ancistrocladus benomensis*. *J. Nat. Prod.* **2004**, *67*, 2058–2062.
274. Bringmann, G.; Ortmann, T.; Zagst, R.; Schöner, B.; Aké Assi, L.; Burschka, C. (±)-Dioncophyllacine A, a Naphthylisoquinoline Alkaloid With a 4-Methoxy Substituent From the Leaves of *Triphyophyllum peltatum*. *Phytochemistry* **1992**, *31*, 4015–4018.
275. Bringmann, G.; Günther, C.; Saeb, W.; Mies, J.; Brun, R.; Aké Assi, L. 8-O-Methyldioncophyllinol B and Revised Structures of Other 7,6′-Coupled Naphthylisoquinoline Alkaloids From *Triphyophyllum peltatum* (Dioncophyllaceae). *Phytochemistry* **2000**, *54*, 337–346.
276. Hallock, Y. F.; Cardellina, J. H., II; Schäffer, M.; Stahl, M.; Bringmann, G.; François, G.; Boyd, M. R. Yaoundamines A and B, New Antimalarial Naphthylisoquinoline Alkaloids From *Ancistrocladus korupensis*. *Tetrahedron* **1997**, *53*, 8121–8128.
277. Bringmann, G.; Teltschik, F.; Michel, M.; Busemann, S.; Rückert, M.; Haller, R.; Bär, S.; Robertson, S. A.; Kaminsky, R. Ancistrobertsonines B, C, and D as well as 1,2-Didehydroancistrobertsonine D from *Ancistrocladus robertsoniorum*. *Phytochemistry* **1999**, *52*, 321–332.
278. Bringmann, G.; Dreyer, M.; Faber, J. H.; Dalsgaard, P. W.; Stærk, D.; Jarozewski, J. W.; Ndangalasi, H.; Mbago, F.; Brun, R.; Reichert, M.; Maksimenka, K.; Christensen, S. B. Ancistrotanzanine A, the First 5,3′-Coupled Naphthylisoquinoline Alkaloid, and Two Further, 5,8′-Linked Related Compounds From the Newly Described Species *Ancistrocladus tanzaniensis*. *J. Nat. Prod.* **2003**, *66*, 1159–1165.
279. Bringmann, G.; Dreyer, M.; Faber, J. H.; Dalsgaard, P. W.; Stærk, D.; Jaroszewski, J. W.; Ndangalasi, H. F.; Brun, R.; Christensen, S. B. Ancistrotanzanine C and Related 5,1′- and 7,3′-Coupled Naphthylisoquinoline Alkaloids From *Ancistrocladus tanzaniensis*. *J. Nat. Prod.* **2004**, *67*, 743–748.

280. Bringmann, G.; Teltschik, F.; Schäffer, M.; Haller, R.; Bär, S.; Robertson, S. A.; Isahakia, M. A. Ancistrobertsonine A and Related Naphthylisoquinoline Alkaloids From *Ancistrocladus robertsoniorum*. *Phytochemistry* **1998**, *47*, 31–35.
281. Govindachari, T. R.; Parthasarathy, P. C.; Rajagopalan, T. G.; Desai, H. K.; Ramachandran, K. S. Hamatine, a New Isoquinoline Alkaloid from *Ancistrocladus hamatus* (Vahl) Gilg. *Indian J. Chem.* **1975**, *13*, 641–643.
282. Govindachari, T. R.; Parthasarathy, P. C.; Desai, H. K.; Saindane, M. T. On the Absolute Stereochemistry of Hamatine. *Indian J. Chem.* **1977**, *15B*, 871–872.
283. Bringmann, G.; Wohlfarth, M.; Rischer, H.; Heubes, M.; Saeb, W.; Diem, S.; Herderich, M.; Schlauer, J. A Photometric Screening Method for Dimeric Naphthylisoquinoline Alkaloids and Complete On-Line Structural Elucidation of a Dimer in Crude Plant Extracts, by the LC-MS/LC-NMR/LC-CD triad. *Anal. Chem.* **2001**, *73*, 2571–2577.
284. Bringmann, G.; Wohlfarth, M.; Rischer, H.; Schlauer, J.; Brun, R. Extract Screening by HPLC Coupled to MS-MS, NMR, and CD: a Dimeric and Three Monomeric Naphthylisoquinoline Alkaloids From *Ancistrocladus griffithii*. *Phytochemistry* **2002**, *61*, 195–204.
285. Bringmann, G.; Schneider, C.; Möhler, U.; Pfeifer, R. M.; Götz, R.; Aké Assi, L.; Peters, E. M.; Peters, K. Two Atropisomeric *N*-Methyldioncophyllines A and *N*-Methylphylline, Their Naphthalene-Free Heterocyclic Moiety, From *Ancistrocladus barteri*. *Z. Naturforsch.* **2003**, *58b*, 577–584.
286. Bringmann, G.; Schneider, C.; Aké Assi, L. Ancistrobarterine A: A New "Mixed" Ancistrocladaceae/Dioncophyllaceae-Type Alkaloid From *Ancistrocladus barteri*. *Planta Med.* **1993**, *59*, A620–A621.
287. Hallock, Y. F.; Manfredi, K. P.; Blunt, J. W.; Cardellina, J. H., II; Schäffer, M.; Gulden, K. P.; Bringmann, G.; Lee, A. Y.; Clardy, J.; François, G.; Boyd, M. R. Korupensamines A-D, Novel Antimalarial Alkaloids From *Ancistrocladus korupensis*. *J. Org. Chem.* **1994**, *59*, 6349–6355.
288. Hallock, Y. F.; Manfredi, K. P.; Dai, J. R.; Cardellina, J. H., II; Gulakowski, R. J.; McMahon, J. B.; Schäffer, M.; Stahl, M.; Gulden, K. P.; Bringmann, G.; François, G.; Boyd, M. R. Michellamines D-F, New HIV Inhibitory Dimeric Naphthylisoquinoline Alkaloids, and Korupensamine E, A New Antimalarial Monomer, From *Ancistrocladus korupensis*. *J. Nat. Prod.* **1997**, *60*, 677–683.
289. Hallock, Y. F.; Cardellina, J. H., II; Schäffer, M.; Bringmann, G.; François, G.; Boyd, M. R. Korundamine A, a Novel HIV-Inhibitory and Antimalarial "Hybrid" Naphthylisoquinoline Alkaloid From *Ancistrocladus korupensis*. *Bioorg. Med. Chem. Lett.* **1998**, *8*, 1729–1734.
290. Hallock, Y. F.; Cardellina, J. H., II; Kornek, T.; Gulden, K. P.; Bringmann, G.; Boyd, M. R. Gentrymine B, the First Quaternary Isoquinoline Alkaloid From *Ancistrocladus korupensis*. *Tetrahedron Lett.* **1995**, *36*, 4753–4756.
291. Bringmann, G.; Dreyer, M.; Kopff, H.; Rischer, H.; Wohlfarth, M.; Hadi, H. A.; Brun, R.; Meimberg, H.; Heubl, G. *ent*-Dioncophylleine A and Related Dehydrogenated Naphthylisoquinoline Alkaloids, the First Asian Dioncophyllaceae-Type Alkaloids, From the "New" Plant Species *Ancistrocladus benomensis*. *J. Nat. Prod.* **2005**, *68*, 686–690.
292. Bringmann, G.; Messer, K.; Wolf, K.; Mühlbacher, J.; Grüne, M.; Louis, A. M. Dioncophylline E From *Dioncophyllum thollonii*, the First 7,3′-Coupled Dioncophyllaceous Naphthylisoquinoline Alkaloid. *Phytochemistry* **2002**, *60*, 389–397.
293. Bringmann, G.; Rübenacker, M.; Vogt, P.; Busse, H.; Aké Assi, L.; Peters, K.; von Schnering, H. G. Dioncopeltine A and Dioncolactone A: Alkaloids From *Triphyophyllum peltatum*. *Phytochemistry* **1991**, *30*, 1691–1696.

294. Bringmann, G.; Saeb, W.; God, R.; Schäffer, M.; François, G.; Peters, K.; Peters, E. M.; Proksch, P.; Hostettmann, K.; Aké Assi, L. 5′-O-Demethyldioncophylline A, a New Antimalarial Alkaloid From *Triphyophyllum peltatum*. *Phytochemistry* **1998**, *49*, 1667–1673.
295. Bringmann, G.; Rübenacker, M.; Koch, W.; Koppler, D.; Ortmann, T.; Schäffer, M.; Aké Assi, L. 5′-O-Demethyl-8-O-Methyl-7-*Epi*-Dioncophylline A and Its 'Regularly' Configured Atropisomer From *Triphyophyllum peltatum*. *Phytochemistry* **1994**, *36*, 1057–1061.
296. Bringmann, G.; Messer, K.; Schwöbel, B.; Brun, R.; Aké Assi, L. Habropetaline A, an antimalarial naphthylisoquinoline alkaloid from *Triphyophyllum peltatum*. *Phytochemistry* **2003**, *62*, 345–349.
297. Bringmann, G.; Messer, K.; Wohlfarth, M.; Kraus, J.; Dumbuya, K.; Rückert, M. HPLC-CD on-Line Coupling in Combination With HPLC-NMR and HPLC-MS/MS for the Determination of the Full Absolute Stereostructure of New Metabolites in Plant Extracts. *Anal. Chem.* **1999**, *71*, 2678–2686.
298. Bringmann, G.; Rübenacker, M.; Geuder, T.; Aké Assi, L. Dioncophylline B, a Naphthylisoquinoline Alkaloid With a New Coupling Type From *Triphyophyllum peltatum*. *Phytochemistry* **1991**, *30*, 3845–3847.
299. Bringmann, G.; Hertlein-Amslinger, B.; Kajahn, I.; Dreyer, M.; Brun, R.; Moll, H.; Stich, A.; Ndjoko Ioset, K.; Schmitz, W.; Ngoc, L. H. Phenolic Analogs of the *N,C*-Coupled Naphthylisoquinoline Alkaloid Ancistrocladinium A, From *Ancistrocladus cochinchinensis* (Ancistrocladaceae), With Improved Antiprotozoal Activities. *Phytochemistry* **2011**, *72*, 89–93.
300. Bringmann, G.; Gulder, T.; Hertlein, B.; Hemberger, Y.; Meyer, F. Total synthesis of the *N,C*-Coupled Naphthylisoquinoline Alkaloids Ancistrocladinium A and B and Related Analogues. *J. Am. Chem. Soc.* **2010**, *132*, 1151–1158.
301. Gulder, T. Neuartige Wirkstoffe Gegen Infektionskrankheiten: *N,C*-Gekuppelte Naphthylisochinolin-Alkaloide. PhD Thesis, University of Würzburg, 2008.
302. Amslinger, B. Isolierung, Totalsynthese, Stereostruktur und -Dynamik Neuartiger Mono- und Dimerer Naphthylisochinoline. PhD Thesis, University of Würzburg, 2012.
303. Seupel, R. Antiinfektive und Antitumorale Naphthylisochinolin-Alkaloide: Isolierung und Strukturaufklärung, Totalsynthese und Untersuchungen zum Wirkmechanismus. PhD Thesis, University of Würzburg, 2018.
304. Domingo, M. P.; Pardo, J.; Cebolla, V.; Galvez, E. M. Berberine: A Fluorescent Alkaloid With a Variety of Applications From Medicine to Chemistry. *Mini-Rev. Org. Chem.* **2010**, *7*, 335–340.
305. Neag, M. A.; Mocan, A.; Echeverria, J.; Pop, R. M.; Bocsan, C. I.; Crişan, G.; Buzoianu, A. D. Berberine: Botanical Occurrence, Traditional Uses, Extraction Methods, and Relevance in Cardiovascular, Metabolic, Hepatic, and Renal Disorders. *Front. Pharmacol.* **2018**, *9*, 557.
306. Ai, X.; Yu, P.; Peng, L.; Luo, L.; Liu, J.; Li, S.; Lai, X.; Luan, F.; Meng, X. Berberine: A Review of Its Pharmacokinetics Properties and Therapeutic Potentials in Diverse Vascular Diseases. *Front. Pharmacol.* **2021**, *12*, 762654.
307. Rauf, A.; Abu-Izneid, T.; Khalik, A. A.; Imran, M.; Shah, Z. A.; Bin Emran, T.; Mitra, S.; Khan, Z.; Alhumaydhi, F. A.; Aljohani, A. S. M.; Khan, I.; Rahman, M. M.; Jeandet, P.; Gondal, T. A. Berberine as a Potential Anticancer Agent: A Comprehensive Review. *Molecules* **2021**, *26*, 7368.
308. McMahon, J. B.; Currens, M. J.; Gulakowski, R. J.; Buckheit, R. W., Jr.; Lackman-Smith, C.; Hallock, Y. F.; Boyd, M. R. Michellamine B, a Novel Plant Alkaloid, Inhibits Human Immunodeficiency Virus-Induced Cell Killing by at Least Two Distinct Mechanisms. *Antimicrob. Agents Chemother.* **1995**, *39*, 484–488.

309. Supko, J. G.; Malspeis, L. Pharmacokinetics of Michellamine B, a Naphthylisoquinoline Alkaloid With In Vitro Activity Against Human Immunodeficiency Virus Types 1 and 2, in the Mouse and Dog. *Antimicrob. Agents Chemother.* **1995**, *39*, 9–14.
310. Nwaka, S.; Ramirez, B.; Brun, R.; Maes, L.; Douglas, F.; Ridley, R. Advancing Drug Innovation for Neglected Diseases—Criteria for Lead Progression. *PLoS Neglected Trop. Dis.* **2009**, *3*, e440.
311. Bringmann, G.; Irmer, A.; Büttner, T.; Schaumlöffel, A.; Zhang, G.; Seupel, R.; Feineis, D.; Fester, K. Axially Chiral Dimeric Naphthalene and Naphthoquinone Metabolites, From Root Cultures of the West African Liana *Triphyophyllum peltatum*. *J. Nat. Prod.* **2016**, *79*, 2094–2103.
312. Bringmann, G.; Ochse, M.; Michel, M. Gentrymine B, an *N*-Quaternary *Ancistrocladus* Alkaloid: Stereoanalysis, Synthesis, and Biomimetic Formation From Gentrymine A. *Tetrahedron* **2000**, *56*, 581–584.
313. Lien, L. Q.; Linh, T. M.; Giang, V. H.; Mai, N. C.; Nhiem, N. X.; Tai, B. H.; Cuc, N. T.; Anh, L. T.; Ban, N. K.; Minh, C. V.; Kiem, P. V. New naphthalene Derivatives and Isoquinoline Alkaloids From *Ancistrocladus cochinchinensis* With Their Anti-Proliferative Activity on Human Cancer Cells. *Bioorg. Med. Chem. Lett.* **2016**, *26*, 3913–3917.
314. Govindachari, T. R.; Parthasarathy, P. C.; Modi, J. D. Chemical Investigation of *Ancistrocladus heyneanus* Wall: Part IV—Structure & Synthesis of Ancistroquinone. *Indian J. Chem.* **1971**, *9*, 1042–1043.
315. Bringmann, G.; Weirich, R.; Reuscher, H.; Jansen, J. R.; Kinzinger, L.; Ortmann, T. The Synthesis of All Possible Isomeric 6,8-Dioxygenated 1,3-Dimethyl-1,2,3,4-Tetrahydroisoquinoline Methyl Ethers—Useful Chiral Building Blocks for Naphthylisoquinoline Alkaloids. *Liebigs Ann.* **1993**, 877–888.
316. Bringmann, G.; Rückert, M.; Schlauer, J.; Herderich, M. Separation and Identification of Dimeric Naphthylisoquinoline Alkaloids by Liquid Chromatography Coupled to Electrospray Ionization Mass Spectrometry. *J. Chromatogr. A* **1998**, *810*, 231–236.
317. Bringmann, G.; Saeb, W.; Peters, K.; Peters, E. M. The Absolute Stereostructure of Dioncophylline A by Anomalous X-ray Dispersion of a 5-Bromo Derivative. *Phytochemistry* **1997**, *45*, 1283–1285.
318. Berova, N.; Nakanishi, K. Exciton Chirality Method: Principles and Application. In *Circular Dichroism: Principles and Application*; Nakanishi, K., Berova, N., Woody, R. W., Eds, 2nd ed.; Wiley-VCH: New York, 2000; pp. 337–376.
319. Berova, N.; Di Bari, L.; Pescitelli, G. Application of Electronic Circular Dichroism in Configurational and Conformational Analysis of Organic Compounds. *Chem. Soc. Rev.* **2007**, *36*, 914–931.
320. Superchi, S.; Scafato, P.; Gorecki, M.; Pescitelli, G. Absolute Configuration Determination by Quantum Mechanical Calculation of Chiroptical Spectra: Basics and Application to Fungal Metabolites. *Curr. Med. Chem.* **2018**, *25*, 287–320.
321. Mándi, A.; Kurtán, T. Application of OR/ECD/VCD to the Structure Elucidation of Natural Products. *Nat. Prod. Rep.* **2019**, *36*, 889–918.
322. Di Bari, L.; Pescitelli, G. Electronic Circular Dichroism. In *Computational Spectroscopy—Methods, Experiments and Applications*; Grunenberg, J., Ed.; Wiley-VCH: Weinheim, 2010; pp. 241–278.
323. Harada, N.; Nakanishi, K.; Berova, N. Electronic CD Exciton Chirality Method: Principles and Application. In *Comprehensive Chiroptical Spectroscopy—Applications in Stereochemical Analysis of Synthetic Compounds, Natural Products, and Biomolecules*; Berova, N., Polavarapu, P. L., Nakanishi, K., Woody, R. W., Eds.; Vol. 2; John Wiley & Sons, Inc., 2012; pp. 115–166.

324. Crawford, T. D. Ab Initio Calculation of Molecular Chiroptical Properties. *Theor. Chem. Acc.* **2006**, *115*, 227–245.
325. Autschbach, J. Computing Chiroptical Properties With First-Principles Theoretical Methods: Background and Illustrative Examples. *Chirality* **2009**, *21*, E116–E152.
326. Li, X. C.; Ferreira, D.; Ding, Y. Determination of Absolute Configuration of Natural Products: Theoretical Calculation of Electronic Circular Dichroism as a Tool. *Curr. Org. Chem.* **2010**, *14*, 1678–1697.
327. Berova, N.; Ellestad, G.; Nakanishi, K.; Harada, N. Recent Advances in the Application of Electronic Circular Dichroism for Studies of Bioactive Natural Products. In *Bioactive Compounds From Natural Sources*; Tringali, C., Ed, 2nd ed.; CRC Press: Boca Raton., 2012; pp. 133–166.
328. Nugroho, A. E.; Morita, H. Circular Dichroism Calculation for Natural Products. *J. Nat. Med.* **2014**, *68*, 1–10.
329. Grauso, L.; Teta, R.; Esposito, G.; Menna, M.; Mangoni, A. Computational Prediction of Chiroptical Properties in Structure Elucidation of Natural Products. *Nat. Prod. Rep.* **2019**, *36*, 1005–1030.
330. Menna, M.; Imperatore, C.; Mangoni, A.; Della Sala, G.; Taglialatela-Scafati, O. Challenges in the Configuration Assignment of Natural Products. A Case-Selective Perspective. *Nat. Prod. Rep.* **2019**, *36*, 476–489.
331. Lightner, D. A. The Octant Rule. In *Circular Dichroism: Principles and Application*; Nakanishi, K., Berova, N., Woody, R. W., Eds, 2nd ed.; Wiley-VCH: New York, 2000; pp. 261–304.
332. Bruhn, T.; Pescitelli, G.; Jurinovich, S.; Schaumlöffel, A.; Witterauf, F.; Ahrens, J.; Bröring, M.; Bringmann, G. Axially Chiral BODIPY DYEmers: an Apparent Exception to the Exciton Chirality Rule. *Angew. Chem., Int. Ed.* **2014**, *53*, 14592–14595.
333. Bringmann, G.; Busemann, S.; Schreier, P.; Herderich, M.; Humpf, H. U.; Schwab, W. The Quantumchemical Calculation of CD Spectra: The Absolute Configuration of Chiral Compounds From Natural or Synthetic Origin. In *Natural Product Analysis*; Vieweg: Braunschweig, 1998; pp. 195–212.
334. Fleischhauer, J.; Koslowski, A.; Kramer, B.; Zobel, E.; Bringmann, G.; Gulden, K. P.; Ortmann, T.; Peter, B. Z. Detection and Calculation of the CD Spectra From the Biaryl Alkaloids Ancistrocladeine and Dioncophylleine A. *Naturforsch.* **1993**, *48b*, 140–148.
335. Bringmann, G.; Mühlbacher, J.; Repges, C.; Fleischhauer, J. MD-Based CD Calculations for the Assignment of the Absolute Axial Configuration of the Naphthylisoquinoline Alkaloid Dioncophylline A. *J. Comput. Chem.* **2001**, *22*, 1273–1278.
336. Dewar, M. J. S.; Zoebisch, E. G.; Healy, E. F.; Stewart, J. J. P. AM1: A New General Purpose Quantum Mechanical Molecular Model. *J. Am. Chem. Soc.* **1985**, *107*, 3902–3909.
337. Stewart, J. J. P. Optimization of Parameters for Semiempirical Methods. I. Method. *J. Comput. Chem.* **1989**, *10*, 209–220.
338. Stewart, J. J. P. Optimization of Parameters for Semiempirical Methods. II. Applications. *J. Comput. Chem.* **1989**, *10*, 221–264.
339. Hostaš, J.; Řezáč, J.; Hobza, P. On the Performance of the Semiempirical Quantum Mechanical PM6 and PM7 Methods for Noncovalent Interactions. *Chem. Phys. Lett.* **2013**, *568–569*, 161–166.
340. Becke, A. D. A New Mixing of Hartree-Fock and Local Density-Functional Theories. *J. Chem. Phys.* **1993**, *98*, 1372–1377.
341. Becke, A. D. Density-Functional Thermochemistry. III. The Role of Exact Exchange. *J. Chem. Phys.* **1993**, *98*, 5648–5652.

342. Hariharan, P. C.; Pople, J. A. The Influence of Polarization Functions on Molecular Orbital Hydrogenation Energies. *Theor. Chim. Acta* **1973**, *28*, 213–222.
343. *SYBYL*, Tripos, Inc., 1699 South Hanley Road, St. Louis, MO 63144, USA n.d.
344. Clark, M.; Crammer, R. D., III; Van Opdenbosch, M. Validation of the General Purpose Tripos 5.2 Force Field. *J. Comput. Chem.* **1989**, *10*, 982–1012.
345. Maple, J. R.; Hwang, M. J.; Stockfisch, T. P.; Dinur, U.; Waldman, M.; Ewig, C. S.; Hagler, A. T. Derivation of Class II Force Fields. I. Methodology and Quantum Force Field for the Alkyl Functional Group and Alkane Molecules. *J. Comput. Chem.* **1994**, *15*, 162–182.
346. Allinger, N. L.; Yuh, Y. H.; Lii, J. H. Molecular Mechanics. The MM3 Force Field for Hydrocarbons. *J. Am. Chem. Soc.* **1989**, *111*, 8551–8566.
347. Lii, J. H.; Allinger, N. L. Molecular Mechanics. The MM3 Force Field for Hydrocarbons. 2. Vibrational Frequencies and Thermodynamics. *J. Am. Chem. Soc.* **1989**, *111*, 8566–8575.
348. Lii, J. H.; Allinger, N. L. Molecular Mechanics. The MM3 Force Field for Hydrocarbons. 3. The Van Der Waals' Potentials and Crystal Data for Aliphatic and Aromatic Hydrocarbons. *J. Am. Chem. Soc.* **1989**, *111*, 8576–8582.
349. Cui, W.; Li, F.; Allinger, N. L. Simulation of Conformational Dynamics With the MM3 Force Field: The Pseudorotation of Cyclopentane. *J. Am. Chem. Soc.* **1993**, *115*, 2943–2951.
350. Del Bene, J.; Jaffé, H. H. Use of the CNDO Method in Spectroscopy. I. Benzene, Pyridine, and the Diazines. *J. Chem. Phys.* **1968**, *48*, 1807–1813.
351. Sandström, J. Determination of Absolute Configurations and Conformations of Organic Compounds by Theoretical Calculations of CD Spectra. *Chirality* **2000**, *12*, 162–171.
352. Ridley, J.; Zerner, M. An Intermediate Neglect of Differential Overlap Technique for Spectroscopy: Pyrrole and the Azines. *Theor. Chim. Acta* **1973**, *32*, 111–134.
353. Zerner, M. C. Semiempirical Molecular Orbital Methods. In *Reviews in Computational Chemistry*; Lipkowitz, K. B., Boyd, D. B., Eds.; Vol. 2; Wiley-VCH: New York, 1991; pp. 313–365.
354. Weber, W.; Thiel, W. Orthogonalization Corrections for Semiempirical Methods. *Theor. Chem. Acc.* **2000**, *103*, 495–506.
355. Sherrill, C. D.; Schaefer, H. F., III. The Configuration Interaction Method: Advances in Highly Correlated Approaches. In *Advances in Quantum Chemistry*; Lowden, P. O., Sabin, J. R., Zerner, M. C., Brändas, E., Eds.; Vol. 34; Academic Press: San Diego, 1999; pp. 143–269.
356. Furche, F.; Ahlrichs, R.; Wachsmann, C.; Weber, E.; Sobanski, A.; Vögtle, F.; Grimme, S. Circular Dichroism of Helicenes Investigated by Time-Dependent Density Functional Theory. *J. Am. Chem. Soc.* **2000**, *122*, 1717–1724.
357. Pescitelli, G.; Bruhn, T. Good Computational Practice in the Assignment of Absolute Configurations by TDDFT Calculations of ECD Spectra. *Chirality* **2016**, *28*, 466–474.
358. Grimme, S.; Waletzke, M. A combination of Kohn-Sham Density Functional Theory and Multi-Reference Configuration Interaction Methods. *J. Chem. Phys.* **1999**, *111*, 5645–5655.
359. Lee, C.; Yang, W.; Parr, R. G. Development of the Colle-Salvetti Correlation-Energy Formula Into a Functional of the Electron Density. *Phys. Rev. B* **1988**, *37*, 785–789.
360. Burke, K. Perspective on Density Functional Theory. *J. Chem. Phys.* **2012**, *136*, 150901.
361. Schäfer, A.; Huber, C.; Ahlrichs, R. Fully Optimized Contracted Gaussian Basis Sets of Triple Zeta Valence Quality for Atoms Li to Kr. *J. Chem. Phys.* **1994**, *100*, 5829–5835.

362. Bringmann, G.; Hamm, A.; Schraut, M. Atroposelective Biaryl Coupling with Chiral Catalysts: Total Synthesis of the Antileishmanial Naphthylisoquinoline Alkaloids Ancistrotanzanine B and Ancistroealaine A. *Org. Lett.* **2003**, *5*, 2805–2808.
363. Bringmann, G.; Jansen, J. R.; Busse, H. An Atropisomer-Differentiating Reaction for the Chemical Analysis of Axial Chirality and Its Computational Investigation: Proof of the Stereostructure of Dioncophylline A. *Liebigs Ann. Chem.* **1991**, 803–809.
364. Bringmann, G.; Jansen, J. R. Stereocontrolled Ring Opening of Axially Prostereogenic Biaryl Lactones With Hydrogen Nucleophiles: Directed Synthesis of a Dioncophylline A Precursor and (Optionally) Its Atropdiastereomer. *Synthesis* **1991**, 825–827.
365. Kushwaha, P. P.; Singh, A. K.; Prajapati, K. S.; Shuaib, M.; Fayez, S.; Bringmann, G.; Kumar, S. Induction of Apoptosis in Breast Cancer Cells by Naphthylisoquinoline Alkaloids. *Toxicol. Appl. Pharmacol.* **2020**, *409*, 115297.
366. Yücer, R.; Fayez, S.; Feineis, D.; Klauck, S. M.; Shan, L.; Bringmann, G.; Efferth, T.; Dawood, M. Cytotoxicity of Dioncophylline A and Related Naphthylisoquinolines in Leukemia Cells, Mediated by NF-κB Inhibition, Angiogenesis Suppression, G2/M Cell Cycle Arrest, and Autophagy Induction. *Phytomedicine* **2024**, *126*, 155267.
367. Soost, D.; Bringmann, G.; Ihmels, H. Towards an Understanding of the Biological Activity of Naphthylisoquinoline Alkaloids: DNA-Binding Properties of Dioncophyllines A, B, and C. *New J. Chem.* **2022**, *46*, 20292–20298.
368. Bringmann, G.; Holzgrabe, U.; Hoerr, V.; Stich, A. Antitrypanosomal Naphthylisoquinoline Alkaloids and Related Compounds. *Pharmazie* **2003**, *58*, 343–346.
369. François, G.; Bringmann, G.; Phillipson, J. D.; Aké Assi, L.; Dochez, C.; Rübenacker, M.; Schneider, C.; Wéry, M.; Warhurst, D. C.; Kirby, G. C. Activity of Extracts and Naphthylisoquinoline Alkaloids From *Triphyophyllum peltatum*, *Ancistrocladus abbreviatus* and *A. barteri* against *Plasmodium falciparum* In Vitro. *Phytochemistry* **1994**, *35*, 1461–1464.
370. François, G.; Bringmann, G.; Dochez, C.; Schneider, C.; Timperman, G.; Aké Assi, L. Activities of Extracts and Naphthylisoquinoline Alkaloids From *Triphyophyllum peltatum*, *Ancistrocladus abbreviatus* and *Ancistrocladus barteri* Against *Plasmodium berghei* (Anka strain) In Vitro. *J. Ethnopharmacol.* **1995**, *46*, 115–120.
371. François, G.; Timperman, G.; Steenackers, T.; Aké Assi, L.; Holenz, J.; Bringmann, G. In Vitro Inhibition of Liver Forms of the Rodent Malaria Parasite *Plasmodium berghei* by Naphthylisoquinoline Alkaloids—Structure-Activity Relationships of Dioncophyllines A and C and Ancistrocladine. *Parasitol. Res.* **1997**, *83*, 673–679.
372. Bringmann, G.; Gampe, C. M.; Reichert, Y.; Bruhn, T.; Faber, J. H.; Mikyna, M.; Reichert, M.; Leippe, M.; Brun, R.; Gelhaus, C. Synthesis and Pharmacological Evaluation of Fluorescent and Photoactivatable Analogues of Antiplasmodial Naphthylisoquinolines. *J. Med. Chem.* **2007**, *50*, 6104–6115.
373. Bringmann, G.; Fayez, S.; Shamburger, W.; Feineis, D.; Winiarczyk, S.; Janecki, R.; Adaszek, Ł. Naphthylisoquinoline Alkaloids and Their Synthetic Analogs as Potent Novel Inhibitors Against *Babesia canis* In Vitro. *Vet. Parasitol.* **2020**, *283*, 109177.
374. Bruneton, J.; Bouquet, A.; Fournett, A.; Cavé, A. La Triphyophylline, Nouvel Alcaloïde Isolé du *Triphyophyllum peltatum*. *Phytochemistry* **1976**, *15*, 817–818.
375. Lavault, M.; Bruneton, J. Sur Trois Alcaloïdes Isolés de Dionchophyllacées. *C. R. Acad. Sci. Paris, Ser. C* **1978**, *287*, 129–131.
376. Bringmann, G.; Zagst, R.; Schöner, B.; Busse, H.; Hemmerling, M.; Burschka, C. Structure of the Naphthylisoquinoline Alkaloid Dioncophylline A. *Acta Crystallogr., Sect. C: Cryst. Struct. Commun.* **1991**, *C47*, 1703–1705.
377. Bringmann, G.; Maksimenka, K.; Bruhn, T.; Reichert, M.; Harada, T.; Kuroda, R. Quantum Chemical CD Calculations of Dioncophylline A in the Solid State. *Tetrahedron* **2009**, *65*, 5720–5728.

378. Rizzacasa, M. A. Total Synthesis of Naphthylisoquinoline Alkaloids. In *Studies in Natural Products Chemistry*; Atta-ur-Rahman, Ed.; Vol. 20; Elsevier Science B.V: Amsterdam, The Netherlands, 1998; pp. 407–455.
379. Lavault, M.; Kouhon, T.; Bruneton, J. O-Méthyl-Triphyophylline et O-methyldéhydro-1,2-Triphyophylline, Noveaux Alcaloïdes du *Triphyophyllum peltatum*. *C. R. Acad. Sci. Paris, Ser. C* **1977**, *285*, 167–169.
380. Neese, F. *ORCA*; Universität Bonn: Germany, 2008.
381. Bringmann, G.; Breuning, M.; Pfeifer, R. M.; Schenk, W. A.; Kamikawa, K.; Uemura, M. The Lactone Concept—A Novel Approach to the Metal-Assisted Atroposelective Construction of Axially Chiral Biaryl Systems. *J. Organomet. Chem.* **2002**, *661*, 31–47.
382. Bringmann, G.; Tasler, S.; Pfeifer, R. M.; Breuning, M. The Directed Synthesis of Axially Chiral Ligands, Reagents, Catalysts, and Natural Products: Through the 'Lactone Methodology'. *J. Organomet. Chem.* **2002**, *661*, 49–65.
383. Bringmann, G.; Jansen, J. R.; Rink, H. P. Regioselective and Atropoisomeric-Selective Aryl Coupling to Give Naphthylisoquinoline Alkaloids: The First Total Synthesis of (-)-Ancistrocladine. *Angew. Chem.* **1986**, *98*, 917–919. *Angew. Chem., Int. Ed. Engl.* **1986**, *25*, 913–915.
384. Bringmann, G.; Reuscher, H. Atropodiastereoselective Ring Opening of Bridged, "Axial-Prostereogenic" Biaryls: Directed Synthesis of (+)-Ancistrocladisine. *Angew. Chem.* **1989**, *101*, 1725–1726. *Angew. Chem., Int. Ed. Engl.* **1989**, *28*, 1672–1673.
385. Bringmann, G.; Holenz, J.; Weirich, R.; Rübenacker, M.; Funke, C.; Boyd, M. R.; Gulakowski, R. J.; François, G. First Synthesis of the Antimalarial Naphthylisoquinoline Alkaloid Dioncophylline C, and Its Unnatural Anti-HIV Dimer, Jozimine C. *Tetrahedron* **1998**, *54*, 497–512.
386. Bringmann, G.; Saeb, W.; Rübenacker, M. Directed Joint Total Synthesis of the Three Naphthylisoquinoline Alkaloids Dioncolactone A, Dioncopeltine A, and 5′-O-Demethyldioncophylline A. *Tetrahedron* **1999**, *55*, 423–432.
387. Bringmann, G.; Ochse, M.; Götz, R. First Atropo-Divergent Total Synthesis of the Antimalarial Korupensamines A and B by the "Lactone Method". *J. Org. Chem.* **2000**, *65*, 2069–2077.
388. Bringmann, G.; Manchala, N.; Büttner, T.; Hertlein-Amslinger, B.; Seupel, R. First Atroposelective Total Synthesis of Enantiomerically Pure Ancistrocladidine and Ancistrotectorine. *Chem. Eur. J.* **2016**, *22*, 9792–9796.
389. Bringmann, G.; Pabst, T.; Henschel, P.; Kraus, J.; Peters, K.; Peters, E. M.; Rycroft, D. S.; Connolly, J. D. Nondynamic and Dynamic Kinetic Resolution of Lactones with Stereogenic Centers and Axes: Stereoselective Total Synthesis of Herbertenediol and Mastigophorenes A and B. *J. Am. Chem. Soc.* **2000**, *122*, 9127–9133.
390. Bringmann, G.; Menche, D. First, Atropo-Enantioselective Total Synthesis of the Axially Chiral Phenylanthraquinone Natural Products Knipholone and 6′-O-Methylknipholone. *Angew. Chem.* **2001**, *113*, 1733–1736. *Angew. Chem., Int. Ed.* **2001**, *40*, 1687–1690.
391. Bringmann, G.; Menche, D.; Kraus, J.; Mühlbacher, J.; Peters, K.; Peters, E. M.; Brun, R.; Bezabih, M.; Abegaz, B. M. Atropo-Enantioselective Total Synthesis of Knipholone and Related Antiplasmodial Phenylanthraquinones. *J. Org. Chem.* **2002**, *67*, 5595–5610.
392. Bringmann, G.; Menche, D.; Mühlbacher, J.; Reichert, M.; Saito, N.; Pfeiffer, S. S.; Lipshutz, B. H. On the Verge of Axial Chirality: Atroposelective Synthesis of the AB-Biaryl Fragment of Vancomycin. *Org. Lett.* **2002**, *4*, 2833–2836.
393. Bringmann, G.; Hinrichs, J.; Henschel, P.; Kraus, J.; Peters, K.; Peters, E. M. Atropo-Enantioselective Synthesis of the Natural Bicoumarin (+)-Isokotanin A via a Configurationally Stable Biaryl Lactone. *Eur. J. Org. Chem.* **2002**, 1096–1106.

394. Kitamura, M.; Ohmori, K.; Kawase, T.; Suzuki, K. *Angew. Chem.* **1999**, *111*, 1308–1311. Total Synthesis of Pradimicinone, the Common Aglycon of the Pradimicin-Benanomicin Antibiotics. *Angew. Chem., Int. Ed.* **1999**, *38*, 1229–1232.
395. Ohmori, K.; Tamiya, M.; Kitamura, M.; Kato, H.; Oorui, M.; Suzuki, K. *Angew. Chem.* **2005**, *117*, 3939–3942. Regio- and Stereocontrolled Total Synthesis of Benanomicin B. *Angew. Chem., Int. Ed.* **2005**, *44*, 3871–3874.
396. Tamiya, M.; Ohmori, K.; Kitamura, M.; Kato, H.; Arai, T.; Oorui, M.; Suzuki, K. General Synthesis Route to Benanomicin-Pradimicin Antibiotics. *Chem. -Eur. J.* **2007**, *13*, 9791–9823.
397. Bringmann, G.; Schneider, S. Improved Methods for Dehydration and Hydroxy/Halogen Exchange Using Novel Combinations of Triphenylphosphine and Halogenated Ethanes. *Synthesis* **1983**, 139–141.
398. Hammond, C. *The Basics of Crystallography and Diffraction*, 4th ed.; International Union of Crystallography, Ed.; Oxford University Press: Oxford, 2015.
399. Albright, A. L.; White, J. M. Determination of Absolute Configuration Using Single Crystal X-ray Diffraction. In *Metabolomics Tools for Natural Product Discovery: Methods and Protocols*; Roessner, U., Dias, A. D., Eds.; Humana Press: Totowa, NJ, 2013; pp. 149–162.
400. Parthasarathy, P. C.; Kartha, G. Rigorous Structural Proof for Ancistrocladisine and Ancistrocladidine, Two Minor Alkaloids of *Ancistrocladus heyneanus* Wall. *Indian J. Chem.* **1983**, *22B*, 590–591.
401. Bijvoet, J. M.; Peerdeman, A. F.; van Bommel, A. J. Determination of the Absolute Configuration of Optically Active Compounds by Means of X-Rays. *Nature* **1951**, *168*, 271–272.
402. Bringmann, G.; Rübenacker, M.; Weirich, R.; Aké Assi, L. Dioncophylline C From the Roots of *Triphyophyllum peltatum*, the First 5,1'-Coupled Dioncophyllaceae Alkaloid. *Phytochemistry* **1992**, *41*, 4019–4024.
403. Anh, N. H.; Porzel, A.; Ripperger, H.; Bringmann, G.; Schäffer, M.; God, R.; Sung, T. V.; Adam, G. Naphthylisoquinoline Alkaloids From *Ancistrocladus cochinchinensis*. *Phytochemistry* **1997**, *45*, 1287–1291.
404. Fayez, S. Isolation, Structural Elucidation, and Biological Evaluation of Naphthylisoquinoline Alkaloids From Two African *Ancistrocladus* Species. PhD Thesis, University of Würzburg, 2019.
405. Hotez, P. J.; Savioli, L.; Fenwick, A. Neglected Tropical Diseases of the Middle East and North Africa: Review of Their Prevalence, Distribution, and Opportunities for Control. *PLoS Neglected Trop. Dis.* **2012**, *6*, e1475.
406. Feasey, N.; Wansbrough-Jones, M.; Mabey, D. C. W.; Solomon, A. W. Neglected tropical Diseases. *Br. Med. Bull.* **2010**, *93*, 179–200.
407. Barakat, R.; El Morshedy, H.; Farghaly, A. Human Schistosomiasis in the Middle East and North Africa Region. In *Neglected Tropical Diseases—Middle East and North Africa*; McDowell, M. A., Rafati, S., Eds.; Springer Verlag: Wien, 2014; pp. 23–58.
408. Rafati, S.; Modabber, F. Cutaneous Leishmaniasis in Middle East and North Africa. In *Neglected Tropical Diseases—Middle East and North Africa*; McDowell, M. A., Rafati, S., Eds.; Springer Verlag: Wien, 2014; pp. 117–140.
409. Ackley, C.; Elsheikh, M.; Zaman, S. Scoping Review of Neglected Tropical Disease Intervention and Health Promotion: A Framework for Successful NTD Interventions as Evidenced by the Literature. *PLoS Neglected Trop. Dis.* **2021**, *15*, e0009278.
410. Phillips, M. A.; Burrows, J. N.; Manyando, C.; van Huijsduijnen, R. H.; van Voorhis, W. C.; Wells, N. C. Malaria. *Nat. Rev. Dis. Primers* **2017**, *3*, 17050.
411. Cowman, A. F.; Healer, J.; Marapana, D.; Marsh, K. Malaria: Biology and Disease. *Cell* **2016**, *167*, 610–624.

412. Moxon, C. A.; Gibbins, M. P.; McGuinness, D.; Milner, D. A., Jr.; Marti, M. New Insights Into Malaria Pathogenesis. *Annu. Rev. Pathol.: Mech. Dis.* **2020**, *15*, 315–343.
413. World Health Organization. *World Malaria Report 2020: 20 Years of Global Progress and Challenges*; World Health Organization: Geneva, 2020.
414. Guyant, P.; Corbel, V.; Guérin, P. J.; Lautissier, A.; Nosten, F.; Boyer, S.; Coosemans, M.; Dondorp, A. M.; Sinou, V.; Yeung, S.; White, N. Past and New Challenges for Malaria Control and Elimination: The Role of Operational Research for Innovation in Designing Interventions. *Malar. J* **2015**, *14*, 279.
415. Wicht, K. J.; Mok, S.; Fidock, D. A. Molecular Mechanisms of Drug Resistance in *Plasmodium falciparum* Malaria. *Annu. Rev. Microbiol.* **2020**, *74*, 431–454.
416. Blasco, B.; Leroy, D.; Fidock, D. A. Antimalarial Drug Resistance: Linking *Plasmodium falciparum* Parasite Biology to the Clinic. *Nat. Med.* **2017**, *23*, 917–928.
417. Duffey, M.; Blasco, B.; Burrows, J. N.; Wells, T. N. C.; Fidock, D. A.; Leroy, D. Assessing Risks of *Plasmodium falciparum* Resistance to Select Next-Generation Antimalarials. *Trends Parasitol.* **2021**, *37*, 709–721.
418. François, G.; Timperman, G.; Holenz, J.; Aké Assi, L.; Geuder, T.; Maes, L.; Dubois, J.; Hanocq, M.; Bringmann, G. Naphthylisoquinoline Alkaloids Exhibit Strong Growth-Inhibiting Activities Against *Plasmodium falciparum* and *P. berghei* In Vitro—Structure-Activity Relationships of Dioncophylline C. *Ann. Trop. Med. Parasitol.* **1996**, *90*, 115–123.
419. Park, W.; Chawla, A.; O'Reilly, E. M. Pancreatic Cancer: A Review. *JAMA* **2021**, *326*, 851–862.
420. Ansari, E.; Tingstedt, B.; Andersson, B.; Holmquist, F.; Sturesson, C.; Williamsson, C.; Sasor, A.; Borg, D.; Bauden, M.; Andersson, R. Pancreatic Cancer: Yesterday, Today and Tomorrow. *Future Oncol.* **2016**, *12*, 1929–1946.
421. Kamisawa, T.; Wood, L. D.; Itoi, T.; Takaori, K. Pancreatic Cancer. *Lancet* **2016**, *388*, 73–85.
422. Kleeff, J.; Korc, M.; Apte, M.; La Vecchia, C.; Johnson, C. D.; Biankin, A. V.; Neale, R. E.; Tempero, M.; Tuveson, D. A.; Hruban, R. H.; Neoptolemos, J. P. Pancreatic Cancer. *Nat. Rev. Dis. Primers* **2016**, *2*, 16022.
423. Kolbeinsson, H. M.; Chandana, S.; Wright, P.; Chung, M. Pancreatic Cancer: A Review of Current Treatment and Novel Therapies. *J. Invest. Surg.* **2023**, *36*, 2129884.
424. Nevala-Plagemann, C.; Hidalgo, M.; Garrido-Laguna, I. From State-of the-Art Treatments to Novel Therapies for Advanced-Stage Pancreatic Cancer. *Nat. Rev. Clin. Oncol.* **2020**, *17*, 108–123.
425. Wang, S.; Zheng, Y.; Yang, F.; Zhu, L.; Zhu, X. Q.; Wang, Z. F.; Wu, X. L.; Zhou, C. H.; Yan, J. Y.; Hu, B. Y.; Kong, B.; Fu, D. L.; Bruns, C.; Zhao, Y.; Qin, L. X.; Dong, Q. Z. The Molecular Biology of Pancreatic Adenocarcinoma: Translational Challenges and Clinical Perspectives. *Signal Transduction Targeted Ther.* **2021**, *6*, 249.
426. Brunner, M.; Wu, Z.; Krautz, C.; Pilarsky, C.; Grützmann, R.; Weber, G. F. Current Clinical Strategies of Pancreatic Cancer Treatment and Open Molecular Questions. *Int. J. Mol. Sci.* **2019**, *20*, 4543.
427. Izuishi, K.; Kato, K.; Ogura, T.; Esumi, H. Remarkable Tolerance of Tumor Cells to Nutrient Deprivation: Possible New Biochemical Target for Cancer Therapy. *Cancer Res.* **2000**, *60*, 6201–6207.
428. Awale, S.; Lu, J.; Kalauni, S. K.; Kurashima, Y.; Tezuka, Y.; Kadota, S.; Esumi, H. Identification of Arctigenin as an Antitumor Agent Having the Ability to Eliminate the Tolerance of Cancer Cells to Nutrient Starvation. *Cancer Res.* **2006**, *66*, 1751–1757.
429. Magolan, J.; Coster, M. J. Targeting the Resistance of Pancreatic Cancer Cells to Nutrient Deprivation: Anti-Austerity Compounds. *Curr. Drug Delivery* **2010**, *7*, 355–369.

430. Gu, Y.; Qi, C.; Sun, X.; Ma, X.; Zhang, H.; Hu, L.; Yuan, J.; Yu, Q. Arctigenin Preferentially Induces Tumor Cell Death Under Glucose Deprivation by Inhibiting Cellular Engergy Metabolism. *Biochem. Pharmacol.* **2012**, *84*, 468–476.
431. Ikeda, M.; Sato, A.; Mochizuki, N.; Toyosaki, K.; Miyoshi, C.; Fujioka, R.; Mitsunaga, S.; Ohno, I.; Hashimoto, Y.; Takahashi, H.; Hasegawa, H.; Nomura, S.; Takahashi, R.; Yomoda, S.; Tsuchihara, K.; Kishino, S.; Esumi, H. Phase I Trial of GBS-01 for Advanced Pancreatic Cancer Refractory to Gemcitabine. *Cancer Sci.* **2016**, *107*, 1818–1824.
432. Kim, I. K.; Zhang, X. H. F. One Microenvironment Does Not Fit All: Heterogeneity Beyond Cancer Cells. *Cancer Metastasis Rev.* **2016**, *35*, 601–629.
433. Feig, C.; Gopinathan, A.; Neesse, A.; Chan, D. S.; Cook, N.; Tuveson, D. A. The Pancreas Cancer Microenvironment. *Clin. Cancer Res.* **2012**, *18*, 4266–4276.
434. Liu, S. L.; Cao, S. G.; Li, Y.; Sun, B.; Chen, D.; Wang, D. S.; Zhou, Y. B. Pancreatic Stellate Cells Facilitate Pancreatic Cancer Cell Viability and Invasion. *Oncol. Lett.* **2019**, *17*, 2057–2062.
435. Ho, W. J.; Jaffee, E. M.; Zheng, L. The Tumour Microenvironment in Pancreatic Cancer—Clinical Challenges and Opportunities. *Nat. Rev. Clin. Oncol.* **2020**, *17*, 527–540.
436. Maneenet, J.; Tawila, A. M.; Omar, S. M.; Phan, N. D.; Ojima, C.; Kuroda, M.; Sato, M.; Mizoguchi, M.; Takahashi, I.; Awale, S. Chemical Constituents of *Callistemon subulatus* and Their Anti-Pancreatic Cancer Activity Against Human PANC-1 Cell Line. *Plants* **2022**, *11*, 11192466.
437. Kohyama, A.; Kim, M. J.; Sun, S.; Omar, A. M.; Phan, N. D.; Meselhy, M. R.; Tsuge, K.; Awale, S.; Matsuya, Y. Structure-Activity Relationship and Mechanistic Study on Guggulsterone Derivatives: Discovery of New Anti-Pancreatic Cancer Candidate. *Bioorg. Med. Chem.* **2022**, *54*, 116563.
438. Sun, S.; Kim, M. J.; Omar, A. M.; Phan, N. D.; Aoike, M.; Awale, S. GDP Induces PANC-1 Human Pancreatic Cancer Cell Death Preferentially Under Nutrient Starvation by Inhibiting PI3K/Akt/mTOR/Autophagy Signaling Pathway. *Chem. Biodiversity* **2021**, *18*, e2100389.
439. Sun, S.; Dibwe, D. F.; Kim, M. J.; Omar, A. M.; Phan, N. D.; Fujino, H.; Pongterdsak, N.; Chaithatwatthana, K.; Phrutivorapongkul, A.; Awale, S. A New Anti-Austerity Agent, 4′-O-Methylgrynullarin From *Derris scandens* Induces PANC-1 Human Pancreatic Cancer Cell Death Under Nutrition Starvation via Inhibition of Akt/mTOR Pathway. *Bioorg. Med. Chem. Lett.* **2021**, *40*, 127967.
440. Tawila, A. M.; Sun, S.; Kim, M. J.; Omar, A. M.; Dibwe, D. F.; Ueda, J. Y.; Toyooka, N.; Awale, S. Highly Potent Antiausterity Agents From *Callistemon citrinus* and Their Mechanism of Action Against PANC-1 Human Pancreatic Cancer Cell Line. *J. Nat. Prod.* **2020**, *83*, 2221–2232.
441. Omar, A. M.; Dibwe, D. F.; Tawila, A. M.; Sun, S.; Phrutivorapongkul, A.; Awale, S. Chemical Constituents of *Anneslea fragans* and Their Antiausterity Activity Against the PANC-1 Human Pancreatic Cancer Cell Line. *J. Nat. Prod.* **2019**, *82*, 3133–3139.
442. Alilou, M.; Dibwe, D. F.; Schwaiger, S.; Khodami, M.; Troppmair, J.; Awale, S.; Stuppner, H. Antiausterity Activity of Secondary Metabolites From the Roots of *Ferula hezarlalehzarica* Against the PANC-1 Human Pancreatic Cancer Cell Line. *J. Nat. Prod.* **2020**, *83*, 1099–1106.
443. Sun, S.; Kim, M. J.; Omar, A. M.; Phan, N. D.; Awale, S. (+)-Panduratin A Induces PANC-1 Human Pancreatic Cancer Cell Death Preferentially Under Nutrient Starvation by Inhibiting PI3K/Akt/mTOR/Autophagy Signaling Pathway. *Phytomedicine Plus* **2021**, *1*, 100101.

444. Longley, D. B.; Harkin, D. P.; Johnston, P. G. 5-Fluorouracil: Mechanisms of Action and Clinical Strategies. *Nat. Rev. Cancer* **2003**, *3*, 330–338.
445. Sethy, C.; Kundu, C. N. 5-Fluorouracil (5-FU) Resistance and the New Strategy to Enhance the Sensivity Against Cancer: Implication of DNA Repair Inhibition. *Biomed. Pharmacother.* **2021**, *137*, 111285.
446. Xie, P.; Mo, J. L.; Liu, J. H.; Li, X.; Tan, L. M.; Zhang, W.; Zhou, H. H.; Liu, Z. Q. Pharmacogenomics of 5-Fluorouracil in Colorectal Cancer: Review and Update. *Cell. Oncol.* **2020**, *43*, 989–1001.
447. Guzmán, C.; Bagga, M.; Kaur, A.; Westermarck, J.; Abankwa, D. ColonyArea: An ImageJ Plugin to Automatically Quantify Colony Formation in Clonogenic Assays. *PLoS One* **2014**, *9*, e92444.
448. Schindelin, J.; Arganda-Carreras, I.; Frise, E.; Kaynig, V.; Longair, M.; Pietzsch, T.; Preibisch, S.; Rueden, C.; Saalfeld, S.; Schmid, B.; Tinevez, J. Y.; White, D. J.; Hartenstein, V.; Eliceiri, K.; Tomancak, P.; Cardon, A. Fiji: An Open-Source Platform for Biological-Image Analysis. *Nat. Methods* **2012**, *9*, 676–682.
449. Hoye, T. R.; Mi, L. Total Syntheses of Korupensamine C and Ancistrobrevine B. *Tetrahedron Lett.* **1996**, *37*, 3097–3098.
450. Hoye, T. R.; Chen, M.; Hoang, B.; Mi, L.; Priest, O. P. Total Synthesis of Michellamines A-C, Korupensamines A-D, and Ancistrobrevine B. *J. Org. Chem.* **1999**, *64*, 7184–7201.
451. Schetters, T. Mechanisms Involved in the Persistence of *Babesia canis* Infection in Dogs. *Pathogens* **2019**, *8*, 94.
452. Solano-Gallego, L.; Sainz, Á.; Roura, X.; Estrada-Peña, A.; Miro, G. A review of Canine Babesiosis: The European Perspective. *Parasites Vectors* **2016**, *9*, 336.
453. Bilić, P.; Kuleš, J.; Barić Rafaj, R.; Mrljak, V. Canine Babesiosis: Where Do We Stand? *Acta Vet. (Belgrade)* **2018**, *68*, 127–160.
454. Baneth, G. Antiprotozoal Treatment of Canine Babesiosis. *Vet. Parasitol.* **2018**, *254*, 58–63.
455. Vial, H. J.; Gorenflot, A. Chemotherapy Against Babesiosis. *Vet. Parasitol.* **2006**, *138*, 147–160.
456. Clark, I. A.; Jacobson, L. S. Do Babesiosis and Malaria Share a Common Disease Process? *Ann. Trop. Med. Parasitol.* **1998**, *92*, 483–488.
457. Krause, P. J.; Daily, J.; Telford, S. R.; Vannier, E.; Lantos, P.; Spielman, A. Shared Features in the Pathobiology of Babesiosis and Malaria. *Trends Parasitol.* **2007**, *23*, 605–610.
458. Rizk, M. A.; El-Salam El-Sayed, S. A.; El-Khodery, S.; Yokoyama, N.; Igarashi, I. Discovering the In Vitro Potent Inhibitors Against *Babesia* and *Theileria* Parasites by Repurposing the Malaria Box: A Review. *Vet. Parasitol.* **2019**, *274*, 108895.
459. Subeki; Matsuura, H.; Takahashi, K.; Yamasaki, M.; Yamato, O.; Maede, Y.; Katakura, K.; Kobayashi, S.; Trimurningsih; Chairul; Yoshihara, T. Anti-Babesial and Anti-Plasmodial Compounds From *Phyllanthus niruri*. *J. Nat. Prod.* **2005**, *68*, 537–539.
460. Kasahara, K.; Nomura, S.; Subeki; Matsuura, H.; Yamasaki, M.; Yamato, O.; Maede, Y.; Katakura, K.; Suzuki, M.; Trimurningsih; Chairul; Yoshihara, T. Anti-Babesial Compounds From *Curcuma zedoaria*. *Planta Med.* **2005**, *71*, 482–484.
461. Murnigsih, T.; Subeki; Matsuura, H.; Takahashi, K.; Yamasaki, M.; Yamato, O.; Maede, Y.; Katakura, K.; Suzuki, M.; Kobayashi, S.; Chairul; Yoshihara, T. Evaluation of the Inhibitory Activities of the Extracts of Indonesian Traditional Medicinal Plants Against *Plasmodium falciparum* and *Babesia gibsoni*. *J. Vet. Med. Sci.* **2005**, *67*, 829–831.

462. Elkhateeb, A.; Subeki; Takahashi, K.; Matsuura, H.; Yamasaki, M.; Yamato, O.; Maede, Y.; Katakura, K.; Yoshihara, T.; Nabeta, K. Anti-Babesial Ellagic Acid Rhamnosides from the Bark of *Elaeocarpus parvifolius*. *Phytochemistry* **2005**, *66*, 2577–2580.
463. Elkhateeb, A.; Yamada, K.; Matsuura, H.; Yamasaki, M.; Maede, Y.; Katakura, K.; Nabeta, K. Anti-Babesial Compounds From *Berberis vulgaris*. *Nat. Prod. Commun.* **2007**, *2*, 173–175.
464. Cowan, A. J.; Green, D. J.; Kwok, M.; Lee, S.; Coffey, D. G.; Holmberg, L. A.; Tuazon, S.; Gopal, A. K.; Libby, E. N. Diagnosis and MANAGEMENT of multiple Myeloma: A Review. *JAMA* **2022**, *327*, 464–477.
465. Rajkumar, S. V. Multiple Myeloma: 2022 Update on Diagnosis, Risk-Stratification and Management. *Am. J. Hematol.* **2022**, *97*, 1086–1107.
466. Van de Donk, N. W. C. J.; Pawlyn, C.; Yong, K. L. Multiple Myeloma. *Lancet* **2021**, *397*, 410–427.
467. Larocca, A.; Mina, R.; Gay, F.; Bringhen, S.; Boccadoro, M. Emerging Drugs and Combinations to Treat Multiple Myeloma. *Oncotarget* **2017**, *8*, 60656–60672.
468. D'Agostinoa, M.; Salvinia, M.; Palumboa, A.; Larocca, A.; Gay, F. Novel Investigational Drugs Active as Single Agents in Multiple Myeloma. *Expert Opin. Invest. Drugs* **2017**, *26*, 699–711.
469. Maruyama, D.; Iida, S.; Ogawa, G.; Fukuhara, N.; Seo, S.; Miyazaki, K.; Yoshimitsu, M.; Kuroda, J.; Tsukamoto, N.; Tsujimura, H.; Hangaishi, A.; Yamauchi, T.; Utsumi, T.; Mizuno, I.; Takamatsu, Y.; Nagata, Y.; Minauchi, K.; Ohtsuka, E.; Hanamura, I.; Yoshida, S.; Yamasaki, S.; Suehiro, Y.; Kamiyama, Y.; Tsukasaki, K.; Nagai, H. Randomised Phase II Study to Optimise Melphalan, Prednisolone, and Bortezomib in Untreated Multiple Myeloma (JCOG1105). *Br. J. Haematol.* **2021**, *192*, 531–541.
470. Logue, S. E.; Elgendy, M.; Martin, S. J. Expression, Purification and Use of Recombinant Annexin V for the Detection of Apoptotic Cells. *Nat. Protoc.* **2009**, *4*, 1383–1395.
471. Bukowski, K.; Kciuk, M.; Kontek, R. Mechanisms of Multidrug Resistance in Cancer Chemotherapy. *Int. J. Mol. Sci.* **2020**, *21*, 3233.
472. Emran, T. B.; Shahriar, A.; Mahmud, A. R.; Rahman, T.; Abir, M. H.; Faijanur, M.; Siddiquee, R.; Ahmed, H.; Rahman, N.; Nainu, F.; Wahyudin, E.; Mitra, S.; Dharma, K.; Habiballah, M. M.; Haque, S.; Islam, A.; Hassan, M. M. Multidrug Resistance in Cancer: Understanding Molecular Mechanisms, Immunoprevention and Therapeutic Approaches. *Front. Oncol.* **2022**, *12*, 891652.
473. Kachalaki, S.; Ebrahimi, M.; Khosroshahi, L. M.; Mohammadinejad, S.; Baradaran, B. Cancer Chemoresistance; Biochemical and Molecular Aspects: A Brief Overview. *Eur. J. Pharm. Sci.* **2016**, *89*, 20–30.
474. Binkhathlan, Z.; Lavasanifar, A. P-Glycoprotein Inhibition as a Therapeutic Approach for Overcoming Multidrug Resistance in Cancer: Current Status and Future Perspectives. *Curr. Cancer Drug Targets* **2013**, *13*, 326–346.
475. Seelig, A. P-Glycoprotein: One Mechanism, Many Tasks and the Consequences for Pharmacotherapy of Cancers. *Front. Oncol.* **2020**, *10*, 576559.
476. Nanayakkara, A. K.; Follit, C. A.; Chen, G.; Williams, N. S.; Vogel, P. D.; Wise, J. G. Targeted Inhibitors of P-Glycoprotein Increase Chemotherapeutic-Induced Mortality of Multidrug Resistant Tumor Cells. *Sci. Rep.* **2018**, *8*, 967.
477. Thorn, C. F.; Oshiro, C.; Marsh, S.; Hernandez-Boussard, T.; McLeod, H.; Klein, T. E.; Altman, R. B. Doxorubicin Pathways: Pharmacodynamics and Adverse Effects. *Pharmacogenet. Genomics* **2011**, *21*, 440–446.
478. Sritharan, S.; Sivalingam, N. A Comprehensive Review on Time-Tested Anticancer Drug Doxorubicin. *Life Sci.* **2021**, *278*, 119527.

479. Kaltschmidt, B.; Greiner, J. F. W.; Kadhim, H. M.; Kaltschmidt, C. Subunit-Specific Role of NF-κB in Cancer. *Biomedicines* **2018**, *6*, 44.
480. Xia, L.; Tan, S.; Zhou, Y.; Lin, J.; Wang, H.; Oyang, L.; Tian, Y.; Liu, L.; Su, M.; Wang, H.; Cao, D.; Liao, Q. Role of the NFκB-Signaling Pathway in Cancer. *OncoTargets Ther.* **2018**, *11*, 2063–2073.
481. Xia, Y.; Shen, S.; Verma, I. M. NF-κB, an Acitve Player in Human Cancers. *Cancer Immunol. Res.* **2014**, *2*, 823–830.
482. Baud, V.; Karin, M. Is NF-κB a Good Target for Cancer Therapy? Hopes and Pitfalls. *Nat. Rev. Drug Discovery* **2009**, *8*, 33–40.
483. Hou, Z. Y.; Tong, X. P.; Peng, Y. B.; Zhang, B. K.; Yan, M. Broad Targeting of Triptolide to Resistance and Sensitization for Cancer Therapy. *Biomed. Pharmacother.* **2018**, *104*, 771–780.
484. Dorsey, J. F.; Dowling, M. L.; Kim, M.; Voong, R.; Solin, L. J.; Kao, G. D. Modulation of the Anti-Cancer Efficacy of Microtubule-Targeting Agents by Cellular Growth Conditions. *Cancer Biol. Ther.* **2010**, *9*, 809–819.
485. Kothari, A.; Hittelman, W. N.; Chambers, T. C. Cell Cycle-Dependent Mechanisms Underlie Vincristine-Induced Death of Primary Acute Lymphoblastic Leukemia Cells. *Cancer Res.* **2016**, *76*, 3553–3561.
486. Lugano, R.; Ramachandran, M.; Dimberg, A. Tumor Angiogenesis: Causes, Consequences, Challenges, and Opportunities. *Cell. Mol. Life Sci.* **2020**, *77*, 1745–1770.
487. Albini, A.; Bruno, A.; Noonan, D. M.; Mortara, L. Contribution to Tumor Angiogenesis From Innate Iummune Cells Within the Tumor Microenvironment: Implications for Immunotherapy. *Front. Immunol.* **2018**, *9*, 527.
488. Jung, S.; Jeong, H.; Yu, S. W. Autophagy as a Decisive Process for Cell Death. *Exp. Mol. Med.* **2020**, *52*, 921–930.
489. Noguchi, M.; Hirata, N.; Tanaka, T.; Suizu, F.; Nakajima, H.; Chiorini, J. A. Autophagy as a Modulator of Cell Death Machinery. *Cell Death Dis.* **2020**, *11*, 517.
490. Blagosklonny, M. V. Cancer Prevention with Rapamycin. *Oncotarget* **2023**, *14*, 345–350.
491. Lin, X.; Han, L.; Wenig, J.; Wang, K.; Chen, T. Rapamycin Inhibits Proliferation and Induces Autophagy in Human Neuroblastoma Cells. *Biosci. Rep.* **2018**, *38*, BSR20181822.
492. Hill, D.; Chen, L.; Snaar-Jagalska, E.; Chaudhry, B. Embryonic Zebrafish Xenograft Assay of Human Cancer Metastasis. *F1000Research* **2018**, *7*, 1682.
493. Gamble, J. T.; Elson, D. J.; Greenwood, J. A.; Tanguay, R. L.; Kolluri, S. K. The Zebrafish Xenograft Models for Investigating Cancer and Cancer Therapeutics. *Biology* **2021**, *10*, 252.
494. Harbeck, N.; Penault-Llorca, F.; Cortes, J.; Gnant, M.; Houssami, N.; Poortmans, P.; Ruddy, K.; Tsang, J.; Cardoso, F. Breast Cancer. *Nat. Rev. Dis. Primers* **2019**, *5*, 66.
495. Hong, R.; Xu, B. Breast Cancer: An Up-to-Date Review and Future Perspectives. *Cancer Commun.* **2022**, *42*, 913–936.
496. Waks, A. G.; Winer, E. P. Breast Cancer Treatment: A Review. *JAMA* **2019**, *321*, 288–300.
497. Sung, H.; Ferlay, J.; Siegel, R. L.; Laversanne, M.; Soerjomataram, I.; Jemal, A.; Bray, F. Global Cancer Statistics 2020: Globocan Estimates of Incidence and Mortality Worldwide for 36 Cancers in 185 countries. *CA-Cancer J. Clin.* **2021**, *71*, 209–249.
498. Siegel, R. L.; Miller, K. D.; Fuchs, H. E.; Jemal, A. Cancer Statistics, 2022. *CA-Cancer J. Clin.* **2022**, *72*, 7–33.
499. Chavez, K. J.; Garimella, S. V.; Lipkowitz, S. Triple Negative Breast Cancer Cell Lines: One Tool in the Search for Better Treatment of Triple Negative Breast Cancer. *Breast Dis.* **2010**, *32*, 35–48.

500. Comşa, Ş.; Cîmpean, A. M.; Raica, M. The Story of MCF-7 Breast Cancer Line: 40 Years of Experience in Research. *Anticancer Res.* **2015**, *35*, 3147–3154.
501. Park, B. K.; Coleman, J. W.; Kitteringham, N. R. Drug Disposition and Drug Hypersensitivity. *Biochem. Pharmacol.* **1987**, *36*, 581–590.
502. Tavsan, Z.; Kayali, H. A. Flavonoids Showed Anticancer Effects on Ovarian Cancer Cells: Involvement of Reactive Oxygen Species, Apoptosis, Cell Cycle and Invasion. *Biomed. Pharmacother.* **2019**, *116*, 1–12.
503. Fu, Y.; Kadioglu, O.; Wiench, B.; Wie, Z.; Gao, C.; Luo, M.; Gu, C.; Zu, Y.; Efferth, T. Cell Cycle Arrest and Induction of Apoptosis by Cajanin Stilbene Acid From *Cajanus cajan* in Breast Cancer Cells. *Phytomedicine* **2015**, *22*, 462–468.
504. Kushwaha, P. P.; Vardhan, P. S.; Kapewangolo, P.; Shuaib, M.; Prajapati, S. K.; Singh, A. K.; Kumar, S. *Bulbine frutescens* Phytochemical Inhibits Notch Signaling Pathway and Induces Apoptosis in Triple Negative and Luminal Breast Cancer Cells. *Life Sci.* **2019**, *234*, 116777.
505. Kapinova, A.; Kubatka, P.; Glolubnitschaja, O.; Kello, M.; Zubor, P.; Solar, P.; Pec, M. Dietary Phytochemicals in Breast Cancer Research: Anticancer Effects and Potential Utility for Effective Chemoprevention. *Environ. Health Prev. Med.* **2018**, *23*, 1–8.
506. Shrivastava, A.; Kuzontkoski, P. M.; Groopman, J. E.; Prasad, A. Cannabidiol Induces Programmed Cell Death in Breast Cancer Cells by Coordinating the Cross-Talk Between Apoptosis and Autophagy. *Mol. Cancer Ther.* **2011**, *10*, 1161–1172.
507. Basu, A.; Kumar, G. S. Nucleic Acids Binding Strategies of Small Molecules: Lessons From Alkaloids. *Biochim. Biophys. Acta* **2018**, *1862*, 1995–2016.
508. Aleksić, M. M.; Kapetanović, V. An Overview of the Optical and Electrochemical Methods for Detection of DNA-Drug Interactions. *Acta Chim. Slov.* **2014**, *61*, 555–573.
509. Berthold, D.; van Otterloo, W. A. L. Unprecedented Direct Asymmetric Total Synthesis of 5,8'-Naphthylisoquinoline Alkaloids From Their Fully Substituted Precursors Employing a Novel Nickel/N,N-Ligand-Catalyzed Atroposelective Cross-Coupling Reaction. *Chem. -Eur. J.* **2023**, *29*, e202302070.
510. Zhang, Y. F.; Shi, Z. J. Upgrading Cross-Coupling Reactions for Biaryl Synthesis. *Acc. Chem. Res.* **2019**, *52*, 161–169.
511. Álvarez-Casao, Y.; Estepa, B.; Monge, D.; Ros, A.; Iglesias-Sigüenza, J.; Álvarez, E.; Fernández, R.; Lassaletta, J. M. Pyridine-Hydrazone Ligands in Enantioselective Palladium-Catalyzed Suzuki-Miyaura Cross-Couplings. *Tetrahedron* **2016**, *72*, 5184–5190.
512. Bringmann, G. Mono- and Dimeric Naphthylisoquinoline Alkaloids—Pharmaceutically and Structurally Exciting Natural Heterocycles With Axial Chirality. *Bull. Soc. Chim. Belg.* **1996**, *105*, 601–613.
513. Bringmann, G.; Tasler, S. Oxidative Aryl Coupling Reactions: A Biomimetic Approach to Configurationally Unstable or Axially Chiral Biaryl Natural Products and Related Bioactive Compounds. *Tetrahedron* **2001**, *57*, 331–343.
514. Bringmann, G.; Saeb, W.; Mies, J.; Messer, K.; Wohlfarth, M.; Brun, R. One-Step Oxidative Dimerization of Genuine, Unprotected Naphthylisoquinoline Alkaloids to Give Michellamines and Other Bioactive Quateraryls. *Synthesis* **2000**, 1843–1847.
515. Bringmann, G.; Saeb, W.; Koppler, D.; François, G. Jozimine A ('Dimeric' Dioncophyllinea), a Non-Natural Michellamine Analog With High Antimalarial Activity. *Tetrahedron* **1996**, *52*, 13409–13418.
516. Bringmann, G.; Saeb, W.; Kraus, J.; Brun, R.; François, G. Jozimine B, a Constitutionally Unsymmetric, Antiplasmodial 'Dimer' of the Naphthylisoquinoline Alkaloid Ancistrocladine. *Tetrahedron* **2000**, *56*, 3523–3531.

517. Bringmann, G.; Saeb, W.; Wohlfarth, M.; Messer, K.; Brun, R. Jozipeltine A, a Novel, Unnatural Dimer of the Highly Hydroxylated Naphthylisoquinoline Alkaloid Dioncopeltine A. *Tetrahedron* **2000**, *56*, 5871–5875.
518. Bringmann, G.; Götz, R.; François, G. Synthesis of Pindikamine A, a Michellamine Related Dimer of a Non-Natural, 'Skew' Naphthylisoquinoline. *Tetrahedron* **1996**, *52*, 13419–13426.
519. Bringmann, G.; Wenzel, M.; Kelly, T. R.; Boyd, M. R.; Gulakowski, R. J. Octadehydromichellamine, a Structural Analog of the Anti-HIV Michellamines Without Centrochirality. *Tetrahedron* **1999**, *55*, 1731–1740.
520. Hemberger, Y.; Zhang, G.; Brun, R.; Kaiser, M.; Bringmann, G. Highly Antiplasmodial Non-Natural Oxidative Products of Dioncophylline A: Synthesis, Absolute Configuration, and Conformational Stability. *Chem. -Eur. J.* **2015**, *21*, 14507–14518.
521. Mokhtari, R. B.; Homayouni, T. S.; Baluch, N.; Morgatskaya, E.; Kumar, S.; Das, B.; Yeger, H. Combination Therapy in Combating Cancer. *Oncotarget* **2017**, *8*, 38022–38043.
522. Tolcher, A. W.; Mayer, L. D. Improving Combination Cancer Therapy: the CombiPlex® Development Platform. *Future Oncol.* **2018**, *14*, 1317–1332.
523. Plana, D.; Palmer, A. C.; Sorger, P. K. Independent Drug Action in Combination Therapy: Implications for Precision Oncology. *Cancer Discovery* **2022**, *12*, 606–624.
524. Bringmann, G.; Harmsen, S.; Holenz, J.; Geuder, T.; Götz, R.; Keller, P. A.; Walter, R.; Hallock, Y. F.; Cardellina, J. H., II; Boyd, M. R. 'Biomimetic' Oxidative Dimerization of Korupensamine A: Completion of the First Total Synthesis of Michellamines A, B, and C. *Tetrahedron* **1994**, *50*, 9643–9648.
525. Bringmann, G.; Götz, R.; Harmsen, S.; Holenz, J.; Walter, R. Biomimetic Total Synthesis of Michellamines A-C. *Liebigs Ann.* **1996**, 2045–2058.
526. Hobbs, P. D.; Upender, V.; Dawson, M. I. Stereospecific Syntheses of Michellamines A and C. *Synlett* **1997**, 965–967.
527. Kelly, T. R.; Garcia, A.; Lang, F.; Walsh, J. J.; Bhaskar, K. V.; Boyd, M. R.; Götz, R.; Keller, P. A.; Walter, R.; Bringmann, G. Convergent Total Synthesis of the Michellamines. *Tetrahedron Lett.* **1994**, *35*, 7621–7624.
528. Bringmann, G.; Götz, R.; Keller, P. A.; Walter, R.; Boyd, M. R.; Lang, F.; Garcia, A.; Walsh, J. J.; Tellitu, I.; Bhaskar, K. V.; Kelly, T. R. A Convergent Total Synthesis of the Michellamines. *J. Org. Chem.* **1998**, *63*, 1090–1097.
529. Wu, J.; Beal, J. L.; Wu, W. N.; Doskotch, R. W. Alkaloids of *Thalictrum*. XXXII. Isolation and Identification of Alkaloids From *Thalictrum revolutum* DC. Fruit. *J. Nat. Prod.* **1980**, *43*, 270–277.
530. Tanahashi, T.; Su, Y.; Nagakura, N.; Nayeshiro, H. Quaternary isoquinoline alkaloids from *Stephania cepharantha*. *Chem. Pharm. Bull.* **2000**, *48*, 370–373.
531. Nishiyama, Y.; Moriyasu, M.; Ichimaru, M.; Iwasa, K.; Kato, Y.; Mathenge, S. G.; Chalo Mutiso, P. B.; Juma, F. D. Quaternary isoquinoline alkaloids from *Xylopia parviflora*. *Phytochemistry* **2004**, *65*, 939–944.
532. Bringmann, G.; Hartung, T.; Göbel, L.; Schupp, O.; Peters, K.; von Schnering, H. G. Synthesis and Structure of Benzonaphthopyrans: Helically Distorted, Bridged Biaryls With Different Steric Hindrance at the Axis. *Liebigs Ann. Chem.* **1992**, 769–775.
533. Bringmann, G.; Breuning, M.; Endress, H.; Vitt, D.; Peters, E. M.; Peters, K. Biaryl Hydroxy Aldehydes as Intermediates in the Metal-Assisted Atropo-Enantioselective Reduction of Biaryl Lactones: Structures and Aldehyde-Lactolequilibria. *Tetrahedron* **1998**, *54*, 10677–10690.
534. This was evident from the "substitution test" (like replacing one or the other methoxy function, e.g., by hydroxy groups, which would lead to atropo-diastereomers) and from the "symmetry test" (due to the fact that the two methoxy groups were

constitutionally identical, but could not be interconverted by a C_2-axis or by an internal mirror plane); see also: Kramer, W. H.; Griesbeck, A. G. The Same and Not the Same: Chirality, Topicity, and Memory of Chirality. *J. Chem. Educ.* **2008**, *85*, 701–709.
535. Bungard, C. J.; Morris, J. C. Total Synthesis of the 7,3′-Linked Naphthylisoquinoline Alkaloid Ancistrocladidine. *J. Org. Chem.* **2006**, *71*, 7354–7363.
536. Rozwadowska, M. D. *seco*-Isoquinoline Alkaloids. In *The Alkaloids: Chemistry and Pharmacology*; Brossi, A., Ed.; Vol. 33; Academic Press: New York, 1988; pp. 231–306.
537. Blasko, G.; Gula, D. J.; Shamma, M. The Phthalideisoquinoline Alkaloids. *J. Nat. Prod.* **1982**, *45*, 105–122.
538. Blanco, O. M.; Castedo, L.; Villaverde, M. C. Alkaloids from *Platycapnos spicata*. *Phytochemistry* **1993**, *32*, 1055–1057.
539. Tojo, E.; Dominguez, D.; Castedo, L. Alkaloids from *Sarcocapnos enneaphylla*. *Phytochemistry* **1991**, *30*, 1005–1010.
540. Lee, K. H.; Chuah, C. H.; Goh, S. H. *seco*-Benzyltetrahydroisoquinolines From *Polyalthia insignis* (Annonaceae). *Tetrahedron Lett.* **1997**, *38*, 1253–1256.
541. Nimgirawath, S. Syntheses of the *seco*-Benzyltetrahydroisoquinoline Alkaloids Polysignine and Methoxypolysignine. *Aust. J. Chem.* **2000**, *53*, 523–525.
542. Desravines, N.; Hsu, C. H.; Mohnot, S.; Sahasrabuddhe, S.; House, M.; Sauter, E.; O'Conner, S.; Bauman, J. E.; Chow, H. H. S.; Rahangdales, L. Feasibility of 5-Fluorouracil and Imiquimod for the Topical Treatment of Cervical Intraepithelial Neoplasias (CIN) 2/3. *Int. J. Gynecol. Obstet.* **2023**, *163*, 862–867.
543. Cohen, P. A.; Jhingran, A.; Oaknin, A.; Denny, L. Cervical CANCER. *Lancet* **2019**, *393*, 169–182.
544. Koh, W.-J.; Abu-Rustum, N. R.; Bean, S.; Bradley, K.; Campos, S. M.; Cho, K. R.; Chon, H. S.; Chu, C.; Clark, R.; Cohn, D.; Crispens, M. A.; Damast, S.; Dorigo, O.; Eifel, P. J.; Fisher, C. M.; Frederick, P.; Gaffney, D. K.; Han, E.; Huh, W. K.; Lurain, J. R., III; Mariani, A.; Mutch, D.; Nagel, C.; Nekhlyudov, L.; Nickles Fader, A.; Remmenga, S. W.; Reynolds, R. K.; Tillmanns, T.; Ueda, S.; Wyse, W.; Yashar, C. M.; McMillian, N. R.; Scavone, J. L. Cervical Cancer, Version 3.2019. *J. Natl. Compr. Cancer Network* **2019**, *17*, 64–84.
545. Small, W., Jr.; Bacon, M. A.; Bajaj, A.; Chuang, L. T.; Fisher, B. J.; Harkenrider, M. M.; Jhingran, A.; Kitchener, H. C.; Mileshkin, L. R.; Viswanathan, A. N.; Gaffney, D. K. Cervical Cancer: A Global Health Crisis. *Cancer* **2017**, *123*, 2404–2412.
546. Burmeister, C. A.; Khan, S. F.; Schäfer, G.; Mbatani, N.; Adams, T.; Moodley, J.; Prince, S. Cervical Cancer Therapies: Current Challenges and Future Perspectives. *Tumour Virus Res.* **2022**, *13*, 200238.
547. Liontos, M.; Kyriazoglou, A.; Dimitriadis, I.; Dimopoulos, M. A.; Bamias, A. Systemic Therapy in Cervical Cancer: 30 Years in Review. *Crit. Rev. Oncol. Hematol.* **2019**, *137*, 9–17.
548. Campello, M. J.; Castedo, L.; Dominguez, D.; De Lera, A. R.; Saá, J. M.; Suau, R.; Tojo, E.; Vidal, M. C. New Oxidized Isocularine Alkaloids From *Sarcocapnos* Plants. *Tetrahedron Lett.* **1984**, *25*, 5933–5936.
549. Speck, K.; Magauer, T. The Chemistry of Isoindole Natural Products. *Beilstein J. Org. Chem.* **2013**, *9*, 2048–2078.
550. Kumar, V.; Poonam; Prasad, A. K.; Parmar, V. S. Naturally Occurring Aristolactams, Aristolochic Acids and Dioxoaporphines and Their Biological Activities. *Nat. Prod. Rep.* **2003**, *20*, 565–583.
551. Michl, J.; Ingrouille, M. J.; Simmonds, M. S. J.; Heinrich, M. Naturally Occurring Aristolochic Acid Analogues and Their Toxicities. *Nat. Prod. Rep.* **2014**, *31*, 676–693.

552. Chanakul, W.; Tuchinda, P.; Anantachoke, N.; Pohmakotr, M.; Piyachaturawat, P.; Jariyawat, S.; Suksen, K.; Jaipetch, T.; Nuntasaen, N.; Reutrakul, V. Cytotoxic Alkaloids From Stems, Leaves and Twigs of *Dasymaschalon blumei*. *Fitoterapia* **2011**, *82*, 964–968.
553. Nayyatip, S.; Thaichana, P.; Buayairaksa, M.; Tuntiwechapikul, W.; Meepowpan, P.; Nuntasaen, N.; Pompimon, W. Aristolactam-Type Alkaloids From *Orophea enterocarpa* and Their Cytotoxicities. *Int. J. Mol. Sci.* **2012**, *13*, 5010–5018.
554. Grimshaw, M. J.; Cooper, L.; Papazisis, K.; Coleman, J. A.; Bohnenkamp, H. R.; Chiapero-Stanke, L.; Taylor-Papdimitriou, J.; Burchel, J. M. Mammosphere Culture of Metastatic Breast Cancer Cells Enriches for Tumorigenic Breast Cancer Cells. *Breast Cancer Res.* **2017**, *10*, R52.
555. Yousefnia, S.; Ghaedi, K.; Forootan, F. S.; Esfahani, M. H. N. Characterization of the Stemness Potency of Mammospheres Isolated From the Breast Cancer Cell Lines. *Tumor Biol.* **2019**, *41*, 1–14.
556. Lombard, Y.; de Giorgio, A.; Coombes, C. R.; Stebbing, J.; Castellano, L. Mammosphere Formation Assay From Human Breast Cancer Tissues and Cell Lines. *J. Visualized Exp.* **2015**, *97*, e52671.
557. Prajapati, K. S.; Shuaib, M.; Gupta, S.; Kumar, S. Withaferin A Mediated Changes of miRNA Expression in Breast Cancer-Derived Mammospheres. *Mol. Carcinog.* **2022**, *61*, 876–889.
558. Jordan, C. T.; Guzman, M. L.; Noble, M. Cancer Stem Cells. *N. Engl. J. Med.* **2006**, *355*, 1253–1261.
559. Schulenburg, A.; Urich-Pur, H.; Thurnher, D.; Eerovic, B.; Florian, S.; Sperr, W. R.; Kalhs, P.; Marian, B.; Wrba, F.; Zielinski, C. C.; Valent, P. Neoplastic Stem Cells: A Novel Therapeutic Target in Clinical Oncology. *Cancer* **2006**, *107*, 2512–2520.
560. Bringmann, G.; Schlauer, J.; Rückert, M.; Wiesen, B.; Ehrenfeld, K.; Proksch, P.; Czygan, F. C. Host-Derived Acetogenins Involved in the Incompatible Parasitic Relationship Between *Cuscuta reflexa* (Convolvulaceae) and *Ancistrocladus heyneanus* (Ancistrocladaceae). *Plant Biol.* **1999**, *1*, 581–584.
561. Bringmann, G.; Kehr, C.; Dauer, U.; Gulden, K. P.; Haller, R. D.; Bär, S.; Isahakia, M. A.; Robertson, S. A.; Peters, K. *Ancistrocladus robertsoniorum* "Produces" Pure Crystalline Droserone When Wounded. *Planta Med.* **1993**, *93*, A622–A623.
562. Peters, K.; Peters, E. M.; von Schnering, H. G.; Bringmann, G.; Kehr, C.; Haller, R. D.; Bär, S.; Isahakia, M. A.; Robertson, S. A. Crystal Structure of Droserone in "Biogenic Crystals" Found Under the Stem Bark of *Ancistrocladus robertsoniorum*. *Z. Kristallogr.* **1995**, *210*, 290–291.
563. Rischer, H.; Hamm, A.; Bringmann, G. *Nepenthes insignis* Uses a C_2-Portion of the Carbon Skeleton of L-Alanine Acquired via Its Carnivorous Organs, to Build Up the Allelochemical Plumbagin. *Phytochemistry* **2002**, *59*, 603–609.
564. Raj, G.; Kurup, R.; Hussain, A. A.; Baby, S. Distribution of Naphthoquinones, Plumbagin, Droserone, and 5-O-Methyldroserone in Chitin-Induced and Uninduced *Nepenthes khasiana*: Molecular Events in Prey Capture. *J. Exp. Bot.* **2011**, *62*, 5429–5436.
565. Dávila-Lara, A.; Rahman-Soad, A.; Reichelt, M.; Mithöfer, A. Carnivorous *Nepenthes x ventrata* Plants Use a Naphthoquinone as Phytoanticipin Against Herbivory. *PLoS One* **2021**, *16*, e0258235.
566. Hook, I.; Mills, C.; Sheridan, H. Bioactive Naphthoquinones From Higher Plants. In *Studies in Natural Products Chemistry*; Atta-ur-Rahman, Ed.; Vol. 41; Elsevier Science B.V.: Amsterdam, The Netherlands, 2014; pp. 119–160.
567. Babula, P.; Adam, V.; Havel, L.; Kizek, R. Noteworthy Secondary Metabolites Naphthoquinones—Their Occurrence, Pharmacological Properties and Analysis. *Curr. Pharm. Anal.* **2009**, *5*, 47–68.

568. Lu, J. J.; Bao, J. L.; Wu, G. S.; Xu, W. S.; Huang, M. Q.; Chen, X. P.; Wang, Y. T. Quinones Derived From Plant Secondary Metabolites as Anti-Cancer Agents. *Anti-Cancer Agents Med. Chem.* **2013**, *13*, 456–463.
569. Padhye, S.; Dandawate, P.; Yusufi, M.; Ahmad, A.; Sarkar, F. H. Perspectives on Medicinal Properties of Plumbagin and Its Analogs. *Med. Res. Rev.* **2012**, *32*, 1131–1158.
570. Liu, Y.; Cai, Y.; He, C.; Chen, M.; Li, H. Anticancer Properties and Pharmaceutical Applications of Plumbagin: A Review. *Am. J. Chin. Med.* **2017**, *45*, 423–441.
571. Badwaik, H. R.; Kumari, L.; Nakhate, K.; Verma, V. S.; Sakure, K. Phytoconstituent Plumbagin: Chemical, Biotechnological and Pharmaceutical Aspects. In *Studies in Natural Products Chemistry*; Atta-ur-Rahman, Ed.; Vol. 63; Elsevier Science B.V.: Amsterdam, The Netherlands, 2019; pp. 415–460.
572. Yin, Z.; Zhang, J.; Chen, L.; Guo, Q.; Yang, B.; Zhang, W.; Kang, W. Anticancer Effects and Mechanisms of Action of Plumbagin: Review of Research Advances. *BioMed. Res. Int.* **2020**, 6940953.
573. Roy, A. Plumbagin: A Potential Anti-Cancer Compound. *Mini-Rev. Med. Chem.* **2021**, *21*, 731–737.
574. Rahman, M.; Islam, R.; Akash, S.; Shohag, S.; Ahmed, L.; Supti, F. A.; Rauf, A.; Aljohani, A. S. M.; Abdulmonem, W. A.; Khalil, A. A.; Sharma, R.; Thiruvengadam, M. Naphthoquinones and Derivatives as Potential Anticancer Agents: An Updated Review. *Chem.-Biol. Interact.* **2022**, *368*, 110198.
575. Navarro-Tovar, G.; Vega-Rodríguez, S.; Leyva, E.; Loredo-Carrillo, S.; de Loera, D.; Vega-Rodríguez, L. I. The Relevance and Insights on 1,4-Naphthoquinones as Antimicrobial and Antitumoral Molecules: A Systematic Review. *Pharmaceuticals* **2023**, *16*, 496.
576. Santos, T. B.; Cavalieri de Moraes, L. G.; Pacheco, P. A. F.; dos Santos, D. G.; de Assis Cabral Ribeiro, R. M.; dos Santo Moreira, C.; da Rocha, D. R. Naphthoquinones as a Promising Class of Compounds for Facing the Challenge of Parkinson's Disease. *Pharmaceuticals* **2023**, *16*, 1577.
577. Veshkurova, O.; Golubenko, Z.; Pshenichnov, E.; Arzanova, I.; Uzbekov, V.; Sultanova, E.; Salikhov, S.; Williams, J. J.; Reibenspies, J. H.; Puckhaber, L. S.; Stipanovic, R. D. Malvone A, Phytoalexin Found in *Malva sylvestris* (Family Malvaceae). *Phytochemistry* **2006**, *67*, 2376–2379.
578. Cannon, J. R.; Lojanapiwatna, V.; Raston, C. L.; Sinchai, W.; White, A. H. The Quinones of *Nepenthes rafflesiana*. The Crystal Structure of 2,5-Dihydroxy-3,8-Dimethoxy-7-Methylnaphtho-1,4-Quinone (Nepenthone E) and a Synthesis of 2,5-Dihydroxy-3-Methoxy-7-Methyl-Naphtho-1,4-Quinone (Nepenthone C). *Aust. J. Chem.* **1980**, *33*, 1073–1093.
579. Brockmann, H.; Zeeck, A. Rubromycine, III. Die Konstitution Von α-Rubromycin, β-Rubromycin, γ-Rubromycin und γ-Iso-Rubromycin. *Chem. Ber.* **1970**, *103*, 1709–1726.
580. Armitage, J. O.; Gascoyne, R. D.; Lunning, M. A.; Cavalli, F. Non-Hodgkin Lymphoma. *Lancet* **2017**, *390*, 298–310.
581. Singh, R.; Shaik, S.; Negi, B. S.; Rajguru, J. P.; Patil, P. B.; Parihar; Sharma, U. Non-Hodgkin's Lymphoma: A Review. *J. Fam. Med. Primary Care* **2020**, *9*, 1834–1840.
582. Kamila, S.; Mukherjee, C.; Mondal, S. S.; De, A. Application of Directed Metallation in Synthesis. Part 3: Studies in the Synthesis of (±)-Semivioxanthin and Its Analogues. *Tetrahedron* **2003**, *59*, 1339–1348.
583. Khaldi, M.; Chrétien, F.; Chapleur, Y. Dramatic Effect of Substituents and TMEDA Additive on the Regioselectivity of Directed Orthometalation of Tetrasubstituted Aromatics. *Tetrahedron Lett.* **1994**, *35*, 401–404.

584. Ohland, R.; Ohlan, S.; Judge, V.; Narang, R.; Ahuja, M.; Narasimhan, B. 2-(2,4-Difluorophenyl)-1,3-Bis(1,2,4-Triazol-1-yl)Propan-2-ol Derivatives: Synthesis, Antifungal Evaluation and QSAR Studies by Hansch Analysis. *Arkivoc* **2007**, *Xiv*, 172–184.
585. Li, H. C.; Wang, C.; Sanchez, T.; Tan, Y. M.; Jiang, C. Y.; Neamati, N.; Zhao, G. Amide-Containing Diketoacids as HIV-1 Integrase Inhibitors: Synthesis, Structure-Activity Relationship Analysis, and Biological Activity. *Bioorg. Med. Chem.* **2009**, *17*, 2913–2919.
586. Cheng, X.; Zhang, G.; Seupel, R.; Feineis, D.; Brünnert, D.; Chatterjee, M.; Schlosser, M.; Bringmann, G. Epoxides Related to Dioncoquinone B: Synthesis, Activity Against Multiple Myeloma Cells, and Search for the Target Protein. *Tetrahedron* **2018**, *74*, 5102–5112.
587. Leslie, B. J.; Hergenrother, P. J. Identification of the Cellular Targets of Bioactive Small Organic Molecules Using Affinity Reagents. *Chem. Soc. Rev.* **2008**, *37*, 1347–1360.
588. Trippier, P. C. Synthetic Strategies for the Biotinylation of Bioactive Small Molecules. *ChemMedChem* **2013**, *8*, 190–203.
589. Ding, C.; Fan, X.; Wu, G. Peroxiredoxin 1—An Antioxidant Enzyme in Cancer. *J. Cell. Mol. Med.* **2017**, *21*, 193–202.
590. Cai, A. L.; Zeng, W.; Cai, W. L.; Liu, J. L.; Zheng, X. W.; Liu, Y.; Yang, X. C.; Long, Y.; Li, J. Peroxiredoxin-1 Promotes Cell Proliferation and Metastasis Through Enhancing Akt/mTOR in Human Osteosarcoma Cells. *Oncotarget* **2017**, *9*, 8290–8302.
591. Chawsheen, H. A.; Ying, Q.; Jiang, H.; Wie, Q. A Critical Role of the Thioredoxin Domain Containing Protein 5 (TXNDC5) in Redox Homeostasis and Cancer Development. *Genes Dis.* **2018**, *5*, 312–322.
592. Wang, X.; Li, H.; Chang, X. The Role and Mechanism of TXNDC5 in Diseases. *Eur. J. Med. Res.* **2022**, *27*, 145.
593. Matsuo, Y. Introducing Thioredoxin-Related Transmembrane Proteins: Emerging Roles of Human TMX and Clinical Implications. *Antioxid. Redox Signaling* **2022**, *36*, 13–15.
594. Borgese, N. Getting Membrane Proteins on and Off the Shuttle Bus Between the Endoplasmic Reticulum and the Golgi Complex. *J. Cell Sci.* **2016**, *129*, 1537–1545.
595. Chen, X.; Cubillos-Ruiz, J. R. Endoplasmic Reticulum Stress Signals in the Tumour and Its Microenvironment. *Nat. Rev. Cancer* **2021**, *21*, 71–88.
596. Cubillos-Ruiz, J. R.; Bettigole, S. E.; Glimcher, L. H. Tumorigenic and Immunosuppressive Effects of Endoplasmic Reticulum Stress in Cancer. *Cell* **2017**, *168*, 692–706.
597. Khmelevskaya, E. A.; Pelageev, D. N. A Convenient Synthetic Approach to Dioncoquinone B and Related Compounds. *Tetrahedron Lett.* **2019**, *60*, 1022–1024.
598. Awale, S.; Baba, H.; Phan, N. D.; Kim, M. J.; Maneenet, J.; Sawaki, K.; Kanda, M.; Okumura, T.; Fujii, T.; Okada, T.; Maruyama, T.; Okada, T.; Toyooka, N. Targeting Pancreatic Cancer With Novel Plumbagin Derivatives: Design, Synthesis, Molecular Mechanism, In Vitro and In Vivo Evaluation. *J. Med. Chem.* **2023**, *66*, 8054–8065.
599. Ji, W.; Sun, X.; Gao, Y.; Lu, M.; Zhu, L.; Wang, D.; Hu, C.; Chen, J.; Cao, P. Natural Compound Shikonin Is a Novel PAK1 Inhibitor and Enhances Efficacy of Chemotherapy Against Pancreatic Cancer Cells. *Molecules* **2022**, *27*, 2747.
600. Gomes, C. L.; de Albuquerque Wanderley Sales, V.; Gomes de Melo, C.; Ferreira da Silva, R. M.; Vicente Nishimura, R. H.; Rolim, L. A.; Rolim Neto, P. J. Beta-Lapachone: Natural Occurrence, Physicochemical Properties, Biological Activities, Toxicity and Synthesis. *Phytochemistry* **2021**, *186*, 112713.
601. Maneenet, J.; Tajuddeen, N.; Nguyen, H. H.; Fujii, R.; Lombe, B. K.; Feineis, D.; Awale, S.; Bringmann, G. Droserone and Dioncoquinone B, and Related

Naphthoquinones as Potent Antiausterity Agents Against Human PANC-1 Pancreatic Cancer Cells. *Results Chem.* **2024**, *7*, 101352.
602. Fares, J.; Fares, M. Y.; Khachfe, H. H.; Salhab, H. A.; Fares, Y. Molecular Principles of Metastasis: A Hallmark of Cancer Revisited. *Signal Transduction Targeted Ther.* **2020**, *5*, 28.
603. Seyfried, T. N.; Huysentruyt, L. C. On the Origin of Cancer Metastasis. *Crit. Rev. Oncog.* **2013**, *18*, 43–73.
604. Lambert, A. W.; Pattabiraman, D. R.; Weinberg, R. A. Emerging Biological Principles of Metastasis. *Cell* **2017**, *168*, 670–691.
605. Franke, T. F.; Kaplan, D. R.; Cantley, L. C. PI3K: Downstream AKTion Blocks Apoptosis. *Cell* **1997**, *88*, 435–437.
606. Burgering, B. M. T.; Coffer, P. J. Protein Kinase B (c-Akt) in Phosphatidylinositol-3-OH Kinase Signal Transduction. *Nature* **1995**, *376*, 599–602.
607. Altomare, D. A.; Khaled, A. R. Homeostasis and the Importance for a Balance Between Akt/mTOR Activity and Intracellular Signaling. *Curr. Med. Chem.* **2012**, *19*, 3748–3762.
608. Roy, H. K.; Olusola, B. F.; Clemens, D. L.; Karolski, W. J.; Ratashak, A.; Lynch, H. T.; Smyrk, T. C. Akt Proto-Oncogene Overexpression Is an Early Event During Sporadic Colon Carcinogenesis. *Carcinogenesis* **2002**, *23*, 201–205.
609. Bringmann, G. Isoquinolines and Naphthalenes From ß-Polyketones: Model Reactions for an Extraordinary Alkaloid Biosynthesis. *Angew. Chem.* **1982**, *94*, 205–206. *Angew. Chem., Int. Ed. Engl.* **1982**, *21*, 200–201.
610. Bringmann, G. A Short Biomimetic Synthesis of the Isoquinoline and the Naphthalene Moieties of *Ancistrocladus* Alkaloids From Common β-Polycarbonyl Precursors. *Tetrahedron Lett.* **1982**, *23*, 2009–2012.
611. Bringmann, G. Biomimetische Synthesen Beider Molekülhälften der *Ancistrocladus*- und der *Triphyophyllum*-Alkaloide Aus Gemeinsamen Vorstufen. *Liebigs Ann. Chem.* **1985**, 2126–2134.
612. Thomas, R. A Biosynthetic Classification of Fungal and Streptomycete Fused-Ring Aromatic Polyketides. *ChemBioChem* **2001**, *2*, 612–627.
613. Kreher, B.; Neszmélyi, A.; Wagner, H. Naphthoquinones from *Dionaea muscipula*. *Phytochemistry* **1990**, 605–606.
614. Schölly, T.; Kapetanidis, I. Flavonol and Naphthoquinone Glycosides of *Drosera rotundifolia*. *Planta Med.* **1989**, *55*, 611–612.
615. Bringmann, G.; Pokorny, F.; Wenzel, M.; Wurm, K.; Schneider, C. Labelled Precursors for Biosynthetic Studies on Naphthylisoquinoline Alkaloids. *J. Labelled Compd. Radiopharm.* **1997**, *39*, 29–38.
616. Hobbs, P. D.; Upender, V.; Liu, J.; Pollart, D. J.; Thomas, D. W.; Dawson, M. I. The First Stereospecific Synthesis of Michellamine B. *Chem. Commun.* **1996**, 923–924.
617. Watanabe, T.; Shakadou, M.; Uemura, M. Stereoselective Synthesis of KORUPENSAMINE A and *ent*-korupensamine b utilizing an Identical Planar Chiral Arene Chromium Complex. *Synlett* **2000**, 1141–1144.
618. Watanabe, T.; Tanaka, Y.; Shoda, R.; Sakamoto, R.; Kamikawa, K.; Uemura, M. Stereoselective Synthesis of Atropisomeric Korupensamines A and B Utilizing Planar Chiral Arene Chromium Complex. *J. Org. Chem.* **2004**, *69*, 4152–4158.
619. Lipshutz, B. H.; Keith, J. M. A Stereospecific, Intermolecular Biaryl-Coupling Approach to Korupensamine A—En Route to the Michellamines. *Angew. Chem., Int. Ed.* **1999**, *38*, 3530–3533.
620. Huang, S.; Petersen, T. B.; Lipshutz, B. H. Total Synthesis of (+)-Korupensamine B via an Atropselective Intermolecular Biaryl Coupling. *J. Am. Chem. Soc.* **2010**, *132*, 14021–14023.

621. Xu, G.; Fu, W.; Liu, G.; Senanayake, C. H.; Tang, W. Efficient Syntheses of Korupensamines A, B and Michellamine B by Asymmetric Suzuki-Miyaura Coupling Reactions. *J. Am. Chem. Soc.* **2014**, *136*, 570–573.
622. Bai, M.; Jia, S.; Zhang, J.; Cheng, H. G.; Cong, H.; Liu, S.; Huan, Z.; Hu, Y.; Chen, X.; Zhou, Q. A Modular Approach for Diversity-Oriented Synthesis of 1,3-*trans*-Disubstituted Tetrahydroisoquinolines: Seven-Step Asymmetric Synthesis of Michellamines B and C. *Angew. Chem., Int. Ed.* **2022**, *61*, e202205245.
623. Toop, H. D.; Brusnahan, J. S.; Morris, J. C. Concise Total Synthesis of Dioncophylline E Through an *Ortho*-Arylation Strategy. *Angew. Chem., Int. Ed.* **2017**, *56*, 8536–8538.
624. Slack, E.; Seupel, R.; Aue, D.; Bringmann, G.; Lipshutz, B. Atroposelective Total Synthesis of the Fourfold *Ortho*-Substituted Naphthyltetrahydroisoquinoline Biaryl O,N-dimethylhamatine. *Chem. -Eur. J.* **2019**, *25*, 14237–14245.
625. Jo, Y. I.; Lee, C. Y.; Cheon, C. H. Asymmetric Total Syntheses of Naphthylisoquinoline Alkaloids via Atroposelective Coupling Reaction Using Central Chirality as Atroposelectivity-Controlling Group. *Org. Lett.* **2020**, *22*, 4653–4658.
626. Jo, Y. I.; Lee, C. Y.; Cheon, C. H. Atroposelective Total Syntheses of Naphthylisoquinoline Alkaloids With (*P*)-Configuration. *J. Org. Chem.* **2020**, *85*, 12770–12776.
627. Kim, G.; Westwood, J. H. Macromolecule Exchange in *Cuscuta*-Host Plant Interactions. *Curr. Opin. Plant Biol.* **2015**, *26*, 20–25.
628. Furuhashi, T.; Nakamura, T.; Iwase, K. Analysis of Metabolites in Stem Parasitic Plant Interactions: Interaction of *Cuscuta–Momordica* Versus *Cassytha–Ipomoea*. *Plants* **2016**, *5*, 43.
629. Kumar, K.; Amir, R. The Effect of a Host on the Primary Metabolic Profiling of *Cuscuta campestris*' Main Organs, Haustoria, Stem and Flower. *Plants* **2021**, *10*, 2098.
630. Smith, J. D.; Mescher, M. C.; De Moraes, C. M. Implications of Bioactive Solute Transfer From Hosts to parasitic plants. *Curr. Opin. Plant Biol.* **2013**, *16*, 464–472.
631. Flores-Sánchez, I. J.; Garza-Ortiz, A. Is There a Secondary/Specialized Metabolism in the Genus *Cuscuta* and Which Is the Role of the Host Plant? *Phytochem. Rev.* **2019**, *18*, 1299–1335.
632. Uebel, T.; Hermes, L.; Haupenthal, S.; Müller, L. α-Asarone, β-Asarone, and γ-Asarone: Current Status of Toxicological Evaluation. *J. Appl. Toxicol.* **2021**, *41*, 1166–1179.
633. Duncan, J. The Toxicology of Molluscicides. *Pharmacol. Ther.* **1981**, *14*, 67–88.
634. Ishida, N.; Koizumi, M.; Kano, H. The NMR Microscope: A Unique Promising Tool for Plant Science. *Ann. Bot.* **2000**, *86*, 259–278.
635. Köckenberger, W.; de Panfilis, C.; Santoro, D.; Dahiya, P.; Rawsthorne, S. High Resolution NMR Microscopy of Plants and Fungi. *J. Microsc.* **2004**, *214*, 182–189.
636. Gussoni, M.; Greco, F.; Pegna, M.; Bianchi, G.; Zeta, L. Solid State and Microscopy NMR Study of the Chemical Constituents of *Afzelia cuanzensis* Seeds. *Magn. Reson. Imaging* **1994**, *12*, 477–486.
637. MacFall, J. S.; Johnson, G. A. Plants, Seeds, Roots, and Soils as Applications of Magnetic Resonance Microscopy. In *Encyclopedia of NMR*; Harris, R. K., Wasylishen, R. E., Eds.; Vol. 6; John Wiley: Chichester, 2012; pp. 3403–3409.
638. Kuchenbrod, E.; Haase, A.; Benkert, R.; Schneider, H.; Zimmermann, U. Quantitative NMR Microscopy on Intact Plants. *Magn. Reson. Imaging* **1995**, *3*, 447–455.
639. Köckenberger, W. Nuclear Magnetic Resonance Micro-Imaging in the Investigation of Plant Cell Metabolism. *J. Exp. Bot.* **2001**, *52*, 641–652.
640. Cheng, Y. C.; Wang, T. T.; Chen, J. H.; Lin, T. T. Spatial-Temporal Analyses of Lycopene and Sugar Contents in Tomatoes During Ripening Using Chemical Shift Imaging. *Postharvest Biol. Technol.* **2011**, *62*, 17–25.

641. Terskikh, V. V.; Feurtado, J. A.; Ren, C.; Abrams, S. R.; Kermode, A. R. Water Uptake and Oil Distribution During Inhibition of Seeds of Western White Pine (*Pinus monticola* Dougl. Ex D. Don) Monitored In Vivo Using Magnetic Resonance Imaging. *Planta* **2005**, *221*, 17–27.
642. Schrader, B.; Klump, H. H.; Schenzel, K.; Schulz, H. Non-Destructive NIR FT Raman Analysis of Plants. *J. Mol. Struct.* **1999**, *509*, 201–212.
643. Gierlinger, N.; Schwanninger, M. The Potential of Raman Microscopy and Raman Imaging in Plant Research. *Spectroscopy* **2007**, *21*, 69–89.
644. Rys, M.; Szaleniec, M.; Skoczowski, A.; Stawoska, I.; Janeczko, A. FT-Raman Spectroscopy as a Tool in Evaluation the Response of Plants to Drought Stress. *Open Chem.* **2015**, *13*, 1091–1100.
645. Saletnik, A.; Saletnik, B.; Puchalski, C. Overview of Popular Techniques of Raman Spectroscopy and Their Potential in the Study of Plant Tissues. *Molecules* **2021**, *26*, 1537.
646. Atanasov, G.; Zotchev, S. B.; Dirsch, V. M.; The International Natural Product Sciences Taskforce; Supuran, C. T. Natural Products in Drug Discovery: Advances and Opportunities. *Nat. Rev. Drug Discovery* **2021**, *20*, 200–216.
647. Newman, D. J.; Cragg, G. M. Natural Products as Sources of New Drugs Over the Nearly Four Decades From 01/1981 to 09/2019. *J. Nat. Prod.* **2020**, *83* (3), 770–803.
648. Newman, D. J. Natural Products and Drug Discovery. *Nat. Sci. Rev.* **2022**, *9*, nwac206.
649. Coy-Barrera, E.; Ogungbe, I. V.; Schmidt, T. J. Natural Products for Drug Discovery in the 21st Century: Innovations for Novel Therapeutics. *Molecules* **2023**, *28*, 3690.
650. Wink, M. Biochemistry, Physiology and Ecological Functions of Secondary Metabolites. *Annu. Plant Rev.* **2010**, *40*, 1–19.
651. Hartmann, T. The Lost Origin of Chemical Ecology in the Late 19th Century. *Proc. Natl. Acad. Sci. U. S. A.* **2008**, *105*, 4541–4546.
652. Kessler, A.; Kalske, A. Plant Secondary Metabolite Diversity and Species Interactions. *Annu. Rev. Ecol. Evol. Syst.* **2018**, *49*, 115–138.
653. Gong, B.; Zhang, G. Interactions Between Plants and Herbivores: A Review of Plant Defense. *Acta Ecol. Sin.* **2014**, *34*, 325–336.
654. Erb, M.; Reymond, P. Molecular Interactions Between Plants and Insect Herbivores. *Annu. Rev. Plant Biol.* **2019**, *70*, 527–557.
655. Desai, H. K.; Gawad, D. H.; Govindachari, T. R.; Joshi, B. S.; Kamat, V. N.; Modi, J. D.; Parthasarathy, P. C.; Radhakrishnan, J.; Shanbhag, M. N.; Sidhaye, A. R.; Viswanathan, N. Chemical Investigations of Indian Plants: Part VII. *Indian J. Chem.* **1973**, *11*, 840–842.
656. Cai, C. H.; Mei, W. L.; Zuo, W. J.; Guo, Z. K.; Wang, H.; Gu, H. G.; Dai, H. F. Antibacterial Components From the Branches of *Ancistrocladus tectorius* (Lour.) Merr. *J. Trop. Subtrop. Bot.* **2013**, *31*, 184–188.
657. Bringmann, G.; Gulden, K. P.; Busse, H.; Fleischhauer, J.; Kramer, B.; Zobel, E. Circular Dichroism of Naphthylisoquinoline Alkaloids: Calculation of CD Spectra by Semiempirical Methods. *Tetrahedron* **1993**, *49*, 3305–3312.
658. Giorgio, E.; Viglione, R. G.; Rosini, C. Assignment of the Absolute Configuration of Large Molecules by *Ab Initio* Calculation of the Rotatory Power Within a Small Basis Set Scheme: The Case of Some Biologically Active Natural Products. *Tetrahedron: Asymmetry* **2004**, *15*, 1979–1986.
659. Bringmann, G.; Jansen, J. R. Chiral Economy with Respect to Rotational Isomerism: Rational Synthesis of Hamatine and (Optionally) Ancistrocladine From Joint Helical Precursors. *Heterocycles* **1989**, *28*, 137–142.
660. Chau, P.; Czuba, I. R.; Rizzacasa, M. A.; Bringmann, G.; Gulden, K. P.; Schäffer, M. Convergent Synthesis of Naphthylisoquinoline Alkaloids: Total Synthesis of (+)-O-Methylancistrocline. *J. Org. Chem.* **1996**, *61*, 7101–7105.

661. Unger, M.; Dreyer, M.; Specker, S.; Laug, S.; Pelzing, M.; Neusüß, C.; Holzgrabe, U.; Bringmann, G. Analytical Characterization of Crude Extracts From an African Ancistrocladus Species Using High-Performance Liquid Chromatography and Capillary Electrophoresis Coupled to Ion Trap Mass Spectrometry. *Phytochem. Anal.* **2004**, *15*, 21–26.
662. Desai, H. K.; Gawad, D. H.; Govindachari, T. R.; Joshi, B. S.; Parthasarathy, P. C.; Ramachandran, K. S.; Ravindranath, K. R.; Sidhaye, A. R.; Viswanathan, N. Chemical Investigation of Some Indian Plants: Part IX. *Indian J. Chem.* **1976**, *14B*, 473–475.
663. Gunatilaka, A. A. L. Alkaloids of Some Plants of Sri Lanka—Chemistry and Pharmacology. *J. Natl. Sci. Counc. Sri Lanka* **1978**, *6*, 39–87.
664. Rizzacasa, M. A.; Sargent, M. V. Synthetic Approaches to the Alkaloids of the Ancistrocladaceae: (−)-O-Methylancistrocladine and (+)-O-Methylhamatine. *J. Chem. Soc., Chem. Commun.* **1990**, 894–896.
665. Rizzacasa, M. A.; Sargent, M. V. Synthetic Approaches to the Naphthylisoquinoline Alkaloids, Part 2: The Total Synthesis of (−)-O-Methylancistrocladine and (+)-O-Methylhamatine and Their Enantiomers. *J. Chem. Soc., Perkin Trans. 1* **1991**, *1*, 845–854.
666. Leighton, B. N.; Rizzacasa, M. A. Formal Synthesis of (−)-O-Methylancistrocladine. *J. Org. Chem.* **1995**, *60*, 5702–5705.
667. Anh, N. H.; Sung, T. V.; Ripperger, H.; Adam, G. Some Results From Chemical Study on Vietnamese *Ancistrocladus cochinchinensis*. *Tap Chi Hoa Hoc* **1996**, *34*, 89–93.
668. Anh, N. H.; Ripperger, H.; Porzel, A.; Sung, T. V.; Adam, G. Tetralones from *Ancistrocladus cochinchinensis*. *Phytochemistry* **1997**, *44*, 549–551.
669. Fournet, A.; Angelo, A.; Muñoz, V.; Roblot, F.; Hocquemiller, R.; Cavé, A. Biological and Chemical Studies of *Pera benensis*, a Bolivian Plant Used in Folk Medicine as a Treatment of Cutaneous Leishmaniasis. *J. Ethnopharmacol.* **1992**, *37*, 159–164.
670. Yue, J.; Lin, Z.; Wang, D.; Feng, Y.; Sun, H. Plumbasides A-C, Three Naphthoquinones Derivatives From *Ceratostigma minus*. *Phytochemistry* **1994**, *35*, 1023–1025.
671. Serrilli, A. M.; Sanfilippo, V.; Ballero, M.; Sanna, C.; Poli, F.; Scartezzini, P.; Serafini, M.; Bianco, A. Polar and Antioxidant Fraction of *Plumbago europaea* L., a Spontaneous Plant of Sardinia. *Nat. Prod. Res.* **2010**, *24*, 633–639.
672. Durand, R.; Zenk, M. H. Biosynthesis of Plumbagin (5-Hydroxy-2-Methyl-1,4-Naphthoquinone) via the Acetate Pathway in Higher Plants. *Tetrahedron Lett.* **1971**, *32*, 3009–3012.
673. Chauhan, R.; Ruby, K.; Shori, A.; Dwivedi, J. *Plumbago zeylanica*, a Potent Herb for Various Ailments: A Review. *Int. J. Pharm. Sci. Rev. Res.* **2012**, *15*, 72–78.
674. Budzianowski, J.; Budzianowska, A.; Kromer, K. Naphthalene Glucoside and Other Phenolics From the Shoot and Callus Cultures of *Drosophyllum lusitanicum*. *Phytochemistry* **2002**, *61*, 421–425.
675. Budzianowski, J. Naphthohydroquinone Glucosides of *Drosera rotundifolia* and *D. intermedia* from *In Vitro* Cultures. *Phytochemistry* **1996**, *42*, 1145–1147.
676. Tokunaga, T.; Dohmura, A.; Takada, N.; Ueda, M. Cytotoxic Antifeedant from *Dionaea muscipula* Ellis: A Defensive Mechanism of Carnivorous Plants Against Predators. *Bull. Chem. Soc. Jpn.* **2004**, *77*, 537–541.
677. Timmers, M. A.; Dias, D. A.; Urban, S. HPLC-NMR Chemical Profiling of the Australian Carnivorous Plant, *Drosera erythrohiza* Subspecies *Magna*. *J. Nat. Prod.* **2013**, *3*, 35–41.
678. Higa, M.; Ogihara, K.; Yogi, S. Bioactive Naphthoquinone Derivatives From *Diospyros maritime* BLUME. *Chem. Pharm. Bull.* **1998**, *46*, 1189–1193.

679. Kayser, O.; Kiderlen, A. F.; Laatsch, H.; Croft, S. L. In Vitro Leishmanicidal Activity of Monomeric and Dimeric Naphthoquinones. *Acta Trop.* **2000**, *77*, 307–314.
680. Ali, A.; Assimopoulou, A. N.; Papageorgiou, V. P.; Kolodziej, H. Structure/Antileishmanial Activity Relationship Study of Naphthoquinones and Dependency of the Mode of Action on the Substitution Patterns. *Planta Med.* **2011**, *77*, 2003–2012.
681. Fournet, A.; Barrios, A. A.; Muñoz, V.; Hocquemiller, R.; Cavé, A. Effect of Natural Naphthoquinones in BALB7c Mice Infected With *Leishmania amazonensis* and *Leishmania venezuelensis*. *Trop. Med. Parasitol.* **1992**, *43*, 219–221.
682. Porras, G.; Chassagne, F.; Lyles, J. T.; Marquez, L.; Dettweiler, M.; Salam, A. M.; Samarakoon, T.; Shabih, S.; Farrokhi, R.; Quave, C. L. Ethnobotany and the Role of Plant Natural Products in Antibiotic Drug Discovery. *Chem. Rev.* **2021**, *121*, 3495–3560.
683. Uc-Cachón, A. H.; Borges-Argáez, R.; Said-Fernández, S.; Vargas-Villarreal, J.; González-Salazar, F.; Méndez-Gonzáles, M.; Cáceres-Farfán, M.; Molina-Salinas, G. M. Naphthoquinones From *Diospyros anisandra* Exhibit Potent Activity Against Pan-Resistant First-Line Drugs *Mycobacterium tuberculosis* Strains. *Pulm. Pharmacol. Ther.* **2014**, *27*, 114–120.
684. Nair, S. V.; Baranwal, G.; Chatterjee, M.; Sachu, A.; Vasudevan, A. K.; Bose, C.; Banerji, A.; Biswas, R. Antimicrobial Activity of Plumbagin, a Naturally Occurring Naphthoquinone From *Plumbago rosea*, Against *Staphylococcus aureus* and *Candida albicans*. *Int. J. Med. Microbiol.* **2016**, *306*, 237–248.
685. Periasamy, H.; Iswarya, S.; Pavithra, N.; Senthilnathan, S.; Gnanamani, A. In Vitro Antibacterial Activity of Plumbagin Isolated from *Plumbago zeylanica* L. Against Methicillin-Resistant *Staphylococcus aureus*. *Lett. Appl. Microbiol.* **2019**, *69*, 41–49.
686. Adusei, E. B. A.; Adosraku, R. K.; Oppong-Kyekyeku, J.; Amengor, C. D. K.; Jibira, Y. Resistance Modulation Action, Time-Kill Kinetics Assay, and Inhibition of Biofilm Formation Effects of Plumbagin from *Plumbago zeylanica* Linn. *J. Trop. Med.* **2019**, 1250645.
687. Jangra, A.; Chadha, V.; Kumar, D.; Kumar, V.; Arora, M. K. Neuroprotective and Acetylcholinesterase Inhibitory Activity of Plumbagin in ICV-LPS-Induced Behavioural Deficits in Rats. *Curr. Res. Behav. Sci.* **2021**, *2*, 100060.
688. Nakhate, K. T.; Bharne, A. P.; Verma, V. S.; Aru, D. N.; Kokare, D. M. Plumbagin Ameliorates Memory Dysfunction in Streptozotocin Induced Alzheimer's Disease via Activation of Nrf2/ARE Pathway and Inhibition of β-Secretase. *Biomed. Pharmacother.* **2018**, *101*, 379–390.
689. Messeha, S. S.; Zarnouh, N. O.; Mendonca, P.; Kolta, M. G.; Soliman, K. F. A. The Attenuating Effects of Plumbagin on Pro-Inflammatory Cytokine Expression in LPS-Activated BV-2 Microglial Cells. *J. Neuroimmunol.* **2017**, *313*, 129–137.
690. Budzianowski, J. Naphthoquinone Glucosides of *Drosera gigantea* From *In Vitro* Cultures. *Planta Med.* **2000**, *66*, 667–669.
691. Lieberherr, C.; Zhang, G.; Grafen, A.; Singethan, K.; Kendl, S.; Vogt, V.; Maier, J.; Bringmann, G.; Schneider-Schaulies, J. The Plant-Derived Naphthoquinone Droserone Inhibits In Vitro Measles Virus Infection. *Planta Med.* **2017**, *83*, 232–238.
692. Likhitwitayawuid, K.; Kaewamatawong, R.; Ruangrungsi, N.; Krungkrai, J. Antimalarial Naphthoquinones from *Nepenthes thorelii*. *Planta Med.* **1998**, *64*, 237–241.
693. Bringmann, G.; Münchbach, M.; Messer, K.; Koppler, D.; Michel, M.; Schupp, O.; Wenzel, M.; Louis, A. M. *Cis-* and *Trans-*Isoshinanolone From *Dioncophyllum thollonii*: Absolute Configuration of Two 'Known', Wide-Spread Natural Products. *Phytochemistry* **1999**, *51*, 693–699.

694. Lavault, M.; Bruneton, J. Alcaloïdes du *Dionocophyllum thollonii*: Isolement de Deux Nouveaux Alcaloïdes, Triphyopeltine et O-Méthyl-5'-Triphyopeltine. *Planta Med.* **1980**, *Suppl*, 17–21.
695. Bringmann, G.; Messer, K.; Saeb, W.; Peters, E. M.; Peters, K. The Absolute Configuration of (+)-Isoshinanolone and In Situ LC-CD Analysis of Its Stereoisomers From Crude Extracts. *Phytochemistry* **2001**, *56*, 387–391.
696. Hanson, S. W.; Crawford, M.; Thanasingh, D. P. J. (+)-Isoshinanolone and 2-Methylbenzofuran-4-Carbaldehyde From the Fish-Stunning Plant *Habropetalum dawei*. *Phytochemistry* **1981**, *20*, 1162–1164.
697. Tezuka, M.; Takahashi, C.; Kuroyanagi, M.; Satake, M.; Yoshihira, K.; Natori, S. New Naphthoquinones From *Diospyros*. *Phytochemistry* **1973**, *12*, 175–183.
698. Gunaherath, G. M. K. B.; Gunatilaka, A. A. L.; Sultanbawa, M. U. S.; Balasubramaniam, S. 1,2(3)-Tetrahydro-3,3'-Biplumbagin: A Naphthalenone and Other Constituents From *Plumbago zeylanica*. *Phytochemistry* **1983**, *22*, 1245–1247.
699. Sreelatha, T.; Hymavathi, A.; Madhusudhana Murthy, J.; Rani, P. U.; Madhusudhana Rao, J.; Suresh Babu, K. Bioactivity-Guided Isolation of Mosquitocidal Constituents From the Rhizomes of *Plumbago capensis* Thunb. *Bioorg. Med. Chem. Lett.* **2010**, *20*, 2974–2977.
700. Bhattacharya, J.; De Carvalho, V. R. *Epi*-Isoshinanolone From *Plumbago scandens*. *Phytochemistry* **1986**, *255*, 764–765.
701. Aung, H. H.; Chia, L. S.; Goh, N. K.; Chia, T. F.; Ahmed, A. A.; Pare, P. W.; Mabry, T. J. Phenolic Constituents From the Leaves of the Carnivorous Plant *Nepenthes gracilis*. *Fitoterapia* **2002**, *73*, 445–447.
702. Kumar, V.; Meepagala, K. M.; Balasubramaniam, S. Quinoid and Other Constituents of *Aristea ecklonii*. *Phytochemistry* **1985**, *24*, 1118–1119.
703. Zhong, S. M.; Waterman, P. G.; Jeffreys, J. A. D. Naphthoquinones and Triterpenes From African *Diospyros* Species. *Phytochemistry* **1984**, *23*, 1067–1072.
704. Bin Zakaria, M.; Jeffreys, J. A. D.; Waterman, P. G.; Zhong, S. M. Naphthoquinones and Triterpenes From Some Asian *Diospyros* Species. *Phytochemistry* **1984**, *23*, 1481–1484.
705. Richomme, P.; Papillon, B.; Cabalion, P.; Bruneton, J. Naphthoquinones de *Diospyros samoensis*. *Pharm. Acta Helv.* **1991**, *66*, 88–89.
706. Chang, C. I.; Chen, C. R.; Chiu, H. L.; Kuo, C. L.; Kuo, Y. H. Chemical Constituents From the Stems of *Diospyros maritima*. *Molecules* **2009**, *14*, 5281–5288.

Cumulative Index of Titles

A

Aconitum alkaloids, **4**, 275 (1954), **7**, 473 (1960), **34**, 95 (1989), **42**, 151 (1992), **87**, 1 (2022)
   C18 diterpenes, **67**, 1 (2009)
   C19 diterpenes, **12**, 1 (1970), **17**, 1 (1979), **69**, 1 (2010)
   C20 diterpenes, **12**, 135 (1970), **18**, 99 (1981), **59**, 1 (2002)
Acridine alkaloids, **2**, 353 (1952)
Acridone alkaloids, **54**, 259 (2000), **78**, 1 (2017)
   experimental antitumor activity of acronycine, **21**, 1 (1983)
Actinomycetes, isoquinolinequinones, **21**, 55 (1983), **53**, 120 (2000)
N-Acyliminium ions as intermediates in alkaloid synthesis, **32**, 271 (1988)
Aerophobins and related alkaloids, **57**, 208 (2001)
Aerothionins, **57**, 219 (2001)
Ajmaline-Sarpagine alkaloids, **8**, 789 (1965), **11**, 41 (1986), **52**, 104 (1999), **55**, 1 (2001)
   enzymes in biosynthesis of, **47**, 116 (1995)
Akuammiline alkaloids, **76**, 171 (2016)
Alkaloid chemistry
   marine cyanobacteria, **57**, 86 (2001)
   synthetic studies, **50**, 377 (1998)
Alkaloid production, plant biotechnology of, **40**, 1 (1991)
Alkaloid structures
   spectral methods, study, **24**, 287 (1985)
   unknown structure, **5**, 301 (1955), **7**, 509 (1960), **10**, 545 (1967), **12**, 455 (1970), **13**, 397 (1971), **14**, 507 (1973), **15**, 263 (1975), **16**, 511 (1977)
   X-ray diffraction, **22**, 51 (1983)
Alkaloids
   as chirality transmitters, **53**, 1 (2000)
   biosynthesis, regulation of, **49**, 222 (1997)
   biosynthesis, molecular genetics of, **50**, 258 (1998)
   biotransformation of, **57**, 3 (2001), **58**, 1 (2002)
   chemical and biological aspects of *Narcissus*, **63**, 87 (2006)
   containing a quinolinequinone unit, **49**, 79 (1997)
   containing a quinolinequinoneimine unit, **49**, 79 (1997)
   containing an isoquinolinoquinone unit, **53**, 119 (2000)
   2,5-diketopiperazine, **90**, 159 (2023)
   ecological activity of, **47**, 227 (1995)
   ent-morphinan, **90**, 1 (2023)
   forensic chemistry of, **32**, 1 (1988)
   histochemistry of, **39**, 165 (1990)
   infrared and raman spectroscopy of, **67**, 217 (2009)
   in the plant, **1**, 15 (1950), **6**, 1 (1960)
   of the Annonaceae, **74**, 233 (2015)
   of the Lauraceae, **82**, 147 (2019)
   of the Menispermaceae, **54**, 1 (2000)
   plant biotechnology, production of, **50**, 453 (1998)
   pyridoacridine, **90**, 97 (2023)

pyridoacridone, **90**, 97 (2023)
pyrroloacridine, **90**, 97 (2023)
toxic to livestock, **67**, 143 (2009)
with antiprotozoal activity, **66**, 113 (2008) 153
Alkaloids from
amphibians, **21**, 139 (1983), **43**, 185 (1993), **50**, 141 (1998)
ants and insects, **31**, 193 (1987)
Chinese traditional medicinal plants, **32**, 241 (1988)
Hernandiaceae, **62**, 175 (2005)
mammals, **21**, 329 (1983), **43**, 119 (1993)
marine bacteria, **53**, 239 (2000), **57**, 75 (2001)
marine organisms, **24**, 25 (1985), **41**, 41 (1992)
medicinal plants of New Caledonia, **48**, 1 (1996)
mushrooms, **40**, 189 (1991)
plants of Thailand, **41**, 1 (1992)
Sri Lankan flora, **52**, 1 (1999)
Alkaloid synthesis
using organocatalysts, **79**, 1 (2018)
Wittig reaction in, **84**, 201 (2020)
Alkyl, aryl, alkylarylquinoline, and related alkaloids, **64**, 139 (2007)
Allelochemical properties of alkaloids, **43**, 1 (1993)
Allo congeners, and tropolonic *Colchicum* alkaloids, **41**, 125 (1992)
Alstonia alkaloids, **8**, 159 (1965), **12**, 207 (1970), **14**, 157 (1973), **81**, 115 (2019)
The alstoscholarisine alkaloids: isolation, structure determination, biogenesis, biological evaluation, and synthesis, **81**, 115 (2019)
Amaryllidaceae
Amaryllidaceae alkaloids, **2**, 331 (1952), **6**, 289 (1960), **11**, 307 (1968), **15**, 83 (1975), **30**, 251 (1987), **51**, 323 (1998), **63**, 87 (2006), **83**, 113 (2020)
Amphibian alkaloids, **21**, 139 (1983), **43**, 185 (1983), **50**, 141 (1998)
Analgesic alkaloids, **5**, 1 (1955)
Anesthetics, local, **5**, 211 (1955)
Annonaceae alkaloids: occurrence and a compilation of their biological activities, **74**, 233 (2015)
Anthranilic acid derived alkaloids, **17**, 105 (1979), **32**, 341 (1988), **39**, 63 (1990)
Antifungal alkaloids, **42**, 117 (1992)
Antimalarial alkaloids, **5**, 141 (1955)
Antiprotozoal alkaloids, **66**, 113 (2008)
Antitumor alkaloids, **25**, 1 (1985), **59**, 281 (2002)
Apocynaceae alkaloids, steroids, **9**, 305 (1967)
Aporphine alkaloids, **4**, 119 (1954), **9**, 1 (1967), **14**, 225 (1973), **24**, 153 (1985), **53**, 57 (1999), **89**, 39 (2023)
Apparicine and related alkaloids, **57**, 235 (2001)
Aristolochia alkaloids, **31**, 29 (1987)
Aristotelia alkaloids, **24**, 113 (1985), **48**, 191 (1996)
Aspergillus alkaloids, **29**, 185 (1986)
Aspidosperma alkaloids, **8**, 336 (1965), **11**, 205 (1968), **17**, 199 (1979), **86**, 1 (2021)
synthesis of, **50**, 343 (1998)
Aspidospermine group alkaloids, **51**, 1 (1998)

Asymmetric catalysis by alkaloids, **53**, 1 (2000)
Azafluoranthene alkaloids, **23**, 301 (1984)

B

Bases
  simple, **3**, 313 (1953), **8**, 1 (1965)
  simple indole, **10**, 491 (1967)
  simple isoquinoline, **4**, 7 (1954), **21**, 255 (1983)
Benzoxazole alkaloids, **79**, 71 (2018)
Benzodiazepine alkaloids, **39**, 63 (1990)
Benzophenanthridine alkaloids, **26**, 185 (1985)
Benzylisoquinoline alkaloids, **4**, 29 (1954), **10**, 402 (1967)
Betalains, **39**, 1 (1990)
Biosynthesis
  C19 diterpene, **69**, 362–374 (2010)
  in *Catharanthus roseus*, **49**, 222 (1997)
  in *Rauwolfia serpentina*, **47**, 116 (1995)
  isoquinoline alkaloids, **4**, 1 (1954)
  jadomycin alkaloids, **84**, 125 (2020)
  pyrrolizidine alkaloids, **46**, 1 (1995)
  quinolizidine alkaloids, **46**, 1 (1995)
  regulation of, **63**, 1 (2006)
  tropane alkaloids, **44**, 116 (1993)
Bisbenzylisoquinoline alkaloids, **4**, 199 (1954), **7**, 439 (1960), **9**, 133 (1967), **13**, 303 (1971), **16**, 249 (1977), **16**, 319 (1977), **30**, 1 (1987), **81**, 1 (2019)
Bisindole alkaloids, **20**, 1 (1981), **63**, 181 (2006), **76**, 259 (2016)
  noniridoid, **47**, 173 (1995)
Bisindole alkaloids of *Catharanthus*
  C-20′ position as a functional hot spot in, **37**, 133 (1990)
  isolation, structure elucidation and biosynthesis of, **37**, 1 (1990), **63**, 181 (2006)
  medicinal chemistry of, **37**, 145 (1990)
  pharmacology of, **37**, 205 (1990)
  synthesis of, **37**, 77 (1990), **59**, 281 (2002)
  therapeutic uses of, **37**, 229 (1990)
Bromotyrosine alkaloids, marine, **61**, 79 (2005)
Buxus alkaloids, steroids, **9**, 305 (1967), **14**, 1 (1973), **32**, 79 (1988)
  chemistry and biology, **66**, 191 (2008)

C

Cactus alkaloids, **4**, 23 (1954)
Calabar bean alkaloids, **8**, 27 (1965), **10**, 383 (1967), **13**, 213 (1971), **36**, 225 (1989)
Calabash curare alkaloids, **8**, 515 (1965), **11**, 189 (1968)
Calycanthaceae alkaloids, **8**, 581 (1965)
Calystegines, **64**, 49 (2007)
Camptothecin and derivatives, **21**, 101 (1983), **50**, 509 (1998)
  clinical studies, **60**, 1 (2003)
Cancentrine alkaloids, **14**, 407 (1973)
Cannabis sativa alkaloids, **34**, 77 (1988)
Canthin-6-one alkaloids, **36**, 135 (1989)

Capsicum alkaloids, **23**, 227 (1984)
Carbazole alkaloids, **13**, 273 (1971), **26**, 1 (1985), **44**, 257 (1993), **65**, 1 (2008)
  biogenesis, **65**, 159 (2008)
  biological and pharmacological activities, **65**, 181 (2008)
  chemistry, **65**, 195 (2008)
Carboline alkaloids, **8**, 47 (1965), **26**, 1 (1985)
β-Carboline congeners and Ipecac alkaloids, **22**, 1 (1983)
Cardioactive alkaloids, **5**, 79 (1955)
Catharanthus alkaloids, **59**, 281 (2002)
Catharanthus roseus, biosynthesis of terpenoid indole alkaloids in, **49**, 222 (1997)
Celastraceae alkaloids, **16**, 215 (1977)
Cephalostatins and Ritterazines, **72**, 153 (2013)
Cephalotaxus alkaloids, **23**, 157 (1984), **51**, 199 (1998), **78**, 205 (2017)
Cevane group of *Veratrum* alkaloids, **41**, 177 (1992)
The chemical synthesis and applications of tropane alkaloids, **81**, 151 (2019)
Chemistry of hapalindoles, fischerindoles, ambiguines, and welwitindolinones, **73**, 65 (2014)
Chemosystematics of alkaloids, **50**, 537 (1998)
Chemotaxonomy of Papaveraceae and Fumariaceae, **29**, 1 (1986)
Chinese medicinal plants, alkaloids from, **32**, 241 (1988)
Chirality transmission by alkaloids, **53**, 1 (2000)
Chromone alkaloids, **31**, 67 (1987)
Cinchona alkaloids, **3**, 1 (1953), **14**, 181 (1973), **34**, 331 (1989), **82**, 29 (2019)
Colchicine, **2**, 261 (1952), **6**, 247 (1960), **11**, 407 (1968), **23**, 1 (1984)
  pharmacology and therapeutic aspects of, **53**, 287 (2000)
Colchicum alkaloids and allo congeners, **41**, 125 (1992)
Configuration and conformation, elucidation by X-ray diffraction, **22**, 51 (1983)
Corynantheine, yohimbine, and related alkaloids, **27**, 131 (1986)
Cularine alkaloids, **4**, 249 (1954), **10**, 463 (1967), **29**, 287 (1986)
Curare-like effects, **5**, 265 (1955)
Cyclic tautomers of tryptamine and tryptophan, **34**, 1 (1988)
Cyclopeptide alkaloids, **15**, 165 (1975), **67**, 79 (2009)
Cylindrospermopsin alkaloids, **70**, 1 (2011)
Cytotoxic alkaloids, modes of action, **64**, 1 (2007)

D

Daphniphyllum alkaloids, **15**, 41 (1975), **29**, 265 (1986), **60**, 165 (2003), **85**, 113 (2021)
Delphinium alkaloids, **4**, 275 (1954), **7**, 473 (1960)
  C10-diterpenes, **12**, 1 (1970)
  C20-diterpenes, **12**, 135 (1970)
Detection of through IR and Raman spectroscopy, **67**, 217 (2009)
Dibenzazonine alkaloids, **35**, 177 (1989)
Dibenzopyrrocoline alkaloids, **31**, 101 (1987)
2,5-Diketopiperazine alkaloids, **90**, 159 (2023)
Diplorrhyncus alkaloids, **8**, 336 (1965)
Diterpenoid alkaloids, **4**, 275 (1954), **7**, 473 (1960), **12**, 1 (1970), **12**, 135 (1970), **17**, 1 (1979), **18**, 99 (1981), **34**, 95 (1989), **42**, 151 (1992), **59**, 1 (2002), **67**, 1 (2009), **69**, 1 (2010), **87**, 1 (2022)
  general introduction, **12**, xv (1970)
Duguetia alkaloids, **68**, 83 (2010)

E

Eburnamine-vincamine alkaloids, **8**, 250 (1965), **11**, 125 (1968), **20**, 297 (1981), **42**, 1 (1992)
Ecological activity of alkaloids, **47**, 227 (1995)
Elaeocarpus alkaloids, **6**, 325 (1960)
Ellipticine and related alkaloids, **39**, 239 (1990), **57**, 235 (2001)
Enamide cyclizations in alkaloid synthesis, **22**, 189 (1983)
Ent-morphinan alkaloids, **90**, 1 (2023)
Enzymatic transformation of alkaloids, microbial and in vitro, **18**, 323 (1981)
Ephedra alkaloids, **3**, 339 (1953)
Epibatidine, **46**, 95 (1995)
Ergot alkaloids, **8**, 726 (1965), **15**, 1 (1975), **38**, 1 (1990), **50**, 171 (1998), **54**, 191 (2000), **63**, 45 (2006), **85**, 1 (2021)
Erythrina alkaloids, **2**, 499 (1952), **7**, 201 (1960), **9**, 483 (1967), **18**, 1 (1981), **48**, 249 (1996), **68**, 39 (2010)
Erythrophleum alkaloids, **4**, 265 (1954), **10**, 287 (1967)
Eupomatia alkaloids, **24**, 1 (1985)

F

Flavoalkaloids, **31**, 67 (1987), **77**, 85 (2017)
Forensic chemistry, alkaloids, **12**, 514 (1970)
   by chromatographic methods, **32**, 1 (1988)

G

Galanthamine
   history and introduction, **68**, 157 (2010)
   production, **68**, 167 (2010)
Galanthus
Galbulimima alkaloids, **9**, 529 (1967), **13**, 227 (1971), **78**, 109 (2017)
Gardneria alkaloids, **36**, 1 (1989)
Garrya alkaloids, **7**, 473 (1960), **12**, 2 (1970), **12**, 136 (1970)
Geissospermum alkaloids, **8**, 679 (1965)
Gelsemium alkaloids, **8**, 93 (1965), **33**, 84 (1988), **49**, 1 (1997)
Glycosides, monoterpene alkaloids, **17**, 545 (1979)
Guatteria alkaloids, **35**, 1 (1989)

H

Halogenated alkaloids
   biosynthesis of, **71**, 167 (2012)
   occurrence of, **71**, 1 (2012)
Haplophyton cimicidum alkaloids, **8**, 673 (1965)
Hasubanan alkaloids, **16**, 393 (1977), **33**, 307 (1988)
Hasubanan and acutumine alkaloids, **73**, 161 (2014)
Hernandiaceae alkaloids, **62**, 175 (2005)
Histochemistry of alkaloids, **39**, 165 (1990)
Holarrhena group, steroid alkaloids, **7**, 319 (1960)
Homalium alkaloids: isolation, synthesis and absolute configuration assignment, **74**, 121 (2015)
Hunteria alkaloids, **8**, 250 (1965)

I

Iboga alkaloids, **8**, 203 (1965), **11**, 79 (1968), **59**, 281 (2002)
Ibogaine alkaloids
    addict self-help, **56**, 283 (2001)
    as a glutamate antagonist, **56**, 55 (2001)
    comparative neuropharmacology, **56**, 79 (2001)
    contemporary history of, **56**, 249 (2001)
    drug discrimination studies with, **56**, 63 (2001)
    effects of rewarding drugs, **56**, 211 (2001)
    gene expression, changes in, **56**, 135 (2001)
    mechanisms of action, **56**, 39 (2001)
    multiple sites of action, **56**, 115 (2001)
    neurotoxicity assessment, **56**, 193 (2001)
    pharmacology of, **52**, 197 (1999)
    review, **56**, 1 (2001)
    treatment case studies, **56**, 293 (2001)
    use in equatorial African ritual context, **56**, 235 (2001)
Imidazole alkaloids, **3**, 201 (1953), **22**, 281 (1983)
Indole alkaloids, **2**, 369 (1952), **7**, 1 (1960), **26**, 1 (1985)
    ajmaline group of, **55**, 1 (2001)
    biomimetic synthesis of, **50**, 415 (1998)
    biosynthesis in *Catharanthus roseus*, **49**, 222 (1997)
    biosynthesis in *Rauvolfia serpentina*, **47**, 116 (1995)
    chippiine class of, **85**, 177 (2021)
    dippinine class of, **85**, 177 (2021)
    distribution in plants, **11**, 1 (1968)
    moschamine-related, **79**, 139 (2018)
    Reissert synthesis of, **31**, 1 (1987)
    sarpagine group of, **52**, 103 (1999)
    simple, **10**, 491 (1967), **26**, 1 (1985)
    tronocarpine class of, **85**, 177 (2021)
Indole-based subincanadine alkaloids, **83**, 187 (2020)
Indole diterpenoid alkaloids, **60**, 51 (2003)
Indolizidine alkaloids, **28**, 183 (1986), **44**, 189 (1993), **75**, 1 (2016)
2,2'-Indolylquinuclidine alkaloids, chemistry, **8**, 238 (1965), **11**, 73 (1968)
Infrared spectroscopy of alkaloids, **67**, 217 (2009)
In vitro and microbial enzymatic transformation of alkaloids, **18**, 323 (1981)
Ipecac alkaloids, **3**, 363 (1953), **7**, 419 (1960), **13**, 189 (1971), **22**, 1 (1983), **51**, 271 (1998)
Isolation of alkaloids, **1**, 1 (1950)
Isoquinoline alkaloids, **7**, 423 (1960)
    biosynthesis, **4**, 1 (1954)
    13C-NMR spectra, **18**, 217 (1981)
    Reissert synthesis of, **31**, 1 (1987)
    simple isoquinoline alkaloids **4**, 7 (1954), **21**, 255 (1983)
Isoquinolinequinones, **21**, 55 (1983), **53**, 120 (2000)
Isoxazole alkaloids, **57**, 186 (2001)

J

Jadomycin alkaloids, **84**, 125 (2020)

K

Khat (Catha edulis) alkaloids, **39**, 139 (1990)
Kopsia alkaloids, **8**, 336 (1965), **66**, 1 (2008)

L

Lamellarin alkaloids, **83**, 1 (2020)
Lauraceae alkaloids, **82**, 147 (2019)
Lead tetraacetate oxidation in alkaloid synthesis, **36**, 70 (1989)
Local anesthetics, **5**, 211 (1955)
Localization in the plant, **1**, 15 (1950), **6**, 1 (1960)
Lupine alkaloids, **3**, 119 (1953), **7**, 253 (1960), **9**, 175 (1967), **31**, 116 (1987), **47**, 1 (1995)
Lycopodium alkaloids, **5**, 295 (1955), **7**, 505 (1960), **10**, 305 (1968), **14**, 347 (1973), **26**, 241 (1985), **45**, 233 (1994), **61**, 1 (2005), **72**, 1 (2013)
Lythraceae alkaloids, **18**, 263 (1981), **35**, 155 (1989)

M

Macrocyclic peptide alkaloids from plants, **26**, 299 (1985), **49**, 301 (1997)
Madangamine group alkaloids, **74**, 159 (2015)
Mammalian alkaloids, **21**, 329 (1983), **43**, 119 (1993)
Manske, R.H.F., biography of, **50**, 3 (1998)
Manzamine alkaloids, **60**, 207 (2003), **84**, 1 (2020)
Marine alkaloids, **24**, 25 (1985), **41**, 41 (1992), **52**, 233 (1999)
  bromotyrosine alkaloids, **61**, 79 (2005)
Marine bacteria, alkaloids from, **53**, 120 (2000)
Marine bi-, bis-, and trisindole alkaloids, **73**, 1 (2014)
Maytansinoids, **23**, 71 (1984)
Melanins, **36**, 254 (1989)
  chemical and biological aspects, **60**, 345 (2003)
Melodinus alkaloids, **11**, 205 (1968)
Mesembrine alkaloids, **9**, 467 (1967)
Metabolic transformation of alkaloids, **27**, 323 (1986)
Microbial and *in vitro* enzymatic transformation of alkaloids, **18**, 323 (1981)
Mitragyna alkaloids, **8**, 59 (1965), **10**, 521 (1967), **14**, 123 (1973)
Molecular modes of action of cytotoxic alkaloids, **64**, 1 (2007)
Monoterpene alkaloids, **16**, 431 (1977), **52**, 261 (1999)
  glycosides, **17**, 545 (1979)
Monoterpenoid bisindole alkaloids, **76**, 259 (2016)
Monoterpenoid indole alkaloids, **77**, 1 (2017)
Morphine alkaloids, **2**, 1 (part 1), **161** (part 2) (1952), **6**, 219 (1960), **13**, 1 (1971), **45**, 127 (1994), **86**, 145 (2021)
Moschamine-related indole alkaloids, **79**, 139 (2018)
Muscarine alkaloids, **23**, 327 (1984)
Mushrooms, alkaloids from, **40**, 190 (1991)
Mydriatic alkaloids, **5**, 243 (1955)

N

α-Naphthophenanthridine alkaloids, **4**, 253 (1954), **10**, 485 (1967)
Naphthylisoquinoline alkaloids, **29**, 141 (1986), **46**, 127 (1995), **91**, 1 (2024)
Narcotics, **5**, 1 (1955)

Narcissus alkaloids, **63**, 87 (2006)
New Caledonia, alkaloids from the medicinal plants of, **48**, 1 (1996)
Nitrogen-containing metabolites from marine bacteria, **53**, 239, (2000), **57**, 75 (2001)
Non-iridoid bisindole alkaloids, **47**, 173 (1995)
Nuclear magnetic resonance imaging, C19 diterpenes, **69**, 381–419 (2010)
Nuphar alkaloids, **9**, 441 (1967), **16**, 181 (1977), **35**, 215 (1989)

O

Ochrosia alkaloids, **8**, 336 (1965), **11**, 205 (1968)
Organocatalysts, **79**, 1 (2018)
Ourouparia alkaloids, **8**, 59 (1965), **10**, 521 (1967)
Oxazole alkaloids, **35**, 259 (1989)
Oxindole alkaloids, **14**, 83 (1973)
Oxoaporphine alkaloids, **14**, 225 (1973)

P

Pancratium alkaloids, **68**, 1 (2010)
Pandanus alkaloids, **66**, 215 (2008), **82**, 1 (2019)
Papaveraceae alkaloids, **10**, 467 (1967), **12**, 333 (1970), **17**, 385 (1979)
    pharmacology, **15**, 207 (1975)
    toxicology, **15**, 207 (1975)
Pauridiantha alkaloids, **30**, 223 (1987)
Pavine and isopavine alkaloids, **31**, 317 (1987)
Pentaceras alkaloids, **8**, 250 (1965)
Peptide alkaloids, **26**, 299 (1985), **49**, 301 (1997)
Phenanthrene alkaloids, **39**, 99 (1990)
Phenanthroindolizidine alkaloids, **19**, 193 (1981)
Phenanthroquinolizidine alkaloids, **19**, 193 (1981)
β-Phenethylamines, **3**, 313 (1953), **35**, 77 (1989)
Phenethylisoquinoline alkaloids, **14**, 265 (1973), **36**, 172 (1989)
Phthalideisoquinoline alkaloids, **4**, 167 (1954), **7**, 433 (1960), **9**, 117 (1967), **24**, 253 (1985)
Picralima alkaloids, **8**, 119 (1965), **10**, 501 (1967), **14**, 157 (1973)
Piperidine alkaloids, **26**, 89 (1985)
Plant biotechnology, for alkaloid production, **40**, 1 (1991), **50**, 453 (1998)
Plant systematics, **16**, 1 (1977)
Pleiocarpa alkaloids, **8**, 336 (1965), **11**, 205 (1968)
Polyamine alkaloids, **22**, 85 (1983), **45**, 1 (1994), **50**, 219 (1998), **58**, 83 (2002)
    analytical aspects of, **58**, 206 (2002)
    biogenetic aspects of, **58**, 274 (2002)
    biological and pharmacological aspects of, **46**, 63 (1995), **58**, 281 (2002)
    catalog of, **58**, 89 (2002)
    synthesis of cores of, **58**, 243 (2002)
Polyhalogenated alkaloids
    in environmental and food samples, **71**, 211 (2012)
Pressor alkaloids, **5**, 229 (1955)
Protoberberine alkaloids, **4**, 77 (1954), **9**, 41 (1967), **28**, 95 (1986), **62**, 1 (2005)
    biotransformation of, **46**, 273 (1955)
    transformation reactions of, **33**, 141 (1988)
Protopine alkaloids, **4**, 147 (1954), **34**, 181 (1988)

Pseudocinchoma alkaloids, **8**, 694 (1965)
Pseudodistomins, **50**, 317 (1998)
Purine alkaloids, **38**, 226 (1990)
Putrescine and related polyamine alkaloids, **58**, 83 (2002)
Pyridine alkaloids, **1**, 165 (1950), **6**, 123 (1960), **11**, 459 (1968), **26**, 89 (1985)
Pyridoacridine alkaloids, **90**, 97 (2023)
Pyridoacridone alkaloids, **90**, 97 (2023)
Pyrimidine-containing alkaloids, synthesis, **88**, 49 (2022)
Pyrrolidine alkaloids, **1**, 91 (1950), **6**, 31 (1960), **27**, 270 (1986)
Pyrrolizidine alkaloids, **1**, 107 (1950), **6**, 35 (1960), **12**, 245 (1970), **26**, 327 (1985), **80**, 1 (2018)
  biosynthesis of, **46**, 1 (1995)
Pyrroloacridine alkaloids, **90**, 97 (2023)
Pyrrolo[2,1-a] isoquinoline alkaloids
  synthesis of **70**, 79 (2011)
Pyrrolo[2,3-*d*]pyrimidine, **79**, 191 (2018)

Q

Quinazolidine alkaloids, *see* Indolizidine alkaloids
Quinazoline alkaloids, **3**, 101 (1953), **7**, 247 (1960), **29**, 99 (1986), **88**, 1 (2022)
Quinazolinocarbolines, **8**, 55 (1965), **21**, 29 (1983)
Quinoline alkaloids, **3**, 65 (1953), **7**, 229 (1960), **9**, 223 (1967), **17**, 105 (1979), **32**, 341 (1988), **64**, 139 (2007), **88**, 1 (2022)
Quinolinequinone alkaloids, **49**, 79 (1997)
Quinolinequinoneimine alkaloids, **49**, 79 (1977)
Quinolizidine alkaloids, **28**, 183 (1986), **55**, 91 (2001), **75**, 1 (2016)
  biological activities of, **89**, 1 (2023)
  biosynthesis of, **46**, 1 (1995)

R

Raman spectroscopy of alkaloids, **67**, 217 (2009)
Rauwolfia alkaloids, **8**, 287 (1965)
  biosynthesis of, **47**, 116 (1995)
Recent studies on the synthesis of strychnine, **64**, 103 (2007)
Regulation of alkaloid biosynthesis in plants, **63**, 1 (2006)
Reissert synthesis of isoquinoline and indole alkaloids, **31**, 1 (1987)
Reserpine, chemistry, **8**, 287 (1965)
Respiratory stimulants, **5**, 109 (1995)
Rhazinilam-leuconoxine-mersicarpine triad, **77**, 1 (2017)
Rhoeadine alkaloids, **28**, 1 (1986)
Rigidins, **79**, 191 (2018)

S

Salamandra group, steroids, **9**, 427 (1967)
Saraine alkaloids, **73**, 223 (2014)
Sarpagan-Ajmalan-type indoles, **76**, 1 (2016)
Sarpagine-type alkaloids, **52**, 104 (1999), **76**, 63 (2016)
Sceletium alkaloids, **19**, 1 (1981)
Secoisoquinoline alkaloids, **33**, 231 (1988)

Securinega alkaloids, **14**, 425 (1973), **74**, 1 (2015)
Senecio alkaloids, *see* Pyrrolizidine alkaloids
Sesquiterpene pyridine alkaloids, **60**, 287 (2003)
Simple indole alkaloids, **10**, 491 (1967)
Simple indolizidine alkaloids, **28**, 183 (1986), **44**, 189 (1993), **75**, 1 (2016)
Simple indolizidine and quinolizidine alkaloids, **55**, 91 (2001), **75**, 1 (2016)
Sinomenine, **2**, 219 (1952)
Solanum alkaloids
   chemistry, **3**, 247 (1953), **74**, 216 (2015)
   steroids, **7**, 343 (1960), **10**, 1 (1967), **19**, 81 (1981)
Sources of alkaloids, **1**, 1 (1950)
Spectral methods, alkaloid structures, **24**, 287 (1985)
Spermidine and related polyamine alkaloids, **22**, 85 (1983), **58**, 83 (2002)
Spermine and related polyamine alkaloids, **22**, 85 (1983), **58**, 83 (2002)
Spider toxin alkaloids, **45**, 1 (1994), **46**, 63 (1995)
Spirobenzylisoquinoline alkaloids, **13**, 165 (1971), **38**, 157 (1990)
Sponges, isoquinolinequinone alkaloids from, **21**, 55 (1983)
Sri Lankan flora, alkaloids, **52**, 1 (1999)
Stemona alkaloids, **9**, 545 (1967), **62**, 77 (2005)
Steroid alkaloids
   Apocynaceae, **9**, 305 (1967), **32**, 79 (1988)
   Buxus group, **9**, 305 (1967), **14**, 1 (1973), **32**, 79 (1988), **66**, 191 (2008)
   chemistry and biology, **50**, 61 (1998), **52**, 233 (1999)
   Holarrhena group, **7**, 319 (1960)
   Salamandra group, **9**, 427 (1967)
   Solanum group, **7**, 343 (1960), **10**, 1 (1967), **19**, 81 (1981), **74**, 204 (2015)
   Veratrum group, **7**, 363 (1960), **10**, 193 (1967), **14**, 1 (1973), **41**, 177 (1992), **74**, 204 (2015)
Stimulants
   respiratory, **5**, 109 (1955)
   uterine, **5**, 163 (1955)
Structure elucidation, by X-ray diffraction, **22**, 51 (1983)
Strychnine, synthesis of, **64**, 104 (2007)
Strychnos alkaloids, **1**, 375 (part 1) (1950), **2**, 513 (part 2) (1952), **6**, 179 (1960), **8**, 515, 592 (1965), **11**, 189 (1968), **34**, 211 (1988), **36**, 1 (1989), **48**, 75 (1996), **86**, 1 (2021)
Subincanadine alkaloids, **83**, 187 (2020)
Sulfur-containing alkaloids, **26**, 53 (1985), **42**, 249 (1992)
Synthesis of alkaloids
   enamide cyclizations for, **22**, 189 (1983)
   lead tetraacetate oxidation in, **36**, 70 (1989)

T

Tabernaemontana alkaloids, **27**, 1 (1983)
Taxoids, **69**, 491-514 (2010)
Taxol, **50**, 509 (1998)
Taxus alkaloids, **10**, 597 (1967), **39**, 195 (1990)
Terpenoid indole alkaloids, **49**, 222 (1997)
Thailand, alkaloids from the plants of, **41**, 1 (1992)
Toxicity to livestock, **67**, 143 (2009)

Toxicology
   Papaveraceae alkaloids, 15, 207 (1975)
Transformation of alkaloids, enzymatic, microbial and in vitro, **18**, 323 (1981)
Tremogenic and non-tremogenic alkaloids, **60**, 51 (2003)
Tropane alkaloids, **1**, 271 (1950), **6**, 145 (1960), **9**, 269 (1967), **13**, 351 (1971), **16**, 83 (1977), **33**, 1 (1988), **44**, 1 (1993), **44**, 115 (1993), **81**, 151 (2019)
Tropoloisoquinoline alkaloids, **23**, 301 (1984)
Tropolonic *Colchicum* alkaloids, **23**, 1 (1984), **41**, 125 (1992)
Tylophora alkaloids, **9**, 517 (1967)

U

Uleine and related alkaloids, **57**, 235 (2001)
Unnatural alkaloid enantiomers, biological activity of, **50**, 109 (1998)
Uterine stimulants, **5**, 163 (1955)

V

Veratrum alkaloids
   cevane group of, **41**, 177 (1992)
   chemistry, **3**, 247 (1952), **74**, 216 (2015)
   steroids, **7**, 363 (1960), **10**, 193 (1967), **14**, 1 (1973)
Veratrum and Solanum alkaloids, **74**, 201 (2015)
Vinca alkaloids, **8**, 272 (1965), **11**, 99 (1968), **20**, 297 (1981)
Voacanga alkaloids, **8**, 203 (1965), **11**, 79 (1968)

W

Wasp toxin alkaloids, **45**, 1 (1994), **46**, 63 (1995)
Wittig reaction
   in alkaloid synthesis **84**, 201 (2020)

X

X-ray diffraction of alkaloids, **22**, 51 (1983)

Y

Yohimbe alkaloids, **8**, 694 (1965), **11**, 145 (1968), **27**, 131 (1986)

Index

Note: Page numbers followed by "*f*" indicate figures and "*t*" indicate tables.

A

Acetate-polymalonate pathway
   catalyze the reaction cascade of, 248–249
   to isoquinoline alkaloids, 238–240, 239*f*
   in plants, 231–249, 232*f*, 234*f*
Acetogenic isoquinoline alkaloids, 235
Acridine orange (AO), 116, 149, 179–180
Acute myeloid leukemia (AML), 131
Aedes aegypti, dioncophylline A as larvicide against, 258
African swine fever (ASF) virus, 32–33
African taxa, synoptic revision of, 19
Aldol cyclization, 233
Alkaloid-producing callus cultures, 37
AML. See Acute myeloid leukemia (AML)
Ancisheynine, 54*f*
5-*epi*-Ancistectorine A$_2$, 110–112, 304–326*t*
   cytotoxic activity of, 111–112
Ancistrobreveine A, 290–293*t*
   HPLC-ECD analysis of, 156, 156*f*
Ancistrobreveine B, 290–293*t*
Ancistrobreveine C, 157*f*, 297–303*t*
ent-Ancistrobreveine C, 297–303*t*
Ancistrobreveine D, 119–120, 332–336*t*
Ancistrobrevidine A, 113–114, 113–114*f*, 329–331*t*
Ancistrobrevidine A–C, 112–117, 113–115*f*
Ancistrobrevidine B, 113–114, 113–114*f*, 329–331*t*
Ancistrobrevidine C, 112, 114–116, 304–326*t*
   impairment of cell viability of human HeLa cervical cancer cells, 114–117, 115*f*
   investigations on, 116
   PANC-1 cells in DMEM by, 117, 118*f*
   treatment of PANC-1 cells with, 116–117
Ancistrobrevine A, 150–153, 156*f*, 297–303*t*
Ancistrobrevine B, 120, 332–336*t*
   and related 5,8'-coupled naphthylisoquinoline alkaloids, 119–122, 120–122*f*
   stereostructure of, 121*f*
   synthesis of, 121–122, 122*f*
Ancistrobrevine C, 142–143, 143*f*, 158, 294*t*
Ancistrobrevine D, 290–293*t*
   and related ancistrocladaceae-type naphthylisoquinoline alkaloids, 140–142, 140–141*f*
Ancistrobrevine E, 110, 304–326*t*
5-*epi*-Ancistrobrevine E, 304–326*t*
Ancistrobrevine F, 304–326*t*
   electronic circular dichroism (ECD) spectra of, 78, 78*f*
5-*epi*-Ancistrobrevine F, 304–326*t*
Ancistrobrevine G, 110, 304–326*t*
Ancistrobrevine H, 150–153, 297–303*t*
Ancistrobrevine I, 150–153, 297–303*t*
Ancistrobrevine J, 150–153, 160*f*, 297–303*t*
Ancistrobrevine K, 106–108, 304–326*t*
Ancistrobrevine L, 304–326*t*
Ancistrobrevine M, 329–331*t*
   absolute stereostructures of, 108, 109*f*
   oxidative degradation of, 108
Ancistrobrevinium A, 6–7, 189–192, 191*f*, 339*t*
Ancistrobreviquinone A, 6–7, 337–338*t*
Ancistrobreviquinone A and B, 185–189, 186–187*f*
Ancistrobreviquinone B, 337–338*t*
Ancistrobrevoline A, 210*f*, 213–215, 346–349*t*
Ancistrobrevoline B, 346–349*t*
   cytotoxic effects of, 213–215, 214*f*
Ancistrobrevoline C, 346–349*t*
Ancistrobrevoline D, 346–349*t*
Ancistrocladaceae, 5, 9–44
   cultivation of, 35–42
   geographic distribution of, 9–10, 10*f*
   from India, Sri Lanka, and Southeast Asia, 33–35
   phylogenetic relationships of, 11–12
   phytochemical studies on, 69–74, 100–101

423

Ancistrocladaceae (*Continued*)
　species-specific production of naphthylisoquinoline alkaloids in, 63–64
　systematic position of, 10–11
　from West, Central, and East Africa, 31–33
Ancistrocladine, 44–46, 45*f*, 304–326*t*
　in situ localization of, 267–269, 268*f*
Ancistrocladinium A, 4–5, 4*f*, 54*f*
Ancistrocladinium B, 54*f*
　atropo-diastereomers of, 86–88, 87*f*
Ancistrocladisine B, 290–293*t*
Ancistrocladus, 18–19
　African species of, 20–23*t*
　characteristics of, 29–30, 29*f*
　cultivation of African and Asian, 42–44, 43*f*
　in evergreen rainforests, 29–30
　Southeast Asian species of, 24–28*t*
Ancistrocladus abbreviatus (West African liana), 7–8, 29*f*, 58, 101, 165–231, 165*f*, 167*f*
　atropo-diastereomeric naphthylisoquinoline dimers, 167–185
　callus cultures of, 42, 43*f*
　clonal propagation of, 41–42, 42*f*
　cultivation and propagation of, 39–42
　dehydrogenated naphthylisoquinoline alkaloids from roots of, 153–161, 154–157*f*, 159–160*f*
　geographic distribution of, 40, 41*f*
　highly oxygenated naphthoquinones from solid callus cultures of, 122*f*, 215–231, 216–218*f*, 220*f*, 221*t*, 222*f*, 223–224*t*, 225–227*f*, 228*t*, 230*f*
　5,1′-linked naphthylisoquinoline alkaloids from leaves of, 112–117, 113–115*f*
　5,1′-linked naphthyltetrahydroisoquinoline alkaloids from roots of, 105–108, 106–107*f*, 109*f*
　oxygenated naphthalene derivatives isolated from, 166–167, 167*f*
　phytochemical studies on, 6, 6*f*
　as rich source of naphthylisoquinoline alkaloids, 161–164, 162–164*f*
　rich source of structurally diverse naphthylisoquinoline alkaloids, 101–164, 102*f*
　secondary metabolites produced by, 165–166, 165*f*
　seed germination and growth of, 264–267, 265–266*f*
　seedlings of, 40
　young plants of, 41
Ancistrocladus benomensis, 5–6, 48–49
Ancistrocladus cochinchinensis, 29–30, 30*f*
Ancistrocladus heyneanus (Indian liana)
　ancistroheynine A in aging parts of, 252–254, 252–253*f*
　cultivation and propagation of, 35–42, 36–37*f*
　induction of callus formation, 37, 37*f*
　influence of different phytohormones, 37–38, 38*f*
　in situ localization of ancistroheynine A and ancistrocladine in, 267–269, 268*f*
　interaction with *Cuscuta reflexa*, 254–255, 255*f*
　investigations on anatomy of, 261–262, 261*f*
　localization of naphthylisoquinoline alkaloids in, 262–264, 263*f*
　regeneration of shoots from callus cultures of, 39, 40*f*
Ancistrocladus ileboensis, 43–44, 58
Ancistrocladus korupensis, 5–6, 47–48
Ancistrocladus robertsoniorum, produces droserone as chemical weapon, 250–252, 250–251*f*
Ancistrocladus tectorius, 34–35, 188–189
　leaf surface of, 29*f*
　twigs and stems of, 55–56
Ancistrocladus uncinatus, 32–33
Ancistrocline, 304–326*t*
Ancistrocyclinones A, 55*f*
Ancistroealaine D, 119–120
Ancistroheynine A
　in aging parts of *Ancistrocladus heyneanus*, 252–254, 252–253*f*
　in situ localization of, 267–269, 268*f*

Index

Ancistrolikokine E3, 4–5, 4f
Ancistroquinone B, 355–368t
Ancistroquinone C, 355–368t
Ancistroquinone D, 355–368t
Ancistroquinone E, 355–368t
Ancistroquinone F, 355–368t
Ancistrosecoline A, 341–345t
Ancistrosecoline A–F, 195–205
Ancistrosecoline B, 341–345t
Ancistrosecoline C, 341–345t
Ancistrosecoline D, 203–205, 341–345t
Ancistrosecoline E, 341–345t
Ancistrosecoline F, 341–345t
Ancistrotanzanine A, absolute axial configuration of, 84–85, 85f
Anisotropic effect, 67–69
Anopheles stephensi, dioncophylline A as larvicide againsts, 258
Antiausterity approach, 111
Antibabesial effects of N-methyldioncophylline A, 126
Anticancer agents, 131–132
Anticancer drug development, dioncophylline A, 127–140, 128t, 130t, 131–133f, 135f, 137f
Anti-MM epoxide, 224
Anti-MM naphthoquinone dioncoquinone B, synthesis of, 222f
Antiplasmodial activities, of 5,1′-linked naphthyltetrahydroisoquinoline alkaloids, 109–110
Antiplasmodial potential, of jozimine-A$_2$-type naphthylisoquinoline dimers, 176–177
AO. *See* Acridine orange (AO)
Apirombandakamine A1, 4–5, 4f
Apoptosis
 in CCRF-CEM cells, 133
 in MDA-MB-231 and MCF-7 breast cancer cells, 138
Aristoyagonine, 212f
Aromatic diamidine compound, 125–126
Aromatic protons
 detection of, 72
 spin pattern of, 67
Aseptic plants, 37
ASF virus. *See* African swine fever (ASF) virus

Atropisomer-specific NOE interactions, 93, 93f
Atropo-diastereomer
 of ancistrocladinium B, 86, 87f
 of 4′-O-demethyl-7-*epi*-dioncophylline, 129
 7-*epi*-dioncophylline A, 4–5, 4f
 electronic circular dichroism (ECD) spectra of, 88
 naphthylisoquinoline dimers, *Ancistrocladus abbreviatus*, 167–185

B

Babesia parasites, 125–127, 126f
B-cell lymphoma, 220
Biaryl chromophore, chiroptical predominance of, 80
Bijvoet analysis, 100
Bioactivities of naphthylisoquinoline alkaloids, 7–8
Biogenic crystals, 251, 251f
Biomphalaria glabrata (Snail)
 molluscicidal activities of alkaloids of *Triphyophyllum peltatum* against, 259–260, 259t
Biosynthesis
 of droserone, 236–238
 of isoquinoline alkaloids, 232f
 of isoshinanolone, 236–238
 of naphthylisoquinoline alkaloids, 231–249, 232f, 234f
 of plumbagin, 236–238
Blood vessel formation (angiogenesis), 132–133
B non-Hodgkin lymphoma, 219–220
Boltzmann statistics, 84–85
Breast cancer, 133–134

C

Cahn-Ingold-Prelog
 formalism, 152–153
 notation, 198–199
Cameroonian species, 5–6
Cancer
 anticancer agents, 131–132
 breast, 133–134
 cervical, 203

Cancer (*Continued*)
  drug development, Jozimine A$_2$, 177–183, 178–179*f*, 179*t*
  pancreatic, 116–117, 227
Cancer cells
  colon, 181
  HeLa cervical, 203–205, 204*f*
  pancreatic, 179–181
  ROS production in, 136
Canine babesiosis, 125–126
Cationic C, 189–192, 189*f*
C-Coupled naphthylisoquinoline alkaloid, 189–192, 189*f*, 191–192*f*
CCRF-CEM leukemia cells, 130*t*
  apoptosis in, 133
Cell-permeable dye, 116
Central Africa, Ancistrocladaceae from, 31–33
Cervical cancer, 203
Chemical-shift imaging (CSI) experiments, 253, 267
Chemodiversity, naphthylisoquinoline alkaloids, 273–274
Chemoecological functions, of naphthylisoquinoline alkaloids, 274–276
Chemotherapeutic agents, 5–6, 125–126
Chinese liana. See *Ancistrocladus tectorius*
Chiral analyte, electronic circular dichroism (ECD) spectrum of, 86
Chiralcel® OD-H (Daicel), 86, 124
Chiroptical behavior of molecule, 82–83
Chiroptical predominance of biaryl chromophore, 80
Cis-configured naphthyltetrahydroisoquinolines, 63–64
Cis-diastereomers, instability of, 63–64
Cis-isoshinanolone, 355–368*t*
  acetogenic origin of, 237*f*
Coat protein complex II (COPII), 225
Colon cancer cells, 181
Column chromatography (CC), 65, 124
Conformational analysis (CA), 82–83
5,1′-Coupled ancistrocladaceae-type naphthylisoquinoline alkaloids, 304–326*t*
5,8′-Coupled ancistrocladaceae-type naphthylisoquinoline alkaloids, 332–336*t*
7,1′-Coupled ancistrocladaceae-type naphthylisoquinoline alkaloids, 290–293*t*
7,8′-Coupled ancistrocladaceae-type naphthylisoquinoline alkaloids, 297–303*t*
3′,3″-Coupled dioncophyllaceae-type dimeric naphthylisoquinoline alkaloids, 350–354*t*
5,1′-Coupled dioncophyllaceae-type naphthylisoquinoline alkaloids, 328*t*
7,1′-Coupled dioncophyllaceae-type naphthylisoquinoline alkaloids, 282–289*t*
7,3′-Coupled dioncophyllaceae-type naphthylisoquinoline alkaloids, 296*t*
7,3′-Coupled dioncophyllaceous, naphthylisoquinoline alkaloid, 145–150, 146–147*f*
5,1′-Coupled hybrid-type naphthylisoquinoline alkaloids, 329–331*t*
7,1′-Coupled hybrid-type naphthylisoquinoline alkaloids, 294*t*
7,1′-Coupled inverse hybrid-type naphthylisoquinoline alkaloid, 295*t*
5,8′-Coupled naphthylisoquinoline alkaloids, ancistrobrevine B and related, 119–122, 120–122*f*
7,1′-Coupled naphthylisoquinoline alkaloids, dioncophylline A and related, 122–145
7,8′-Coupled naphthyltetra- and-dihydroisoquinoline alkaloids, 150–153, 150*f*, 152–153*f*
5,1′-Coupling site and structural diversity, naphthylisoquinoline alkaloids with, 104–119, 105*f*
Cultivation of ancistrocladaceae plants, 35–42
Cuscuta reflexa interaction with *Ancistrocladus heyneanus*, 254–255, 255*f*
Cymes, 30–31

Index

Cytotoxic activity
  of dioncophylline A, 133, 133f
  of 5-*epi*-ancistectorine A$_2$, 111–112
  of 1,4-naphthoquinone, 230f
Cytotoxic effects
  of ancistrobrevoline B, 213–215, 214f
  against dioncophylline A, 134, 135f
  of dioncophylline A, 136–138, 137f

D

Defense-mediating processes, 7–8
Degradation, oxidative, 79–80, 79f, 120–121, 151–152, 200
  of ancistrobrevine M, 108
  of *N*-methylated naphthylisoquinoline alkaloids, 79–80, 79f
  ruthenium-mediated, 188, 198, 252–253
Dehydrogenated naphthylisoquinoline alkaloids
  from roots of *Ancistrocladus abbreviatus*, 153–161, 154–157f, 159–160f
Dehydrogenated naphthylisoquinoline ancistrobreveine D, 119–120
1-*nor*-8-*O*-Demethylancistrobrevine H, 192–195, 193–194f, 340t
Density functional theory (DFT), 83
Dimeric naphthylisoquinoline alkaloids, 8–9
1,3-Dimethyl-1,2,3,4-tetrahydroisoquinolines, 74
Dioncoline A, 144–145, 144f, 295t
Dioncophyllaceae, 5, 9–44
  geographic distribution of, 9–10, 10f
  phylogenetic relationships of, 11–12
  species-specific production of naphthylisoquinoline alkaloids in, 63–64
  systematic position of, 10–11
Dioncophyllaceae plants, phytochemical studies on, 69–74, 100–101
Dioncophyllaceae-type naphthylisoquinoline alkaloids, 167–175, 169f, 171f, 174–175f
  dioncophylline A and related, 122–127, 123f, 125–126f
Dioncophylleine A, absolute configuration of, 80–85, 82f

Dioncophyllidine E, 145–150, 146–147f, 296t
Dioncophylline A, 4–5, 4f, 282–289t
  absolute axial configuration of, 93–97, 95–96f
  anticancer drug development, 127–140, 128t, 130t, 131–133f, 135f, 137f
  antiproliferative activities of, 134
  biosynthesis from dihydro- and tetrahydroisoquinoline precursors, 240–249, 241–243f, 245f, 247–248f
  biosynthesis of, 238–240, 239f
  constitution and stereostructure of, 99–100
  cytotoxic activities of, 133, 133f
  cytotoxic effects against, 134, 135f
  cytotoxic effects of, 136–138, 137f
  exemplified for, 88–100, 89f
  as larvicide against *Anopheles stephensi* and *Aedes aegypti*, 258
  and related 7,1'-coupled naphthylisoquinoline alkaloids, 122–145
  and related dioncophyllaceae-type naphthylisoquinoline alkaloids, 122–127, 123f, 125–126f
  stereoselective total synthesis of, 97–99, 98f
  structural elucidation of, 91–93, 92–93f
  in vivo localization of, 269–273, 270–271f
ent-Dioncophylleine A, 282–289t
7-*epi*-Dioncophylline A, 282–289t
5-*epi*-Dioncophyllidine C$_2$, 117–119, 118–119f, 328t
Dioncophylline C, 117–119
Dioncophyllum tholIonii, 12–14, 147–148
Dioncoquinone B, 216–217, 217f, 220, 227f, 355–368t
  antiproliferative properties of, 220, 220f
  synthetic approach to, 226f
Dioncotetralone A, 282–289t
Dioncotetralone B, 282–289t
Directed ortho-metalation (DOM) reaction, 221
DMEM. *See* Dulbecco's modified Eagle's medium (DMEM)

Droserone, 215–216, 227f, 355–368t
  biosynthesis of, 236–238
  as chemical weapon, 250–252, 250–251f
Dulbecco's modified Eagle's medium (DMEM), 111, 114–116, 115f, 179–180
  nutrient-rich, 149, 227

E

Ealamine A, 4–5, 4f
East Africa, Ancistrocladaceae from, 31–33
EB. *See* Ethidium bromide (EB)
ECD. *See* Electronic circular dichroism (ECD)
Ecological functions of naphthylisoquinoline alkaloids, 273–274
Electronic circular dichroism (ECD)
  behavior, 78
  computations, 96–97
  spectra, 80–81, 85–86
    of ancistrobrevine F, 78, 78f
    of atropo-diastereomers, 88
    of chiral analyte, 86
Electronic circular dichroism (ECD) spectroscopy
  absolute axial configuration
    by experimental, 80–85, 82f, 85f
    of dioncophylline A, 93–97, 95–96f
Electrospray ionization (ESI), 66
Enzyme-catalyzed pathway, 233–235
Estrogen receptors (ER), 134
Ethidium bromide (EB), 116, 149, 179–180
Exciton Chirality Method, 81–82

F

Fast centrifugal partition chromatography (FCPC), 65
Fibrosarcoma, 181
Friedel-Crafts acylation of pyrocatechol, 226
FT-Raman microspectroscopy, 267
FT-Raman spectra, 269

G

Gas chromatography with mass-selective detection (GC-MSD), 79–80, 79f
Gellan gum, 215
Gel permeation chromatography, 65

Gelrite®, 215
Giemsa-stained blood smears of parasites, 126, 126f
Guanine-cytosine (GC) units, 139–140

H

Habropetalum dawei, 243
Hamatine, 304–326t
H$_2$DCFDA (2',7'-dichlorodihydrofluorescein diacetate), 138
Heavy-atom method, 100
HeLa cervical cancer cells, 203–205, 204f
  impairment of cell viability of human, 114–117, 115f
Herbivores, activities against, 255–258
Heterodimeric naphthylisoquinoline alkaloids, 73
Heteronuclear Multiple Bond Correlation (HMBC)
  experiments, 219
  interactions, 67, 73, 244–246
High-speed countercurrent chromatography (HSCCC), 65
HI virus, 5–6
Hoechst 33342, 136
Homo naphthylisoquinoline alkaloids, 73
Hormone-positive MCF-7 cells, 134
Hormone-unresponsive, MDA-MB-231 breast cancer cells, 107f, 136–138
HPLC-ECD analysis, of ancistrobreveine A, 156, 156f
Human epidermal growth factor 2 (HER2), 134

I

Imidocarb dipropionate, 125–126
India, Ancistrocladaceae from, 33–35
In situ localization
  of ancistrocladine, 267–269, 268f
  of ancistroheynine A, 267–269, 268f
Inter-simple sequence repeat (ISSR) fingerprint, 19
In vivo localization
  of dioncophylline A, 269–273, 270–271f
  of naphthylisoquinoline alkaloids, 267–273, 268f, 270–271f
Ionic chromatography, 65

Isoindolinones, naphthylisoquinoline-
derived, 346–349t
Isoquinoline, 7
non-coupled, 61–63
Isoquinoline alkaloids, 231–232
acetate-polymalonate pathway to,
231–249, 232f, 234f, 239f
acetogenic, 235
biosynthesis of, 232f
biosynthetic origin of, 8–9
Isoquinoline building block, stereogenic
centers in, 121–122
Isoquinolinemoiety, non-hydrogenated,
155–156
Isoshinanolone, biosynthesis of, 236–238

J

Jozibrevine A, 350–354t
semisynthesis of, 183–185
Jozibrevine B, 350–354t
Jozibrevine C, 350–354t
Jozibrevines A and B
cytotoxicity of
against fibrosarcoma and colon cancer cells, 181
against leukemia cells, 181–183
Jozibrevines A–D, 167–175, 174–175f
Jozibrevines C and D, cytotoxicity of, 181–183
Jozimine A, 282–289t
Jozimine A$_2$, 4–5, 4f, 167–175, 169f, 171f, 350–354t
absolute configuration of, 168–170, 169f
binaphthalene core of, 168–170
cancer drug development, 177–183, 178–179f, 179t
cytotoxicity of
against cervical and pancreatic cancer cells, 179–181
against fibrosarcoma and colon cancer cells, 181
against leukemia cells, 181–183
semisynthesis of, 183–185, 184f
Jozimine-A$_2$-type naphthylisoquinoline
dimers, antiplasmodial potential of,
176–177

L

Lactone method, 97–99, 98f
Lactone strategy, 141–142
LC-ECD coupling, naphthylisoquinoline
alkaloids, 85–88, 87f
LC-UV, 69
Leukemia cells, 181–183
CCRF-CEM, 130t
Linear dichroisms (LD), 139–140
7,1′-Linked ancistrocladaceae-type alkaloid,
156–157
5,1′-Linked dioncophyllaceous
naphthylisoquinoline alkaloid,
117–119, 118–119f
7,1′-Linked hybrid-type
naphthylisoquinoline alkaloids,
142–143, 143f
5,1′-Linked naphthylisoquinoline alkaloids
from leaves of *Ancistrocladus abbreviatus*,
112–117, 113–115f
5,1′-Linked naphthyltetrahydroisoquinoline
alkaloids
antiplasmodial activities of, 109–110
preferential cytotoxicities of, 110–112, 112f
from roots of *Ancistrocladus abbreviatus*,
105–108, 106–107f, 109f
5,1′-Linked naphthyltetrahydroisoquinoline
alkaloids
from roots of *Ancistrocladus abbreviatus*,
105–108, 106–107f, 109f
5,1′-Linked 6-O-methylhamateine, 155
Lux® Cellulose-1 (Phenomenex), 86,
157–158
Lymphoma
B-cell, 220
B non-Hodgkin, 219–220

M

Malignant cells, 132–133
Mass spectrometry (MS), 66
affinity capture experiments, 225
Mbandakamine A, 58–60, 59f
MCF-7 breast cancer cells
apoptosis in, 138
hormone-positive, 134

Index

MCF-7 breast cancer cells (*Continued*)
  reactive oxygen species (ROS) production in, 138
MDA-MB-231 breast cancer cells
  apoptosis in, 138
  hormone-unresponsive, 107f, 136–138
  reactive oxygen species (ROS) production in, 138
MDA-MB-231 cancer cells, 134
MDR. *See* Multi-drug resistance (MDR)
Me-3 group, 74
Metabolites
  enriched fractions, 65
  naphthalene, 61–63
Michellamine A, regioselective synthesis of, 183, 183f
Michellamine B, 4–6, 4f
Micro-FT Raman studies, 267
Mitochondrial membrane potential (MMP), 136–138
MM. *See* Multiple myeloma (MM)
MMP. *See* Mitochondrial membrane potential (MMP)
Molecular-dynamics (MD)
  approach, 82–83, 93–94
  simulation, 83–85
Molecular rDNA, nuclear internal transcribed spacers (ITS) of, 19
Molluscicidal activities of alkaloids of *Triphyophyllum peltatum*, 259–260, 259t
Monomeric alkaloids, 6
Mosher derivatives of alkaloid sample, 79–80
Multi-drug resistance (MDR), 129
Multiple myeloma (MM), 127, 220

N

N-Acetyldioncophylline A, 129
Naphthalene
  metabolites, 61–63
  moiety, 75
1-Naphthalene acetic acid (NAA), 17–18
Naphthoquinones, 228
  with different oxygenation patterns, 215–216, 216f
  related to naphthylisoquinoline alkaloids, 355–368t

1,4-Naphthoquinones, 217f, 235
  cytotoxic activity of, 230f
Naphthyldihydroisoquinoline, 64
  with five aromatic protons, 197f
Naphthylisoindolinones ancistrobrevolines A–D, 205–215, 205–206f, 208f, 210–212f, 214f
Naphthylisoquinoline alkaloids, 5
  absolute stereostructure of, 88–100, 89f
  axial configuration of, 80–81
  biosynthesis of, 231–249, 232f, 234f
  C,C-coupled monomeric, 44–53, 66f
    5,3'- and 5,6'-coupling, 51f, 52
    7,3'- and 7,8'-coupling, 50–52
    5,1'-coupling, 50
    5,8'-coupling, 50
    7,1'-coupling, 52
    7,6'-coupling, 53
  chemodiversity, storage, and ecological functions of, 273–274
  chemo-ecological functions of, 7–8, 274–276
  chemo-ecological interactions and localization of, 249–274, 250–253f, 255–257f
  5,1'-coupled ancistrocladaceae-type, 304–326t
  5,8'-coupled ancistrocladaceae-type, 332–336t
  7,1'-coupled ancistrocladaceae-type, 290–293t
  7,8'-coupled ancistrocladaceae-type, 297–303t
  5,1'-coupled dioncophyllaceae-type, 328t
  7,1'-coupled dioncophyllaceae-type, 282–289t
  7,3'-coupled dioncophyllaceae-type, 296t
  3',3''-coupled dioncophyllaceae-type dimeric, 350–354t
  7,3'-coupled dioncophyllaceous, 145–150, 146–147f
  5,1'-coupled hybrid-type, 329–331t
  7,1'-coupled hybrid-type, 294t
  7,1'-coupled inverse hybrid-type, 295t
  with 5,1'-coupling site and structural diversity, 104–119, 105f
  determination of relative configuration of, 75–78, 76f, 78f

Index

dimeric, 8–9
discovered in roots of *Ancistrocladus abbreviatus*, 195–205, 196–197*f*, 199–200*f*, 202*f*, 204*f*
diversity of bioactivities of, 7–8
electronic circular dichroism (ECD) spectra of two atropo-diastereomeric, 78
elucidation of absolute configuration, 79–80, 79*f*
with intriguing structural motifs, 275*f*
inverse hybrid-type, 144–145, 144*f*
isolation of, 5–7, 64–101, 66*f*, 68*f*, 70–71*f*
lacking a methyl substituent at C-1, 340*t*
LC-ECD coupling, 85–88, 87*f*
localization, in *Ancistrocladus heyneanus*, 262–264, 263*f*
molecular halves of, 7
naphthoquinones and tetralones related to, 355–368*t*
N,C-coupled naphthylisoquinolines, 53–56, 54*f*
NMR imaging of organs and seeds of *Ancistrocladus* species and localization of, 260–267
non-hydrogenated, detection of, 63
ortho-quinoid, 337–338*t*
oxidative degradation, 79–80, 79*f*
rich source of structurally diverse, 101–164, 102*f*
ring-contracted, 205–215, 205–206*f*, 208*f*, 210–212*f*, 214*f*
secondary metabolites related to hetero- or isocyclic molecular portions of, 61–63, 62*f*
seco-type, 341–345*t*
species-specific production of, 63–64
structural diversity and classification of, 44–64
structural elucidation of, 7
structural features of, 64–65
structural variety of, 4–5, 4*f*
structures of selected, 279*f*
subclasses of, 47*f*
suspended callus cultures of *A. heyneanus*, 39
tentative classification of fully dehydrogenated, 48–49, 49*f*
tetrahydroisoquinoline portions of, 74, 74*f*
with unusual molecular scaffolds, 165–231, 165*f*, 167*f*
in vivo localization of, 267–273, 268*f*, 270–271*f*
Naphthylisoquinoline-derived isoindolinones, 346–349*t*
Naphthylisoquinolines, 233
conformational analysis of, 83
dimers and subclasses, 56–61, 57*f*, 59*f*
Naphthyltetrahydroisoquinoline ancistrobrevine M, 113
Natural tetrahydroisoquinoline phylline, 246
N,C-coupled Alkaloids, 72
N,C-coupled Naphthylisoquinoline alkaloids, 8–9
N,C-coupled Naphthylisoquinolines, 53–56, 54*f*
NDM. *See* Nutrient-deprived medium (NDM)
Near-infrared Fourier transform (NIR-FT) Raman microspectroscopy
inclusions of fresh stem material of *Triphyophyllum peltatum* by, 269–273, 270–271*f*
Nepenthone A, 355–368*t*
N-Methylancistrocladisine A, 290–293*t*
N-Methylated naphthylisoquinoline alkaloids, 63–64
oxidative degradation of, 79–80, 79*f*
N-Methyldioncophylline A, 124, 125–126*f*, 144–145, 282–289*t*
antibabesial effects of, 126
N-Methyl-7-*epi*-dioncophylline A, 282–289*t*
N,N-Ligand, 161
Non-coupled isoquinoline, 61–63
Non-hydrogenated alkaloids, 158
Non-hydrogenated isoquinolinemoiety, 155–156
Non-hydrogenated naphthylisoquinoline alkaloids, detection of, 63
Non-invasive FT-Raman microspectroscopy
in vivo localization of naphthylisoquinoline alkaloids by, 267–273, 268*f*, 270–271*f*

Non-invasive NMR microscopy, investigations on anatomy of *Ancistrocladus heyneanus* by, 261–262, 261f
Novel-type naphthylisoquinoline alkaloids, 185–189
Nuclear factor kappa B (NF-κB) signaling pathway, 131
Nuclear internal transcribed spacers (ITS) of molecular rDNA, 19
Nuclear magnetic resonance (NMR) chemical-shift imaging (CSI) measurements, 249–250
  microscopy, 260
    and chemical shift imaging (CSI) experiments, 264–267, 265–266f
    and radial spectroscopic imaging, 262–264, 263f
  spectroscopy, fruitful interplay of, 141–142
Nuclear Overhauser and Exchange (NOE) correlation between Me-1 and H-8 0, 77
  effects between isoquinoline and naphthalene, 77
  interactions, 75–78, 120–121
  measurements, 77
Nuclear Overhauser and Exchange Spectroscopy (NOESY), 67
  correlation between Me-1 and H-8', 77
  interactions, 76f, 92–93
  measurement, naphthylisoquinoline alkaloids, 75–78, 76f, 78f
Nucleus-specific Hoechst 33342 staining method, 136
Nutrient-deprived medium (NDM), 111, 114–117, 115f
Nutrient-rich Dulbecco's modified Eagle's medium (DMEM), 114–116, 115f, 149, 179–180, 227

O

Octant Rule, 80–82
6-O-Demethylancistrobrevine A, 297–303t
5'-O-Demethylancistrobrevine B, 119–120, 332–336t
6-O-Demethylancistrobrevine H, 297–303t
4'-O-Demethylancistrocladinium A, 54f
4'-O-Demethyldioncophylline A, 124, 127–128
4'-O-Demethyldioncophylline A, 282–289t
5'-O-Demethyldioncophylline A, 127–128
4'-O-Demethyl-7-*epi*-dioncophylline A, 124, 282–289t
4'-O-Demethyl-7-*epi*-dioncophylline, atropo-diastereomer of, 129
6-O-Methylancistectorine A$_3$, 112, 113f, 304–326t
6-O-Methylancistrocladine, 304–326t
O-Methylation, 111–112
6-O-Methylhamateine, 304–326t
6-O-Methylhamatine, 111–112, 304–326t
6-O-Methylhamatinine, 112, 113f, 304–326t
6-O-Methyl-4'-O-demethyl-ancistrocladine, 304–326t
6-O-Methyl-4'-O-demethyl-hamatine, 111–112, 304–326t
1D nuclear magnetic resonance (NMR) spectroscopy, 74, 74f
O-Permethylated alkaloid 6-O-methylhamatinine, 112, 113f
8-O-(*p*-nitrobenzyl)Dioncophylline A, in silico binding study of, 131–132, 132f
Orange-colored crystals, 250–251
Ortho-quinoid naphthylisoquinoline alkaloids, 337–338t
Oxidative degradation, 79–80, 79f, 120–121, 151–152, 200
  of ancistrobrevine M, 108
  of *N*-methylated naphthylisoquinoline alkaloids, 79–80, 79f
  ruthenium-mediated, 188, 198, 252–253

P

PANC-1 cells in DMEM, 117, 118f
Pancreatic cancer, 116–117, 227
Pancreatic cancer cells, 149, 179–181, 227
  naphthyltetrahydroisoquinoline alkaloids against, 110–112, 112f
  preferential cytotoxic activity against PANC-1 human, 114–116, 115f
Parasitemia of cells, 126, 126f
Parasites, giemsa-stained blood smears of, 126, 126f

Peripheral blood of healthy donors, 127–128
Peripheral mononuclear blood cells (PMBCs), 220, 225–226
P-glycoprotein (P-gp), 129
Pictet-Spengler-type condensation, 231–232
PKS. See Polyketide synthase (PKS)
Plant-pathogenic fungi
  fungicidal activities of alkaloids of *Triphyophyllum peltatum* against, 260
Plasmodium parasites, 126–127
Plastid *matK*, single-gene analysis of, 11*f*
Plumbagin, 215–216, 253–254, 253*f*, 355–368*t*
  biosynthesis of, 236–238
PMBCs. See Peripheral mononuclear blood cells (PMBCs)
Polyketide synthase (PKS), 248–249
Positive chirality, 81–82
P,R-Atropisomer, 209
Preferential cytotoxicities, of 5,1′-linked naphthyltetrahydroisoquinoline alkaloids, 110–112, 112*f*
Progesterone receptors (PR), 134
Pyrocatechol, Friedel-Crafts acylation of, 226

Q

Quantum-chemical electronic circular dichroism (ECD) calculations, 82

R

Raman spectra, 267
Reactive oxygen species (ROS) production
  in cancer cells, 136
  in MDA-MB-231 and MCF-7 breast cancer cells, 138
Ring-contracted naphthylisoquinoline alkaloids, 205–215, 205–206*f*, 208*f*, 210–212*f*, 214*f*
Ring-opened naphthylisoquinoline alkaloids, 195–205, 196–197*f*, 199–200*f*, 202*f*, 204*f*
ROESY spectroscopy, 75
Ruthenium-mediated oxidative degradation, 188, 198, 246, 252–253

S

Secondary gymnosperms, 14
Secondary metabolites, structural diversity of, 280, 281*f*
Seco-type naphthylisoquinoline alkaloids, 341–345*t*
Seramis® substrate, 18, 41–42
Serine/threonine kinase Akt (protein kinase B), 231
Slice-selective one-dimensional NMR spectrum, 262–263
Snail *Biomphalaria glabrata*
  molluscicidal activities of alkaloids of *Triphyophyllum peltatum* against, 259–260, 259*t*
Soft-ionization techniques, 66
Solid-phase extraction (SPE), 65
Solid-state electronic circular dichroism (ECD) spectroscopy, 95
Southeast Asia, Ancistrocladaceae from, 33–35
SPE. See Solid-phase extraction (SPE)
Species-specific production of naphthylisoquinoline alkaloids, 63–64
Spirombandakamines, 61
Spodoptera littoralis (Noctuidae), 257–258, 257*f*
Sri Lanka, Ancistrocladaceae from, 33–35
Stereogenic *N*-iminium-*C*-aryl axis, 53–56
Stereoisomer, 85–86

T

TDDFT. See Time-dependent density functional theory (TDDFT)
Tetralones, related to naphthylisoquinoline alkaloids, 355–368*t*
Thidiazuron, 17–18
Thioredoxin domain-containing 5 (TXNDC5), 225
Thioredoxin-related transmembrane protein 1 (TMX1), 225
Tick-borne disease, 125–126
Time-dependent density functional theory (TDDFT), 83, 87*f*

Triphyophyllum peltatum, 5–6, 12, 117–119
   cultivation of, 15*f*
   fungicidal activities of alkaloids of, 260
   geographic distribution of, 12, 13*f*
   germination, 16–17
   greenhouse conditions, 16–17
   insect-growth retarding and antifeedant activities of alkaloids of, 255–258, 256*f*
   in juvenile phase, 14–16
   metabolite of, 127–128
   molluscicidal activities of alkaloids of, 259–260, 259*t*
   NIR-FT Raman microspectroscopy inclusions of fresh stem material of, 269–273, 270–271*f*
   rooting of, 17–18
   seedling of, 15*f*, 17
   viable seeds of, 15*f*, 16–17

Triptolide, 131–132, 132*f*
Tropical infectious diseases, 109
2D nuclear magnetic resonance (NMR) spectroscopy, 74, 74*f*
TXNDC5. *See* Thioredoxin domain-containing 5 (TXNDC5)

W

West Africa, Ancistrocladaceae from, 31–33
Western blotting, 138, 231

X

X-ray diffraction, 99–100

Z

Zebrafish model, 133
Zymogen pro-caspase-3 (caspase-3 precursor protein), 138

9780443295546